Statistical Methods, Experimental Design, and Scientific Inference

Sir Ronald A. Fisher
(Photograph courtesy of Antony Barrington-Brown)

Statistical Methods, Experimental Design, and Scientific Inference

A Re-issue of
Statistical Methods for Research Workers,
The Design of Experiments,
and
Statistical Methods and Scientific Inference

R. A. FISHER

Edited by
J. H. BENNETT

With a Foreword by
F. YATES

OXFORD UNIVERSITY PRESS · OXFORD

This book has been printed digitally and produced in a standard specification in order to ensure its continuing availability

OXFORD
UNIVERSITY PRESS

Great Clarendon Street, Oxford OX2 6DP

Oxford University Press is a department of the University of Oxford.
It furthers the University's objective of excellence in research, scholarship,
and education by publishing world-wide in

Oxford New York

Auckland Bangkok Buenos Aires Cape Town Chennai
Dar es Salaam Delhi Hong Kong Istanbul Karachi Kolkata
Kuala Lumpur Madrid Melbourne Mexico City Mumbai Nairobi
São Paulo Shanghai Taipei Tokyo Toronto

Oxford is a registered trade mark of Oxford University Press
in the UK and in certain other countries

Published in the United States
by Oxford University Press Inc., New York

© The University of Adelaide 1990
Original volumes published by
Hafner Publishing Company, New York

The moral rights of the author have been asserted

Database right Oxford University Press (maker)

Reprinted 2003

All rights reserved. No part of this publication may be reproduced,
stored in a retrieval system, or transmitted, in any form or by any means,
without the prior permission in writing of Oxford University Press,
or as expressly permitted by law, or under terms agreed with the appropriate
reprographics rights organization. Enquiries concerning reproduction
outside the scope of the above should be sent to the Rights Department,
Oxford University Press, at the address above

You must not circulate this book in any other binding or cover
And you must impose this same condition on any acquirer

ISBN 978-0-19-852229-4

CONTENTS

Foreword

Statistical Methods for Research Workers

The Design of Experiments

Statistical Methods and Scientific Inference

FOREWORD

F. Yates

When Henry Bennett told me that the Oxford University Press were preparing to issue a set of Fisher's three statistical textbooks, *Statistical Methods for Research Workers*, *The Design of Experiments*, and *Statistical Methods and Scientific Inference*, and would like me to write a foreword I was somewhat daunted. However, in view of my long association and friendship with Fisher, and our mutual interest, not only in practical statistics, but also in genetics and evolution, I felt I owed it to his memory to accept the invitation.

Statistical Methods for Research Workers was first published in 1925, and at Fisher's death in 1962 a further 12 editions had appeared. Fisher had planned to publish a 14th edition, but his sudden death in July 1962 prevented this. However, E. A. Cornish was able to complete the task from notes Fisher had already prepared. In the present set, the text of the 14th edition of *Statistical Methods* is reproduced along with Fisher's Preface to the 13th edition and with the companion volumes of *The Design of Experiments* and *Statistical Methods and Scientific Inference*, to mark the centenary of his birth on 17th February 1890. In what follows I will refer to the first and last of these books by the short titles *Statistical Methods* and *Scientific Inference*, respectively.

As the full title indicates, *Statistical Methods* was originally written for scientists, particularly biologists, engaged in active research, many of whom had little or no training in the statistics of the time. It therefore came to be used by many as a laboratory notebook rather than as a textbook for students, and for this purpose it was admirably suited, as apart from the so-called introductory chapter, every topic in the first edition was based on one or more specific worked numerical examples employing real data which themselves illustrated the problems they posed.

Fisher's general practice was to include the preface to the first edition of his books in all subsequent editions. He did this with all the editions of *Statistical Methods* until the eighth in 1941. However, it was omitted from the ninth edition in 1944 and from all later editions. I reproduce it here because it expresses in vivid terms his motives when he wrote the first edition.

PREFACE TO THE FIRST EDITION

FOR several years the author has been working in somewhat intimate cooperation with a number of biological research departments; the present book is in every sense the product of this circumstance. Daily contact with the statistical problems which present themselves to the laboratory worker has stimulated the purely mathematical researches upon which are based the methods here presented. Little experience is sufficient to show that the traditional machinery of statistical processes is wholly unsuited to the needs of practical research. Not only does it take a cannon to shoot a sparrow, but it misses the

FOREWORD ix

sparrow! The elaborate mechanism built on the theory of infinitely large samples is not accurate enough for simple laboratory data. Only by systematically tackling small sample problems on their merits does it seem possible to apply accurate tests to practical data. Such at least has been the aim of this book.

I owe more than I can say to Mr W. S. Gosset, Mr E. Somerfield, and Miss W. A. Mackenzie, who have read the proofs and made many valuable suggestions. Many small but none the less troublesome errors have been removed; I shall be grateful to readers who will notify me of any further errors and ambiguities they may detect.

ROTHAMSTED EXPERIMENTAL STATION,
February 1925.

To appreciate fully the revolutionary nature of *Statistical Methods* we must recall the state of the subject at the time. Karl Pearson was then at the height of his fame, and *Biometrika* the leading journal of mathematical statistics. It was the age of correlation and curve fitting; in *Tables for statisticians and biometricians*, 37 per cent of the tabular matter was devoted to curve fitting, and a further 18 per cent to various forms of correlation. It was also the age of coefficients of all kinds. In attempts to assess the degree of association in 2 × 2 contingency tables, for example, such measures as the coefficient of association, the coefficient of mean square contingency, the coefficient of tetrachoric correlation, equiprobable tetrachoric correlation, and the coefficient of colligation, were proposed. The way in which these coefficients were used revealed considerable confusion between the problem of estimating the degree of association, and that of testing the significance of the existence of an association.

This confusion permeated the whole of the statistical writing and thinking of the Pearsonian school. It is perhaps not surprising, therefore, that the paramount need in statistics for methods which would provide the most accurate possible estimates from the available data had been largely lost sight of. This need had been very apparent to earlier thinkers, in particular Gauss, who had developed the method of least squares—accepted by astronomers and geodetic surveyors as the standard method for "the adjustment of observations"—with just this end in view.

A further weakness of the Pearsonian school was their failure to consider the need of experimenters for methods appropriate to small samples involving quantitative observations.

The exact solution for the test of the significance of the mean of a small sample that could be assumed to be normally distributed, but with unknown standard deviation, was in fact provided by Gosset ("Student"), a research chemist, in 1908. It was published in *Biometrika*, and the relevant table (the equivalent of the t table) was reproduced in *Tables for statisticians and biometricians*. But as Fisher said in his obituary of "Student" (1939), this new contribution to the theory of errors was received with "weighty apathy". It was left to Fisher to draw attention to its importance, and to extend the solution to more complex problems. This he did in *Statistical Methods*.

Apart from the addition of Chapter 9 in the second edition there have been no major changes in the structure of the book, but many insertions of new matter. Inserted sections, examples, and tables have been distinguished by the dot notation, so that references to earlier editions are unchanged. The reader coming new to the book should first read the preface to the present edition, which gives a detailed account of the changes and insertions made over the years.

One useful detail, introduced with the eighth edition in 1941, is the printing of the current section numbers at the head of each page. For this reprint edition, Henry Bennett has prepared a list in the Contents of section headings and the examples with their titles and page numbers. This is invaluable to the serious reader. My bête noire in the earlier editions was example 19 on p. 121 on paired comparisons of two sleeping drugs, which I often wanted to recommend to novices. A section index would have given me the necessary clue.

This is a minor matter. A greater difficulty for the reader is the development in parallel of problems involving quantitative variates based on the normal distribution, and those involving integers, based on the binomial and Poisson distributions (Chapter 3).

The first chapter, "Introductory", is not, as may be thought, merely an introduction to the book. The first two sections are a masterly exposition of the scope and general method of statistics, the third an explanation of what is meant by consistent, efficient, and sufficient statistics and the method of maximum likelihood. Only in Section 4 do we have an account of the scope of the book, and the considerations Fisher had in mind when writing it. The fifth section, formerly occupied by a list of tables available for testing significance, was replaced in the fourth edition by a historical note on Bayes, Laplace, and Gauss, and their relation to modern workers. This provides an introduction to the logic of inductive inference, which is fully argued in *Scientific Inference*.

The exposition in Section 3 of the introductory chapter is very useful for an understanding of Chapter 9, the Principles of statistical estimation. To those not conversant with the genetical terminology of cross-breeding, this chapter may at first sight look forbidding, but with careful reading they will soon become familiar with it. Indeed Section 52 gives a full specification. I strongly recommend the chapter to statisticians who are well acquainted with the more elementary parts of the book.

Following the introductory chapter comes a chapter on diagrams. This must have been an innovation surprising to statisticians of that time. Now that computers are available the use of diagrams has become much more common. Chapter 3 I have already commented on above. Chapter 4 on the use of χ^2 draws attention to Karl Pearson's error on degrees of freedom. It is indeed surprising that this error persisted for so long. Yule had suspected it for many years and had used the correct value of 1 d.f. for 2×2 tables in the first edition of his textbook in 1914 and had done further tests since then, but was reluctant to publish until Fisher launched his attack. Even then the correction was not accepted in certain quarters, as is shown by a remarkable review in the *British Medical Journal* of the first edition:

> The trained statistician interested in Mr. Fisher's researches will miss a detailed justification of his conclusions. . . . Even if the statement that Professor Pearson's treatment of a fundamental problem contained a "serious error" had not been disputable, and therefore improper in a work addressed to elementary students, it would have reminded anyone of Macaulay's remark on a similar occasion—"just so we have heard a baby, mounted on the shoulders of his father, cry out, 'how much taller I am than Papa!'"

Actually, the point was discussed by Fisher in a passage (Example 8), which for clarity of statement

FOREWORD   xiii

and convincingness of argument would be difficult to better.

Although in the present preface Fisher refers to "weathering the hostile criticisms inevitable to such a venture", this review was the only one that I know of that was really hostile.

Chapter 5 mainly deals with regression. The only curved regression lines that Fisher considers are polynomial functions for single variables. For these he usually used orthogonal polynomials. Although he does not mention it, this provides a useful safeguard; if, for instance, one is fitting a third degree polynomial, terms for x and x^2 are automatically included. If powers of x are fitted there is always the temptation, particularly now that computers are available, to retain only those powers which produce a significant reduction in the residuals.

Chapter 6 deals with the correlation coefficient. The purely statistical defects of this coefficient are due to the limited range (-1 to $+1$), which results in a much smaller variance at the ends of the range than at the centre. This Fisher overcame by introducing the transformation

$$z = \tfrac{1}{2}\{\log_e(1+r) - \log_e(1-r)\}$$

A table of this transformation is provided in *Statistical Methods*. In the first sentence of the opening paragraph (Section 30), Fisher made significant changes in the 4th and 5th editions, in particular deleting the word "successfully". He also added a paragraph on the next page: "That this method was once centred on the correlation coefficient gives to this statistic a certain importance, even to those who prefer to develop their analysis in other terms."

Chapter 7 starts with intraclass correlation. This is replaced in Section 40 by the analysis of variance. This section opens with the sentence: "A very great simplification is introduced into questions involving intra-class correlation when we recognise that in such cases the correlation merely measures the relative importance of two groups of factors causing variation."

Because Fisher was thinking in terms of correlation, or addressing those who were so thinking, he took four pages to demonstrate this simplification. In components of variance terms it can be demonstrated in a few lines. The analysis of variance for n families, each with k members, can in fact be written:

	d.f.	MS	Components
Between families	$n-1$	s_2^2	$W+kB$
Within families	$n(k-1)$	s_1^2	W

$$W = s_1^2; \qquad kB = s_2^2 - s_1^2$$

Note that when there is little difference between families, but intense competition within families, B may be negative. Surprisingly, this contingency was not considered by Fisher. In the extreme case, when all the family totals are equal, $s_2^2 = 0$, and kB therefore has a lower limit of $-W$.

Unfortunately there has been a tendency to denote variance components by σ^2 with a suffix, such as σ_W^2 and σ_{kB}^2. This has led to the erroneous conclusion that variance components cannot be negative, and to replacing negative values by 0. This practice must result in serious positive bias when combining estimates of such components.

The first example of the analysis of variance of an actual field experiment given in *Statistical Methods* occurs in Chapter 7, and indeed is the only such analysis; all the analyses in Chapter 8 are on data from uniformity trials. This experiment is particularly interesting in that it exhibits an analysis of variance of what are now known as split-plot designs, and therefore have two levels of error.

The experiment must have been one of the earliest of this complexity done on the Rothamsted farm. As is stated in *Statistical Methods*, the object was to test whether different varieties of potatoes differed in their manurial needs. Twelve varieties differing greatly in yield were chosen, and were tested for their potash requirements by applications of chloride and sulphate of potash and no potash on subdivisions of the varietal plots. The experiment was in two parts, one of which was on dunged land, the other on land without dung, with three replicates of each variety on each part.

The varietal plots were described as "scattered" over the area, but were not properly randomized, and the potash treatments were in the same order on each plot. The half with dung was chosen for inclusion in *Statistical Methods*.

The means of the three replicates of all the 60 variety–potash combinations and a plan of the experiment were reported separately for the dunged and undunged parts of the experiment in a paper by Fisher and Mackenzie (1923) listed in the bibliography in *Statistical Methods*. The means of these over all 12 varieties (in lb) are:

	Dunged	Undunged
Sulphate	20.1	15.3
Choride	19.8	15.5
No potash	18.6	4.5

The undunged land was clearly very short of potash. There are also substantial differences in the yields of the varieties, ranging on the undunged series from 8.9 lb to 21.5 lb. The consistency of the results for the different varieties in the body of the table shows, without any formal analysis, that the results are of reasonable accuracy. The analysis of variance in the paper, however, is very defective. I suspect that Fisher did not give his full attention to it.

These defects were eliminated from the analysis set out in *Statistical Methods*. Indeed, if Tables 48 and 49 were combined into one table the resultant form would be very similar to that now customarily adopted for split-plot experiments.

The one thing that this experiment lacked was proper randomization of the treatments assigned to the plots and to the sub-plots. This concept is developed in Sections 48 and 49 of Chapter 8, using data from uniformity trials. Completely randomized designs, randomized blocks, and Latin squares, were included.

An important addition to the techniques of experimental design and analysis is the use of covariance on preliminary data obtained before the treatments are applied. This is developed in Section 49.1 of the 4th edition and illustrated by data provided by T. Eden from a tea plantation in Ceylon. The text following

Table 61.7 up to and including Table 61.71 was added in the 5th edition in response to criticisms by Wishart, who had perceived the need for these auxiliary estimates of error, but had been unable to provide a solution.

A full list of the additions, some of them of great importance and novelty, but which do not impinge directly on experimental design, will be found in the Preface.

An example of the subtleties that Fisher sometimes introduced into relatively straightforward problems is shown by Section 24.1 on the comparison between two means. In the example in Section 23 of the first edition two sleeping drugs were tested on each of 10 subjects, thus giving 10 paired comparisons, so that the standard error of the difference of the means is obtained from the 10 differences. Someone must have asked him, what do I do if the tests of the two drugs are made on different subjects? He gave the obvious answer: calculate the variance of each set and use the sum of the variances to test the difference of the means. This advice remained until the seventh edition, but in the eighth a further refinement, which I would not myself recommend, had been added; namely that if the two samples are in fact from the same population, the difference between the observed means will give an additional degree of freedom for error, and this can be included in the test conditions at the expense of some algebraic gymnastics, but would need further extension if three or more separate groups are included.

That a little clear thinking on this problem is very necessary is shown by an early paper by Wishart. Confronted by some paired comparison data, he first calculated the standard error ignoring the pairing and then recalculated it taking the pairing into account. He claimed that this had greatly increased the accuracy of the results. But as the mean difference is the same by both methods it should have been clear to him that it was not his ingenuity which had given a better estimate, but that his first calculation of the error was wrong.

In the second paragraph of the preface to the 13th edition Fisher states: "Today exact tests of significance need no apology. The demand, steadily increasing over a long period, for a book designed for a much smaller public has justified at least some of the innovations in its plan which at first must have seemed questionable. The recognition of degrees of freedom; the use of fixed probability levels in tabulating the functions used in tests of significance; the analysis of variance; the need for randomization in experimental design, etc."

The use of chosen probability levels for the tabulation of t, χ^2, and z (or F) is the obvious course for printed tables, but should not be taken to imply that these levels are more than convenient points on a continuous scale. In fact, t and χ^2 have been fairly fully tabulated and present no problems, but F, because it is dependent on a pair of degrees of freedom, requires much more computation and more extensive tabulation. Although rough interpolation between tabulated levels of significance is possible, it is tiresome, and *, **, and *** were customary for indicating 5 per cent, 1 per cent, and 0.1 per cent levels in analysis of variance tables.

The situation has now been entirely changed by advances in electronics. The desk calculator I have used for the last ten years can expeditiously provide probabilities corresponding to any observed values of

t, χ^2, or F without the use of any tables. The same is, or should be, true of computer programs for the analysis of variance.

A Comedy of Misconceptions

In the late 1920s, an epidemic of doubt developed, chiefly amongst the theoreticians, on the reliability of the z-test when applied to data that deviated markedly from the normal distribution. Eden, at Fisher's suggestion, started an investigation designed to test its performance on some very skew data. The data selected comprised 256 heights of wheat shoots taken from an observational scheme on the growth of wheat. It was planned that this should be used to construct a uniformity trial of 8 blocks, each with 4 treatments, but preliminary tests to get a measure of the skewness of the data were made on the 256 individual shoots, and numerical and graphical evidence on this is shown in the paper by Eden and Yates (*J. Agric. Sci.*, Vol. 23, 1933). The planned uniformity trial, however, required only 32 measurements, and these were obtained by taking the means of the groups of 8 of the original measurements. A random selection of one unit out of each group should have been taken, as taking the means eliminated most of the skewness from the data; unfortunately this was not noticed at the time.

Eden found that the work was taking much longer than he expected and asked if I had any suggestions. With the aid of a printer-adder, a 10 × 10 Millionaire, and some cardboard, I devised a method of speeding up the computations, and 1000 samples were successfully taken. But the method can scarcely be described as "obtaining rapidly the results" (*Design,* end of Section 23), nor did it "demonstrate how closely the theoretical distribution was verified in material that was far from normal".

It was some years later, on thinking over this experiment, that I realized that the effect of skewness is always reduced by randomization, as any observation will affect the relevant treatment total positively or negatively, depending on the luck of the draw. Most of the effect, in fact, is to transfer the increase in the third moment to the fourth moment.

With modern computers, problems of this sort can readily be solved by simulation. In fact, the standard t and z tables could have been generated directly from first principles without even establishing the necessary algebraic expressions. In 1972 (*Biometrika*, Vol. 59), I used simulation to test the behaviour of Tukey's one degree of freedom, which singles out the largest of a set of two-factor interactions, mainly to try out the performance of a new computer we had acquired.

The Design of Experiments

The first edition of this book was published in 1935, ten years after *Statistical Methods* first appeared and two years after Fisher's appointment to the Galton Chair. Much of the book is devoted to factorial designs for field experiments on agricultural crops, which Fisher developed in the years immediately following the potato experiment described in Chapter 7 of *Statistical Methods*, work which I joined in when I came to Rothamsted in 1931.

The first three chapters are delightful. Chapter 1 starts with a humorous section on criticisms of experiments by superior persons, and recalls to my mind an occasion when Fisher and I were sitting together near the front in a lecture by an eminent speaker, when he said to me in a loud aside "What a feast of incompetent biometrics." The rest of the chapter gives what I think is the best criticism I have read of inverse probability.

The second chapter gives an account of the world-famous tea-tasting experiment, designed to test the claim of a lady that she could discriminate whether the milk or the tea infusion was first added to the cup. Fisher discusses possible refinements of technique in this experiment, and suggests that only those that are practically convenient need be adopted. This seems to be going too far. I think it will be admitted that in this experiment the lady may confidently and correctly divide the cups into two groups of four, but she may be guided by the proportion of milk in the tea, not by difference in taste when the proportions are equal. The server is probably in the habit of pouring the milk first, and may easily get a different proportion when pouring last. This suggests that the milk should be accurately measured for both types of tea, in spite of the practical inconvenience. I would also adopt six cups of each kind, rather than four, as is suggested in Section 11, particularly if there are several competitors.

Chapter 3 is particularly interesting. It gives a very full account of Galton's attempts to interpret Darwin's data on the growth of crossed and self-fertilized seedlings of *Zea mays*. Galton, who founded the Galton Laboratory, had a passion for numerical data, particularly data on human populations, but had little conception of how best to analyse them. By juggling with the figures, he greatly exaggerated the difference between the crossed and self-fertilized plants, whereas Darwin, by considering only simple averages, had got the right answer. The one defect in his experimental procedures, as Fisher pointed out, was not to randomize the orientation of the pots in the greenhouse.

This omission was of interest to me, as very early in my career at Rothamsted, Winifred Brenchley, the head botanist, came to me with a problem with an experiment on boron. This was a replicated pot experiment on beans, comparing 0, 1, 2, 3, 4, 5 levels of boron dressings. The analysis revealed a slight but significant depression with dressing 1, an expected benefit over 0 for dressing 2, and then a decline to a very low level for dressing 5. In discussion she revealed she had not randomized the pots within replicates. This had the consequence that the 5 pot in each replicate came next to the 0 pot in the next replicate, which had consequently more light and air than the 1 pot. This presumably accounted for the anomaly.

The chapter ends with a criticism of "non-parametric tests", containing the typical Fisherian sentence: "In inductive logic, however, an erroneous assumption of ignorance is not innocuous; it often leads to manifest absurdities."

The rest of the book is very different in style to *Statistical Methods*. In the chapters dealing with field experiments, the data of actual experiments are presented as exercises for the reader without further comment, and without any results against which the reader can check his own analysis and conclusions. Thus in Chapter 4 on simple randomized blocks, a hypothetical experiment on 5 varieties with 8 replicates is discussed. The only numerical table is that for the degrees of freedom. A further table, showing how the observed yields and their totals and means should be set out, is also given. The reader is told to calculate the sums of squares from the sums of the products of totals and means, less the product of the grand total by its mean. (Surely sums of squares with the appropriate divisors would be simpler.) At the end of the chapter we are given, as an example for the reader, a 5 × 12 table of yields, which in the first edition was said to be an experiment on 5 varieties of barley. At my insistence Fisher added a note in the second edition on what the data actually represented, namely varietal totals from 12 experiments located at six centres in each of two years. I was anxious that this should be done, as the experimental error was considerably larger than would be expected from a well-conducted single trial.

One of Fisher's major contributions to experimental techniques was his recognition of the great gains that are obtainable by well-designed factorial experiments. Surprisingly, however, there were considerable doubts in certain quarters. In 1926, Sir John Russell, the Director of Rothamsted, published a paper on experimental design in the *Journal of the Ministry of Agriculture* which put forward this thesis (i.e., investigating one question at a time), and showed how little the subject was then generally understood. It was of value, however, in that it stimulated Fisher to set out his own ideas, which characteristically he did not hesitate to publish in the same journal at the earliest possible moment. Equally characteristically, Sir John bore him no ill-will for this.

In this paper Fisher made a very strong recommendation in favour of factorial design: "In most experiments involving manuring or cultural treatment, the comparisons involving single factors, e.g. with or without phosphate, are of far higher interest and practical importance than the much more numerous possible comparisons involving several factors. This circumstance, through a process of reasoning which can best be illustrated by a practical example, leads to the remarkable consequence that large and complex experiments have a much higher efficiency than simple ones. No aphorism is more frequently repeated in connection with field trials, than that we must ask Nature few questions, or ideally, one question, at a time. The writer is convinced that this view is wholly mistaken. Nature, he suggests, will best respond to a logical and carefully thought out questionnaire; indeed, if we ask her a single question, she will often refuse to answer until some other topic has been discussed."

Of course, the levels of the factors must be chosen sensibly. If it is known, for example, that some nitrogen is certainly necessary, levels 1, 2, 3 of N might be included, with an additional no-nitrogen treatment with medium P and K.

One unwelcome feature of factorial design is that it gives rise to a large number of treatment combinations. Ways of dealing with this problem, partial and total confounding, and estimation of error from high-order interactions, are discussed in the last part of Chapter 6 and in Chapters 7 and 8.

The last three chapters, 9, 10, and 11, are of more general interest. Chapter 11, in particular, is worth study by all readers. Amongst other things it contains sections on percentage mortality at different concentrations of insecticides, etc. (probits), and estimation of linkage.

The reader may well wonder why there is little in

the book on experiments on farm animals. There are several reasons. The main one is that Rothamsted, although a mixed farm, has never been concerned with animal nutrition or breeding. I have myself later been much involved in these matters, but that is another story.

There is a general point of some importance on which I am not in agreement with Fisher. At the beginning of Section 24 (Chapter 4) he writes, "If the yields of the different varieties in the experiment fail to satisfy the test of significance they will not often need to be considered further, for the results, as so far tested, are compatible with the view that all differences observed in the experiment are due to variations in the fertility of the experimental area, and this is the simplest interpretation to put upon them." But the lack of significance may be due either to small differences between the varieties, or to high experimental error, or both. Rejection should only be made on the criterion of unacceptably high error, or better, if similar experiments are made at several places or in several seasons, all the results can be reported with their errors, ready for a combined analysis if one is required.

Scientific Inference

Scientific Inference was first published in 1956, 31 years after *Statistical Methods*. A second edition followed in 1959, but with only minor alterations, mostly textual. The section numbers start from 1 in each chapter, giving a convenient form of reference, e.g. II.4.

The first chapter consists mainly of comments on the work of Galton and Karl Pearson, similar to those in the earlier books, and draws attention to the need for further development of estimation procedures.

A full understanding of the book requires more mathematical expertise than *Statistical Methods* or *The Design of Experiments*, but the first four chapters should be comprehensible to those who have mastered these books. Chapters V and VI, dealing with likelihood and estimation, are more mathematical. I will comment here on only a few specific items which strike me as of particular interest. The rest I leave to the reader.

In Chapter II.4 there is mention, in connection with gaming, of *recognizable sub-sets*. This is a very important concept. If such a sub-set can be recognized (in this case before the die is cast) then the relevant ratio for the gambler is that of the sub-set, not the entire set.

Note that sub-division into sub-sets necessarily reduces the range of probability levels when there is a finite number of observations, and we are dealing with integral data. In the 2 × 2 table, discussed at length in Section IV.4, there are seven possible lower margins, of which only the 3,3 margin gives any significant values at the 5 per cent level.

Other sections that I found particularly interesting are IV.1, Tests of significance and acceptance decisions, essentially what is known as quality control, and IV.5, Composite hypotheses, where it is demonstrated that the frequency of rejection is not to be equated with the level of significance.

The last two sections of this chapter, IV.7 and IV.8, record some of Fisher's disagreements with Bartlett, and with Neyman and Pearson. Fisher in fact was remarkably forbearing in his criticisms of Neyman in *Scientific Inference*. For a fuller account of the conflict between them I would refer the reader to my review of a biography of Neyman by Constance Reid (*J. Roy. Stat. Soc., A*, Vol. 147, 1984).

Barnard, who had at first advocated the "much more powerful" unconditional test for 2 × 2 tables, accepted Fisher's arguments, but much later came to the conclusion that there might be a little information in the marginal totals, which could be extracted by maximum likelihood. This, however, is a secondary matter. Tables published by Pearson and Hartley (1954) are based on Fisher's arguments, though without acknowledgements.

The subject is still a live issue, and papers on it are continually appearing. In 1984, I presented a lengthy paper to the Royal Statistical Society (*J. Roy. Stat Soc., A*, Vol. 147, 1984) in the hope of clearing up some of the confusion. The summary reads as follows:

> Fisher's exact test, and the approximation to it by the continuity-corrected χ^2 test, have repeatedly been attacked over the past 40 years, recently with the support of extensive computer exercises. The present paper argues, on commonsense grounds, supported by simple examples, that these attacks are misconceived, and are mainly due to uncritical acceptance of the Neyman–Pearson approach to tests of significance, the use of nominal levels, and refusal to accept the arguments for conditioning on the margins.
>
> Two-sided tests have also added to the confusion; it is argued that the best definition of a two-sided probability is twice the observed one-tail probability.

Further developments

As the above re-examination of *Statistical Methods* indicates, the basic ideas of the analysis of variance and its application to experiments were all included in the first edition in 1925.

As I wrote in a paper to mark the 25th anniversary of the publication of *Statistical Methods*, "The new methods have been completely accepted by biologists and agricultural research workers, and are rapidly spreading through other branches of scientific and technical research in which the variability of the experimental material necessitates refined techniques. Their introduction has resulted in an immense gain in the accuracy and certainty of experimental results. As examples we may instance work on long-term change-over trials in agriculture, biological assay, and industrial experimentation and quality control."

As is inevitable with any new ideas, misrepresentation and criticism were not slow in making their appearance. In part this was due to not fully understanding the new methods, in part due to rivalry. M. G. Kendall is an example of the former, Neyman of the latter.

Unfortunately, as so often happens, mathematical extensions, conceived as improvements, appeared. The first was one by Churchill Eisenhart, who proposed two different models, the fixed and the random effects models, for Fisher's analysis of variance, depending on whether factors had levels chosen by the experimenter, or are randomly selected, as might happen with varieties. This gives rise to different components of variance, but if these are rightly interpreted there is no real difference between them. My short answer, therefore, is "follow Fisher's example and always use the random effects model", but geneticists interested in components of variance might read the paper I gave at the 5th Berkeley Symposium, which is reproduced together with an added note in *Experimental Design: Selected Papers* (Yates 1970).

FOREWORD

Statistical Tables for Biological, Agricultural and Medical Research

By the mid-1930s it became interestingly obvious that a book of tables, containing properly bound copies of those included in *Statistical Methods*, would be of great benefit to practical workers. When I first suggested this Fisher was averse to it, but eventually he changed his mind. I then discovered, somewhat to my surprise, that he had indeed been thinking about this for some time.

The first paragraph of the Preface, which he dictated to me without any modification, well states the aims of the book:

> The problems with which an active statistical department may have to deal require, if their solutions are to be made widely accessible, a great variety of special tables. A number of those in the present book are familiar to statisticians, and are already widely used. In presenting them in a convenient form, the opportunity has been taken to supplement them with a selection of others, chosen as likely also to be of value and not accessible elsewhere. The volume is completed with a number of tables of standard functions of general utility. The experience of the authors of the problems arising in practical research is the basis of the selection, from among those tables which from time to time have been computed for special purposes.

Part of the success of the venture was that we had the opportunity of publishing tables for new procedures at an early date. The first tables of balanced incomplete blocks, for example, were included in the first edition, and their preparation stimulated Fisher to consider and incorporate combinatorial aspects of these designs, although at that time the idea of recovering interblock information had not occurred to me. This in turn stimulated other combinatorial experts to further research.

United Nations Sub-Commission on Statistical Sampling

Although there is nothing in *Statistical Methods* on sampling, Fisher, while at Rothamsted, in co-operation with the botanists, developed powerful improved techniques for the sampling of experimental plots, both for intermediate measurements on individual plants, and for estimates of final yields. This was extended to estimates of the yields of commercial crops, with which I became involved. I also, before the war, advised the Forestry Commission on sampling problems arising in their Census of Woodlands, and during the war became much more heavily involved in sampling problems of all kinds.

In 1947, the United Nations set up a Sub-Commission on Statistical Sampling to prepare for their projected 1950 World Censuses of Agriculture and Population. In all, the Sub-Commission held seven sessions at yearly intervals, the last being held at Mahalanobis's Indian Statistical Institute, in Calcutta.

The members of the Sub-Commission were selected as leading statistical experts from various countries, some seven members in all, including Darmois from France, Linder from Switzerland, and Deming from the United States. Mahalanobis was Chairman and Fisher was selected as the leading statistical expert on sampling of agricultural crops. It must have been he who suggested that I should be a member.

The Sub-Commission was particularly concerned

with the need for wider use of sampling in the less-developed areas and it was originally intended that only sampling problems in these areas should be dealt with. At their first meeting in 1947 at Lake Success, it was suggested that a manual be prepared for the surveys and the choice to do this fell on me. On consideration, however, I came to the conclusion that conditions differed so greatly in different areas that it would be necessary to cover a wide variety of methods which in essentials differed little from the methods appropriate for fully developed areas. It therefore seemed best to take the opportunity of writing a more general book, referred to in the foreword to *Scientific Inference*; this appeared in 1949.

Sampling of human populations was in a terrible mess in European countries, including the UK, at that time. One effect of Fisher's participation in the Sub-Commission was that a much more rigorous terminology, which has stood the test of time, was defined at the first session. Fisher, of course, was very insistent that all good sampling should have a random component. I maintained that for some purposes systematic location of the chosen elements would be permissible. I won my point on this (Yates, F., *Phil. Trans. A*, Vol. 241).

Computers

Fisher had little interest in computers, then in their infancy, and referred to their calculations as "Meccano arithmetic". He did, however, solve one problem using a computer, whereupon some unkindly soul pointed out that a desk calculator would do the job equally well. So far as I know Fisher made no further attempts.

To me this is evidence of Fisher's phenomenal computing ability. He was not only a fast and accurate computer, he also stored in his memory the whole sequence of computations done years before.

Statistical Methods for Research Workers

Statistical Methods for Research Workers

BY

SIR RONALD A. FISHER, SC.D., F.R.S.

D.Sc. (Adelaide, Ames, Chicago, Harvard, Indian Statistical Institute, Leeds, London)
LL.D. (Calcutta, Glasgow)

Honorary Research Fellow, Division of Mathematical Statistics, C.S.I.R.O., University of Adelaide; Foreign Associate, United States National Academy of Sciences; Foreign Honorary Member, American Academy of Arts and Sciences; Foreign Member, American Philosophical Society; Honorary Member, American Statistical Association; Honorary President International Statistical Institute; Foreign Member, Royal Swedish Academy of Sciences; Member, Royal Danish Academy of Sciences; Member, Pontifical Academy; Member, Imperial German Academy of Natural Science; formerly Fellow of Gonville and Caius College, Cambridge; formerly Galton Professor, University of London; and formerly Balfour Professor of Genetics, University of Cambridge.

FOURTEENTH EDITION—REVISED AND ENLARGED

HAFNER PUBLISHING COMPANY
New York
1973

FIRST PUBLISHED	1925
SECOND EDITION	1928
THIRD EDITION	1930
FOURTH EDITION	1932
FIFTH EDITION	1934
SIXTH EDITION	1936
SEVENTH EDITION	1938
EIGHTH EDITION	1941
NINTH EDITION	1944
TENTH EDITION	1946
TENTH EDITION REPRINTED	1948
ELEVENTH EDITION	1950
TWELFTH EDITION	1954
THIRTEENTH EDITION	1958
THIRTEENTH EDITION REPRINTED	1963
THIRTEENTH EDITION REPRINTED	1967
FOURTEENTH EDITION	1970
FOURTEENTH EDITION REPRINTED	1973

COPYRIGHT © 1970 University of Adelaide

PUBLISHED BY
HAFNER PUBLISHING COMPANY
866 THIRD AVE.
NEW YORK, NEW YORK

PREFACE TO THIRTEENTH EDITION

For several years prior to the preparation of this book, the author had been working in somewhat intimate co-operation with a number of biological research departments at Rothamsted; the book was very decidedly the product of this circumstance. Daily contact with statistical problems as they presented themselves to laboratory workers stimulated the purely mathematical researches upon which the new methods were based. It was clear that the traditional machinery inculcated by the biometrical school was wholly unsuited to the needs of practical research. The futile elaboration of innumerable measures of correlation, and the evasion of the real difficulties of sampling problems under cover of a contempt for small samples, were obviously beginning to make its pretentions ridiculous. These procedures were not only ill-aimed, but, for all their elaboration, not sufficiently accurate. Only by tackling small sample problems on their merits, in the author's view, did it seem possible to apply accurate tests to practical data. With the encouragement of my colleagues, and the valued help of the late W. S. Gosset ("Student"), his assistant Mr E. Somerfield, and Miss W. A. Mackenzie, the first edition was prepared and weathered the hostile criticisms inevitable to such a venture.

To-day exact tests of significance need no apology. The demand, steadily increasing over a long period, for a book designed originally for a much smaller public has justified at least some of the innovations in its plan which at first must have seemed questionable. (The recognition of degrees of freedom; the use of fixed probability levels in tabulating the functions used in tests of significance; the analysis of variance; the need for randomisation in experimental design, etc.) The author was impressed with the practical importance of many recent mathematical advances, which to others seemed to be merely academic refinements. He felt sure, too, that workers with research experience would appreciate a book which, without entering into the mathematical theory of statistical methods, should embody the latest results of that theory, presenting them in the form of practical procedures appropriate to those types of data with which research workers are actually concerned. The practical application of general theorems is a different art from their establishment by mathematical proof. It requires fully as deep an understanding of their meaning, and is, moreover, useful to many to whom the other is unnecessary. To carry out this plan new matter has had to be added with each new edition, to illustrate extensions and improvements, the value of which had in the meantime been established by experience.

In most cases the new methods actually simplify the handling of the data. The conservatism of some university courses in elementary statistics, in stereotyping unnecessary approximations and inappropriate conventions, still hinders many students in the use of exact methods. In reading this book they should try to remember that departures from tradition have not been made capriciously, but only when they have been found to be definitely helpful.

Especially in the order of presentation, the book bears traces of the state of the subject when it first

PREFACE TO THIRTEENTH EDITION

appeared. More recent books have, rightly from the teacher's standpoint, introduced the analysis of variance earlier, and given it more space. They have thus carried further than I the process of abstracting from the field formerly embraced by the correlation coefficient, problems capable of a more direct approach. In excusing myself from the difficult task of a fundamental rearrangement, I may plead that it is of real value to understand the problems discussed by earlier writers, and to .be able to translate them into the system of ideas in which they may be more simply or more comprehensively resolved. I have therefore contented myself with indicating the analysis of variance procedure as an alternative approach in some early examples, as in Sections 24 and 24·1.

With a class capable of mastering the whole book, I should now postpone the matter of Sections 30 to 40, dealing with correlation, until further experience has been gained of the applications of the Analysis of Variance, but should later give time to the ideas of correlation and partial correlation, for their importance in understanding the literature of quantitative biology, which has been so largely influenced by them.

In the second edition the importance of providing a striking and detailed illustration of the principles of statistical estimation led to the addition of a ninth chapter. The subject had received only general discussion in the first edition, and, in spite of its practical importance, had not yet won sufficient attention from teachers to drive out of practice the demonstrably defective procedures which were still unfortunately taught to students. The new chapter superseded Section 6 and Example 1 of the first edition; in the third edition it was enlarged by two new sections (57.1 and 57.2) illustrating further the applicability of the method of maximum likelihood, and of the quantitative evaluation of information. Later K. Mather's admirable book *The Measurement of Linkage in Heredity* has illustrated the appropriate procedures for a wider variety of genetical examples.

In Section 27 a general method of constructing the series of orthogonal polynomials was added to the third edition, in response to the need which is felt, with respect to some important classes of data, to use polynomials of higher degree than the fifth. Simple and direct algebraic proofs of the methods of Sections 28 and 28·1 have been published by Miss F. E. Allan.

In the fourth edition the Appendix to Chapter III, on technical notation, was entirely rewritten, since the inconveniences of the moment notation seemed by that time definitely to outweigh the advantages formerly conferred by its familiarity. Section 5, formerly occupied by an account of the tables available for testing significance, was given to a historical note on the principal contributors to the development of statistical reasoning. The principal new matter in that edition was added in response to the increasing use of the analysis of covariance, which is explained in Section 49·1. Since several writers had found difficulty in applying the appropriate tests of significance to deviations from regression formulæ, this section was further enlarged in the fifth edition.

Other new sections in the fifth edition were 21·01, giving a correction for continuity recently introduced by F. Yates, and 21·02 giving the exact test of significance for 2×2 tables. Workers who are accustomed to handle regression equations with a large number of variates will be interested in Section 29·1, which provides the relatively simple adjustments to be made when, at a late stage, it is decided that one or more of the variates used may with advantage be omitted. The possibility of doing this without laborious recalculations should encourage workers to make the list of independent variates included more comprehensive than has, in the past, been thought advisable.

In the sixth edition Example 15·1, Section 22, gave a new test of homogeneity for data with hierarchical subdivisions. Attention was also called to Working and Hotelling's formula for the sampling error of values estimated by regression, and in Section 29·2 to an extended use of successive summation in fitting polynomials.

I am indebted to Dr W. E. Deming for the extension of the Table of z to the 0·1 per cent. level of significance. Such high levels of significance are especially useful when the test we make is the most favourable out of a number which *a priori* might equally well have been chosen.

Two changes in the seventh edition may be mentioned. Section 27 was expanded so as to give a fuller introduction to the theory of orthogonal polynomials, by way of orthogonal comparisons between observations, which most practical workers find easier to grasp. The arithmetical construction is simpler by this path, and the full generality of the original treatment can be retained without very complicated algebraic expressions. A useful range of tables giving the serial values to the fifth degree is now available in *Statistical Tables*.

Section 49·2 was added to give an outline of the important new subject of the use of multiple measurements to form the best discriminant functions of which they are capable. The tests of significance appropriate to this process are approximate and deserve further study. The diversity of problems which yield to this method is very striking.

A section new in the ninth edition is given to the test of homogeneity of evidence used in estimation, since this subject is the natural and logical complement to the methods of combining independent evidence illustrated in the previous examples. In the tenth edition is an extension of the t-test to find fiducial limits for the ratio of means or regression coefficients (Section 26·2).

The sections of Chapter VIII, the Principles of Experimentation, which have always been too short to do justice to aspects of the subject other than the purely statistical, have since developed into an independent book, *The Design of Experiments* (Oliver and Boyd, 1935, 1937, 1942, 1947, 1949, 1951, 1953). The tables of this book, together with a number of others calculated for a variety of statistical purposes, with illustrations of their use, are now available under the title of *Statistical Tables* (Oliver and Boyd, 1938, 1943, 1948, 1953, 1957). Both of these publications relieve the present work of claims for expansion in directions which threatened to obstruct its usefulness as a single course of study. The serious student should make sure that these volumes also are accessible to him.

Since the middle of this century a flood of literature has appeared bearing on statistical methods. The authors are largely in mathematical teaching departments, and better trained as mathematicians than some of their predecessors. Too often, however, their experience has not included the training and mental

PREFACE TO THIRTEENTH EDITION

discipline of the natural sciences, and much space is given to the trivial and the irrelevant. Competition also has led to methods of publicity with propagandist zeal. Though the methods of this book, and of *The Design of Experiments* have been widely used, the underlying logic has been often misapprehended, and erroneous numerical tables have been published. In 1956, therefore, the previous books were supplemented by one devoted to the logic of induction, under the title of *Statistical Methods and Scientific Inference*. Detailed and explicit demonstration is there given of the logical concepts such as fiducial probability, taken for granted in this book, as distinct from "Decision Functions," "Inverse Probability" and like approaches.

It should be noted that numbers of sections, tables and examples have been unaltered by the insertion of fresh material, so that references to them, though not to pages, will be valid irrespective of the edition used.

DEPARTMENT OF GENETICS, CAMBRIDGE
1958

NOTE TO THIS REPRINT EDITION

THE FOURTEENTH EDITION, 1970, which is reprinted here, included new material introduced from notes written for this purpose by Sir Ronald Fisher some time before his death on 29th July, 1962. The two main changes were an extension of Section 29 and the addition of Section 57.4. In the former, attention is directed to methods available for testing hypotheses involving all p regression coefficients simultaneously, and the latter describes the combination of efficient scores derived from different sets of data all of which are relevant to the same question.

J. H. Bennett

CONTENTS

I INTRODUCTORY

1. The Scope of Statistics — 1
2. General Method, Calculation of Statistics — 6
3. The Qualifications of Satisfactory Statistics — 11
4. Scope of this Book — 16
5. Historical Note — 20

II DIAGRAMS

7. Diagrams — 24
8. Time Diagrams, Growth Rate, and Relative Growth Rate — 24
9. Correlation Diagrams — 29
10. Frequency Diagrams — 33
    10.1 Transformed Frequencies — 37

III DISTRIBUTIONS

11. Distributions — 41
12. The Normal Distribution — 43
13. Fitting the Normal Distribution (Ex. 2) — 45
14. Test of Departure From Normality (Ex. 3) — 52
15. Discontinuous Distributions — 54
16. Small Samples of a Poisson Series (Ex. 4) — 57
17. Presence and Absence of Organisms in Samples — 61
18. The Binomial Distribution (Ex. 5, 6) — 63
19. Small Samples of the Binomial Series (Ex. 7) — 68
    Appendix on Technical Notation and Formulae — 70

IV TESTS OF GOODNESS OF FIT, INDEPENDENCE AND HOMOGENEITY; WITH TABLE OF χ^2

20. The χ^2 Distribution (Ex. 8, 9) — 78
21. Tests of Independence, Contingency Tables (Ex. 10, 11, 12, 13) — 85
    21.01 Yates' Correction for Continuity (Ex. 13.1) — 92
    21.02 The Exact Treatment of 2 × 2 Tables — 96

CONTENTS

21.03	Exact Tests based on the χ^2 Distribution (Ex. 14)	97
21.1	The Combination of Probabilities from Tests of Significance (Ex. 14.1)	99
22.	Partition of χ^2 into its Components (Ex. 15, 15.1)	101

V TESTS OF SIGNIFICANCE OF *MEANS*, DIFFERENCES OF MEANS, AND REGRESSION COEFFICIENTS

23.	The Standard Error of the Mean (Ex. 16, 17, 18)	114
24.	The Significance of the Mean of a Unique Sample (Ex. 19)	119
24.1	Comparison of Two Means (Ex. 20, 21)	122
25.	Regression Coefficients	129
26.	Sampling Errors of Regression Coefficients (Ex. 22)	132
26.1	The Comparison of Regression Coefficients (Ex. 23)	140
26.2	The Ratio of Means and Regression Coefficients (Ex. 23.1, 23.2)	142
27.	The Fitting of Curved Regression Lines	147
28.	The Arithmetical Procedure of Fitting	151
28.1	The Calculation of the Polynomial Values	154
29.	Regression with Several Independent Variates (Ex. 24)	156
29.1	The Omission of an Independent Variate	166
29.2	Polynomial Fitting when the Frequencies are Unequal	168

VI THE CORRELATION COEFFICIENT

30.	The Correlation Coefficient	177
31.	The Statistical Estimation of the Correlation (Ex. 25)	185
32.	Partial Correlations (Ex. 26)	189
33.	Accuracy of the Correlation Coefficient	194
34.	The Significance of an Observed Correlation (Ex. 27, 28)	195
35.	Transformed Correlations (Ex. 29, 30, 31, 32, 33)	199
36.	Systematic Errors	207
37.	Correlation between Series	208

VII INTRACLASS CORRELATIONS AND THE ANALYSIS OF VARIANCE

38.	Intraclass Correlations	213
39.	Sampling Errors of Intraclass Correlations (Ex. 34, 35, 36)	217
40.	Intraclass Correlation as an Example of the Analysis of Variance	223
41.	Test of Significance of Difference of Variance (Ex. 37, 38, 39)	227

CONTENTS

42. Analysis of Variance into more than Two Portions (Ex. 40, 41) — 235

VIII FURTHER APPLICATIONS OF THE ANALYSIS OF VARIANCE

43. Analysis of Variance — 250
44. Fitness of Regression Formulae (Ex. 42) — 250
45. The "Correlation Ratio" η — 257
46. Blakeman's Criterion — 259
47. Significance of the Multiple Correlation Coefficient (Ex. 43) — 260
48. Technique of Plot Experimentation (Ex. 44, 45) — 263
49. The Latin Square (Ex. 46) — 269
    49.1 The Analysis of Covariance (Ex. 46.1) — 272
    49.2 The Discrimination of Groups by Means of Multiple Measurements; Appropriate Scores (Ex. 46.2) — 287
    49.3 The Precision of Estimated Scores — 297

IX THE PRINCIPLES OF STATISTICAL ESTIMATION

50. The Estimation of Genetic Linkage (Ex. 47) — 301
51. The Significance of Evidence for Linkage — 302
52. The Specification of the Progeny Population for Linked Factors — 303
53. The Multiplicity of Consistent Statistics — 305
54. The Comparison of Statistics by Means of the Test of Goodness of Fit — 307
55. The Sampling Variance of Statistics — 308
56. Comparison of Efficient Statistics — 316
57. The Interpretation of the Discrepancy χ^2 — 318
    57.1 Fragmentary Data (Ex. 48) — 322
    57.2 The Amount of Information: Design and Precision — 327
    57.3 Test of Homogeneity of Evidence used in Estimation — 331
    57.4 Compiling and Summarizing Data of use in Estimation — 335
58. Summary of Principles — 337

Sources Used for Data and Methods — 340

Bibliography — 345

Index — 359

TABLES

I. and II. Normal Distribution — 77
III. Table of χ^2 — 112

CONTENTS

IV.	Table of t	176
V.A.	Correlation Coefficient—Significant Values	211
V.B.	Correlation Coefficient—Transformed Values	212
VI.	Table of z	244–249

WORKED EXAMPLES

2.	Fitting a normal distribution to a large sample	46
3.	Use of higher powers to test normality	52
4.	Test of agreement with Poisson series of a number of small samples	58
5.	Binomial distribution given by dice records	63
6.	Comparison of sex ratio in human families with binomial distribution	66
7.	The accuracy of estimates of infestation	68
8.	Comparison with expectation of Mendelian class frequencies	82
9.	Comparison with expectation of the Poisson series and binomial series	83
10.	Test of independence in a 2×2 classification	85
11.	Test of independence in a $2 \times n'$ classification	87
12.	Test of independence in a 4×4 classification	88
13.	Homogeneity of different families in respect of ratio black:red	91
13.1	Frequency of criminality among the twin brothers and sisters of criminals	94
14.	Agreement with expectation of normal variance	98
14.1	Significance of the product of a number of independent probabilities	100
15.	Partition of observed discrepancies from Mendelian expectation	101
15.1	Complex test on homogeneity in data with hierarchical subdivisions	106
16.	Significance of mean of a large sample	115
17.	Standard error of difference of means from large samples	116
18.	Standard error of mean of differences	117
19.	Significance of mean of a small sample	121
20.	Significance of difference of means of small samples	125
21.	Significance of change in bacterial numbers	127
22.	Effect of nitrogenous fertilisers in maintaining yield	136
23.	Comparison of relative growth rate of two cultures of an alga	140
23.1	[Limits for the ratio of the potencies of two drugs]	142
23.2	The age at which girls become taller than boys	145
24.	Dependence of rainfall on position and altitude	160
25.	Parental correlation in stature	186

26.	Elimination of age in organic correlations with growing children	189
27.	Significance of a correlation coefficient between autumn rainfall and wheat crop	196
28.	Significance of a partial correlation coefficient	198
29.	Test of the approximate normality of the distribution of z	203
30.	Further test of the normality of the distribution of z	204
31.	Significance of deviation from expectation of an observed correlation coefficient	204
32.	Significance of difference between two observed correlations	205
33.	Combination of values from small samples	206
34.	Accuracy of an observed intraclass correlation	220
35.	Extended use of Table V.B.	221
36.	Significance of intraclass correlation from large samples	222
37.	Sex difference in variance of stature	229
38.	Homogeneity of small samples	230
39.	Comparison of intraclass correlations	231
40.	Diurnal and annual variation of rain frequency	235
41.	Analysis of variation in experimental field trials	238
42.	Test of straightness of regression line	253
43.	Significance of a multiple correlation	262
44.	Accuracy attained by random arrangement	264
45.	Restrictions upon random arrangement	265
46.	Doubly restricted arrangements	269
46.1	Covariance of tea yields in successive periods	274
46.2	The derivation of an additive scoring system from serological readings	291
47.	[Estimation of linkage from the progeny of self-fertilised heterozygotes]	301
48.	[Estimation of linkage from fragmentary data]	323

I

INTRODUCTORY

1. The Scope of Statistics

THE science of statistics is essentially a branch of Applied Mathematics, and may be regarded as mathematics applied to observational data. As in other mathematical studies, the same formula is equally relevant to widely different groups of subject-matter. Consequently the unity of the different applications had usually been overlooked, the more naturally because the development of the underlying mathematical theory had been much neglected. We shall therefore consider the subject-matter of statistics under three different aspects, and then show in more mathematical language that the same types of problems arise in every case. Statistics may be regarded as (i) the study of **populations,** (ii) as the study of **variation,** (iii) as the study of methods of the **reduction of data.**

The original meaning of the word "statistics" suggests that it was the study of populations of human beings living in political union. The methods developed, however, have nothing to do with the political unity of the group, and are not confined to populations of men or of social insects. Indeed, since no observational record can completely specify a human being, the populations studied are always to some extent abstractions. If we have records of the stature of 10,000 recruits, it is rather the population of statures than the population of recruits that is open to study. Nevertheless, in a real sense, statistics is the study of populations, or aggregates of individuals, rather than of individuals. Scientific theories which involve the properties of large aggregates of individuals, and not necessarily the properties of the individuals themselves, such as the Kinetic Theory of Gases, the Theory of Natural Selection, or the chemical Theory of Mass Action, are essentially statistical arguments, and are liable to misinterpretation as soon as the statistical nature of the argument is lost sight of. In Wave Mechanics this is now clearly recognised. Statistical methods are essential to social studies, and it is principally by the aid of such methods that these studies may be raised to the rank of sciences. This particular dependence of social studies upon statistical methods has led to the unfortunate misapprehension that statistics is to be regarded as a branch of economics, whereas in truth methods adequate to the treatment of economic data, in so far as these exist, have mostly been developed in the study of biology and the other sciences.

The idea of a population is to be applied not only to living, or even to material, individuals. If an observation, such as a simple measurement, be repeated indefinitely, the aggregate of the results is a population of measurements. Such populations are the particular field of study of the **Theory of Errors,** one of the oldest and most fruitful lines of statistical investigation. Just as a single observation may be regarded as an individual, and its repetition as generating a population, so the entire result of an extensive experiment may be regarded as but one of a possible population of such experiments. The salutary habit of repeating important experiments, or of carrying out original observations in replicate, shows a tacit appreciation of the fact that the object of our study is not the individual result, but the population of possibilities of which we do our best to make our experiments representative. The calculation of means and standard errors shows a deliberate attempt to learn something about that population.

The conception of statistics as the study of variation is the natural outcome of viewing the subject as the study of populations; for a population of individuals in all respects identical is completely described by a description of any one individual, together with the number in the group. The populations which are the object of statistical study always display variation in one or more respects. To speak of statistics as the study of variation also serves to emphasise the contrast between the aims of modern statisticians and those of their predecessors. For until comparatively recent times, the vast majority of workers in this field appear to have had no other aim than to ascertain aggregate, or average, values. The variation itself was not an object of study, but was recognised rather as a troublesome circumstance which detracted from the value of the average. The error curve of the *mean* of a normal sample has been familiar for a century, but that of the *standard deviation* was the object of researches up to 1915. Yet, from the modern point of view, the study of the causes of variation of any variable phenomenon, from the yield of wheat to the intellect of man, should be begun by the examination and measurement of the variation which presents itself.

The study of variation leads immediately to the concept of a **frequency distribution**. Frequency distributions are of various kinds; the number of classes in which the population is distributed may be finite or infinite; again, in the case of quantitative variates, the intervals by which the classes differ may be finite or infinitesimal. In the simplest possible case, in which there are only two classes, such as male and female births, the distribution is simply specified by the proportion in which these occur, as for example by the statement that 51 per cent. of the births are of males and 49 per cent. of females. In other cases the variation may be discontinuous, but the number of classes indefinite, as with the number of children born to different married couples; the frequency distribution would then show the frequency with which 0, 1, 2 . . . children were recorded, the number of classes being sufficient to include the largest family in the record. The variable quantity, such as the number of children, is called the **variate,** and the frequency distribution specifies how frequently the variate takes each of its possible values. In the third group of cases, the variate, such as human stature, may take any intermediate value within its range of variation; the variate is then said to vary continuously, and the frequency distribution may be expressed by stating, as a mathematical function of the variate, either (i) the proportion of the population for which the variate is less than any given value, or (ii) by the mathematical device of differentiating this function, the (infinitesimal) proportion of the population for which the variate falls within any infinitesimal element of its range.

The idea of a frequency distribution is applicable either to populations which are finite in number, or to infinite populations, but it is more usefully and more simply applied to the latter. A finite population can only be divided in certain limited ratios, and cannot in any case exhibit continuous variation. Moreover, in most cases only an infinite population can exhibit accurately, and in their true proportion, the whole of the possibilities arising from the causes actually at work, and which we wish to study. The actual observations can only be a sample of such possibilities. With an infinite population the frequency distribution specifies the fractions of the population assigned to the several classes; we may have (i) a finite number of fractions adding up to unity as in the Mendelian frequency distributions, or (ii) an infinite series of finite fractions adding up to unity, or (iii) a mathematical function expressing the fraction of the total in each of the infinitesimal elements in which the range of the variate may be divided. The last possibility may be represented by a frequency curve; the values of the variate are set out along a horizontal axis, the fraction of the total population, within any limits of the variate, being represented by the area of the curve standing on the corresponding length of the axis. It should be noted that the familiar concept of the frequency curve is only applicable to an infinite population with a continuous variate.

The study of variation has led not merely to measurement of the amount of variation present, but to the study of the qualitative problems of the type, or form, of the variation. Especially important is the study of the simultaneous variation of two or more variates. This study, arising principally out of the work of Galton and Pearson, is generally known under the name of **Correlation**, or, more descriptively, as **Covariation**.

The third aspect under which we shall regard the scope of statistics is introduced by the practical need to reduce the bulk of any given body of data. Any investigator who has carried out methodical and extensive observations will probably be familiar with the oppressive necessity of reducing his results to a more convenient bulk. No human mind is capable of grasping in its entirety the meaning of any considerable quantity of numerical data. We want to be able to express all the *relevant* information contained in the mass by means of comparatively few numerical values. This is a purely practical need which the science of statistics is able to some extent to meet. In some cases at any rate it *is* possible to give the whole of the relevant information by means of one or a few values. In all cases, perhaps, it is possible to reduce to a simple numerical form the main issues which the investigator has in view, in so far as the data are competent to throw light on such issues. The number of independent facts supplied by the data is usually far greater than the number of facts sought, and in consequence much of the information supplied by any body of actual data is irrelevant. It is the object of the statistical processes employed in the reduction of data to exclude this irrelevant information, and to isolate the whole of the relevant information contained in the data.

2. General Method, Calculation of Statistics

The discrimination between the irrelevant information and that which is relevant is performed as follows. Even in the simplest cases the values (or sets of values) before us are interpreted as a random sample of a hypothetical infinite population of such values as might have arisen in the same circumstances. The distribution of this population will be capable of some kind of mathematical specification, involving a certain number, usually few, of **parameters,** or " constants " entering into the mathematical formula. These parameters are the characters of the population. If we could know the exact values of the parameters, we should know all (and more than) any sample from the population could tell us. We cannot in fact know the parameters exactly, but we can make estimates of their values, which will be more or less inexact. These estimates, which are termed **statistics,** are of course calculated from the observations. If we can find a mathematical form for the population which adequately represents the data, and then calculate from the data the best possible estimates of the required parameters, then it would seem that there is little, or nothing, more that the data can tell us; we shall have extracted from it all the available relevant information.

A recent practice, or affectation, is to call these estimates in preference " estimators ". This innovation appears to arise from, and to lead to, confusion of thought. It is difficult in any particular case to know whether by " estimator " is meant a method of estimation, or the algebraic specification of the estimate reached by that method, or the particular value in a single instance. To speak of the " estimate " is unambiguous, whether its value is expressed arithmetically or algebraically; the word is not easily mistaken as meaning " Method of estimation ".

The value of such estimates as we can make is enormously increased if we can calculate the magnitude and nature of the errors to which they are subject. If we can rely upon the specification adopted, this presents the purely mathematical problem of deducing from the nature of the population what will be the behaviour of each of the possible statistics which can be calculated. This type of problem, with which until recent years comparatively little progress had been made, is the basis of the tests of significance by which we can examine whether or not the data are in harmony with any suggested hypothesis. In particular, it is necessary to test the adequacy of the hypothetical specification of the population upon which the method of reduction was based.

The problems which arise in the reduction of data may thus conveniently be divided into three types :

(i) Problems of **Specification**, which arise in the choice of the mathematical form of the population. This is not arbitrary, but requires an understanding of the way in which the data are supposed to, or did in fact, originate. Its further discussion depends on such fields as the theory of Sample Survey, or that of Experimental Design.

(ii) When a specification has been obtained, problems of **Estimation** arise. These involve the choice among the methods of calculating, from our sample, statistics fit to estimate the unknown parameters of the population.

(iii) Problems of **Distribution** include the mathematical deduction of the exact nature of the distributions in random samples of our estimates of the parameters, and of other statistics designed to test the validity of our specification (tests of **Goodness of Fit**).

The statistical examination of a body of data is thus logically similar to the general alternation of inductive and deductive methods throughout the sciences. A hypothesis is conceived and defined with all necessary exactitude ; its logical consequences are ascertained by a deductive argument ; these consequences are compared with the available observations ; if these are completely in accord with the deductions, the hypothesis is justified at least until fresh and more stringent observations are available. The author has attempted a fuller examination of the logic of planned experimentation in his book, *The Design of Experiments*; and of rational induction in *Statistical Methods and Scientific Inference*.

The deduction of inferences respecting samples, from assumptions respecting the populations from which they are drawn, shows us the position in Statistics of the classical **Theory of Probability**. For a given population we may calculate the probability with which any given sample will occur, and if we can solve the purely mathematical problem presented, we can calculate the probability of occurrence of any given statistic calculated from such a sample. The problems of distribution may in fact be regarded as applications and extensions of the theory of probability. Three of the distributions with which we shall be concerned, Bernoulli's binomial distribution, Laplace's normal distribution, and Poisson's series, were developed by writers on probability. For many years, extending over a century and a half, attempts were made to extend the domain of the idea of probability to the deduction of inferences respecting populations from assumptions (or observations) respecting samples. Such inferences were formerly distinguished under the heading of **Inverse Probability**, and have at times gained wide acceptance. This is not the place to enter into the subtleties of a prolonged controversy; it will be sufficient in this general outline of the scope of Statistical Science to reaffirm my personal conviction, which I have sustained elsewhere, that the theory of inverse probability is founded upon an error, and must be wholly rejected. Inferences respecting populations, from which known samples have been drawn, cannot by this method be expressed in terms of probability, save in those cases in which there is an observational basis for making exact probability statements in advance about the population in question.

The probabilities arising from such tests of significance, as those we shall later designate by t and z, are, however, entirely distinct from statements of inverse probability, and are free from the objections which apply to these latter. Their interpretation as probability statements respecting populations constitutes an application unknown to the classical writers on probability. The method of reasoning which is demonstrated explicitly in *Scientific Inference*, is distinguished from that of inverse probability, as the fiducial argument. The statements arrived at are often called statements of **Fiducial Probability**, though indeed the "probability" concerned is in perfect strictness the Mathematical Probability of such old masters as Fermat, Pascal, Bernoulli, De Moivre and Bayes.

The rejection of the theory of inverse probability was for a time wrongly taken to imply that we cannot draw, from knowledge of a sample, inferences respecting the corresponding population. Such a view would entirely deny validity to all experimental science. What has now appeared is that the mathematical concept of probability is, in cases in which fiducial probability is not available, inadequate to express our mental confidence or diffidence in making such inferences, and that the mathematical quantity which usually appears to be appropriate for measuring our order of preference among different possible populations does not in fact obey the laws of probability. To distinguish it from probability, I have used the term "**Likelihood**" to designate this quantity *; since both the words "likelihood" and "probability" are loosely used in common speech to cover both kinds of relationship.

3. The Qualifications of Satisfactory Statistics

The solutions of problems of distribution (which may be regarded as purely deductive problems in the theory of probability) not only enable us to make critical tests of the significance of statistical results, and of the adequacy of the hypothetical distributions upon which our methods of numerical inference are based, but afford real guidance in the choice of appropriate statistics for purposes of estimation. Such statistics may be divided into classes according to the behaviour of their distributions in large samples.

If we calculate a statistic, such, for example, as the mean, from a very large sample, we are accustomed to ascribe to it great accuracy; and indeed it will usually, but not always, be true, that if a number of such statistics can be obtained and compared, the discrepancies among them will grow less and less, as the samples from which they are drawn are made larger and larger. In fact, as the samples are made larger without limit, the statistic will usually tend to some fixed value characteristic of the population, and, therefore, expressible in terms of the parameters of the population. If, therefore, such a statistic is to be used to estimate these parameters, there is only one parametric function to which it can properly be equated. If it be equated to some other parametric function, we

* A more specialised application of the likelihood is its use, under the name of "power function," for comparing the sensitiveness, in some chosen respect, of different possible tests of significance.

shall be using a statistic which even from an infinite sample does not give the correct value; it tends indeed to a fixed value, but to a value which is erroneous from the point of view with which it was used. Such statistics are termed **Inconsistent** Statistics; except when the error is extremely minute, as in the use of Sheppard's adjustments, inconsistent statistics should be regarded as outside the pale of decent usage.

Consistent statistics, on the other hand, all tend more and more nearly to give the correct values, as the sample is more and more increased; at any rate, if they tend to any fixed value it is not to an incorrect one. In the simplest cases, with which we shall be concerned, they not only tend to give the correct value, but the errors, for samples of a given size, tend to be distributed in a well-known distribution (of which more in Chap. III) known as the Normal Law of Frequency of Error, or more simply as the **normal distribution**. The liability to error may, in such cases, be expressed by calculating the mean value of the squares of these errors, a value which is known as the **variance**; and in the class of cases with which we are concerned, the variance falls off with increasing samples, in inverse proportion to the number in the sample.

The foregoing paragraphs specify the notion of consistency in terms suitable to the theory of Large Samples, *i.e.* by means of the properties required as the sample is increased without limit. Logically it is important that consistency can also be defined strictly for small (*i.e.* finite) samples by the stipulation that if for each frequency observed its expectation were substituted, then consistent statistics would be equal identically to the parameters of which they are estimates. The method is illustrated in Section 53.

For the purpose of estimating any parameter, such as the centre of a normal distribution, it is usually possible to invent any number of statistics such as the arithmetic mean, or the median, etc., which shall be consistent in the sense defined above, and each of which has in large samples a variance falling off inversely with the size of the sample. But for large samples of a fixed size the variance of these different statistics will generally be different. Consequently, a special importance belongs to a smaller group of statistics, the error distributions of which tend to the normal distribution, as the sample is increased, with the least possible variance. We may thus separate off from the general body of consistent statistics a group of especial value, and these are known as **efficient** statistics.

The reason for this term may be made apparent by an example. If from a large sample of (say) 1000 observations we calculate an efficient statistic, A, and a second consistent statistic, B, having twice the variance of A, then B will be a valid estimate of the required parameter, but one definitely inferior to A in its accuracy. Using the statistic B, a sample of 2000 values would be required to obtain as good an estimate as is obtained by using the statistic A from a sample of 1000 values. We may say, in this sense, that the statistic B makes use of 50 per cent. of the relevant information available in the observations; or, briefly, that its **efficiency** is 50 per cent. The term " efficient " in its absolute sense is reserved for statistics the efficiency of which is 100 per cent.

Statistics having efficiency less than 100 per cent.

may be legitimately used for many purposes. It is conceivable, for example, that it might in some cases be less laborious to increase the number of observations than to apply a more elaborate method of calculation to the results. It may often happen that an inefficient statistic is accurate enough to answer the particular questions at issue. None the less, so much teaching time is wasted on these inexpensive, but inefficient, methods that the student is often found to have learnt no others ; and it is often overlooked that if we are to make accurate tests of significance, the methods of fitting employed must not introduce **errors of fitting** comparable to the **errors of random sampling** ; when this requirement is investigated, it appears that when tests of significance or of goodness of fit are required, the statistics employed in fitting must be not only consistent, but must be of 100 per cent. efficiency. This is a very serious limitation to the use of inefficient statistics, since in the examination of any body of data it is desirable to be able at any time to test the validity of one or more of the provisional assumptions which might be made.

Numerous examples of the calculation of statistics will be given in the following chapters, and, in these illustrations of method, efficient statistics have been chosen. The discovery of efficient statistics in new types of problem may require some mathematical investigation. The researches of the author have led him to the conclusion that an efficient statistic can in all cases be found by the **Method of Maximum Likelihood** ; that is, by choosing statistics so that the estimated population should be that for which the likelihood is greatest. In view of the mathematical difficulty of some of the problems which arise it is also

§ 3] INTRODUCTORY 15

useful to know that *approximations* to the maximum likelihood solution are also in most cases efficient statistics. Some simple examples of the application of the method of maximum likelihood, and other methods, to genetical problems are developed in the final chapter.

For practical purposes it is not generally necessary to press refinement of methods further than the stipulation that the statistics used should be efficient. With large samples it may be shown that all efficient statistics tend to equivalence, so that little inconvenience arises from diversity of practice. There is, however, one class of statistics, including some of the most frequently recurring examples, which is of theoretical interest for possessing the remarkable property that, even in small samples, a statistic of this class alone includes the whole of the relevant information which the observations contain. Such statistics are distinguished by the term **sufficient** and, in the use of small samples, sufficient statistics, when they exist, are definitely superior to other efficient statistics. Examples of sufficient statistics are the **arithmetic mean** of samples from the normal distribution, or from the Poisson series; it is the fact of providing sufficient statistics for these two important types of distribution which gives to the arithmetic mean its theoretical importance. The method of maximum likelihood leads to these sufficient statistics when they exist. By a further extension, also depending on a special, but not uncommon, functional relationship, the advantage of sufficient statistics, namely **exhaustive** estimation, may be gained by using **ancillary** statistics, even when no statistic sufficient by itself exists.

While diversity of practice within the limits of

efficient statistics will not with large samples lead to inconsistencies, it is, of course, of importance in all cases to distinguish clearly the parameter of the population, of which it is desired to estimate the value from the actual statistic employed as an estimate of its value; and to inform the reader by which of the considerable variety of processes which exist for the purpose the estimate was actually obtained.

4. Scope of this Book

The prime object of this book is to put into the hands of research workers, and especially of biologists, the means of applying statistical tests accurately to numerical data accumulated in their own laboratories or available in the literature. Such tests are the result of solutions of problems of distribution, most of which are but recent additions to our knowledge and have previously only appeared in specialised mathematical papers. The mathematical complexity of these problems has made it seem undesirable to do more than (i) to indicate the kind of problem in question, (ii) to give numerical illustrations by which the whole process may be checked, (iii) to provide numerical tables by means of which the tests may be made without the evaluation of complicated algebraical expressions.

It would have been impossible to give methods suitable for the great variety of kinds of tests which are required but for the unforeseen circumstance that each mathematical solution appears again and again in questions which at first sight appeared to be quite distinct. For example, Helmert's solution in 1875 of the distribution of the sum of the squares of deviations from a mean, is in reality equivalent to the distribution of χ^2 given by K. Pearson in 1900. It was again discovered independently by "Student" in 1908, for the distribution of the variance of a normal sample. The same distribution was found by the author for the index of dispersion derived from small samples from a Poisson series. What is even more remarkable is that, although Pearson's paper of 1900 contained a serious error, which vitiated most of the tests of goodness of fit made by this method until 1921, yet the correction of this error, when efficient methods of estimation are used, leaves the form of the distribution unchanged, and only requires that some few units should be deducted from one of the variables with which the Table of χ^2 is entered.

It is equally fortunate that the distribution of t, first established by "Student" in 1908, in his study of the probable error of the mean, should be applicable, not only to the case there treated, but to the more complex, but even more frequently needed problem of the comparison of two mean values. It further provides an exact solution of the sampling errors of the enormously wide class of statistics known as regression coefficients.

In studying the exact theoretical distributions in a number of other problems, such as those presented by intraclass correlations, the goodness of fit of regression lines, the correlation ratio, and the multiple correlation coefficient, the author has been led repeatedly to a third distribution, which may be called the distribution of z, and which is intimately related to, and indeed a natural extension of, the distributions introduced by Pearson and "Student." It has thus been possible to classify the necessary distributions covering a very great variety of cases, under these three main groups; and, what is equally important, to make some provision for the need for numerical values by means of a few tables only. Tables needed for a wider range of problems, with illustrations of their use, have since been published separately.

The book has been arranged so that the student may make acquaintance with these three main distributions in a logical order, and proceeding from more simple to more complex cases. Methods developed in later chapters are frequently seen to be generalisations of simpler methods developed previously. Studying the work methodically as a connected treatise, the student will, it is hoped, not miss the fundamental unity of treatment under which such very varied material has been brought together; and will prepare himself to deal competently and with exactitude with the many analogous problems which cannot be individually exemplified. On the other hand, it is recognised that many will wish to use the book for laboratory reference, and not as a connected course of study. This use would seem desirable only if the reader will be at the pains to work through, in all numerical detail, one or more of the appropriate examples, so as to assure himself, not only that his data are appropriate for a parallel treatment, but that he has obtained a critical grasp of the meaning to be attached to the processes and results.

It is necessary to anticipate one criticism, namely, that in an elementary book, without mathematical proofs, and designed for readers without special mathematical training, so much has been included which from the teacher's point of view is advanced; and indeed much that has not previously appeared in print. By way of apology the author would like to put forward the following considerations.

(1) For non-mathematical readers, numerical

tables are in any case necessary; accurate tables are no more difficult to use, though more laborious to calculate, than inaccurate tables embodying the approximations formerly current.

(2) The process of calculating a probable or standard error from one of the established formulæ gives no real insight into the random sampling distribution, and can only supply a test of significance by the aid of a table of deviations of the normal curve, and on the assumption that the distribution is in fact very nearly normal. Whether this procedure should, or should not, be used must be decided, not by the mathematical attainments of the investigator, but by discovering whether it will or will not give a sufficiently accurate answer. The fact that such a process has been used successfully by eminent mathematicians in analysing very extensive and important material does not imply that it is sufficiently accurate for the laboratory worker anxious to draw correct conclusions from a small group of perhaps preliminary observations.

(3) The exact distributions, with the use of which this book is chiefly concerned, have been in fact developed in response to the practical problems arising in biological and agricultural research; this is true not only of the author's own contribution to the subject, but from the beginning of the critical examination of statistical distributions in " Student's " paper of 1908.

The greater part of the book is occupied by numerical examples; and these have steadily increased in number as fresh points needed illustration. In choosing them it has appeared to the author a hopeless task to attempt to exemplify the great variety of subject-matter to which these processes may be usefully applied. There are no examples from astronomical statistics, in which important work has been done in recent years, few from social studies, and the biological applications are scattered unsystematically. The examples have rather been chosen each to exemplify a particular process, and seldom on account of the importance of the data used, or even of similar examinations of analogous data. By a study of the processes exemplified, the student should be able to ascertain to what questions, in his own material, such processes are able to give a definite answer ; and, equally important, what further observations would be necessary to settle other outstanding questions. In conformity with the purpose of the examples the reader should remember that they do not pretend to be discussions of general scientific questions, which would require the examination of much more extended data, and of other evidence, but are solely concerned with the critical examination of the particular batch of data presented.

5. Historical Note

Since much interest has been evinced in the historical origin of the statistical theory underlying the methods of this book, and as some misapprehensions have occasionally gained publicity, ascribing to the originality of the author methods well known to some previous writers, or ascribing to his predecessors modern developments of which they were quite unaware, it is hoped that the following notes on the principal contributors to statistical theory will be of value to students who wish to see the modern work in its historical setting.

Thomas Bayes' celebrated essay published in 1763 is well known as containing the first attempt to use the theory of probability as an instrument of inductive reasoning; that is, for arguing from the particular to the general, or from the sample to the population. It was published posthumously, and we do not know what views Bayes would have expressed had he lived to publish on the subject. We do know that the reason for his hesitation to publish was his dissatisfaction with the postulate associated with "Bayes' Theorem." While we must reject this postulate, we should also recognise Bayes' greatness in perceiving the problem to be solved, in illustrating the possibility of its experimental solution, and finally in realising more clearly than many subsequent writers the weakness of the axiomatic method.

Whereas Bayes excelled in logical penetration, Laplace (1820) was unrivalled for his mastery of analytic technique. He admitted the principle of inverse probability, quite uncritically, into the foundations of his exposition. On the other hand, it is to him we owe the principle that the distribution of a quantity compounded of independent parts shows a whole series of features—the mean, variance, and other cumulants (p. 73)—which are simply the sums of like features of the distributions of the parts. These seem to have been later discovered independently by Thiele (1889), but mathematically Laplace's methods were more powerful than Thiele's and far more influential on the development of the subject in France and England. A direct result of Laplace's study of the distribution of the resultant of numerous independent causes was the recognition of the normal law of error, a law more usually ascribed, with some reason, to his great contemporary, Gauss.

Gauss, moreover, approached the problem of

statistical estimation in an empirical spirit, raising the question of the estimation not only of probabilities but of other quantitative parameters. He perceived the aptness for this purpose of the Method of Maximum Likelihood, although he attempted to derive and justify this method from the principle of inverse probability. The method has been attacked on this ground, but it has no real connection with inverse probability. Gauss, further, perfected the systematic fitting of regression formulæ, simple and multiple, by the method of least squares, which, in the cases to which it is appropriate, is a particular example of the method of maximum likelihood.

The first of the distributions characteristic of modern tests of significance, though originating with Helmert, was rediscovered by K. Pearson in 1900, for the measure of discrepancy between observation and hypothesis, known as χ^2. This, I believe, is the great contribution to statistical methods by which the unsurpassed energy of Prof. Pearson's work will be remembered. It supplies an exact and objective measure of the joint discrepancy from their expectations of a number of normally distributed, and mutually correlated, variates. In its primary application to frequencies, which are discontinuous variates, the distribution is necessarily only an approximate one, but when small frequencies are excluded the approximation is satisfactory. The distribution is exact for other problems solved later. With respect to frequencies, the apparent goodness of fit is often exaggerated by the inclusion of vacant or nearly vacant classes which contribute little or nothing to the observed χ^2, but increase its expectation, and by the neglect of the effect on this expectation of adjusting the parameters of the population to fit those of the sample. The need for correction on this score was for long ignored, and later disputed, but is now, I believe, admitted. The chief cause of error tending to lower the apparent goodness of fit is the use of inefficient methods of fitting (Chapter IX). This limitation could scarcely have been foreseen in 1900, when the very rudiments of the theory of estimation were unknown.

The study of the exact sampling distributions of statistics commences in 1908 with " Student's " paper *The Probable Error of a Mean*. Once the true nature of the problem was indicated, a large number of sampling problems were within reach of mathematical solution. " Student " himself gave in this and a subsequent paper the correct solutions for three such problems—the distribution of the estimate of the variance, that of the mean divided by its estimated standard deviation, and that of the estimated correlation coefficient between independent variates. These sufficed to establish the position of the distributions of χ^2 and of t in the theory of samples, though further work was needed to show how many other problems of testing significance could be reduced to these same two forms, and to the more inclusive distribution of z. " Student's " work was not quickly appreciated (it had, in fact, been totally ignored in the journal in which it had appeared), and from the first edition it has been one of the chief purposes of this book to make better known the effect of his researches, and of mathematical work consequent upon them, on the one hand, in refining the traditional doctrine of the theory of errors and mathematical statistics, and on the other, in simplifying the arithmetical processes required in the interpretation of data.

II

DIAGRAMS

7. The preliminary examination of most data is facilitated by the use of diagrams. Diagrams prove nothing, but bring outstanding features readily to the eye; they are therefore no substitute for such critical tests as may be applied to the data, but are valuable in suggesting such tests, and in explaining the conclusions founded upon them.

8. Time Diagrams, Growth Rate, and Relative Growth Rate

The type of diagram in most frequent use consists in plotting the values of a variable, such as the weight of an animal or of a sample of plants against its age, or the size of a population at successive intervals of time. Distinction should be drawn between those cases in which the same group of animals, as in a feeding experiment, is weighed at successive intervals of time, and the cases, more characteristic of plant physiology, in which the same individuals cannot be used twice, but a parallel sample is taken at each age. The same distinction occurs in counts of micro-organisms between cases in which counts are made from samples of the same culture, or from samples of parallel cultures. If it is of importance to obtain the general form of the growth curve, the second method has the advantage that any deviation from the expected curve may be confirmed from independent evidence at the next measurement, whereas using the same material no such independent confirmation is obtainable. On the other hand, if interest centres on the growth rate, there is an advantage in using the same material, for only so are actual increases in weight measurable. Both aspects of the difficulty can be got over only by replicating the observations; by carrying out measurements on a number of animals under parallel treatment it is possible to test, from the individual weights, though not from the means, whether their growth curve corresponds with an assigned theoretical course of development, or differs significantly from it or from a series differently treated. Equally, if a number of plants from each sample are weighed individually, growth rates may be obtained with known probable errors, and so may be used for critical comparisons. Care should of course be taken that each is strictly a random sample.

Fig. 1 represents the growth of a baby weighed to the nearest ounce at weekly intervals from birth. Table 1 indicates the calculation from these data of the absolute growth rate in ounces per day and the relative growth rate per day. The absolute growth rates, representing the average actual rates at which substance is added during each period, are found by subtracting from each value that previously recorded, and dividing by the length of the period. The relative growth rates measure the rate of increase not only per unit of time, but also per unit of weight already attained; using the mathematical fact, that

$$\frac{1}{m}\frac{dm}{dt} = \frac{d}{dt}(\log_e m),$$

it is seen that the true average value of the relative

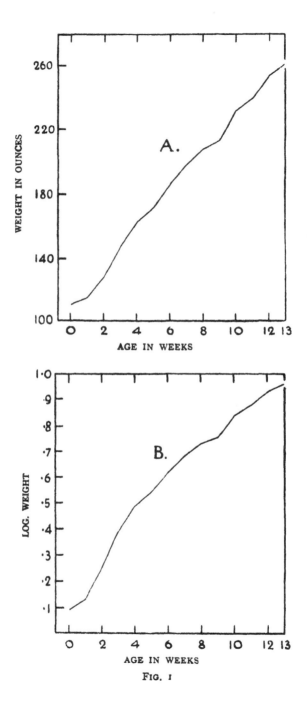

FIG. 1

growth rate for any period is obtained from the natural logarithms of the successive weights, just as the actual rates of increase are from the weights themselves.

TABLE 1

Age in Weeks.	Weight in Ounces.	Increase.	Growth Rate per Day (Oz.).	Natural Log of Weight.	Increase.	Relative Growth Rate per cent. per Day.
$\dfrac{t}{7}$	m	δm	$\dfrac{\delta m}{\delta t}$	$\log \dfrac{m}{100}$	$\delta \log m$	$\dfrac{\delta}{\delta t} \log m$
0	110			·0953		
		4·	·57		·0357	·51
1	114			·1310		
		14	2·00		·1159	1·66
2	128			·2469		
		19	2·71		·1384	1·98
3	147			·3853		
		16	2·29		·1033	1·47
4	163			·4886		
		9	1·29		·0537	·77
5	172			·5423		
		14	2·00		·0783	1·12
6	186			·6206		
		12	1·71		·0625	·89
7	198			·6831		
		10	1·43		·0493	·70
8	208			·7324		
		5	·71		·0237	·34
9	213			·7561		
		19	2·71		·0855	1·22
10	232			·8416		
		8	1·14		·0339	·48
11	240			·8755		
		14	2·00		·0567	·81
12	254			·9322		
		7	1·00		·0272	·39
13	261			·9594		

Such relative rates of increase are conveniently multiplied by 100, and thereby expressed as the percentage rate of increase per day. If these percentage rates of increase had been calculated on the principle of simple interest, by dividing the actual increase by the weight at the beginning of the period, somewhat higher values would have been obtained; the reason for this is that the actual weight of the baby at any time during each period is usually somewhat higher than its weight at the beginning. The error introduced by the simple interest formula becomes exceedingly great when the percentage increases between successive weighings are large.

Fig. 1 A shows the course of the increase in absolute weight; the average slope of such a diagram shows the absolute rate of increase. In this diagram the points fall approximately on a straight line, showing that the absolute rate of increase was nearly constant at about 1·66 oz. per diem. Fig. 1 B shows the course of the increase in the natural logarithm of the weight; the slope at any point shows the relative rate of increase, which, apart from the first week, falls off perceptibly with increasing age. The features of such curves are best brought out if the scales of the two axes are so chosen that the graph makes with them approximately equal angles; with nearly vertical, or nearly horizontal lines, changes in the slope are not so readily perceived.

A rapid and convenient way of displaying the line of increase of the logarithm is afforded by the use of graph paper in which the horizontal rulings are spaced on a logarithmic scale, with the actual values indicated in the margin (see Fig. 5). The horizontal scale can then be adjusted to give the line an appropriate slope. This method avoids the use of a logarithm table, which, however, will still be required if the values of the relative rate of increase are needed.

In making a rough examination of the agreement of the observations with any law of increase, it is desirable so to manipulate the variables that the law to be tested will be represented by a straight line. Thus Fig. 1 A is suitable for a rough test of the law that the absolute rate of increase is constant; if it were suggested that the relative rate of increase were constant, Fig. 1 B would show clearly that this was not so. With other hypothetical growth curves other transformations may be used; for example, in the so-called "autocatalytic" or "logistic" curve the relative growth rate falls off in proportion to the actual weight attained at any time. If, therefore, the relative growth rate be plotted against the actual weight, the points should fall on a straight line if the "autocatalytic" curve fits the facts. For this purpose it is convenient to plot against each observed weight the mean of the two adjacent relative growth rates. To do this for the above data for the growth of an infant may be left as an exercise to the student; twelve points will be available for weights 114 to 254 ounces. The relative growth rates, even after averaging adjacent pairs, will be very irregular, so that no clear indications will be found from these data. If a straight line is found to fit the data, the weight at which growth will cease, supposing the law of growth continues unchanged, is found by producing the line to meet the axis.

9. Correlation Diagrams

Although most investigators make free use of diagrams in which an uncontrolled variable is plotted against the time, or against some controlled factor such as concentration of solution, or temperature, much more use might be made of correlation diagrams in which one uncontrolled factor is plotted against another. When this is done as a dot diagram, a number of dots are obtained, each representing a single experiment, or pair of observations, and it is usually clear from such a diagram whether or not any close connection exists between the variables. When the observations are few a dot diagram will often tell us whether or not it is worth while to accumulate observations of the same sort; the range and extent of our experience is visible at a glance; and associations may be revealed which are worth while following up.

If the observations are so numerous that the dots cannot be clearly distinguished, it is best to divide up the diagram into squares, recording the frequency in each; this semi-diagrammatic record is a correlation table.

Fig. 2 shows in a dot diagram the yields obtained from an experimental plot of wheat (dunged plot, Broadbalk field, Rothamsted) in years with different total rainfall. The plot was under uniform treatment during the whole period 1854-1888; the 35 pairs of observations, indicated by 35 dots, show well the association of high yield with low rainfall. Even when few observations are available a dot diagram may suggest associations hitherto unsuspected, or what is equally important, the absence of associations which would have been confidently predicted. Their value lies in giving a simple conspectus of the experience hitherto gathered, and in bringing to the mind suggestions which may be susceptible of more exact statistical or experimental examination.

Instead of making a dot diagram the device is sometimes adopted of arranging the values of one variate in order of magnitude, and plotting the values of a second variate in the same order. If the line so obtained shows any perceptible slope, or general trend, the variates are taken to be associated. Fig. 3 represents the line obtained for rainfall, when the

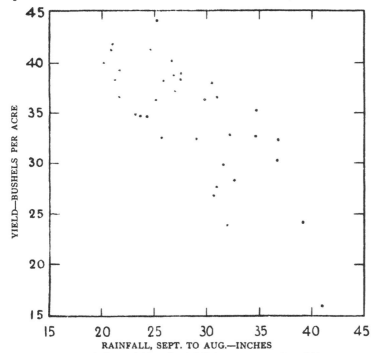

FIG. 2.—Wheat yield and rainfall for 35 years, 1854-1888.

years are arranged in order of wheat yield. Such diagrams are usually far less informative than the dot diagram, and often conceal features of importance brought out by the former. In addition, the dot diagram possesses the advantage that it is easily used as a correlation table if the number of dots is small, and easily transformed into one if the number of dots is large.

In the correlation table the values of both variates are divided into classes, and the class intervals should be equal for all values of the same variate. Thus we might divide the value for the yield of wheat throughout at intervals of one bushel per acre, and the values of the rainfall at intervals of an inch. The diagram is thus divided into squares, and the number

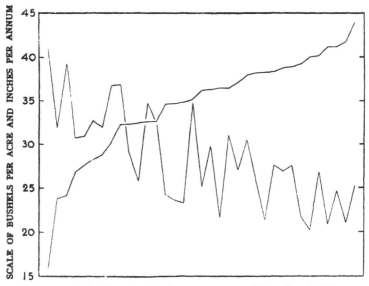

FIG. 3.—Rainfall and yield of 35 years arranged in order of yield.

of observations falling into each square is counted and recorded. The correlation table is useful for three distinct purposes. It affords a valuable visual representation of the whole of the observations, which with a little experience is as easy to comprehend as a dot diagram; it serves as a compact record of extensive data, which, as far as the two variates are concerned, is complete. With more than two variates correlation tables may be given for every pair. This will not indeed enable the reader to reconstruct the original data in its entirety, but it is a fortunate fact that for the great majority of statistical purposes a set of such twofold distributions provides complete information. Original data involving more than two variates are most conveniently recorded for reference on cards, each case being given a separate card with the several variates entered in corresponding positions upon them. The publication of such complete data presents difficulties but it is not yet sufficiently realised how much of the essential information can be presented in a compact form by means of correlation tables. The third feature of value about the correlation table is that the data so presented form a convenient basis for the immediate application of methods of statistical reduction. The most important statistics which the data provide, means, variances, and covariance, can be most readily calculated from the correlation table. An example of a correlation table is shown in Table 31, p. 180.

10. Frequency Diagrams

When a large number of individuals are measured in respect of physical dimensions, weight, colour, density, etc., it is possible to describe with some accuracy the *population* of which our experience may be regarded as a sample. By this means it may be possible to distinguish it from other populations differing in their genetic origin, or in environmental circumstances. Thus local races may be very different as populations, although individuals may overlap in all characters; or, under experimental conditions, the aggregate may show environmental effects, on size, death-rate, etc., which cannot be detected in the individual. A visible representation of a large number of measurements of any one feature is afforded by a frequency diagram. The feature measured is used as abscissa, or measurement along the horizontal axis, and as ordinates are set off vertically the *frequencies*, corresponding to each range.

Fig. 4 is a frequency diagram illustrating the distribution in stature of 1375 women (Pearson and Lee's data modified). The whole sample of women is divided up into successive height ranges of 1 inch.

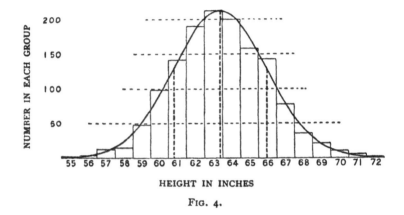

FIG. 4.

Equal areas on the diagram represent equal frequency; if the data be such that the ranges into which the individuals are subdivided are not equal, care should be taken to make the areas correspond to the observed frequencies, so that the area standing upon any interval of the base line shall represent the actual frequency observed in that interval.

The class containing the greatest number of observations is technically known as the modal class. In Fig. 4 the modal class indicated is the class whose central value is 63 inches. When, as is very frequently the case, the variate varies continuously, so that all intermediate values are possible, the choice of the grouping interval and limits is arbitrary and will make a perceptible difference to the appearance of the diagram. Usually, however, the possible limits of grouping will be governed by the smallest units in which the measurements are recorded. If, for example, measurements of height were made to the nearest quarter of an inch, so that all values between $66\frac{7}{8}$ inches and $67\frac{1}{8}$ were recorded as 67 inches, all values between $67\frac{1}{8}$ and $67\frac{3}{8}$ were recorded as $67\frac{1}{4}$, then we have no choice but to take as our unit of grouping 1, 2, 3, 4, etc., quarters of an inch, and the limits of each group must fall on some odd number of eighths of an inch. For purposes of calculation the smaller grouping units are more accurate, but for diagrammatic purposes coarser grouping is often preferable. Fig. 4 indicates a unit of grouping suitable in relation to the total range for a large sample; with smaller samples a coarser grouping is usually necessary in order that sufficient observations may fall in each class.

In all cases where the variation is continuous the frequency diagram should be in the form of a histogram, rectangular areas standing on each grouping interval showing the frequency of observations in that interval. The alternative practice of indicating the frequency by a single ordinate raised from the centre of the interval is sometimes preferred, as giving to the diagram a form more closely resembling a continuous curve. The advantage is illusory, for not only is the form of the curve thus indicated somewhat misleading, but the utmost care should always be taken to distinguish the infinitely large hypothetical population from which our sample of observations is drawn, from the actual sample of observations which we possess; the conception of a continuous frequency curve is applicable only to the former, and in illustrating the latter no attempt should be made to slur over this distinction.

This consideration should in no way prevent a frequency curve fitted to the data from being superimposed upon the histogram (as in Fig. 4); the contrast between the histogram representing the sample, and the continuous curve representing an estimate of the form of the hypothetical population, is well brought out in such diagrams, and the eye is aided in detecting any serious discrepancy between the observations and the hypothesis. No eye observation of such diagrams, however experienced, is really capable of discriminating whether or not the observations differ from expectation by more than we should expect from the circumstances of random sampling. Accurate methods of making such tests will be developed in later chapters.

With discontinuous variation, when, for example, the variate is confined to whole numbers, the reasons given for insisting on the histogram form have little weight, for there are, strictly speaking, no ranges of variation within each class. On the other hand, there is no question of a frequency curve in such cases. Representation of such data by means of a histogram is usual and not inconvenient; it is especially appropriate if we regard the discontinuous variation as due to an underlying continuous variate, which can, however, express itself only to the nearest whole number.

10.1. Transformed Frequencies

It is, of course, possible to treat the values of the frequency like any other variable, by plotting the value of its logarithm, or its actual value on loga-

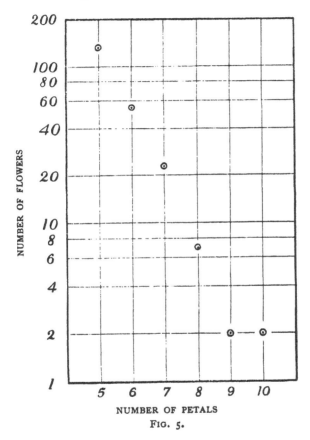

FIG. 5.

rithmic paper, when it is desired to illustrate the agreement of the observations with any particular law of frequency. Fig. 5 shows in this way the number of flowers (buttercups) having 5 to 10 petals (Pearson's data), plotted upon logarithmic paper, to facilitate comparison with the hypothesis that the frequency, for petals above five, falls off in geometric progression. Such illustrations are not, properly speaking, frequency diagrams, although the frequency is one of the variables employed, because they do not adhere to the convention that equal frequencies are represented by equal areas.

A useful form, similar to the above, is used to compare the death-rates, throughout life, of different populations. The logarithm of the number of survivors at any age is plotted against the age attained. Since the death-rate is the rate of decrease of the logarithm of the number of survivors, equal gradients on such curves represent equal death-rates. They therefore serve well to show the increase of death-rate with increasing age, and to compare populations with different death-rates. Such diagrams are less sensitive to small fluctuations than would be the corresponding frequency diagrams showing the distribution of the population according to age at death; they are therefore appropriate when such small fluctuations are due principally to errors of random sampling, which in the more sensitive type of diagram might obscure the larger features of the comparison. It should always be remembered that the choice of the appropriate methods of statistical treatment is quite independent of the choice of methods of diagrammatic representation.

A need which is felt frequently in Genetics and occasionally in other studies is to survey the evidence on some particular frequency ratio provided by a number of different samples, which may or may not be homogeneous in this respect. The classification of samples, such as progenies of plants or animals, according to the frequency-ratio they exhibit, and the homogeneity of the samples classified alike, are in such studies of critical importance, and the explicit tests of Chapter IV will usually be needed. A graphical survey of the evidence gives useful guidance as to what particular points should be tested, and is of further value, as a means of presenting the evidence most simply to the reader.

The frequencies observed of the two alternatives in each sample may be used as co-ordinates of a point, so that just so many points are shown as there are samples. In Fig. 5·1 the useful device has been adopted of plotting not the absolute frequencies, but their square roots. Points representing samples of n observations will then fall on a quadrant of a circle of radius $\sqrt{n}$. Samples showing a frequency ratio $p:q$, where $p+q=1$, will fall on a radius vector making an angle ϕ with the axis, such that

$$\sin^2\phi = p, \quad \cos^2\phi = q.$$

The device thus allows the diagram to exhibit a wider range of sample size, and a wider range of frequency ratio, than would otherwise be possible. Graph paper embodying this principle has been designed by F. Mosteller and J. W. Tukey and is now available. Since 1951, also, a similar chart, designed by M. Nasuyama has been available in Japan.

Since, moreover, the standard error of random sampling of ϕ, for given n, is proportional to $1/\sqrt{n}$, and is independent of ϕ, it follows that the scatter of the observation points on either side of the radii to which they approximate is nearly equal in all parts of the diagram, and the eye is thus materially aided in recognising homogeneous groups.

In the material for *Lythrum salicaria* illustrated in Fig. 5·1, three classes represented by 1, 19 and 7 families respectively, appeared according to expectation. The one family of 41 plants all mid-styled, which evidently belongs to a fourth class, was un-

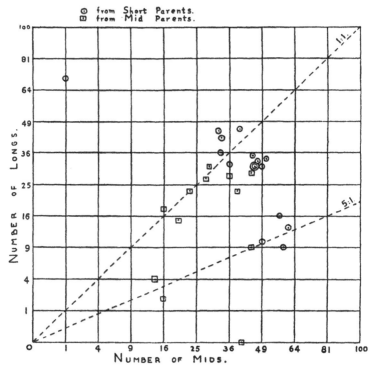

Fig. 5·1.—Frequencies plotted on square-root chart.

expected; later experiments showed it to contain three dominant genes for Mid, due to double reduction having occurred in the preceeding meiosis, and that by the same process it gave about 2 per cent. Longs in a more extensive test.

III

DISTRIBUTIONS

11. The idea of an infinite **population** distributed in a **frequency distribution** in respect of one or more characters is fundamental to all statistical work. From a limited experience, for example, of individuals of a species, or of the weather of a locality, we may obtain some idea of the infinite hypothetical population from which our sample is drawn, and so of the probable nature of future samples to which our conclusions are to be applied. If a second sample belies this expectation we infer that it is, in the language of statistics, drawn from a different population; that the treatment to which the second sample of organisms had been exposed did in fact make a material difference, or that the climate (or the methods of measuring it) had materially altered. Critical tests of this kind may be called tests of significance, and when such tests are available we may discover whether a second sample is or is not significantly different from the first.

A **statistic** is a value calculated from an observed sample with a view to characterising the population from which it is drawn. For example, the *mean* of a number of observations $x_1, x_2 \ldots x_n$, is given by the equation

$$\bar{x} = \frac{1}{n} S(x),$$

where S stands for summation over the whole sample

(this symbol is the one regularly used in our subject), and n for the number of observations. Such statistics are of course variable from sample to sample, and the idea of a frequency distribution is applied with especial value to the variation of such statistics. If we know exactly how the original population was distributed it is theoretically possible, though often a matter of great mathematical difficulty, to calculate how any statistic derived from a sample of given size will be distributed. The utility of any particular statistic, and the nature of its distribution, both depend on the original distribution, and appropriate and exact methods have been worked out for only a few cases. The application of these cases is greatly extended by the fact that the distribution of many statistics tends to the **normal** form as the size of the sample is increased. For this reason it is customary to apply to many cases what is called " the theory of large samples " which is to assume that such statistics are normally distributed, and to limit consideration of their variability to calculations of the standard error.

In the present chapter we shall give some account of three principal distributions—(i) the normal distribution, (ii) the Poisson series, (iii) the binomial distribution. It is important to have a general knowledge of these three distributions, the mathematical formulæ by which they are represented, the experimental conditions upon which they occur, and the statistical methods of recognising their occurrence. On the latter topic we shall be led to some extent to anticipate methods developed more systematically in Chapters IV and V.

12. The Normal Distribution

A variate is said to be normally distributed when it takes all values from $-\infty$ to $+\infty$, with frequencies given by a definite mathematical law, namely, that the logarithm of the frequency at any distance d from the centre of the distribution is less than the logarithm of the frequency at the centre by a quantity proportional to d^2. The distribution is therefore symmetrical, with the greatest frequency at the centre; although

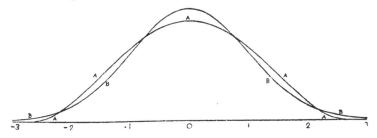

FIG. 6.—Showing a way in which a symmetrical frequency curve may depart from the normal distribution. A, flat-topped curve (γ_2 negative); B, normal curve ($\gamma_2 = 0$).

the variation is unlimited, the frequency falls off to exceedingly small values at any considerable distance from the centre, since a large negative logarithm corresponds to a very small number. Fig. 6 B represents a normal curve of distribution. The frequency in any infinitesimal range dx may be written as

$$df = \frac{1}{\sigma\sqrt{2\pi}} e^{-\frac{1}{2} \cdot \frac{(x-\mu)^2}{\sigma^2}} dx,$$

where $x-\mu$ is the distance of the observation, x, from the centre of the distribution, μ; and σ, called the **standard deviation**, measures in the same units the extent to which the individual values are scattered.

Geometrically σ is the distance, on either side of the centre, of the points at which the slope is steepest, or the points of inflexion of the curve (Fig. 4).

In practical applications we do not so often want to know the frequency at any distance from the centre as the total frequency beyond that distance; this is represented by the area of the tail of the curve cut off at any point. Tables of this total frequency, or probability integral, have been constructed from which, for any value of $(x-\mu)/\sigma$, we can find what fraction of the total population has a larger deviation; or, in other words, what is the probability that a value so distributed, chosen at random, shall exceed a given deviation. Tables I and II have been constructed to show the deviations corresponding to different values of this probability. The rapidity with which the probability falls off as the deviation increases is well shown in these tables. A deviation exceeding the standard deviation occurs about once in three trials. Twice the standard deviation is exceeded only about once in 22 trials, thrice the standard deviation only once in 370 trials, while Table II shows that to exceed the standard deviation sixfold would need nearly a thousand million trials. The value for which $P = \cdot 05$, or 1 in 20, is 1·96 or nearly 2; it is convenient to take this point as a limit in judging whether a deviation is to be considered significant or not. Deviations exceeding twice the standard deviation are thus formally regarded as significant. Using this criterion we should be led to follow up a false indication only once in 22 trials, even if the statistics were the only guide available. Small effects will still escape notice if the data are insufficiently numerous to bring them out, but no lowering of the standard of significance would meet this difficulty.

Some little confusion is sometimes introduced by the fact that in some cases we wish to know the probability that the deviation, known to be positive, shall exceed an observed value, whereas in other cases the probability required is that a deviation, which is equally frequently positive and negative, shall exceed an observed value; the latter probability is always half the former. For example, Table I shows that the normal deviate falls outside the range $\pm 1 \cdot 598193$ in 11 per cent. of cases, and consequently that it exceeds $+1 \cdot 598193$ in 5·5 per cent. of cases.

The value of the deviation beyond which half the observations lie is called the **quartile** distance, and bears to the standard deviation the ratio ·67449. It was formerly a common practice to calculate the standard error and then, multiplying it by this factor, to obtain the **probable error**. The probable error is thus about two-thirds of the standard error, and as a test of significance a deviation of three times the probable error is effectively equivalent to one of twice the standard error. The common use of the probable error is its only recommendation; when any critical test is required the deviation must be expressed in terms of the standard error in using the tables of normal deviates (Tables I and II).

Further tables of the normal distribution are given in *Statistical Tables* IX and X, and in *Sheppard's Tables*, 1938.

13. Fitting the Normal Distribution

From a sample of n individuals of a normal population the mean and the standard deviation of the population may be **estimated** by using two easily calculated statistics. The best estimate of μ is $\bar{x}$ where

$$\bar{x} = \frac{1}{n} S(x),$$

while for the best estimate of σ, we calculate s from

$$s^2 = \frac{1}{n-1} S(x-\bar{x})^2;$$

these two statistics are calculated from the sums of the first two powers of the observations (*see* Appendix, p. 70), and are specially related to the normal distribution, in that they summarise the whole of the information which the sample provides as to the distribution from which it was drawn, provided the latter was normal. Fitting by sums of powers, and especially by the particular system of statistics known as **moments**, has also been widely applied to skew (asymmetrical) distributions, and others which are not normal; but such distributions have not generally the peculiar properties which make the first two powers especially appropriate, and where the distributions differ widely from the normal form the two statistics defined above may be of little or no use.

Ex. 2. *Fitting a normal distribution to a large sample.*—In calculating the statistics from a large sample it is not necessary to calculate individually the squares of the deviations from the mean of each measurement. The measurements are grouped together in equal intervals of the variate, and the whole of the calculation may be carried out rapidly as shown in Table 2, where the distribution of the stature of 1164 men is analysed.

The first column shows the central height in inches of each group, followed by the corresponding frequency. A central group (68·5″) is chosen as " working mean." To form the next column the frequencies are multiplied by 1, 2, 3, etc., according to their distance from the working mean; this process being repeated to form the fourth column, which is

§ 13] DISTRIBUTIONS 47

TABLE 2.

Central Height (Inches).	Men (Frequency).	Frequency × Deviation.	Frequency × (Deviation)².	Women.
52·5	...	...	...	·5
53·5	...	...	...	·5
54·5	...	...	...	...
55·5	...	...	...	1
56·5	...	...	...	5
57·5	...	...	...	15
58·5	...	...	...	15·5
59·5	1	− 9	81	52
60·5	2·5	− 20	160	101
61·5	1·5	− 10·5	73·5	150
62·5	9·5	− 57	342	199
63·5	31	−155	775	223
64·5	56	−224	896	215
65·5	78·5	−235·5	706·5	169·5
66·5	127	−254	508	151·5
67·5	178·5	−178·5	178·5	81·5
68·5	189	−1143·5		40·5
69·5	137	137	137	19·5
70·5	137	274	548	10
71·5	93	279	837	5
72·5	52·5	210	840	...
73·5	39	195	975	1
74·5	17	102	612	...
75·5	6·5	45·5	318·5	...
76·6	3·5	28	224	...
77·5	1	9	81	...
78·5	2	20	200	...
79·5	1	11	121	...
	1164	1310·5 +167	8614	1456

Mean		+·1435		Estimated	
Correction for mean		167²÷1164	23·96	Variance.	S.D.
Corrected sum of squares			8590·04	7·3861	2·7177
Sampling variance of mean				·006345	·0797
Sampling variance of variance				·09382	·3063
Adjustment for grouping				·0833	
Adjusted variance				7·3028	2·7024

summed from top to bottom in a single operation; in the third column, however, the upper portion, representing negative deviations, is summed separately and subtracted from the sum of the lower portion. The difference, in this case positive, shows that the whole sample of 1164 individuals has in all 167 inches more than if every individual were 68·5″ in height. This balance divided by 1164 gives the amount by which the mean of the sample exceeds 68·5″. The mean of the sample is therefore 68·6435″.

From the sum of the fourth column is subtracted a correction to give the value we should have obtained had the working mean been the true mean. This correction is the product of the total, 167″, and the mean 0·1435″ derived from it. The corrected sum of squares divided by 1163, one less than the sample number, provides the estimate of the variance, 7·3861 square inches, which is the basis of all subsequent calculations.

Corresponding to any estimate of a variance, we have, by taking the square root, the corresponding estimate of the standard deviation. Thus from the value 7·3861 square inches, we obtain at once the estimate 2·7177 inches for the standard deviation. This, however, represents the standard deviation of the population as grouped. The process of grouping may be represented as the addition to any true value of a grouping error, positive or negative, which takes all values from $-\frac{1}{2}$ to $\frac{1}{2}$ of a grouping unit with equal frequency. The effect of this on the population, and its average effect upon samples, is to add a constant quantity $\frac{1}{12}$ (= ·0833) to the variance. Sheppard's adjustment for grouping consists in deducting this quantity from the estimate of variance of the population as grouped. This gives 7·3028 square inches for the adjusted variance, and 2·702 for the corresponding estimate of the standard deviation.

Any interval may be used as a unit of grouping; and the whole calculation is carried through in such units, the final results being transformed into other units if required, just as we might wish to transform the mean and standard deviation from inches to centimetres by multiplying by the appropriate factor. It is advantageous that the units of grouping should be exact multiples of the units of measurement; so that if the above sample had been measured to tenths of an inch, we might usefully have grouped them at intervals of 0·6″ or 0·7″.

Regarded as estimates of the mean and the standard deviation of a normal population of which the above is regarded as a sample, the values found are affected by errors of random sampling; that is, we should not expect a second sample to give us exactly the same values. The values for different (large) samples of the same size would, however, be distributed very accurately in normal distributions, so the accuracy of any one such estimate may be satisfactorily expressed by its standard error. These standard errors may be calculated from the variance of the grouped population, and in treating large samples we take our estimate of this variance as the basis of the calculation.

The formulæ for the variances of random sampling of estimates of the mean and of the variance of a normal population are (as given in Appendix, p. 75)

$$\frac{\sigma^2}{n}, \quad \frac{2\sigma^4}{n-1}.$$

Putting our value for k_2 7·3861, in place of σ^2 in these formulæ, we find that our estimate of the mean has a sampling variance ·006345 square inches, or, taking the square root, a standard error ·0797 inches. From this value it is seen that our sample shows significant aberration (± twice standard error) from any population whose mean lay outside the limits 68·48″ to 68·80″. It is therefore probable, in the fiducial sense, that the mean of the population from which our sample was drawn lay between these limits. Similarly, our value for the variance of the population is seen to have a sampling variance ·09382, or a standard error ·3063 ; we have therefore equally good evidence that the variance of the grouped population from which our sample was drawn lay between 6·773 and 7·999 square inches. For the ungrouped population we should deduct ·083 from both limits.

It may be asked, Is nothing lost by grouping ? Grouping in effect replaces the actual data by fictitious data placed arbitrarily at the central values of the groups ; evidently a very coarse grouping might be very misleading. It has been shown that as regards obtaining estimates of the parameters of a normal population, the loss of information caused by grouping is less than 1 per cent., provided the group interval does not exceed one-quarter of the standard deviation ; the grouping of the sample above in whole inches is thus somewhat too coarse ; the loss in the estimation of the standard deviation is 2·28 per cent., or about 27 observations out of 1164 ; the loss in the estimation of the mean is half as great. With suitable group intervals, however, little is lost by grouping, and much labour is saved.

Another way of regarding the loss of information involved in grouping is to consider how near the estimates obtained for the mean and the standard deviation will be to the estimates obtained without grouping. From this point of view we may calculate a standard **error of grouping**, not to be confused with the standard error of random sampling which measures the deviation of the sample values from the population value. In grouping units, the standard error due to grouping of both the mean and the standard deviation is

$$\frac{1}{\sqrt{12n}},$$

or in this case ·0085″. For sufficiently fine grouping this should not exceed one-tenth of the standard error of random sampling.

In the analysis of a large sample the estimate of the variance often employed is

$$\frac{1}{n} S(x - \bar{x})^2,$$

which differs from the formula given previously (p. 46) in that we have divided by n instead of by $(n-1)$. In large samples the difference between these formulæ is small, and that using n may claim some theoretical advantage if we wish for an estimate to be used in conjunction with the estimate of the mean from the same sample, as in fitting a frequency curve to the data; in general it is best to use $(n-1)$. In small samples the difference is still small compared to the probable error, but becomes important if a variance is estimated by averaging estimates from a number of small samples. Thus if a series of experiments is carried out each with six parallels and we have reason to believe that the variation is in all cases due to the operation of analogous causes, we may take the average of such quantities as

$$\frac{1}{n-1} S(x-\bar{x})^2 = \frac{1}{5} S(x-\bar{x})^2$$

to obtain an unbiased estimate of the variance, whereas we should underestimate it were we to divide by 6.

14. Test of Departure from Normality

It is sometimes necessary to test whether an observed sample does or does not depart significantly from normality. For this purpose the third, and sometimes the fourth powers, are used; from each of these it is possible to calculate a quantity, g, the average value of which is zero for a normal distribution, and which is distributed normally for large samples—the standard error being calculable from the size of the sample. The quantity g_1, which is calculated from the third powers, is essentially a measure of asymmetry; the parameter γ_1, of which it provides an estimate, may be equated to $\pm \sqrt{\beta_1}$ of Pearson's notation, though Pearson also used β_1 to designate a statistic which is not the equivalent of $g_1{}^2$; g_2, calculated from the fourth powers, is in like manner a measure of departure from normality, in this case of a symmetrical type, by which the apex and the two tails of the curve are increased at the expense of the intermediate portion, or when negative, the top and tails are depleted and the shoulders filled out, making a relatively flat-topped curve. (See Fig. 6, p. 43.)

Ex. 3. *Use of higher powers to test normality.*—Departures from normal form, unless very strongly marked, can only be detected in large samples; conversely, they make little difference to statistical tests on other questions. We give an example (Table 3) of the calculation for 90 values of the yearly rainfall at Rothamsted; the process of calculation is similar to that of finding the mean and standard deviation, but it is carried two stages further, in the summation of the 3rd and 4th powers. The formulæ by which the sums are reduced to the true mean and the statistics

TABLE 3
Test of Normality of Yearly Rainfall

Year's rain in inches.	Frequency.				
16	1	−12	144	−1728	20736
17	...	...	...	...	...
18	...	...	...	...	...
19	3	−27	243	−2187	19683
20	2	−16	128	−1024	8192
21	3	−21	147	−1029	7203
22	...	...	...	...	...
23	3	−15	75	−375	1875
24	2	−8	32	−128	512
25	12	−36	108	−324	972
26	4	−8	16	−32	64
27	7	−7	7	−7	7
28	4	...	...	...	...
29	8	8	8	8	8
30	9	18	36	72	144
31	6	18	54	162	486
32	7	28	112	448	1792
33	4	20	100	500	2500
34	4	24	144	864	5184
35	4	28	196	1372	9604
36	3	24	192	1536	12288
37	3	27	243	2187	19683
38	...	...	...	...	...
39	1	11	121	1331	14641
s	90	56	2106	1646	125574
Corrections to Mean			−34·84	−3931·2 +43·4	−4096·7 +4892·2 −40·5
S	90	...	2071·16	−2241·8	126329·0
k	...	·62	23·2715	−25·761	−162·487
Adjustment	...	...	−·0833		+·008
k'		·62	23·1882	−25·761	−162·479
g				−·231	−·302
Standard error				±·254	±·503

k and g are calculated, are gathered in an Appendix, p. 70. For the k statistics we obtain in terms of group intervals

$$k_1 = \cdot 62, \quad k_2 = 23 \cdot 2715, \quad k_3 = -25 \cdot 76, \quad k_4 = -162 \cdot 49,$$

whence are calculated

$$g'_1 = k'_3/k'_2{}^{3/2} = -\cdot 231, \quad g'_2 = k'_4/k'_2{}^2 = -\cdot 302.$$

For samples from a normal distribution the sampling variances of g_1 and g_2 are given exactly by the formulæ in the Appendix, and the numerical values of the standard error have been appended in Table 3. It will be seen that neither is significant, or even exceeds its standard error. A negative value of γ_1, which is suggested but not established by the data, would indicate an asymmetry of the distribution in the sense that moderately dry and very wet years are respectively less frequent than moderately wet and very dry years.

15. Discontinuous Distributions

Frequently a variable is not able to take all possible values, but is confined to a particular series of values, such as the whole numbers. This is obvious when the variable is a frequency, obtained by counting, such as the number of cells on a square of a hæmacytometer, or the number of colonies on a plate of culture medium. The normal distribution is the most important of the continuous distributions; but among discontinuous distributions the Poisson series is of the first importance. If a variate can take the values 0, 1, 2, . . ., x, . . ., and the relative frequencies with which the values occur are given by the series

$$e^{-m}\left(1, m, \frac{m^2}{2!}, \ldots, \frac{m^x}{x!}, \ldots\right)$$

(where $x!$ stands for "factorial x" $= x(x-1)(x-2) \ldots 1$), then the number is distributed in the Poisson series. The total frequency is unity, since

$$e^m = 1 + m + \frac{m^2}{2!} + \frac{m^3}{3!} + \ldots$$

Whereas the normal curve has two unknown parameters, μ and σ, the Poisson series has only one. This value may be estimated from a series of observations, by taking their mean, the mean being a statistic as appropriate to the Poisson series as it is to the normal curve. It may be shown theoretically that if the probability of an event is exceedingly small, but a sufficiently large number of independent cases are taken to obtain a number of occurrences, then this number will be distributed in the Poisson series. For example, the chance of a man being killed by horse-kick on any one day is exceedingly small, but if an army corps of men are exposed to this risk for a year, often one or more of them will be killed in this way. The following data (Bortkewitch's data) were obtained from the records of ten army corps for twenty years, supplying 200 such observations.

TABLE 4

Deaths.	Frequency observed.	Expected.
0	109	108·67
1	65	66·29
2	22	20·22
3	3	4·11
4	1	·63
5	...	·08
6	...	·01

The average, $\bar{x}$, is 0·61, and taking this as an estimate of m the numbers calculated agree excellently with those observed.

The importance of the Poisson series in biological research was first brought out in connexion with the accuracy of counting with a hæmacytometer. It was shown that when the technique of the counting process was effectively perfect, the number of cells on each square should be theoretically distributed in a Poisson series; it was further shown that this distribution was, in favourable circumstances, actually realised

TABLE 5

Number of Cells.	Frequency observed.	Frequency expected.
0	...	3·71
1	20	17·37
2	43	40·65
3	53	63·41
4	86	74·19
5	70	69·44
6	54	54·16
7	37	36·21
8	18	21·18
9	10	11·02
10	5	5·16
11	2	2·19
12	2	·86
13	...	·31
14	...	·10
15	...	·03
16	...	·01
Total . .	400	400·00

in practice. Thus the preceding table ("Student's" data) shows the distribution of yeast cells in the 400 squares into which one square millimetre was divided.

The total number of cells counted is 1872, and the mean number is therefore 4·68. The expected frequencies calculated from this mean agree well with those observed. The methods of testing the agreement are explained in Chapter IV.

When a number is the sum of several components, each of which is independently distributed in a Poisson series, then the total number is also so distributed. Thus the total count of 1872 cells may be regarded as a sample of one individual from a series, for which m is not far from 1872. The variance of a Poisson series, like its mean is equal to m; and for such large values of m the distribution of numbers approximates closely to the normal form; we may therefore attach to the number counted, 1872, the standard error $\pm\sqrt{1872} = \pm 43\cdot 26$, to represent the standard error of random sampling of such a count. The density of cells in the original suspension is therefore estimated with a standard error of 2·31 per cent. If, for instance, a parallel sample differed by 7 per cent., the technique of sampling would be suspect.

16. Small Samples of a Poisson Series

Exactly the same principles as govern the accuracy of a hæmacytometer count would also govern a count of bacterial or fungal colonies in estimating the numbers of those organisms by the dilution method, if it could be assumed that the technique of dilution afforded a perfectly random distribution of organisms, and that these could develop on the plate without mutual interference. Agreement of the observations with the Poisson distribution thus affords in the dilution method of counting a test of the suitability of the technique and medium similar to the test afforded of the technique of hæmacytometer counts. The great practical difference between these cases is that from the hæmacytometer we can obtain a record of a large number of squares with only a few organisms on each, whereas in a bacterial count we may have only 5 parallel plates, bearing perhaps 200 colonies apiece. From a single sample of 5 it would be impossible to demonstrate that the distribution followed the Poisson series; however, when a large number of such samples have been obtained under comparable conditions, it is possible to utilise the fact that for all Poisson series the variance is numerically equal to the mean.

For each set of parallel plates with $x_1, x_2, \ldots, x_n$ colonies respectively, after finding the mean $\bar{x}$, an index of dispersion may be calculated by the formula

$$\chi^2 = \frac{S(x-\bar{x})^2}{\bar{x}}.$$

It has been shown that for true samples of a Poisson series, χ^2 calculated in this way will be distributed in a known manner; Table III (p. 112) shows the principal values of χ^2 for this distribution; entering the table with n equal to one less than the number of parallel plates. For small samples the permissible range of variation of χ^2 is wide; thus for five plates with $n=4$, χ^2 will be less than 1·064 in 10 per cent. of cases, while the highest 10 per cent. will exceed 7·779; a single sample of 5 thus gives us little information; but if we have 50 or 100 such samples, we are in a position to verify with accuracy if the expected distribution is obtained.

Ex. 4. *Test of agreement with Poisson series of a number of small samples.*—From 100 counts of bacteria in sugar refinery products the following values were obtained (Table 6); there being 6 plates in each case, the values of χ^2 were taken from the χ^2 table for $n = 5$.

TABLE 6

χ^2.	Expected.	Observed.	Expected 43 per cent.
0			
	1	26	·43
·554			
	1	6	·43
·752			
	3	11	1·29
1·145			
	5	7	2·15
1·610			
	10	7	4·3
2·343			
	10	2	4·3
3·000			
	20	12	8·6
4·351			
	20	7	8·6
6·064			
	10	3	4·3
7·289			
	10	4	4·3
9·236			
	5	1	2·15
11·070			
	3	3	1·29
13·388			
	1	0	·43
15·086			
	1	11	·43
Total	100	100	43·00

It is evident that the observed series differs strongly from expectation; there is an enormous excess in the first class, and in the high values over 15; the relatively few values from 2 to 15 are not far from the expected proportions, as is shown in the last column by taking 43 per cent. of the expected values. It is possible then that even in this case nearly half of the samples were satisfactory, but about 10 per cent. were excessively variable, and in about 45 per cent. of the cases the variability was abnormally depressed.

It is often desirable to test if the variability is of the right magnitude when we have not accumulated a large number of counts, all with the same number of parallel plates, but where a certain number of counts is available with various numbers of parallels. In this case we cannot indeed verify the theoretical distribution with any exactitude, but can test whether or not the general level of variability conforms with expectation. The sum of a number of independent values of χ^2 is itself distributed in the manner shown in the Table of χ^2, provided we take for n the number $S(n)$, calculated by adding the several values of n for the separate experiments. Thus for six sets of 4 plates each the total value of χ^2 was found to be 13·85, the corresponding value of n is $6 \times 3 = 18$, and the χ^2 table shows that for $n = 18$ the value 13·85 is exceeded in between 70 and 80 per cent. of cases; it is therefore not an abnormal value to obtain. In another case the following values were obtained:

TABLE 7

Number of Plates in Set.	Number of Sets.	$S(n)$.	Total χ^2
4	8	24	27·31
5	36	144	133·96
9	1	8	8·73
Total	...	176	170·00

We have therefore to test if $\chi^2 = 170$ is an unreasonably small or great value for $n = 176$. The χ^2 table has not been calculated beyond $n = 30$, but for higher values we make use of the fact that the distribution of χ becomes nearly normal. A good approximation is given by assuming that $(\sqrt{2\chi^2} - \sqrt{2n-1})$ is normally distributed about zero with unit standard deviation. If this quantity exceeds 2, or even 1·645 for the 5 per cent. level, the value of χ^2 significantly exceeds expectation. In the example before us

$$2\chi^2 = 340, \qquad \sqrt{2\chi^2} = 18\cdot 44$$
$$2n-1 = 351, \qquad \sqrt{2n-1} = 18\cdot 73$$
$$\text{Difference} = -\cdot 29$$

The set of 45 counts thus shows variability between parallel plates, very close to that to be expected theoretically. The internal evidence thus suggests that the technique was satisfactory.

17. Presence and Absence of Organisms in Samples

When the conditions of sampling justify the use of the Poisson series, the number of samples containing 0, 1, 2, . . . organisms is, as we have seen, connected by a calculable relation with the mean number of organisms in the sample. With motile organisms, or in other cases which do not allow of discrete colony formation, the mean number of organisms in the sample may be inferred from the proportion of fertile cultures, provided a single organism is capable of developing. If m is the mean number of organisms in the sample, the proportion of samples containing none, that is the proportion of sterile samples, is e^{-m}, from which relation we can calculate, as in the following table, the mean number of organisms corresponding to 10 per cent., 20 per cent., etc., fertile samples.

TABLE 8

Percentage of fertile samples	10	20	30	40	50	60	70	80	90
Mean number of organisms	·1054	·2231	·3567	·5108	·6932	·9163	1·2040	1·6094	2·3026

In connexion with the use of the table above it is worth noting that for a given number of samples tested the frequency ratio of fertile to sterile is most accurately determined at 50 per cent. fertile, but for the minimum percentage error in the estimate of the number of organisms, nearly 80 per cent. fertile or 1·6 organism per sample is most accurate. At this point the standard error of sampling may be reduced to 10 per cent. by taking about 155 samples, whereas at 50 per cent., to obtain the same accuracy, 208 samples would be required. (See *Design of Experiments*, Section 68.)

The Poisson series also enables us to calculate what percentage of the fertile cultures obtained have been derived from a single organism, for the percentage of impure cultures, *i.e.* those derived from 2 or more organisms, can be calculated from the percentage of cultures which proved to be fertile. If e^{-m} are sterile, me^{-m} will be pure cultures, and the remainder impure. The following table gives representative values of the percentage of cultures which are fertile, and the percentage of fertile cultures which are impure:

TABLE 9

	·1	·2	·3	·4	·5	·6	·7
Mean number of organisms in sample							
Percentage fertile	9·52	18·13	25·92	32·97	39·35	45·12	50·34
Percentage of fertile cultures impure	4·92	9·67	14·25	18·67	22·92	27·02	30·95

If it is desired that the cultures should be pure with high probability, a sufficiently low concentration must be used to render at least nine-tenths of the samples sterile.

18. The Binomial Distribution

The binomial distribution is well known as the first example of a theoretical distribution to be established. It was found by Bernoulli, about the end of the seventeenth century, that if the probability of an event occurring were p and the probability of it not occurring were $q(=1-p)$, then if a random sample of n trials were taken, the frequencies with which the event occurred 0, 1, 2, ..., n times were given by the expansion of the binomial

$$(q+p)^n.$$

This rule is a particular case of a more general theorem dealing with cases in which not only a simple alternative is considered, but in which the event may happen in s ways with probabilities $p_1, p_2, \ldots, p_s$; then it can be shown that the chance of a random sample of n giving a_1 of the first kind, a_2 of the second, ..., a_s of the last is

$$\frac{n!}{a_1! a_2! \ldots a_s!} p_1^{a_1} p_2^{a_2} \ldots p_s^{a_s},$$

which is the general term in the multinomial expansion of

$$(p_1+p_2+ \ldots +p_s)^n.$$

Ex. 5. *Binomial distribution given by dice records.*—In throwing a true die the chance of scoring more than 4 is 1/3, and if 12 dice are thrown together the number of dice scoring 5 or 6 should be distributed with frequencies given by the terms in the expansion of

$$(\tfrac{2}{3}+\tfrac{1}{3})^{12}.$$

If, however, one or more of the dice were not true, but if all retained the same bias throughout the experiment, the frequencies should be given approximately by

$$(q+p)^{12},$$

where p is a fraction to be determined from the data. The following frequencies were observed (Weldon's data) in an experiment of 26,306 throws.

TABLE 10

Number of Dice with 5 or 6.	Observed Frequency.	Expected True Dice.	Expected Biased Dice.	Measure of Divergence $\frac{x^2}{m}$.	
				True Dice.	Biased Dice.
0	185	202·75	187·38	1·554	·030
1	1149	1216·50	1146·51	3·745	·005
2	3265	3345·37	3215·24	1·931	·770
3	5475	5575·61	5464·70	1·815	·019
4	6114	6272·56	6269·35	4·008	3·849
5	5194	5018·05	5114·65	6·169	1·231
6	3067	2927·20	3042·54	6·677	·197
7	1331	1254·51	1329·73	4·664	·001
8	403	392·04	423·76	·306	1·017
9	105	87·12	96·03	3·670	·838
10	14	13·07	14·69⎫		
11	4	1·19	1·36⎬	·952	·222
12	...	·05	·06⎭		
	26306	26306·02	26306·00	35·491	8·179
				$n = 10$	$n = 9$

It is apparent that the observations are not compatible with the assumption that the dice were unbiased. With true dice we should expect more cases than have been observed of 0, 1, 2, 3, 4, and fewer cases than have been observed of 5, 6, . . ., 11 dice scoring more than four. The same conclusion is more clearly brought out in the fifth column, which shows the values of the measure of divergence

$$\frac{x^2}{m},$$

where m is the expected value and x the difference between the expected and observed values. The aggregate of these values is χ^2, which measures the deviation of the whole series from the expected series of frequencies, and the actual chance of χ^2 exceeding 35·49, the value for the hypothesis that the dice are true, is ·0001. (See Section 20.)

The total number of times in which a die showed 5 or 6 was 106,602, out of 315,672 trials, whereas the number expected with true dice is 105,224; from the former number, the value of p can be calculated, and proves to be ·337,698,6, and hence the expectations of the fourth column were obtained. These values are much more close to the observed series, and indeed fit them satisfactorily, showing that the conditions of the experiment were really such as to give a binomial series.

The variance of the binomial series is pqn. Thus with true dice and 315,672 trials the expected number of dice scoring more than 4 is 105,224 with variance 70149·3 and standard error 264·9; the observed number exceeds expectation by 1378, or 5·20 times its standard error; this is the most sensitive test of the bias, and it may be legitimately applied, since for such large samples the binomial distribution closely approaches the normal. From the table of the probability integral it appears that a normal deviation only exceeds 5·2 times its standard error once in 5 million times.

The reason why this last test gives so much higher odds than the test for goodness of fit, is that the latter is testing for discrepancies of any kind, such, for example, as copying errors would introduce. The actual discrepancy is almost wholly due to a single item, namely, the value of p, and when that point is tested separately its significance is more clearly brought out.

Ex. 6. *Comparison of sex ratio in human families with binomial distribution.*—Biological data are rarely so extensive as this experiment with dice; Geissler's data on the sex ratio in German families will serve as an example. It is well known that male births are slightly more numerous than female births, so that if a family of 8 is regarded as a random sample of 8 from the general population, the number of boys in such families should be distributed in the binomial
$$(q+p)^8,$$
where p is the proportion of boys. If, however, families differ not only by chance, but by a tendency on the part of some parents to produce males or females, then the distribution of the number of boys should show an excess of unequally divided families, and a deficiency of equally or nearly equally divided families. The data in Table 11 show that there is evidently such an excess of very unequally divided families.

The observed series differs from expectation markedly in two respects: one is the excess of unequally divided families; the other is the irregularity of the central values, showing an apparent bias in favour of even values. No biological reason is suggested for the latter discrepancy, which therefore detracts from the value of the data. The excess of the extreme types of family may be treated in more detail by

§ 18] DISTRIBUTIONS 67

comparing the observed with the expected variance. The expected variance, npq, is 1·998,28, while that calculated from the data is 2·067,45, showing an excess of ·06917, or 3·46 per cent. The sampling variance of this estimate of variance is (p. 75)

$$\frac{2\kappa_2^2}{N-1} + \frac{\kappa_4}{N}$$

where N is the number of families, and κ_2 and κ_4 are the second and fourth cumulants of the theoretical distribution, namely,

$$\kappa_2 = npq = 1\cdot 99828$$
$$\kappa_4 = npq(1-6pq) = -\cdot 99656.$$

The values given are calculated from the value of p as estimated from the frequency of boys in the sample. The standard error of the variance, which as the values show is nearly $\sqrt{7/N}$, is found to be ·01141. The excess of the observed variance over that appropriate to a binomial distribution is thus over six times its standard error.

TABLE 11

Number of Boys.	Number of Families Observed.	Expected.	Excess (x).	$\frac{x^2}{m}$.
0	215	165·22	+ 49·78	14·998
1	1485	1401·69	+ 83·31	4·952
2	5331	5202·65	+128·35	3·166
3	10649	11034·65	−385·65	13·478
4	14959	14627·60	+331·40	7·508
5	11929	12409·87	−480·87	18·633
6	6678	6580·24	+ 97·76	1·452
7	2092	1993·78	+ 98·22	4·839
8	342	264·30	+ 77·70	22·843
	53680	53680·00		91·869

One possible cause of the excessive variation lies in the occurrence of multiple births, for it is known that children of the same birth tend to be of the same sex. The multiple births are not separated in these data, but an idea of the magnitude of this effect may be obtained from other data for the German Empire. These show about 12 twin births per thousand, of which $\frac{5}{8}$ are of like sex and $\frac{3}{8}$ of unlike, so that one-quarter of the twin births, 3 per thousand, may be regarded as "identical" or necessarily alike in sex. Six children per thousand would therefore probably belong to such "identical" twin births, the additional effect of triplets, etc., being small. Now with a population of identical twins it is easy to see that the theoretical variance is doubled; consequently, to raise the variance by 3·46 per cent. we require that 3·46 per cent. of the children should be "identical" twins; this is more than five times the general average; and, although it is probable that the proportion of twins is higher in families of 8 than in the general population, we cannot reasonably ascribe more than a fraction of the excess variance to multiple births.

19. Small Samples of the Binomial Series

With small samples, such as ordinarily occur in experimental work, agreement with the binomial series cannot be tested with much precision from a single sample. It is, however, possible to verify that the variation is approximately what it should be, by calculating an index of dispersion similar to that used for the Poisson series.

Ex. 7. *The accuracy of estimates of infestation.*—The proportion of barley ears infested with gout-fly may be ascertained by examining 100 ears, and counting the infested specimens; if this is done repeatedly, the numbers obtained, if the material is homogeneous, should be distributed in the binomial

$$(q+p)^{100},$$

where p is the proportion infested, and q the proportion free from infestation. The following are the data from 10 such observations made on the same plot (J. G. H. Frew's data):

16, 18, 11, 18, 21, 10, 20, 18, 17, 21.   Mean 17·0.

Is the variability of these numbers ascribable to random sampling; *i.e.* Is the material apparently homogeneous? Such data differ from those to which the Poisson series is appropriate, in that a fixed total of 100 is in each case divided into two classes, infested and not infested, so that in taking the variability of the infested series we are equally testing the variability of the series of numbers not infested. The modified form of χ^2, the index of dispersion, appropriate to the binomial is

$$\chi^2 = \frac{S(x-\bar{x})^2}{npq} = \frac{S(x-\bar{x})^2}{\bar{x}q},$$

differing from the form appropriate to the Poisson series in containing the divisor q, or in this case, ·83. The value of χ^2 is 9·21, which, as the χ^2 table shows, is a perfectly reasonable value for $n = 9$, one less than the number of values available.

Such a test of the single sample is, of course, far from conclusive, since χ^2 may vary within wide limits. If, however, a number of such small samples are available, though drawn from plots of very different infestation, we can test, as with the Poisson series, if the general trend of variability accords with the binomial distribution. Thus from 20 such plots the total χ^2 is 193·64, while S(n) is 180. Testing as before (p. 61), we find

$$\sqrt{387\cdot 28} = 19\cdot 68$$
$$\sqrt{359} = 18\cdot 95$$
Difference $+\cdot 73$.

The difference being less than one, we conclude that the variance shows no sign of departure from that of the binomial distribution. The difference between the method appropriate for this case, in which the samples are small (10), but each value is derived from a considerable number (100) of observations, and that appropriate for the sex distribution in families of 8, where we had many families, each of only 8 observations, lies in the omission of the term

$$\kappa_4 = npq(1-6pq)$$

in calculating the standard error of the variance. When n is 100 this term is very small compared to $2n^2p^2q^2$, and in general the χ^2 method is highly accurate if the number in all the observational categories is as high as 10.

Appendix on Technical Notation and Formulæ

A. *Statistics derived from sums of powers.*

If we have n observations of a variate x, it is easy to calculate for the sample the sums of the simpler powers of the values observed; these we may write

$$s_1 = S(x) \qquad s_2 = S(x^2)$$
$$s_3 = S(x^3) \qquad s_4 = S(x^4)$$

and so on.

It is convenient arithmetically to calculate from

these the sums of powers of deviations from the mean defined by the equations

$$S_2 = s_2 - \frac{1}{n} s_1^2$$

$$S_3 = s_3 - \frac{3}{n} s_2 s_1 + \frac{2}{n^2} s_1^3$$

$$S_4 = s_4 - \frac{4}{n} s_3 s_1 + \frac{6}{n^2} s_2 s_1^2 - \frac{3}{n^3} s_1^4.$$

Many statistics in frequent use are derived from these values.

(i) Moments about the arbitrary origin, $x = 0$; these are derived simply by dividing the corresponding sum by the number in the sample; in general if p stand for 1, 2, 3, 4, . . ., they are defined by the formula

$$m'_p = \frac{1}{n} s_p.$$

Clearly m'_1 is the arithmetic mean, usually written $\bar{x}$.

(ii) In order to obtain values independent of the arbitrary origin, and more closely related to the intrinsic characteristics of the population sampled, values called " moments about the mean " are widely used, which are found by dividing the sums of powers about the mean by the sample number; thus if $p = 2, 3, 4, \ldots$

$$m_p = \frac{1}{n} S_p;$$

these are the moments which would have been obtained if, as would usually be inconvenient arithmetically, the arithmetic mean had been chosen as origin.

(iii) A more recent system which has been shown to have great theoretical advantages is to replace the mean and the moments about the mean by the single series of k-statistics

$$k_1 = \frac{1}{n} s_1$$

$$k_2 = \frac{1}{n-1} S_2$$

$$k_3 = \frac{n}{(n-1)(n-2)} S_3$$

$$k_4 = \frac{n}{(n-1)(n-2)(n-3)} \left\{ (n+1)S_4 - 3\frac{n-1}{n} S_2^2 \right\}.$$

It is easy to verify the following relations:

$$m'_1 = k_1$$

$$m_2 = \frac{n-1}{n} k_2$$

$$m_3 = \frac{(n-1)(n-2)}{n^2} k_3$$

$$m_4 = \frac{n-1}{n^2(n+1)} \left\{ (n-2)(n-3)k_4 + 3(n-1)^2 k_2^2 \right\},$$

by which the moment statistics, when they are wanted, may be obtained from the k-statistics.

(iv) It is of historical interest to note that a series of statistics, termed half-invariants, were defined by Thiele, which are related to the moment statistics m' and m in exactly the same way as the cumulants (see B below) are related to the moments μ' and μ of the population. Thus if $h_1, h_2, h_3, \ldots$ stand for the half-invariants, we have

$$h_1 = m'_1 \qquad h_2 = m_2 \qquad h_3 = m_3$$
$$h_4 = m_4 - 3m_2^2 \qquad h_5 = m_5 - 10 m_3 m_2$$

and so on. Thiele used the same term "half-invariants" also to designate the population parameters of which these statistics may be regarded as estimates, just as the single term " moments " has been used in both senses by Pearson and his followers, so that the cumulants have been frequently referred to as half-invariants or semi-invariants of the population, and even the k-statistics have been mistakenly called semi-invariants of the sample. The half-invariants as originally defined by Thiele are not now of importance, and are only mentioned here to clear up the confusion of terminology.

B. *Moments and cumulants of theoretical distributions.*

Either of the systems of statistics derived from sums of powers may be regarded as estimates of corresponding parameters of theoretical distributions, to which they would usually tend if the sample were increased indefinitely. These true, or population, values are designated by Greek letters; thus m'_4 is an estimate of μ'_4, the fourth moment of the population about an arbitrary origin, m_4 is an estimate of μ_4, the fourth moment of the population about its mean, and k_4 is an estimate of κ_4, the fourth cumulant of the population. The relations between these population values are simpler than those between m and k, thus

$$\mu'_1 = \kappa_1 \qquad \mu_2 = \kappa_2 \qquad \mu_3 = \kappa_3$$
$$\mu_4 = \kappa_4 + 3\kappa_2^2 \qquad \mu_5 = \kappa_5 + 10\kappa_3\kappa_2$$

and so on. The general rule for the formation of the coefficients may be seen from the facts that three is the number of ways of dividing four objects into two sets of two each, while ten is the number of ways of dividing five objects into sets of two and three respectively.

In respect of the relationship between the estimates

and the corresponding parameters, the only elementary point to be noted is that whereas the mean value of any m' from samples of n is equal to the corresponding μ' and the mean value of any k equal to the corresponding κ, this property is not enjoyed by the series of moments about the mean $m_2, m_3, m_4, \ldots$, for

$$\overline{m}_2 = \frac{n-1}{n} \mu_2$$

$$\overline{m}_3 = \frac{(n-1)(n-2)}{n^2} \mu_3$$

$$\overline{m}_4 = \frac{n-1}{n^3} \left\{ (n^2 - 3n + 3)\mu_4 + 3(2n-3)\mu_2^2 \right\},$$

a series of formulæ which sufficiently exhibits the practical inconvenience of using the moments about the mean, and which is typical of the much heavier algebra to which the use of these statistics leads, in comparison with the k-statistics.

The half-invariants, h, of Thiele suffer from the same drawback; for, though they may be regarded as estimates of the cumulants κ, their mean values from the aggregate of finite samples are not equal to the corresponding values κ. In fact,

$$\overline{h}_2 = \frac{n-1}{n} \kappa_2$$

$$\overline{h}_3 = \frac{(n-1)(n-2)}{n^2} \kappa_3$$

$$\overline{h}_4 = \frac{n-1}{n^3} \left\{ (n^2 - 6n + 6)\kappa_4 - 6n\kappa_2^2 \right\},$$

showing that the higher members of this series suffer from the same degree of troublesome complexity as do the moments about the mean.

§ 19]   DISTRIBUTIONS   75

The table below gives the first four cumulants of the three distributions considered in this chapter in terms of the parameters of the distribution:

	Symbol.	Normal.	Poisson.	Binomial.
Mean	κ_1	μ	m	np
Variance	κ_2	σ^2	m	npq
Third cumulant	κ_3	0	m	$-npq(p-q)$
Fourth cumulant	κ_4	0	m	$npq(1-6pq)$

C. *Sampling variance of statistics derived from samples of* N.

Sampling variances are needed primarily for tests of significance. The principal use so far developed for sums of powers higher than the second is in testing normality. The two simplest measures of departure from normality are those dependent from the statistics of the 3rd and 4th degree, defined as

$$g_1 = k_3/k_2^{3/2} \qquad g_2 = k_4/k_2^2.$$

It should be noted that these do not exactly correspond to the statistics γ_1 and γ_2 defined in the first three editions. These Greek symbols are best used not for statistics, but for the parameters of which g_1 and g_2 are estimates. The sampling variances are shown below.

Variance of	General Form.	Normal.
k_1	$\dfrac{\kappa_2}{N}$	$\dfrac{\sigma^2}{N}$
k_2	$\dfrac{\kappa_4}{N} + \dfrac{2\kappa_2^2}{N-1}$	$\dfrac{2\sigma^4}{N-1}$
g_1	...	$\dfrac{6N(N-1)}{(N-2)(N+1)(N+3)}$
g_2	...	$\dfrac{24N(N-1)^2}{(N-3)(N-2)(N+3)(N+5)}$

D. *Adjustments for grouping.*

When the sums of powers are calculated from grouped data, it is desirable for some purposes to introduce an adjustment designed to annul the average effect of the grouping process. These adjustments were worked out for the moment notation by Sheppard, and affect the sums of even powers about the mean. Using unit grouping interval, the adjusted values of the second and fourth k-statistics, represented by k'_2 and k'_4, may be obtained from the formulæ

$$k'_2 = k_2 - \tfrac{1}{12} \qquad k'_4 = k_4 + \tfrac{1}{120}.$$

These adjustments should be used for purposes of estimation, but not usually for tests of significance. Thus k'_2 will be a better estimate of the variance than k_2, but the sampling variance, or standard error, both of the mean and of the variance, should be calculated from the unadjusted value, k_2.

TABLE I
TABLE OF x

The deviation in the normal distribution in terms of the standard deviation.

	·01.	·02.	·03.	·04.	·05.	·06.	·07.	·08.	·09.	·10.
·00	2·575829	2·326348	2·170090	2·053749	1·959964	1·880794	1·811911	1·750686	1·695398	1·644854
·10	1·598193	1·554774	1·514102	1·475790	1·439521	1·405072	1·372204	1·340755	1·310579	1·281552
·20	1·253565	1·226528	1·200359	1·174987	1·150349	1·126391	1·103063	1·080319	1·058122	1·036433
·30	1·015222	·994458	·974114	·954165	·934589	·915365	·896473	·877896	·859617	·841621
·40	·823894	·806421	·789192	·772193	·755415	·738847	·722479	·706303	·690309	·674490
·50	·658838	·643345	·628006	·612813	·597760	·582841	·568051	·553385	·538836	·524401
·60	·510073	·495850	·481727	·467699	·453762	·439913	·426148	·412463	·398855	·385320
·70	·371856	·358459	·345125	·331853	·318639	·305481	·292375	·279319	·266311	·253347
·80	·240426	·227545	·214702	·201893	·189118	·176374	·163658	·150969	·138304	·125661
·90	·113039	·100434	·087845	·075270	·062707	·050154	·037608	·025069	·012533	0

The value of P for each entry is found by adding the column heading to the value in the left-hand margin. The corresponding value of x is the deviation such that the probability of an observation falling outside the range from $-x$ to $+x$ is P. For example, P = ·03 for x = 2·170090; so that 3 per cent. of normally distributed values will have positive or negative deviations exceeding the standard deviation in the ratio 2·170090 at least.

TABLE II
VALUES OF x FOR SMALL VALUES OF P

P	·001	·000,1	·000,01	·000,001	·000,000,1	·000,000,01	·000,000,001
x	3·29053	3·89059	4·41717	4·89164	5·32672	5·73073	6·10941

IV

TESTS OF GOODNESS OF FIT, INDEPENDENCE AND HOMOGENEITY; WITH TABLE OF χ^2

20. The χ^2 Distribution

In the last chapter some use has been made of the χ^2 distribution as a means of testing the agreement between observation and hypothesis; in the present chapter we shall deal more generally with the very wide class of problems which may be solved by means of the same distribution.

The element common to these tests is the comparison of the numbers actually observed to fall into any number of classes with the numbers which upon some hypothesis are expected. If m is the number expected, and $m+x$ the number observed, in any class, we calculate

$$\chi^2 = S\left(\frac{x^2}{m}\right),$$

the summation extending over all the classes. This formula gives the value of χ^2, and it is clear that the more closely the observed numbers agree with those expected the smaller will χ^2 be; in order to utilise the table it is necessary to know also the value of n with which the table is to be entered. The rule for finding n is that n is equal to the number of degrees of freedom in which the observed series may differ from the hypothetical; in other words, it is equal to the number of classes the frequencies in which may be filled up arbitrarily, without altering the expectations. Several examples will be given to illustrate this rule.

For any value of n, which must be a whole number,

the form of distribution of χ^2 was established by Pearson in 1900 ; it is therefore possible to calculate in what proportion of cases any value of χ^2 will be exceeded. This proportion is represented by P, which is therefore the probability that χ^2 shall exceed any specified value. To every value of χ^2 there thus corresponds a certain value of P ; as χ^2 is increased from 0 to infinity, P diminishes from 1 to 0. Equally, to any value of P in this range there corresponds a certain value of χ^2. Algebraically the relation between these two quantities is a complex one, so that it is necessary to have a table of corresponding values, if the χ^2 test is to be available for practical use.

An important table of this sort was prepared by Elderton, and is known as Elderton's Table of Goodness of Fit. Elderton gives the values of P to six decimal places corresponding to each integral value of χ^2 from 1 to 30, and thence by tens to 70. In place of n, the quantity n' ($= n+1$) was used, since it was then believed that this could be equated to the number of frequency classes. Values of n' from 3 to 30 were given, these corresponding to values of n from 2 to 29. A table for $n' = 2$, or $n = 1$, was subsequently supplied by Yule. Owing to copyright restrictions we have not reprinted Elderton's table, but have given a new table (Table III, p. 112) in a form which experience has shown to be more convenient. Instead of giving the values of P corresponding to an arbitrary series of values of χ^2, we have given the values of χ^2 corresponding to specially selected values of P. We have thus been able in a compact form to cover those parts of the distributions which have hitherto not been available, namely, the values of χ^2 less than unity, which frequently occur for small values of n, and the values exceeding 30, which for larger values of n become of importance.

It is of interest to note that the measure of dispersion, Q, introduced by the German economist Lexis, is, if accurately calculated, equivalent to χ^2/n of our notation. In the many references in English to the method of Lexis, it has not, I believe, been noted that the discovery of the distribution of χ^2 in reality completed the method of Lexis. If it were desired to use Lexis' notation, our table could be transformed into a table of Q merely by dividing each entry by n.

In preparing this table we have borne in mind that in practice we do not always want to know the exact value of P for any observed χ^2, but, in the first place, whether or not the observed value is open to suspicion. If P is between ·1 and ·9 there is certainly no reason to suspect the hypothesis tested. If it is below ·02 it is strongly indicated that the hypothesis fails to account for the whole of the facts. Belief in the hypothesis as an accurate representation of the population sampled is confronted by the logical disjunction : *Either* the hypothesis is untrue, *or* the value of χ^2 has attained by chance an exceptionally high value. The actual value of P obtainable from the table by interpolation indicates the strength of the evidence against the hypothesis. A value of χ^2 exceeding the 5 per cent. point is seldom to be disregarded.

To compare values of χ^2, or of P, by means of a " probable error " is merely to substitute an inexact (normal) distribution for the exact distribution given by the χ^2 table.

The term Goodness of Fit has caused some to fall into the fallacy of believing that the higher the value of P the more satisfactorily is the hypothesis verified.

§ 20] GOODNESS OF FIT, ETC. 81

Values over ·999 have sometimes been reported which, if the hypothesis were true, would only occur once in a thousand trials. Generally such cases are demonstrably due to the use of inaccurate formulæ, but occasionally small values of χ^2 beyond the expected range do occur, as in Ex. 4 with the colony numbers obtained in the plating method of bacterial counting. In these cases the hypothesis considered is as definitely disproved as if P had been ·001.

When a large number of values of χ^2 are available for testing, it may be possible to reveal discrepancies which are too small to show up in a single value; we may then compare the observed distribution of χ^2 with that expected. This may be done immediately by simply distributing the observed values of χ^2 among the classes bounded by values given in the χ^2 table, as in Ex. 4, p. 58. The expected frequencies in these classes are easily written down, and, if necessary, the χ^2 test may be used to test the agreement of the observed with the expected frequencies.

It is useful to remember that the sum of any number of quantities, χ^2, is distributed in the χ^2 distribution, with n equal to the sum of the values of n corresponding to the values of χ^2 used. Such a test is sensitive, and will often bring to light discrepancies which are hidden or appear obscurely in the separate values.

The table we give has values of n up to 30; some higher values, and an asymptotic method, are given in *Statistical Tables*, ordinarily it will be found sufficient to assume that $\sqrt{2\chi^2}$ is distributed normally with unit standard deviation about a mean $\sqrt{2n-1}$. The values of P obtained by applying this rule to the values of χ^2 given for $n = 30$, may be worked out as an exercise. The errors are small for $n = 30$, and become progressively smaller for higher values of n.

Ex. 8. *Comparison with expectation of Mendelian class frequencies.*—In a cross involving two Mendelian factors we expect by interbreeding the hybrid (F_1) generation to obtain four classes in the ratio 9 : 3 : 3 : 1; the hypothesis in this case is that the two factors segregate independently, and that the four classes of offspring are equally viable. Are the following observations on *Primula* (de Winton and Bateson) in accordance with this hypothesis?

TABLE 12

	Flat Leaves.		Crimped Leaves.		Total.
	Normal Eye.	Primrose Queen Eye.	Lee's Eye.	Primrose Queen Eye.	
Observed ($m+x$)	328	122	77	33	560
Expected (m)	315	105	105	35	560
x^2/m	·537	2·752	7·467	·114	10·870

The expected values are calculated from the observed total, so that the four classes must agree in their sum, and if three classes are filled in arbitrarily the fourth is therefore determinate; hence $n = 3$; $\chi^2 = 10\cdot 87$, the chance of exceeding which value is between ·01 and ·02; if we take P = ·05 as the limit of significant deviation, we shall say that in this case the deviations from expectation are clearly significant.

Let us consider a second hypothesis in relation to the same data, differing from the first in that we suppose that the plants with crimped leaves are to some extent less viable than those with flat leaves. Such a hypothesis could of course be tested by means of additional data; we are here concerned only with the question whether or no it accords with the values before us. The hypothesis tells us nothing of what degree of relative viability to expect; we therefore take the totals of flat and crimped leaves observed, and divide each class in the ratio 3 : 1.

TABLE 13

	Flat Leaves.		Crimped Leaves.		χ^2.
	Normal Eye.	Primrose Queen Eye.	Lee's Eye.	Primrose Queen Eye.	
Observed	328	122	77	33	...
Expected	337·5	112·5	82·5	27·5	...
x^2/m	·267	·802	·367	1·100	2·536

The value of n is now 2, since only two entries can be made arbitrarily; the value of χ^2, however, is so much reduced that P exceeds ·2, and the departure from expectation is no longer significant. The significant part of the original discrepancy lay in the proportion of flat to crimped leaves.

It was formerly believed that, in entering the χ^2 table, n was always to be equated to one less than the number of frequency classes; this view led to many discrepancies, and has since been disproved with the establishment of the rule stated above. On the old view, any elaboration of the hypothesis such as that which in the instance above admitted differential viability, was bound to give an apparent improvement in the agreement between observation and hypothesis. When the change in n is allowed for, this bias disappears, and if the value of P, rightly calculated, is many fold increased, as in this instance, the increase may safely be ascribed to an improvement in the hypothesis, and not to a mere increase in the number of parameters which may be adjusted to suit the observations.

Ex. 9. *Comparison with expectation of the Poisson*

series and Binomial series.—In Table 5, p. 56, we give the observed and expected frequencies in the case of a Poisson series. In applying the χ^2 test to such a series it is desirable that the number expected should in no group be less than 5, since the calculated distribution of χ^2 is not very closely realised for very small classes. We therefore pool the numbers for 0 and 1 cells, and also those for 10 and more, and obtain the following comparison:

TABLE 14

	0 and 1	2	3	4	5	6	7	8	9	10 and more	Total
Observed	20	43	53	86	70	54	37	18	10	9	400
Expected	21·08	40·65	63·41	74·19	69·44	54·16	36·21	21·18	11·02	8·66	400
x^2/m	·055	·136	1·709	1·880	·005	·000	·017	·477	·093	·013	4·385

Using 10 frequency classes we have $\chi^2 = 4\cdot385$; in ascertaining the value of n we have to remember that the expected frequencies have been calculated, not only from the total number of values observed (400), but also from the observed mean; there remain, therefore, 8 degrees of freedom, and $n = 8$. For this value the χ^2 table shows that P is between ·8 and ·9, showing a close, but not an unreasonably close, agreement with expectation.

Similarly, in Table 10, p. 64, we have given the value of χ^2 based upon 11 classes for the two hypotheses of "true dice" and "biased dice"; with "true dice" the expected values are calculated from the total number of observations alone, and $n = 10$, but in allowing for bias we have brought also the means into agreement so that n is reduced to 9. In the first case χ^2 is far outside the range of the table showing a highly significant departure from expectation; in the second it appears that P lies between ·5 and ·7, so that the value of χ^2 is within the expected range.

21. Tests of Independence, Contingency Tables

A special and important class of cases where the agreement between expectation and observation may be tested comprises the tests of independence. If the same group of individuals is classified in two (or more) different ways, as persons may be classified as inoculated and not inoculated, and also as attacked and not attacked by a disease, then we may require to know if the two classifications are independent.

In the simplest case, when each classification comprises only two classes, we have a 2×2 table, or, as it is often called, a fourfold table.

Ex. 10. The following table is taken from Greenwood and Yule's data for Typhoid:

TABLE 15
OBSERVED

	Attacked.	Not Attacked.	Total.
Inoculated . .	56	6,759	6,815
Not inoculated .	272	11,396	11,668
Total .	328	18,155	18,483

TABLE 16
EXPECTED

	Attacked.	Not Attacked.	Total.
Inoculated .	120·94	6,694·06	6,815
Not Inoculated .	207·06	11,460·94	11,668
Total .	328	18,155	18,483

In testing independence we must compare the observed values with values calculated so that the four frequencies are *in proportion*; since we wish to test independence only, and not any hypothesis as to the total numbers attacked, or inoculated, the "expected" values are calculated from the marginal totals observed, so that the numbers expected agree with the numbers observed in the margins; only one value need be calculated, *e.g.*

$$\frac{328 \times 6815}{18483} = 120\cdot 94;$$

the others are written down at once by subtraction from the margins. It is thus obvious that the observed values can differ from those expected in only 1 degree of freedom, so that in testing independence in a fourfold table, $n = 1$. Since $\chi^2 = 56\cdot 234$ the observations are clearly opposed to the hypothesis of independence. Without calculating the expected values, χ^2 may, for fourfold tables, be directly calculated by the formula

$$\chi^2 = \frac{(ad-bc)^2(a+b+c+d)}{(a+b)(c+d)(a+c)(b+d)},$$

where a, b, c, and d are the four observed numbers.

When only one of the classifications is of two classes, the calculation of χ^2 may be simplified to some extent, if it is not desired to calculate the expected numbers. If a, a' represent any pair of observed frequencies, and n, n' the corresponding totals, we may, following Pearson, calculate from each pair

$$\frac{1}{a+a'}(an'-a'n)^2,$$

and the sum of these quantities divided by nn' will be χ^2.

An alternative formula, which besides being quicker, has the advantage of agreeing more closely with the general method used in the Analysis of Variance, has been developed by Brandt and Snedecor. From each pair of frequencies the fraction,

$$p = a/(a+a'),$$

is calculated, and from the totals

$$\bar{p} = n/(n+n');$$

then

$$\chi^2 = \frac{1}{\bar{p}\bar{q}}\left\{ S(ap) - n\bar{p} \right\},$$

where $\bar{q} = 1 - \bar{p}$. It is a further advantage of this method of calculation that it shows the actual fractions observed in each class; where there is any great difference between the two rows it is usually convenient to use the smaller series of fractions.

Ex. 11. *Test of independence in a $2 \times n'$ classification.*—From the pigmentation survey of Scottish children (Tocher's data) the following are the numbers of boys and girls from the same district (No. 1) whose hair colour falls into each of five classes:

TABLE 17
HAIR COLOUR

	Fair.	Red.	Medium.	Dark.	Jet Black.	Total.
Boys	592	119	849	504	36	2100
Girls	544	97	677	451	14	1783
Total	1136	216	1526	955	50	3883
Sex Ratio	·52113	·55093	·55636	·52775	·72000	·54082

The sex ratio, proportion of boys, is given under the total for each hair colour; multiplying each by the number of boys, and deducting the corresponding product for the total, there remains 2·603, which on dividing by $\bar{p}\bar{q}$ gives $\chi^2 = 10\cdot48$.

In this table 4 values could be filled in arbitrarily without conflicting with the marginal totals, so that $n = 4$. The value of P is between ·02 and ·05, so that sex difference in the classification by hair colours is probably significant as judged by this district alone. It is to be noticed that, with this method, the ratios must be calculated with somewhat high precision. Using five decimal places, the value of χ^2 given is not quite correct in the second decimal, and to avoid doubts as to the precision of calculation two more places would have been desirable. It is evident from the ratios that the principal discrepancy is due to the excess of boys in the " Jet Black " class.

Ex. 12. *Test of independence in a* 4×4 *classification.*—As an example of a more complex contingency table we may take the results of a series of back-crosses in mice, involving the two factors Black-Brown, Self-Piebald (Wachter's data) :

TABLE 18

	Black Self.	Black Piebald.	Brown Self.	Brown Piebald.	Total.
Coupling—					
F_1 Males .	88 (85·37)	82 (75·24)	75 (70·93)	60 (73·46)	305
F_1 Females	38 (34·43)	34 (30·34)	30 (28·60)	21 (29·63)	123
Repulsion—					
F_1 Males .	115 (117·00)	93 (103·11)	80 (97·21)	130 (100·68)	418
F_1 Females	96 (100·20)	88 (88·31)	95 (83·26)	79 (86·23)	358
Total .	337	297	280	290	1204

The back-crosses were made in four ways, according as the male or female parents were heterozygous (F_1) in the two factors, and according to whether the

§ 21] GOODNESS OF FIT, ETC. 89

two dominant genes were received both from one (Coupling) or one from each parent (Repulsion).

The simple Mendelian ratios may be disturbed by differential viability, by linkage, or by linked lethals. Linkage is not suspected in these data, and if the only disturbance were due to differential viability of the four genotypes, these should always appear in the same proportion; to test if the data show significant departures we may apply the χ^2 test to the whole 4×4 table. The values expected on the hypothesis that the proportions are independent of the matings used, or that the four series are homogeneous, are given above in brackets. The contributions to χ^2 made by each cell are given below (Table 19).

TABLE 19

·081	·607	·234	2·466	3·388
·370	·442	·069	2·514	3·395
·034	·991	3·047	8·539	12·611
·176	·001	1·655	·606	2·438
·661	2·041	5·005	14·125	21·832

The value of χ^2 is therefore 21·832; the value of n is 9, for we could fill up a block of three rows and three columns and still adjust the remaining entries to check with the margins. In general for a contingency table of r rows and c columns $n = (r-1)(c-1)$. For $n = 9$, the value of χ^2 shows that P is less than ·01, and therefore the departures from proportionality are not fortuitous; it is apparent that the discrepancy is due to the exceptional number of Brown Piebalds in the F_1 males Repulsion series.

It should be noted that the methods employed in this chapter are not designed to measure the *degree* of association between one classification and another, but solely to test whether the observed departures from independence are or are not of a magnitude ascribable to chance. The same degree of association may be significant for a large sample but insignificant for a small one ; if it is insignificant we have no reason on the data present to suspect any degree of association at all, and it is useless to attempt to measure it. If, on the other hand, it is significant the value of χ^2 indicates the fact, but does not measure the degree of association. Provided the deviation is clearly significant, it is of no practical importance whether P is ·01 or ·000,001, and it is for this reason that we have not tabulated the value of χ^2 beyond ·01. To measure the degree of association it is necessary to have some hypothesis as to the nature of the departure from independence to be measured. With Mendelian frequencies, for example, the recombination percentage may be used to measure the degree of association of two factors, and the significance of evidence for linkage may be tested by comparing the difference between the recombination percentage and 50 per cent. (the value for unlinked factors), with its standard error. Such a comparison, if accurately carried out, must agree absolutely with the conclusion drawn from the χ^2 test. To take a second example, the values in a fourfold table may be sometimes regarded as due to the partition of a normally correlated pair of variates, according as the values are above or below arbitrarily chosen dividing-lines ; as if a group of stature measurements of fathers and sons were divided between those above and those below 68 inches. In this case the departure from independence may be properly measured by the correlation in stature between father and son; this quantity can be estimated from the observed frequencies, and a comparison between the value obtained and its standard error, if accurately carried out, will agree with the χ^2 test as to the significance of the association; the significance will become more and more pronounced as the sample is increased in size, but the correlation obtained will tend to a fixed value. The χ^2 test does not attempt to measure the degree of association, but as a test of significance it is independent of all additional hypotheses as to the nature of the association.

Tests of **homogeneity** are mathematically identical with tests of independence; the last example may equally be regarded in either light; in Chapter III the tests of agreement with the Binomial series were essentially tests of homogeneity; the ten samples of 100 ears of barley (Ex. 7, p. 68) might have been represented as a 2 × 10 table. The χ^2 index of dispersion would then be equivalent to the χ^2 obtained from the contingency table. The method of this chapter is more general, and is applicable to cases in which the successive samples are not all of the same size.

Ex. 13. *Homogeneity of different families in respect of ratio black : red.*—The following data show in 33 families of *Gammarus* (Huxley's data) the numbers with black and red eyes respectively:

TABLE 20

Black	79	120	24	117	62	79	66	45	61	64	208	154	31	158	21	105	28
Red	14	31	6	29	17	20	12	11	14	13	52	45	4	45	4	28	7
Total	93	151	30	146	79	99	78	56	75	77	260	199	35	203	25	133	35
Black	58	81	25	95	47	67	30	70	139	179	129	44	24	19	45	91	2565
Red	19	27	8	29	16	21	11	28	57	62	44	17	9	8	23	41	772
Total	77	108	33	124	63	88	41	98	196	241	173	61	33	27	68	132	3337

The totals 2565 black and 772 red are distinctly not in the ratio 3 : 1 ; the discrepancy is ascribed to linkage. The question before us is whether or not all the families indicate the same ratio between black and red, or whether the discrepancy is due to a few families only. For the whole table $\chi^2 = 35 \cdot 620$, $n = 32$. This is beyond the range of the table, so we apply the method explained on p. 81 :

$$\sqrt{2\chi^2} = 8 \cdot 44 ;$$
$$\sqrt{2n-1} = 7 \cdot 94 ;$$
$$\text{Difference} = + \cdot 50 \pm 1.$$

The series is therefore not significantly heterogeneous ; effectively all the families agree and confirm each other in indicating the black-red ratio observed in the total.

Exactly the same procedure would be adopted if the black and red numbers represented two samples distributed according to some character or characters each into 33 classes. The question " Are these samples of the same population ? " is in effect identical with the question " Is the proportion of black to red the same in each family ? " To recognise this identity is important, since it has been very widely disregarded.

21·01. Yates' Correction for Continuity

The distribution of χ^2, tabulated as in Table III, is a continuous distribution. The distribution of frequencies must, however, always be discontinuous. Consequently, the use of χ^2 in the comparison of observed with expected frequencies can only be of approximate accuracy, the continuous distribution being in fact the limit towards which the true discontinuous distribution tends as the sample is made ever larger. It was in order to avoid the irregularities produced by small numbers that we have stipulated above that in no group shall the expected number be less than five. This safeguard generally ensures that the number of possible sets of observations shall be large, each occurring with only a small frequency, so giving to χ^2 a distribution closely simulating the continuous distribution of the table.

A case of special interest arises, however, when there is only 1 degree of freedom, and when the value of χ^2 can, consequently, be calculated from the number observed in a single class. If the number in this class is small, *e.g.* 3, the probability of this number may be by no means negligible compared with the sum of the probabilities of the more extreme deviations represented by 2, 1, or 0 occurrences in the class. If we want to know whether the observed number, 3, is so small as to indicate a significant departure from expectation, we require to know whether the sum of the probabilities of 3, 2, 1 or 0 together is less than a standard value, such as ·05; or, in other words, whether the total probability of obtaining our observed deviation, or any deviation more extreme, is so small that we should be unwilling to ascribe the deviation observed to mere chance.

Our actual problem, therefore, when stated exactly, concerns a limited number of finite probabilities, which in simple cases it may be convenient to calculate directly. The Table of χ^2, on the other hand, gives the area of the tail of a continuous curve. Inasmuch, however, as this curve supplies a close approximation to the actual distribution, the area between the values of χ^2 corresponding to observed frequencies of $3\frac{1}{2}$ and $2\frac{1}{2}$ will be a good approximation to the actual probability of observing 3 ; and the area of the tail beyond the value of χ^2 corresponding to $3\frac{1}{2}$ will be a good approximation to the sum of the probabilities of observing 3 or less. Thus our actual problem will best be resolved by entering the table of χ^2, not with the value calculated from the actual frequencies, but with the value it would have if our observed frequencies had been less extreme than they really are each by half a unit. This useful adjustment is due to F. Yates.

Ex. 13.1. *Frequency of criminality among the twin brothers or sisters of criminals.*—Among 13 criminals who were monozygotic twins Lange reports that 10 had twin brothers or sisters who had also been convicted, while in 3 cases the twin brother had, apparently, not taken to crime. Among 17 criminals who were dizygotic twins (of like sex), 2 had convicted twin brothers or sisters, while those of the other 15 were not known to be criminals. It is argued that the environmental circumstances are as much alike for dizygotic twins of like sex as for monozygotic twins, and that if the latter are more alike in their social reactions these reactions must be largely conditioned by genetic factors. Do Lange's data show that criminality is significantly more frequent among the monozygotic twins of criminals than among the dizygotic twins of criminals?

Our data consist of the four-fold table :—

GOODNESS OF FIT, ETC.

TABLE 20·1

	Convicted.	Not Convicted.	Total.
Monozygotic . .	10	3	13
Dizygotic . .	2	15	17
Total . .	12	18	30

The difference $(ad-bc)$ is 144 and

$$\chi^2 = \frac{144^2 \cdot 30}{12 \cdot 18 \cdot 13 \cdot 17} = 13.032$$

a very significant value, equivalent to a normal deviate 3·61 times its standard error. The probability of exceeding such a deviation in the right direction is about 1 in 6500.

Using Yates' adjustment we should rewrite the table with the larger frequencies 10 and 15 reduced by a half, and the smaller frequencies 2 and 3 increased by half.

The difference between the cross products $ad-bc$ is now reduced to 129, which, it may be noted, is just 15, or half the total number of observations, less than its previous value 144. In other respects the calculation is unchanged. The new value of χ^2 is 10·458, still a very significant value for 1 degree of freedom, but now corresponding to a normal deviation of 3·234 times its standard error, or to odds of 1 in 1638. The exact odds in this case are 1 in 2150, as will be shown in the next section. The adjustment has slightly over-corrected the exaggeration of significance due to using a table of a continuous distribution.

21·02. The Exact Treatment of 2×2 Tables

The treatment of frequencies by means of χ^2 is an approximation, which is useful for the comparative simplicity of the calculations. The exact treatment is somewhat more laborious, though necessary in cases of doubt, and valuable as displaying the true nature of the inferences which the method of χ^2 is designed to draw.

If p is the probability of any event, the probability that it will occur a times in $(a+b)$ independent trials is the term of the binomial expansion,

$$\frac{(a+b)!}{a!\,b!}\, p^a q^b,$$

where $q = 1-p$. The probability that in a sample of $(c+d)$ trials it will occur c times is

$$\frac{(c+d)!}{c!\,d!}\, p^c q^d.$$

So that the probability of the observed frequencies a, b, c, and d in a 2 × 2 table is the product

$$\frac{(a+b)!\,(c+d)!}{a!\,b!\,c!\,d!}\, p^{a+c} q^{b+d},$$

and this in general must be unknown if p is unknown. The unknown factor involving p and q will, however, be the same for all tables having the same marginal frequencies $a+c$, $b+d$, $a+b$, $c+d$, so that among possible sets of observations having the same marginal frequencies, the probabilities are in proportion to

$$\frac{1}{a!\,b!\,c!\,d!},$$

whatever may be the value of p, or, in other words, for all populations in which the four frequencies are in proportion.

Now the sum of the quantities $1/a!\,b!\,c!\,d!$ for all samples having the same margins is found to be

$$\frac{n!}{(a+b)!\,(c+d)!\,(a+c)!\,(b+d)!}$$

where $n = a+b+c+d$; so that, given the marginal frequencies, the probability of any observed set of entries is

$$\frac{(a+b)!\,(c+d)!\,(a+c)!\,(b+d)!}{n!} \cdot \frac{1}{a!\,b!\,c!\,d!}.$$

In the case considered in Ex. 13·1, we have therefore

$$\frac{18!\,12!\,17!\,13!}{30!}\left\{\frac{1}{2!\,3!\,10!\,15!},\frac{1}{2!\,11!\,16!},\frac{1}{1!\,12!\,17!}\right\}$$

for the probabilities of the set of frequencies observed, and the two possible more extreme sets of frequencies which might have been observed. Without any assumption or approximation, therefore, the table observed may be judged significantly to contradict the hypothesis of proportionality if

$$\frac{18!\,13!}{30!}(2992+102+1)$$

is a small quantity. This amounts to 619/1330665, or about 1 in 2150, showing that for any case in which the hypothesis of proportionality were true, observations of the kind recorded would be highly exceptional.

21·03. Exact Tests based on the χ^2 Distribution

In its primary purpose of the comparison of a series of observed frequencies with those expected on the hypothesis to be tested, the χ^2 test is an approximate one, though validly applicable in an immense range of important cases. For other cases where the observations are measurements, instead of frequencies, it provides exact tests of significance. Of these the two most important are:—

(i) its use to test whether a sample from a normal distribution confirms or contradicts the variance which this distribution is expected on theoretical grounds to have, and

(ii) its use in combining the indications drawn from a number of independent tests of significance.

Ex. 14. *Agreement with expectation of normal variance.*—If x_1, x_2, ..., are a sample of a normal population, the standard deviation of which population is σ, then

$$\frac{1}{\sigma^2} S(x-\bar{x})^2$$

is distributed in random samples as is χ^2, taking n one less than the number in the sample. J. W. Bispham gives three series of experimental values of the partial correlation coefficient, each based on thirty observations of the values of three variates, which he assumes should be distributed so that $1/\sigma^2 = 29$, but which properly should have $1/\sigma^2 = 28$. The values of $S(x-\bar{x})^2$ for the three samples of 1000, 200, 100 respectively are, as judged from the grouped data,

35·0279,   7·4573,   3·6146,

whence the values of χ^2 on the two theories are those given in Table 21.

GOODNESS OF FIT, ETC.

TABLE 21

	Exp. 1.	2.	3.	Total.	$\sqrt{2\chi^2}$.	Difference.
29 $S(x-\bar{x})^2$	1015·81	216·26	104·82	1336·89	51·71	+·79
28 $S(x-\bar{x})^2$	980·78	208·80	101·21	1290·79	50·81	−·11
Expectation (n)	999	199	99	1297	50·92	

It will be seen that the true formula for the variance gives slightly the better agreement. That the difference is not significant may be seen from the last two columns. About 6000 observations would be needed to discriminate experimentally, with any certainty, between the two formulæ.

21·1. The Combination of Probabilities from Tests of Significance

When a number of quite independent tests of significance have been made, it sometimes happens that although few or none can be claimed individually as significant, yet the aggregate gives an impression that the probabilities are on the whole lower than would often have been obtained by chance. It is sometimes desired, taking account only of these probabilities, and not of the detailed composition of the data from which they are derived, which may be of very different kinds, to obtain a single test of the significance of the aggregate, based on the product of the probabilities individually observed.

The circumstance that the sum of a number of values of χ^2 is itself distributed in the χ^2 distribution with the appropriate number of degrees of freedom, may be made the basis of such a test. For in the particular case when $n = 2$, the natural logarithm of the probability is equal to $-\tfrac{1}{2}\chi^2$. If therefore we take the natural logarithm of a probability, change its sign and double it, we have the equivalent value of χ^2 for 2 degrees of freedom. Any number of such values may be added together, to give a composite test, using the Table of χ^2 to examine the significance of the result.

Ex. 14·1. *Significance of the product of a number of independent probabilities.*—Three tests of significance have yielded the probabilities ·145, ·263, ·087; test whether the aggregate of these three tests should be regarded as significant. We have

P	$-\log_e P$	Degrees of Freedom.
·145	1·9310	2
·263	1·3356	2
·087	2·4419	2
	5·7085	6

$$\chi^2 = 11\cdot4170$$

For 6 degrees of freedom we have found a value 11·417 for χ^2. The 5 per cent. value is 12·592 while the 10 per cent. value is 10·645. The probability of the aggregate of the three tests occurring by chance therefore exceeds ·05, and is not far from ·075.

In applying this method it will be noticed that we require to know from the individual tests not only whether they are or are not individually significant, but also, to two or three figure accuracy, what are the actual probabilities indicated. For this purpose it is convenient and sufficiently accurate for most purposes to interpolate in the table given (Table III), using the logarithms of the values of P shown. Either natural or common logarithms may equally be employed. We may exemplify the process by applying it to find the probability of χ^2 exceeding 11·417, when $n = 6$.

Our value of χ^2 exceeds the 10 per cent. point by ·772, while the 5 per cent. point exceeds the 10 per cent. point by 1·947; the fraction

$$\frac{\cdot 772}{1\cdot 947} = \cdot 397.$$

The difference between the common logarithm of 5 and of 10 is ·3010, which multiplied by ·397 gives ·119; the negative logarithm of the required probability is thus found to be 1·119, and the probability to be ·076. For comparison, the value calculated by exact methods is ·07631.

22. Partition of χ^2 into its Components

Just as values of χ^2 may be aggregated together to make a more comprehensive test, so in some cases it is possible to separate the contributions to χ^2 made by the individual degrees of freedom, and so to test the separate components of a discrepancy.

Ex. 15. *Partition of observed discrepancies from Mendelian expectation.*—The table on p. 102 (de Winton and Bateson's data) gives the distribution of sixteen families of Primula in the eight classes obtained from a back-cross with the triple recessive.

The theoretical expectation is that the eight classes should appear in equal numbers, corresponding to the hypothesis that in each factor the allelomorphs occur with equal frequency, and that the three factors are unlinked. This expectation is fairly realised in the totals of the sixteen families, but the individual families are somewhat irregular. The values of χ^2 obtained by comparing each family with expectation are given in the lowest line. These values each correspond to 7 degrees of freedom, and it appears that in 6 cases out of 16, P is less than ·1, and of these

TABLE 22

Type.	Family Number.																Total.
	54.	55.	58.	59.	107.	110.	119.	121.	122.	127.	129.	131.	132.	133.	135.	178.	
Ch G W	5	18	17	2	12	17	9	10	24	9	3	16	20	9	11	10	192
Ch G w	10	13	11	12	20	16	10	7	23	3	6	24	18	2	13	12	200
Ch g W	4	10	17	3	14	10	6	8	19	5	5	23	18	10	7	12	171
Ch g w	9	17	11	11	13	13	9	8	9	6	3	12	18	1	9	12	161
ch G W	13	22	20	10	5	5	16	2	30	3	8	21	19	4	9	12	199
ch G w	14	16	18	9	12	6	14	3	16	5	7	13	14	4	13	10	174
ch g W	10	11	12	6	7	3	18	2	11	5	4	14	23	4	6	13	149
ch g w	7	12	16	6	10	8	10	4	23	5	4	22	23	7	8	16	181
Total	72	119	122	59	93	78	92	44	155	41	40	145	153	41	76	97	1427
χ^2	9·78	7·86	5·48	13·00	12·55	19·23	10·09	12·36	18·06	4·86	4·80	9·21	3·18	14·22	5·05	2·05	151·78

§ 22]   GOODNESS OF FIT, ETC.   103

2 are less than ·02. This confirms the impression of irregularity, and the total value of χ^2 (not to be confused with χ^2 derived from the totals), which corresponds to 112 degrees of freedom, is 151·78.

Now
$$\sqrt{223} = 14\cdot93;$$
$$\sqrt{303\cdot56} = 17\cdot42;$$
$$\text{Difference} = +2\cdot49;$$

so that, judged by the total χ^2, the general evidence for departures from expectation in individual families is clear.

Each family is free to differ from expectation in seven independent ways. To carry the analysis further, we must separate the contribution to χ^2 of each of these 7 degrees of freedom. Mathematically the subdivision may be carried out in more than one way, but the only way which appears to be of biological interest is that which separates the parts due to inequality of the allelomorphs of the three factors, and the three possible linkage connexions. If we separate the frequencies into positive and negative values according to the following seven ways:—

TABLE 23

	Ch.	G.	W.	G W.	Ch W.	Ch G.	Ch G W.
Ch G W	+	+	+	+	+	+	+
Ch G w	+	+	−	−	−	+	−
Ch g W	+	−	+	−	+	−	−
Ch g w	+	−	−	+	−	−	+
ch G W	−	+	+	+	−	−	−
ch G w	−	+	−	−	+	−	+
ch g W	−	−	+	−	−	+	+
ch g w	−	−	−	+	+	+	−

then it will be seen that all seven subdivisions are wholly independent, since any two of them agree in four signs and disagree in four. The first 3 degrees of freedom represent the inequalities in the allelomorphs of the three factors **Ch**, **G**, and **W**; the next are the degrees of freedom involved in an inquiry into the linkage of the three pairs of factors, while the 7th degree of freedom has no simple biological meaning but is necessary to complete the analysis. If we take in the first family, for example, the difference between the numbers of the **W** and **w** plants, namely 8, then the contribution of this degree of freedom to χ^2 is found by squaring the difference and dividing by the number in the family, *e.g.* $8^2 \div 72 = \cdot 889$. In this way the contribution of each of the 112 degrees of freedom in the sixteen families is found separately, as shown in the following table :—

TABLE 24

Family.	Ch.	G.	W.	G W.	Ch W.	Ch G.	Ch GW.	Total.
54	3·556	2·000	·889	·222	2·000	·889	·222	9·778
55	·076	3·034	·076	3·034	·412	1·017	·210	7·859
58	·820	·820	·820	·295	1·607	·820	·295	5·477
59	·153	·831	4·898	·017	6·119	·831	·153	13·002
107	6·720	·269	3·108	1·817	·097	·269	·269	12·549
110	14·821	1·282	·821	·821	·205	1·282	0	19·232
119	6·261	·391	·391	·174	2·130	·043	·696	10·086
121	11·000	0	0	·364	·818	·091	·091	12·364
122	·161	6·200	1·090	1·865	·523	·316	7·903	18·058
127	·610	·024	·220	·610	1·195	·220	1·976	4·855
129	·900	1·600	0	·400	·100	·900	·900	4·800
131	·172	·062	·062	·062	·062	·338	8·448	9·206
132	·163	·791	·320	·320	·059	1·471	·059	3·183
133	·220	·220	4·122	·024	8·805	·220	·610	14·221
135	·211	3·368	1·316	·053	·053	0	·053	5·054
178	·258	·835	·093	·093	·010	·258	·505	2·052
Total	46·102	21·727	18·226	10·171	24·195	8·965	22·390	151·776

Looking at the total values of χ^2 for each column, since n is 16 for these, we see that all except the first have values of P between ·05 and ·95, while the contribution of the 1st degree of freedom is very clearly significant. It appears then that the greater part, if not the whole, of the discrepancy is ascribable to the behaviour of the Sinensis-Stellata factor, **Ch**, and its behaviour strongly suggests close linkage with a recessive lethal gene of one of the familiar types. In four families, 107-121, the only high contribution is in the first column. If these four families are excluded $\chi^2 = 97\cdot545$, and this exceeds the expectation for $n = 84$ by only just over the standard error; the total discrepancy cannot therefore be regarded as significant.

There does, however, appear to be an excess of very large entries, and it is noticeable of the seven largest, that six appear in pairs belonging to the same family. The distribution of the remaining 12 families according to the value of P is as follows :—

TABLE 25

P . .	1·0	·9	·8	·7	·5	·3	·2	·1	·05	·02	·01	0	Total
Families	1	1	0	4	1	2	0	1	1	1	0		12

from which it would appear that there is some slight evidence of an excess of families with high values of χ^2. This effect, like other non-significant effects, is only worth further discussion in connexion with some plausible hypothesis capable of explaining it.

The general procedure to follow in analysing χ^2 into its components will be developed in Section 55.

Ex. 15·1. *Complex test on homogeneity in data with hierarchical subdivisions.*—Table 25·1 shows the total number of offspring and the number of recom-

TABLE 25·1

TOTAL PLANTS (T) AND RECOMBINATIONS (C) IN 22 PROGENIES AND IN THE AGGREGATES OF PROGENIES IN WHICH THEY ARE GROUPED.

Descendants of									
Single Plants.		Fraternities.		Parents F_2.		Grandparents F_1.		Total.	
T	c	T	c	T	c	T	c	T	c
34	3	77	11	171	21				
43	8								
94	10	94	10						
73	3					427	42		
20	2	130	8						
16	2								
21	1			256	21				
31	6								
51	4	126	13						
29	2								
15	1							922	105
64	9	119	22	119	22	119	22		
55	13								
55	7	89	12	89	12				
34	5								
37	3	37	3	37	3				
45	7					376	41		
28	0	108	9						
35	2			250	26				
68	8								
44	5	142	17						
30	4								

binations found in 22 progenies of the garden pea, grown by Rasmusson. Each progeny was derived from a single plant, tested by back-crossing. Unequal numbers of these plants belonged, as shown by the

GOODNESS OF FIT, ETC.

table, to 9 different fraternities, for each of which the total number of offspring and the total recom-

TABLE 25·2
VALUES OF $\dfrac{c^2}{T}$ FOR ALL GROUPS AND SUBGROUPS

Individual Plants.	Fraternities.	Parents F_2.	Grandparents F_1.	Total.
0·26471 1·48837	1·57143	2·57895		
1·06383	1·06383			
0·12329 0·20000 0·25000	0·49231		4·13115	
0·04762 1·16129		1·72266		
0·31373 0·13793 0·06667	1·34127			11·95770
1·26562 3·07273	4·06723	4·06723	4·06723	
0·89091 0·73529	1·61798	1·61798		
0·24324	0·24324	0·24324	4·47074	
1·08889 0·00000 0·11429	0·75000	2·70400		
0·94118 0·56818 0·53333	2·03521			
14·57110	13·18250	12·93406	12·66912	11·95770
1·38860 13·760 13	0·24844 2·462 3	0·26494 2·625 3	0·71142 7·050 2	Differences. x^2 n

binations are shown in the table. In three cases, moreover, 2 fraternities had been derived by different matings of the same parent plants, which had been bred with a view to this linkage test, so that the 9 fraternities were the offspring of 6 F_2 parents, which in turn were derived from 3 F_1 grandparents. In all, the final generation yielded 922 plants, of which 105 were of the types recognised as recombinations. It is required to test whether heterogeneity in the fraction of recombinations obtained occurs at any of the four stages represented by the groups and subgroups of progenies.

The method of Brandt and Snedecor is of great value when adapted to the analysis of data of this kind. If in any progeny, or group of progenies, we have c recombinations out of T plants, we may at once calculate c^2/T for each group in the record. These ratios appear in Table 25·2 arranged to show the affiliations of the different groups and subgroups. Whenever the proportion of recombinations observed is different for different subgroups of the same group, the values for these subgroups will together exceed the value for the corresponding group; thus the five totals shown in Table 25·2 form a diminishing series, the successive differences between the terms of which afford measures of the heterogeneity observable.

If such a process were applied to completely homogeneous material, it would only be necessary to divide each of these differences by the same quantity, pq, where p is the proportion of recombinations and q of old combinations, to obtain values distributed in χ^2 distributions. The numbers of degrees of freedom appropriate to each are the differences between the numbers of entries in the successive columns.

It is apparent from the values at the foot of the table that the only apparent heterogeneity in the linkage values occurs among the F_2 plants, or at the earliest stage at which segregation might appear. Here the value of χ^2 is 7·050 for 2 degrees of freedom, a value which lies between the 5 per cent. and the 2 per cent. points. It is probable, therefore, that segregations affecting linkage occurred at this stage, and, in consequence, the values we have obtained for later stages must be revised with this heterogeneity in view.

TABLE 25·3
DIFFERENCES AMONG SISTER PLANTS AND SUBGROUPS, WITH DEGREES OF FREEDOM CORRESPONDING

Sister Plants.	Half-sister Progenies.	F_3 Plants.	F_2 Plants.
·18165 (1)	·05631 (1)	·17046 (1)	
...	·11092 (1)		
·12860 (3)			
·33835 (3)			·71142 (2)
·27112 (1)	...	...	
·00822 (1)	...		
...	...	·09448 (2)	
·45318 (2)	·08121 (1)		
·00748 (2)			

A generally applicable procedure would be to recalculate the divisor, pq, for each of the three F_2 plants. In this case, however, it is evident that the first and third of these differ but little, having recombination fractions 9·836 per cent. and 10·904 per cent. respectively, while both show closer linkage than is shown by the descendants of the second plant, which gave 18·487 per cent. We shall, therefore, in recalculating χ^2 use the same fraction, 83/803, for the descendants of the first and third F_2 plants, and the fraction 22/119 for the descendants of the second plant.

When different factors are to be applied in different parts of the table a convenient first step is to take the differences between the total value for the subgroups, and the value for the group to which they belong, in each available case, as is shown in Table 25·3. In this table the whole set of 21 degrees of freedom has been partitioned among 13 entries. The values of χ^2 to which these parts correspond depend on the values of pq, by which they are divided. In the case of the 2 degrees of freedom among the F_2 plants we must use $p = 105/922$, obtaining as before $\chi^2 = 7\cdot0498$ for 2 degrees of freedom. For the descendants of the first and third F_2 plants we divide by ·0926786, and for the descendants of the second plant by ·1506956, so obtaining the values of χ^2 shown in Table 25·4.

TABLE 25·4
χ^2 FOR 13 RELEVANT SUBDIVISIONS

Sister Plants.	Half-sister Progenies.	F_3 Plants.	F_2 Plants.
1·9600 (1)	·6076 (1)	1·8393 (1)	
1·3876 (3)	1·1968 (1)		
3·6508 (3)			
1·7991 (1)	...	...	7·0498 (2)
·0887 (7)	...		
...	...	1·0194 (2)	
4·8898 (2)	·8763 (1)		
·0807 (2)			
13·8567	2·6807	2·8587	7·0498
13	3	3	2

The totals for the different subdivision stages do not differ greatly from those shown in Table 25·2, but afford a better test for heterogeneity in these later stages, once it is suspected that the F_2 plants were not homogeneous. It will be observed, in fact, that the value of χ^2 for the 3 degrees of freedom representing differences between the pairs of half-sister progenies from the same F_3 plants, and that for differences among the F_3 plants, are slightly raised, through using a smaller divisor for the descendants of the first and third F_2 plants, since in these columns there is no compensation due to using a larger divisor for the descendants of the second plant.

The absence of significant values in the first three columns of Table 25·4 shows that no further modifications of the divisors are necessary, since there is no further evidence of heterogeneity.

[TABLE

TABLE III—

n.	P = ·99.	·98.	·95.	·90.	·80.	·70.
1	·000157	·000628	·00393	·0158	·0642	·148
2	·0201	·0404	·103	·211	·446	·713
3	·115	·185	·352	·584	1·005	1·424
4	·297	·429	·711	1·064	1·649	2·195
5	·554	·752	1·145	1·610	2·343	3·000
6	·872	1·134	1·635	2·204	3·070	3·828
7	1·239	1·564	2·167	2·833	3·822	4·671
8	1·646	2·032	2·733	3·490	4·594	5·527
9	2·088	2·532	3·325	4·168	5·380	6·393
10	2·558	3·059	3·940	4·865	6·179	7·267
11	3·053	3·609	4·575	5·578	6·989	8·148
12	3·571	4·178	5·226	6·304	7·807	9·034
13	4·107	4·765	5·892	7·042	8·634	9·926
14	4·660	5·368	6·571	7·790	9·467	10·821
15	5·229	5·985	7·261	8·547	10·307	11·721
16	5·812	6·614	7·962	9·312	11·152	12·624
17	6·408	7·255	8·672	10·085	12·002	13·531
18	7·015	7·906	9·390	10·865	12·857	14·440
19	7·633	8·567	10·117	11·651	13·716	15·352
20	8·260	9·237	10·851	12·443	14·578	16·266
21	8·897	9·915	11·591	13·240	15·445	17·182
22	9·542	10·600	12·338	14·041	16·314	18·101
23	10·196	11·293	13·091	14·848	17·187	19·021
24	10·856	11·992	13·848	15·659	18·062	19·943
25	11·524	12·697	14·611	16·473	18·940	20·867
26	12·198	13·409	15·379	17·292	19·820	21·792
27	12·879	14·125	16·151	18·114	20·703	22·719
28	13·565	14·847	16·928	18·939	21·588	23·647
29	14·256	15·574	17·708	19·768	22·475	24·577
30	14·953	16·306	18·493	20·599	23·364	25·508

For larger values of n, the expression $\sqrt{2\chi^2} - \sqrt{2n-1}$

TABLE OF χ^2

·50.	·30.	·20.	·10.	·05.	·02.	·01.
·455	1·074	1·642	2·706	3·841	5·412	6·635
1·386	2·408	3·219	4·605	5·991	7·824	9·210
2·366	3·665	4·642	6·251	7·815	9·837	11·345
3·357	4·878	5·989	7·779	9·488	11·668	13·277
4·351	6·064	7·289	9·236	11·070	13·388	15·086
5·348	7·231	8·558	10·645	12·592	15·033	16·812
6·346	8·383	9·803	12·017	14·067	16·622	18·475
7·344	9·524	11·030	13·362	15·507	18·168	20·090
8·343	10·656	12·242	14·684	16·919	19·679	21·666
9·342	11·781	13·442	15·987	18·307	21·161	23·209
10·341	12·899	14·631	17·275	19·675	22·618	24·725
11·340	14·011	15·812	18·549	21·026	24·054	26·217
12·340	15·119	16·985	19·812	22·362	25·472	27·688
13·339	16·222	18·151	21·064	23·685	26·873	29·141
14·339	17·322	19·311	22·307	24·996	28·259	30·578
15·338	18·418	20·465	23·542	26·296	29·633	32·000
16·338	19·511	21·615	24·769	27·587	30·995	33·409
17·338	20·601	22·760	25·989	28·869	32·346	34·805
18·338	21·689	23·900	27·204	30·144	33·687	36·191
19·337	22·775	25·038	28·412	31·410	35·020	37·566
20·337	23·858	26·171	29·615	32·671	36·343	38·932
21·337	24·939	27·301	30·813	33·924	37·659	40·289
22·337	26·018	28·429	32·007	35·172	38·968	41·638
23·337	27·096	29·553	33·196	36·415	40·270	42·980
24·337	28·172	30·675	34·382	37·652	41·566	44·314
25·336	29·246	31·795	35·563	38·885	42·856	45·642
26·336	30·319	32·912	36·741	40·113	44·140	46·963
27·336	31·391	34·027	37·916	41·337	45·419	48·278
28·336	32·461	35·139	39·087	42·557	46·693	49·588
29·336	33·530	36·250	40·256	43·773	47·962	50·892

may be used as a normal deviate with unit variance.

V

TESTS OF SIGNIFICANCE OF *MEANS*, DIFFERENCES OF MEANS, AND REGRESSION COEFFICIENTS

23. The Standard Error of the Mean

THE fundamental proposition upon which the statistical treatment of mean values is based is that—If a quantity be normally distributed with variance σ^2, then the mean of a random sample of n such quantities is normally distributed with variance σ^2/n.

The utility of this proposition is somewhat increased by the fact that even if the original distribution were not exactly normal, that of the mean usually tends to normality, as the size of the sample is increased; the method is therefore applied widely and legitimately to cases in which we have not sufficient evidence to assert that the original distribution was normal, but in which we have reason to think that it does not belong to the exceptional class of distributions for which the distribution of the mean does not tend to normality.

If, therefore, we know the variance of a population, we can calculate the variance of the mean of a random sample of any size, and so test whether or not it differs significantly from any fixed value. If the difference is many times greater than the standard error, it is certainly significant, and it is a convenient convention to take twice the standard error as the limit of significance; this is roughly equivalent to

§ 23]   SIGNIFICANCE OF *MEANS*, ETC.   115

the corresponding limit P = ·05, already used for the χ^2 distribution. The deviations in the normal distribution corresponding to a number of values of P are given in the lowest line of the Table of *t* at the end of this chapter (p. 176). More detailed information has been given in Table I.

Ex. 16. *Significance of mean of a large sample.*— We may consider from this point of view Weldon's die-casting experiment (Ex. 5, p. 63). The variable quantity is the number of dice scoring " 5 " or " 6 " in a throw of 12 dice. In the experiment this number varies from zero to eleven, with an observed mean of 4·0524 ; the expected mean, on the hypothesis that the dice were true, is 4, so that the deviation observed is ·0524. If now we estimate the variance of the whole sample of 26,306 values as explained in Ex. 2, without using Sheppard's correction (for the data are not grouped, and even with grouped data, since the mean is affected by grouping errors, its variance should be estimated without this adjustment), we find

$$\sigma^2 = 2\cdot 69826,$$
whence $\quad \sigma^2/n = \cdot 0001026,$
and $\quad \sigma/\sqrt{n} = \cdot 01013.$

The standard error of the mean is therefore about ·01, and the observed deviation is nearly 5·2 times as great ; thus by a slightly different path we arrive at the same conclusion as that of p. 65. The difference between the two methods is that our treatment of the mean does not depend upon the hypothesis that the distribution is of the binomial form, but on the other hand we do assume the correctness of the value of σ derived from the observations. This assumption breaks down for small samples, and the principal purpose of this chapter is to show how accurate allowance can be made in these tests of significance for the errors in our estimates of the standard deviation.

To return to the cruder theory, we may often, as in the example above, wish to compare the observed mean with the value appropriate to a hypothesis which we wish to test; but equally or more often we wish to compare two experimental values and to test their agreement. In such cases we require the variance of the difference between two quantities whose variances are known; to find this we make use of the proposition that the variance of the difference of two *independent* variates is equal to the sum of their variances. Thus, if the standard deviations are σ_1, σ_2, the variances are σ_1^2 and σ_2^2; consequently the variance of the difference is $\sigma_1^2+\sigma_2^2$, and the standard error of the difference is $\sqrt{\sigma_1^2+\sigma_2^2}$.

Ex. 17. *Standard error of difference of means from large samples.*—In Table 2 is given the distribution in stature of a group of men, and also of a group of women; the means are 68·64 and 63·87 inches, giving a difference of 4·77 inches. The variance obtained for the men was 7·3861 square inches. Dividing this by 1164, we find the variance of the mean is ·006345. Similarly, the variance for the women is 6·7832, which divided by 1456 gives the variance of the mean of the women as ·004659. To find the variance of the difference between the means, we must add together these two contributions, and find in all ·011004; the standard error of the difference between the means is therefore ·1049 inches. The sex difference in stature may therefore be expressed as

$$4\cdot77\pm\cdot105 \text{ inches.}$$

§ 23] SIGNIFICANCE OF *MEANS*, ETC. 117

It is manifest that this difference is significant, the value found being over 45 times its standard error. In this case we can not only assert a significant difference, but place its value with some confidence at between $4\frac{1}{2}$ and 5 inches. It should be noted that we have treated the two samples as *independent*, as though they had been given by different authorities; as a matter of fact, in many cases brothers and sisters appeared in the two groups; since brothers and sisters tend to be alike in stature, we have overestimated the probable error of our estimate of the sex difference. Whenever possible, advantage should be taken of such facts in designing experiments. In the common phrase, sisters provide a better "control" for their brothers than do unrelated women. (See *Design of Experiments*, Chap. III.) The sex difference could therefore be more accurately estimated from the comparison of each brother with his own sister. In the following example (Pearson and Lee's data), taken from a correlation table of stature of brothers and sisters, the material is nearly of this form; it differs from it in that in some instances the same individual has been compared with more than one sister, or brother.

Ex. 18. *Standard error of mean of differences.*— The following table gives the distribution of the excess in stature of a brother over his sister in 1401 pairs.

TABLE 26

Stature difference in inches	-5	-4	-3	-2	-1	0	1	2	3	4	5
Frequency	·25	1·5	1·25	4·5	11·25	27·5	71·75	122·75	171·75	209·75	220·5

Stature difference in inches	6	7	8	9	10	11	12	13	14	15	16	Total
Frequency	205·5	148·75	95·75	57	26	11·25	8·5	2·75	1		·75	1401

Treating this distribution as before, we obtain: mean = 4·895, estimate of variance = 6·5480, variance of mean = ·004674, standard error of mean = ·0684; showing that we may estimate the mean sex difference as 4¾ to 5 inches.

In the examples given above, which are typical of the use of the standard error applied to mean values, we have assumed that the variance of the population is determined with exactitude. It was pointed out by "Student" in 1908, with small samples, such as are of necessity usual in field and laboratory experiments, where the variance of the population can only be roughly estimated from the sample, that the errors of estimation are calculable, and that accurate allowance can be made for them.

If x (for example the mean of a sample) is a value normally distributed about zero, and σ is its true standard error, then the probability that x/σ exceeds any specified value may be obtained from the appropriate table of the normal distribution; but if we do not know σ, but in its place have s, an estimate of the value of σ, the distribution required will be that of x/s, and this is not normal. The true value has been divided by a factor, s/σ, which introduces an error. We have seen in the last chapter that the distribution in random samples of s^2/σ^2 is that of χ^2/n, when n is equal to the number of degrees of freedom, in the group (or groups) of which s^2 is the mean square deviation. Consequently, the distribution of s/σ is calculable, and although σ is unknown, we can use in its place the fiducial distribution of σ given s to find the probability of x exceeding a given multiple of s. Hence the true distribution of x/s is all that is required. The only modification required in these cases depends solely on the number n, representing the number of degrees of freedom available for the estimation of σ. The necessary distributions were given by "Student" in 1908; fuller tables have since been given by the same author, and at the end of this chapter (p. 176) we give the distributions in a similar form to that used for our Table of χ^2.

24. The Significance of the Mean of a Unique Sample

If $x_1, x_2, \ldots x_{n'}$ is a sample of n' values of a variate x, and if this sample constitutes the whole of the information available on the point in question, then we may test whether the mean of x differs significantly from zero by calculating the statistics

$$\bar{x} = \frac{1}{n'} S(x),$$

$$\frac{s^2}{n'} = \frac{1}{n'(n'-1)} S(x-\bar{x})^2,$$

$$t = \bar{x} \div \sqrt{\frac{s^2}{n'}},$$

$$n = n' - 1.$$

Arithmetically, the calculations depend on the simple fact that the sum of squares of deviations from the mean may be obtained from the sum of squares of deviations from zero by deducting the product of the total and the mean. Thus,

$$S(x^2) = S(x-\bar{x})^2 + \bar{x} S(x).$$

This is a sub-division of the sum of squares of x into two portions, the first of which represents variation within the sample, while the second is due only to the deviation of the observed mean from zero. The first part has $n'-1$ degrees of freedom, and the second part only 1. The more complex cases treated in later chapters are greatly simplified by setting out these two sub-divisions, of the sum of squares and of the degrees of freedom, in parallel columns and comparing the mean squares in each class. Thus in this case we have

	Degrees of Freedom.	Sum of Squares.	Mean Square.
Deviation	1	$\bar{x}S(x)$	t^2s^2
Within sample	$n'-1$	$S(x-\bar{x})^2$	s^2
Total	n'	$S(x^2)$	

The mean squares are obtained in each class by dividing the sum of squares by the corresponding degrees of freedom. The observed ratio of the mean squares is, in this case, t^2. This useful form of arrangement is of much wider application than the algebraical expressions by which the calculations can be expressed, and is known as the Analysis of Variance.

The distribution of t for random samples of a normal population distributed about zero as mean is given in the Table of t for each value of n. The successive columns show, for each value of n, the values of t for which P, the probability of falling outside the range $\pm t$, takes the values ·9, . . ., ·01, at the head of the columns. Thus the last column shows that, when $n = 10$, just 1 per cent. of such random samples will give values of t exceeding $+3\cdot169$, or less than $-3\cdot169$. If it is proposed to consider the chance of exceeding the given values of t, in a positive (or negative) direction only, then the values of P should be halved. It will be seen from the table that for any degree of certainty we require higher values of t, the smaller the value of n. The bottom line of the table, corresponding to infinite values of n, gives the values of a normally distributed variate, in terms of its standard deviation, for the same values of P.

Ex. 19. *Significance of mean of a small sample.*—The following figures (Cushny and Peebles' data), which I quote from "Student's" paper, show the result of an experiment with ten patients on the effect of two supposedly soporific drugs, A and B, in producing sleep.

TABLE 27

ADDITIONAL HOURS OF SLEEP GAINED BY THE USE OF TWO TESTED DRUGS.

Patient.	A.	B.	Difference (B−A).
1	+0·7	+1·9	+1·2
2	−1·6	+0·8	+2·4
3	−0·2	+1·1	+1·3
4	−1·2	+0·1	+1·3
5	−0·1	−0·1	0·0
6	+3·4	+4·4	+1·0
7	+3·7	+5·5	+1·8
8	+0·8	+1·6	+0·8
9	0·0	+4·6	+4·6
10	+2·0	+3·4	+1·4
Mean ($\bar{x}$)	+·75	+2·33	+1·58

The last column gives a controlled comparison of the efficacy of the two drugs as soporifics, for the same patients were used to test each; from the series of differences we find

$$\bar{x} = +1·58,$$
$$\frac{s^2}{10} = ·1513,$$
$$s/\sqrt{10} = ·3890,$$
$$t = 4·06$$

For $n = 9$, only one value in a hundred will exceed 3·250 by chance, so that the difference between the results is clearly significant. By the methods of the previous chapters we should, in this case, have been led to the same conclusion with almost equal certainty; for if the two drugs had been equally effective, positive and negative signs would occur in the last column with equal frequency. Of the 9 values other than zero, however, all are positive, and it appears from the binomial distribution,

$$(\tfrac{1}{2}+\tfrac{1}{2})^9,$$

that all will be of the same sign, by chance, only twice in 512 trials. The method of the present chapter differs from that in taking account of the actual values and not merely of their signs, and is consequently the more sensitive method when the actual values are available.

24·1. Comparison of Two Means

In experimental work it is even more frequently necessary to test whether two samples differ significantly in their means, or whether they may be regarded as belonging to the same population. In the latter case any difference in treatment which they may have received will have shown no significant effect.

If $x_1, x_2, \ldots, x_{n_1+1}$ and $x'_1, x'_2, \ldots, x'_{n_2+1}$ be two samples, the significance of the difference between their means may be tested by calculating the following statistics:

$$\bar{x} = \frac{1}{n_1+1} S(x), \quad \bar{x}' = \frac{1}{n_2+1} S(x'),$$

$$s^2 = \frac{1}{n_1+n_2}\left\{S(x-\bar{x})^2 + S(x'-\bar{x}')^2\right\},$$

$$t = \frac{\bar{x}-\bar{x}'}{s}\sqrt{\frac{(n_1+1)(n_2+1)}{n_1+n_2+2}}$$

$$n = n_1+n_2$$

§ 24·1] SIGNIFICANCE OF *MEANS*, ETC.   123

The means are calculated as usual; the standard deviation is estimated by pooling the sums of squares from the two samples and dividing by the total number of the degrees of freedom contributed by them; if σ were the true standard deviation, the variance of the first mean would be $\sigma^2/(n_1+1)$, of the second mean $\sigma^2/(n_2+1)$, and therefore that of the difference would be $\sigma^2 \{1/(n_1+1)+1/(n_2+1)\}$; t is therefore found by dividing $\bar{x}-\bar{x}'$ by its standard error as estimated, and the error of the estimation is allowed for by entering the table with n equal to the number of degrees of freedom available for estimating s; that is $n=n_1+n_2$. It is thus possible to extend "Student's" treatment of the error of a mean to the comparison of the means of two samples.

The method of building the corresponding analysis of variance for this case should be studied. If we put down the analyses for the two samples separately and add their items, we have

	Degrees of Freedom.	Sum of Squares.
Deviations	2	$\bar{x}S(x)+\bar{x}'S(x')$
Within samples	n_1+n_2	$S(x-\bar{x})^2+S(x'-\bar{x}')^2$
Total	n_1+n_2+2	$S(x^2)+S(x'^2)$

But if we had treated all the observations as a single sample with mean m, we should have

	Degrees of Freedom.	Sum of Squares.
Deviations	1	$mS(x)+mS(x')$
Within samples	n_1+n_2+1	$S(x-m)^2+S(x'-m)^2$
Total	n_1+n_2+2	$S(x^2)+S(x'^2)$

These are two different analyses of the same total, and since all comparisons within the separate samples are also comparisons within the grand sample made by throwing them together, we may subtract one from the other, obtaining

	Degrees of Freedom.	Sum of Squares.
Difference	1	$\bar{x}S(x)+\bar{x}'S(x')-mS(x)-mS(x')$
Within samples .	n_1+n_2	$S(x-\bar{x})^2+S(x'-\bar{x}')^2$
Total .	n_1+n_2+1	$S(x^2)+S(x'^2)-mS(x)-mS(x')$

Each item is now easily calculated. The student will do well to verify that t^2 obtained from the procedure first set out is in fact the ratio of the mean squares obtained from the analysis of variance.

It may be noted in connexion with this method, and with later developments, which also involve a pooled estimate of the variance, that a difference in variance between the populations from which the samples are drawn will tend sometimes to enhance the value of t obtained. The test, therefore, is decisive, if the value of t is significant, in showing that the samples could not have been drawn from the same population; but it might conceivably be claimed that the difference indicated lay in the variances and not in the means. The theoretical possibility, that a significant value of t should be produced by a difference between the variances only, seems to be unimportant in the application of the method to experimental data; as a supplementary test, however, the significance of the difference between the variances may always be tested directly by the method of Section 41.

It has been repeatedly stated, perhaps through a

§ 24·1]   SIGNIFICANCE OF *MEANS*, ETC.   125

misreading of the last paragraph, that our method involves the " assumption " that the two variances are equal. This is an incorrect form of statement; the equality of the variances is a necessary part of the hypothesis to be tested, namely that the two samples are drawn from the same normal population. The validity of the t-test, as a test of this hypothesis, is therefore absolute, and requires no assumption whatever. It would, of course, be legitimate to make a different test of significance appropriate to the question: Might these samples have been drawn from different normal populations having the same mean ? This problem has, in fact, been solved, but in relation to the real situations arising in biological research, the question it answers appears to be somewhat academic. Numerical tables of this test were first calculated by W. V. Behrens (1929) and much more completely by P. V. Sukhatmé. These are of use, when there is reason to suspect unequal variances, in removing any doubt from the interpretation of the test of significance. (*Statistical Tables*, V1 and V2; from the 5th edition fuller tables are given as VI, VI1 and VI2.)

Ex. 20 *Significance of difference of means of small samples.*—Let us suppose that the figures of Table 27 had been obtained using different patients for the two drugs; the experiment would have been less well controlled, and we should expect to obtain less certain results from the same number of observations, for it is *a priori* probable, and the above figures suggest, that personal variations in response to the drugs will be to some extent similar.

Taking, then, the figures to represent two different sets of patients, we have

$\bar{x} - \bar{x}' = +1\cdot58$, $t = +1\cdot861$,
$s^2(\frac{1}{10} + \frac{1}{10}) = \cdot7210$, $n = 18$.

The value of P is, therefore, between ·1 and ·05, and cannot be regarded as significant. This example shows clearly the value of design in small scale experiments, and that the efficacy of such design is capable of statistical measurement.

The use of " Student's " distribution enables us to appreciate the value of observing a sufficient number of parallel cases ; their value lies, not only in the fact that the standard error of a mean decreases inversely as the square root of the number of parallels, but in the fact that the accuracy of our estimate of the standard error increases simultaneously. The need for duplicate experiments is sufficiently widely realised ; it is not so widely understood that in some cases, when it is desired to place a high degree of confidence (say $P = ·01$) on the results, triplicate experiments will enable us to detect differences as small as one-seventh of those which, with a duplicate experiment, would justify the same degree of confidence.

The confidence to be placed in a result depends not only on the magnitude of the mean value obtained, but equally on the agreement between parallel experiments. Thus, if in an agricultural experiment a first trial shows an apparent advantage of 8 bushels to the acre, and a duplicate experiment shows an advantage of 9 bushels, we have $n = 1$, $t = 17$, and the results would justify some confidence that a real effect had been observed ; but if the second experiment had shown an apparent advantage of 18 bushels, although the mean is now higher, we should place not more but less confidence in the conclusion that the treatment was beneficial, for t has fallen to 2·6, a value which for $n = 1$ is often exceeded by chance. The apparent paradox may be explained by pointing out that the

§ 24·1] SIGNIFICANCE OF *MEANS*, ETC. 127

difference of 10 bushels between the experiments indicates the existence of uncontrolled circumstances so influential that in both cases the apparent benefit may be due to chance, whereas in the former case the relatively close agreement of the results suggests that the uncontrolled factors were not so very influential. Much of the advantage of further replication lies in the fact that when few tests are made, and these only duplicated, our estimate of the importance of the uncontrolled factors is extremely hazardous.

In cases in which each observation of one series corresponds in some respects to a particular observation of the second series, it is always legitimate to take the differences and test them, as in Ex. 19, as a single sample; but it is not always desirable to do so. A more precise comparison is obtainable by this method only if the corresponding values of the two series are positively correlated, and only if they are correlated to a sufficient extent to counterbalance the loss of precision due to basing our estimate of variance upon fewer degrees of freedom. An example will make this plain.

Ex. 21. *Significance of change in bacterial numbers.* — The following table shows the mean number of bacterial colonies per plate obtained by

TABLE 28

Method.	4 P.M.	8 P.M.	Difference.
A	29·75	39·20	+9·45
B	27·50	40·60	+13·10
C	30·25	36·20	+5·95
D	27·80	42·40	+14·60
Mean	28·825	39·60	+10·775

four slightly different methods from soil samples taken at 4 P.M. and 8 P.M. respectively (H. G. Thornton's data).

From the series of differences we have $\bar{x} = +10\cdot775$, $\frac{1}{4}s^2 = 3\cdot756$, $t = 5\cdot560$, $n = 3$, whence the table shows that P is between $\cdot01$ and $\cdot02$. If, on the contrary, we use the method of Ex. 20, and treat the two separate series, we find $\bar{x} - \bar{x}' = +10\cdot775$, $\frac{1}{2}s^2 = 2\cdot188$, $t = 7\cdot285$, $n = 6$; this is not only a larger value of n but a larger value of t, which is now far beyond the range of the table, showing that P is extremely small. In this case the differential effects of the different methods are either negligible, or have acted quite differently in the two series, so that precision was lost in comparing each value with its counterpart in the other series. In cases like this it sometimes occurs that one method shows no significant difference, while the other brings it out; if either method indicates a definitely significant difference, its testimony cannot be ignored, even if the other method fails to show the effect; for the tests of significance are used as an aid to judgement, and should not be confused with automatic acceptance tests, or "decision functions". When no correspondence exists between the members of one series and those of the other, the second method only is available. Pairing arbitrarily, or at random, will certainly supply a "decision," as, indeed will tossing a coin without reference to the data, but such methods ignore the statistician's obligation to summarise adequately the weight of evidence actually supplied by the data, against any theoretical view, and to do this is the only valid purpose of tests of significance.

25. Regression Coefficients

The methods of this chapter are applicable not only to mean values, in the restricted sense of the word, but to the very wide class of statistics known as regression coefficients. The idea of regression used usually to be introduced in connexion with the theory of correlation, but it is in reality a more general, and a simpler idea ; moreover, the regression coefficients are of interest and scientific importance in many classes of data where the correlation coefficient, if used at all, is an artificial concept of no real utility. The following qualitative discussion is intended to familiarise the student with the concept of regression, and to prepare the way for the accurate treatment of numerical examples.

It is a commonplace that the height of a child depends on his age, although, knowing his age, we cannot accurately calculate his height. At each age the heights are scattered over a considerable range in a frequency distribution characteristic of that age; any feature of this distribution, such as the mean, will be a continuous function of age. The function which represents the mean height at any age is termed the regression function of height on age ; it is represented graphically by a regression curve, or regression line. In relation to such a regression line *age* is termed the **independent** variate, and *height* the **dependent** variate.

The two variates bear very different relations to the regression line. If errors occur in the heights, this will not influence the regression of height on age, provided that at all ages positive and negative errors are equally frequent, so that they balance in the averages. On the contrary, errors in age will in general alter the regression of height on age, so that from a record with ages subject to error, or classified in broad age groups, we should not obtain the true physical relationship between mean height and age. A second difference should also be noted : the regression function does not depend on the frequency distribution of the independent variate, so that a true regression line may be obtained even when the age groups are arbitrarily selected, as when an investigation deals with children of "school age." On the other hand, a selection of the dependent variate may change the regression line altogether.

It is clear from these two instances that the regression of height on age is quite different from the regression of age on height ; and that one may have a definite physical meaning in cases in which the other has only the conventional meaning given to it by mathematical definition. In certain cases both regressions are of equal standing ; thus, if we express in terms of the height of the father the average adult height of sons of fathers of a given height, observation shows that each additional inch of the fathers' height corresponds to about half an inch in the mean height of the sons. Equally, if we take the mean height of the fathers of sons of a given height, we find that each additional inch of the sons' height corresponds to half an inch in the mean height of the fathers. No selection has been exercised in the heights either of fathers or of sons ; each variate is distributed normally, and the aggregate of pairs of values forms a normal correlation surface. Both regression lines are straight, and it is consequently possible to express the facts of regression in the simple rules stated above.

When the regression line with which we are concerned is straight, or, in other words, when the regression function is linear, the specification of regression is much simplified, for in addition to the general means we have only to state the ratio which the increment of the mean of the dependent variate bears to the corresponding increment of the independent variate. Such ratios are termed regression coefficients. The regression function takes the form
$$Y = a + b(x - \bar{x}),$$
where b is the regression coefficient of y on x, and Y is the predicted value of y for each value of x. The physical dimensions of the regression coefficient depend on those of the variates; thus, over an age range in which growth is uniform we might express the regression of height on age in inches per annum, in fact as an average growth rate, while the regression of father's height on son's height is half an inch per inch, or simply $\frac{1}{2}$. Regression coefficients may, of course, be positive or negative.

Curved regression lines are of common occurrence; in such cases we may have to use such a regression function as
$$Y = a + bx + cx^2 + dx^3,$$
in which all four coefficients of the regression function may, by an extended use of the term, be called regression coefficients. More elaborate functions of x may be used, but their practical employment offers difficulties in cases where we lack theoretical guidance in choosing the form of the regression function, and at present the simple power series (or polynomial in x) is alone in frequent use. By far the most important case in statistical practice is the straight regression line.

26. Sampling Errors of Regression Coefficients

The linear regression formula contains two parameters which are to be estimated from the data. If we use the form

$$Y = a + b(x - \bar{x})$$

then the value chosen for *a* will be simply the mean, $\bar{y}$, of the observed values of the dependent variate. This ensures that the sum of the residuals $y - Y$ shall be zero, for the sum of the values of $b(x - \bar{x})$ must be zero, whatever may be the value of *b*.

The value given to *b*, our estimate of the regression coefficient of *y* on *x*, is obtained from the sum of the products of *x* and *y*. Just as with a single variate we estimate the variance from the sum of squares, first by deducting $n\bar{x}^2$, so as to obtain the sum of the squares of deviations from the mean, in accordance with the formula,

$$S\{(x - \bar{x})^2\} = S(x^2) - n\bar{x}^2,$$

and then dividing by $(n - 1)$ to obtain an estimate of the variance; so with any two variates *x* and *y*, we may obtain the sum of the products of deviations from the means by deducting $n\bar{x}\bar{y}$; for

$$S\{(x - \bar{x})(y - \bar{y})\} = S(xy) - n\bar{x}\bar{y}.$$

The mean product of two variates, thus measured from their means, is termed their **covariance**, and, just as in the case of the variance of a single variate, we estimate its value by dividing the sum of products by $n - 1$. The sum of products from which the covariance is estimated may evidently be written equally in the forms

$$S\{y(x - \bar{x})\}, \qquad S\{x(y - \bar{y})\}.$$

Our estimate of *b* is simply the ratio of the covariance of the two variates, to the variance of the independent variate; or, since we may ignore the factor $(n-1)$ which appears in both terms of the ratio, our method of estimation may be expressed by the formula

$$b = \frac{S\{y(x-\bar{x})\}}{S\{(x-\bar{x})^2\}}.$$

We thus have estimates calculable from the observations, of the two parameters, needed to specify the straight line. The true regression formula, which we should obtain from an infinity of observations, may be represented by

$$Y = a + \beta(x-\bar{x})$$

and the differences $a-a$, $b-\beta$, are the errors of random sampling of our statistics.

To ascertain the magnitude of the sampling errors to which they are subject consider a population of samples having the same values for x. The variations from sample to sample in our statistics will be due only to the fact that for a given value of x the values of y in the population sampled are not all equal. If σ^2 represent the variance of y for a given value of x, then clearly the error of a is merely the mean of n' independent errors each having a variance σ^2, so that the variance of a is σ^2/n'. The second statistic b is also a linear function of the values, y, and its sampling variance may be obtained by an extension of the same reasoning. In this case each deviation of y from the true regression formula is multiplied by $x-\bar{x}$; the variance of the product is therefore $\sigma^2(x-\bar{x})^2$, and that of the sum of the products, which is the numerator of the expression for b, must be

$$\sigma^2 S(x-\bar{x})^2.$$

To find b we divide this numerator by $S\{(x-\bar{x})^2\}$ so that the variance of b is found by dividing the variance of the numerator by $S^2\{(x-\bar{x})^2\}$ which gives us the expression

$$\frac{\sigma^2}{S(x-\bar{x})^2}$$

for the sampling variance of the statistic b.

It will be noticed that the value stated for the sampling variance of a is not merely the sampling variance of our estimate of the mean of y, but of our estimate of the mean of y for a given value of x, this value being chosen at, or near to, the mean of our sample, and supposed invariable from sample to sample. The distinction, which at first sight appears somewhat subtle, is worth bearing in mind. From a set of measurements of school children we may make estimates of the mean stature at age ten, and of the mean stature of the school, and these estimates will be equal if the mean age of the school children is exactly ten. Nevertheless, the former will usually be the more accurate estimate, for it eliminates the variation in mean school age, which will doubtless contribute somewhat to the variation in mean school stature.

In order to test the significance of the difference between b, and any hypothetical value, β, to which it is to be compared, we must estimate the value of σ^2; the best estimate for the purpose is

$$s^2 = \frac{1}{n'-2} S(y-Y)^2,$$

found by summing the squares of the deviations of y from its calculated value Y, and dividing by $(n'-2)$. The reason the divisor is $(n'-2)$ is that from the n' values of y two statistics have already been calculated which enter into the formula for Y, consequently the group of differences, $y-Y$, represent in reality only $n'-2$ degrees of freedom.

When n' is small, the estimate of s^2 obtained above is somewhat uncertain, and in comparing the difference $b-\beta$ with its standard error, in order to test its significance we shall have to use "Student's" method, with $n = n'-2$. When n' is large the t-distribution tends to normality. The value of t with which the table must be entered is found by dividing $(b-\beta)$ by its standard error as estimated, and is therefore,

$$t = \frac{(b-\beta)\sqrt{S(x-\bar{x})^2}}{s}.$$

Similarly, to test the significance of the difference between a and any hypothetical value a, the table is entered with
$$t = \frac{(a-a)\sqrt{n'}}{s}, \quad n = n'-2;$$

this test for the significance of a will be more sensitive than that ignoring the regression, if the variation in y is to any considerable extent expressible in terms of that of x, for the value of s obtained from the regression line will then be smaller than that obtained from the original group of observations. On the other hand, 1 degree of freedom is always lost, so that if b is small, no greater precision is obtained.

In general, when the mean value of the dependent variate is estimated for values other than the mean of the independent variate, we need, as was shown by Working and Hotelling, to know the sampling variance of the estimate

$$Y = a + b(x-\bar{x}).$$

Since the sampling errors of a and b are independent, this is given by $V(a)+(x-\bar{x})^2 V(b)$,

where $V(a)$ and $V(b)$ stand for the sampling variances of our estimates a and b.

We have, therefore,
$$V(Y) = \sigma^2\left(\frac{1}{n'} + \frac{(x-\bar{x})^2}{S(x-\bar{x})^2}\right)$$
where σ^2 is the true variance of y for given x. For values of x near the mean, that is, where $(x-\bar{x})$ is small, this variance will not greatly exceed that at the mean of the observed sample, but for values more remote from the centre of our experience the precision of the estimate is naturally lower, and the second component of error, due to the estimation of b, becomes predominant.

Ex. 22. *Effect of nitrogenous fertilisers in maintaining yield.*—The yields of dressed grain in bushels per acre shown in Table 29 were obtained from two plots on Broadbalk wheat field during thirty years; the only difference in manurial treatment was that "9 a" received nitrate of soda, while "7 b" received an equivalent quantity of nitrogen as sulphate of ammonia. In the course of the experiment plot "9 a" appears to be gaining in yield on plot "7 b." Is this apparent gain significant?

A great part of the variation in yield from year to year is evidently similar in the two plots; in consequence, the series of differences will give the clearer result. In one respect these data are especially simple, for the thirty values of the independent variate form a series with equal intervals between the successive values, with only one value of the dependent variate corresponding to each. In such cases the work is simplified by using the formula
$$S(x-\bar{x})^2 = \tfrac{1}{12}n'(n'^2 - 1),$$
where n' is the number of terms, or 30 in this case.

To evaluate b it is necessary to calculate the sum of products
$$S\{y(x-\bar{x})\};$$

§ 26] SIGNIFICANCE OF *MEANS*, ETC.

which bears the same relation to the covariance of two variates as does the sum of squares to the variance of a single variate; this may be done in several ways.

TABLE 29

Harvest Year.	9 a.	7 b.	9 a—7 b.	
1855	29·62	33·00	−3·38	
1856	32·38	36·91	−4·53	
1857	43·75	44·84	−1·09	
1858	37·56	38·94	−1·38	
1859	30·00	34·66	−4·66	
1860	32·62	27·72	+4·90	$S(x-\bar{x})^2 = \dfrac{n'(n'^2-1)}{12} = 2247·5$
1861	33·75	34·94	−1·19	
1862	43·44	35·88	+7·56	
1863	55·56	53·66	+1·90	$b = ·26679$
1864	51·06	45·78	+5·28	
1865	44·06	40·22	+3·84	$S(y-\bar{y})^2 = 1020·56$
1866	32·50	29·91	+2·59	$b^2 S(x-\bar{x})^2 = 159·97$
1867	29·13	22·16	+6·97	
1868	47·81	39·19	+8·62	$S(y-Y)^2 = 860·59$
1869	39·00	28·25	+10·75	
1870	45·50	41·37	+4·13	$s^2 = 30·74$
1871	34·44	22·31	+12·13	
1872	40·69	29·06	+11·63	$s^2/S(x-\bar{x})^2 = ·013675$
1873	35·81	22·75	+13·06	
1874	38·19	39·56	−1·37	$= (·11694)^2$
1875	30·50	26·63	+3·87	
1876	33·31	25·50	+7·81	$t = 2·2814$
1877	40·12	19·12	+21·00	
1878	37·19	32·19	+5·00	$n = 28$
1879	21·94	17·25	+4·69	
1880	34·06	34·31	−·25	
1881	35·44	26·13	+9·31	
1882	31·81	34·75	−2·94	
1883	43·38	36·31	+7·07	
1884	40·44	37·75	+2·69	
Mean	37·50	33·03	+4·47	

We may multiply the successive values of y by -29, -27, ... $+27$, $+29$, add, and divide by 2. This is the direct method suggested by the formula. The same result is obtained by multiplying by 1, 2, ..., 30 and subtracting $15\frac{1}{2}$ $\left(=\frac{n'+1}{2}\right)$ times the sum of values of y; the latter method may be conveniently carried out by successive addition. Starting from the bottom of the column, the successive sums 2·69, 9·76, 6·82, ... are written down, each being found by adding a new value of y to the total already accumulated; the sum of the new column, less $15\frac{1}{2}$ times the sum of the previous column, will be the value required. In this case we find the value 599·615, and dividing by 2247·5, the value of b is found to be ·26679. The yield of plot "9 a" thus appears to have gained on that of "7 b" at a rate somewhat over a quarter of a bushel per acre per annum.

To estimate the standard error of b, we require the value of the sum of squares of the deviations, or residuals, from the regression formula,

$$S(y-Y)^2;$$

knowing the value of b, it is easy to calculate the thirty values of Y from the formula

$$Y = \bar{y} + (x-\bar{x})b;$$

for the first value, $x-\bar{x} = -14\cdot 5$, and the remaining values of Y may be found in succession by adding b each time. By subtracting each value of Y from the corresponding y, squaring, and adding, the required quantity may be calculated directly. This method is laborious, and it is preferable in practice to utilise the algebraical fact that

$$\begin{aligned}S(y-Y)^2 &= S(y-\bar{y})^2 - b^2 S(x-\bar{x})^2 \\ &= S(y^2) - n'\bar{y}^2 - b^2 S(x-\bar{x})^2.\end{aligned}$$

The work then consists in squaring the values of y and adding, then subtracting the two quantities, which can be directly calculated from the mean value of y and the value of b. In using this shortened method it should be noted that small errors in $\bar{y}$ and b may introduce considerable errors in the result, so that it is necessary to be sure that these are calculated accurately to as many significant figures as are needed in the quantities to be subtracted. Errors of arithmetic which would have little effect in the first method may altogether vitiate the results if the second method is used. The subsequent work in calculating the standard error of b may best be followed in the scheme given beside the table of data; the estimated standard error is ·1169, so that in testing the hypothesis that $\beta = 0$, that is that plot "9 a" has not been gaining on plot "7 b," we divide b by this quantity and find $t = 2·2814$. Since s was found from 28 degrees of freedom $n = 28$, and the result of t shows that P is between ·02 and ·05.

The result must be judged significant, though barely so; in view of the data we cannot ignore the possibility that on this field, and in conjunction with the other manures used, nitrate of soda has conserved the fertility better than sulphate of ammonia; the data do not, however, demonstrate this point beyond possibility of doubt.

The standard error of $\bar{y}$, calculated from these data, is 1·012, so that there can be no doubt that the difference in mean yields is significant; if we had tested the significance of the mean, without regard to the order of the values, that is calculating s^2 by dividing 1020·56 by 29, the standard error would have been 1·083. The value of b was therefore high enough to have reduced the standard error. This suggests the possibility that if we had fitted a more complex regression line to the data the probable errors would be further reduced to an extent which would put the significance of *b* beyond doubt. We shall deal later with the fitting of curved regression lines to this type of data. (Sections 27, 28).

26·1. The Comparison of Regression Coefficients

Just as the method of comparison of means is applicable when the samples are of different sizes, if we obtain an estimate of the error by combining the sums of squares derived from the two different samples, so we may compare regression coefficients when the series of values of the independent variate are not identical ; or if they are identical we can ignore the fact in comparing the regression coefficients.

Ex. 23. *Comparison of relative growth rate of two cultures of an alga.*—Table 30 shows the logarithm (to the base 10) of the volumes occupied by algal cells on successive days, in parallel cultures, each taken over a period during which the relative growth rate was approximately constant. In culture A nine values are available, and in culture B eight (Dr M. Bristol-Roach's data).

The method of finding $Sy(x-\bar{x})$ by summation is shown in the second pair of columns : the original values are added up from the bottom, giving successive totals from 6·087 to 43·426 ; the final value should, of course, tally with the total below the original values. From the sum of the column of totals is subtracted the sum of the original values multiplied by 5 for A and by 4½ for B. The differences are $Sy(x-\bar{x})$; these must be divided by the respective values of $S(x-\bar{x})^2$,

§ 26.1] SIGNIFICANCE OF *MEANS*, ETC.

namely, 60 and 42, to give the values of b, measuring the relative growth rates of the two cultures. To test if the difference is significant we calculate in the two cases $S(y^2)$, and subtract successively the product of the mean with the total, and the product of b with $Sy(x-\bar{x})$; this process leaves the two values of $S(y-Y)^2$, which are added as shown in the table, and

TABLE 30

	Log Values.		Summation Values.				
	A.	B.	A.	B.			
	3·592	3·538	43·426	38·358	$S(y-Y)^2$, A		·05089
	3·823	3·828	39·834	34·820	,, B		·07563
	4·174	4·349	36·011	30·992			
	4·534	4·833	31·837	26·643	ns^2		·12652
	4·956	4·911	27·303	21·810	s^2		·009732
	5·163	5·297	22·347	16·899	$s^2/60$		·0001622
	5·495	5·566	17·184	11·602	$s^2/42$		·0002317
	5·602	6·036	11·689	6·036			
	6·087	...	6·087	...			·0003939
Total	43·426	38·358	235·718	187·160	Standard error		·01985
Mean	4·8251	4·7947	217·130	172·611	$b'-b$		·0366
			$Sy(x-\bar{x})$ 18·588	14·549	t		1·844
			b ·3098	·3464	n		13

the sum divided by n, to give s^2. The value of n is found by adding the 7 degrees of freedom from series A to the 6 degrees from series B, and is therefore 13. Estimates of the variance of the two regression coefficients are obtained by dividing s^2 by 60 and 42, and that of the variance of their difference is the sum of these. Taking the square root we find the standard error to be ·01985, and $t = 1·844$. The difference between the regression coefficients, though relatively large, cannot be regarded as significant. There is not sufficient evidence to assert that culture B was growing more rapidly than culture A.

26·2. The Ratio of Means and Regression Coefficients

Ex. 23·1. When pairs of observations are available, such as those shown in Table 27 (page 121), showing as these do a decidedly significant difference between the means, we have gained some idea of the magnitude of the true difference between the means, which we may expect to lie between limits given by the observed value plus or minus an appropriate multiple of its standard error. This multiple specifies the level of significance chosen and, in a well defined sense, the probability that the true difference should lie between the limits assigned. Thus this probability is 95 per cent. if we choose as the appropriate multiplier the 5 per cent. value of t for the number of degrees of freedom available, or 2·262 for the 9 degrees of freedom in that example.

It may well be that the difference between the average effects of two treatments is of less intrinsic interest than the ratio of their effects. This will be so if the effect of each drug is proportional to the quantity used, but in other cases also the ratio of the effects may be constant, so that at any dosage the ratio of the effects provides an estimate of the potency ratio; while the difference between the average effects of a chosen dose will depend greatly on the experimental material used, on the conditions of the experiment, and on the actual amount of the dose. It is useful, therefore, to be able to assign similar limits for a presumed constant ratio between the effects in place of those for a presumed constant difference.

Now, if x and y are the observed effects of treat-

§ 26·2] SIGNIFICANCE OF *MEANS*, ETC.   143

ments A and B in any particular case, and a stands for the potency of A relative to that of B (in the simplest case, the weight of B equivalent to unit weight of A), then we may consider the quantity

$$z = x - ay$$

as an observed value, variations of which from case to case may be estimated from the experimental data. The arithmetic required is nearly the same as that of Example 20, namely the means and sums of squares of the variates x and y, with the addition of their sum of products.

Thus for x we have

$S(x^2)$	34·43
$\bar{x}S(x)$	5·625
$S(x-\bar{x})^2$	28·805 ;

for y

$S(y^2)$	90·37
$\bar{y}S(y)$	54·289
$S(y-\bar{y})^2$	36·081 ;

and, for the product,

$S(xy)$	43·11
$\bar{x}S(y) = \bar{y}S(x)$	17·475
$S(x-\bar{x})(y-\bar{y})$	25·635.

Then, leaving a still undetermined, it is clear that

$$S(z) = 7\cdot5 - 23\cdot3\,a$$

and

$$S(z-\bar{z})^2 = 28\cdot805 - 2a(25\cdot635) + a^2(36\cdot081).$$

Moreover, the data will show a significant deviation from the value of a adopted if

$$\frac{9}{10}S^2(z) > t^2 S(z-\bar{z})^2.$$

Taking, for the 5 per cent. point,
$$t = 2·262, \quad t^2 = 5·116644,$$
then the equation for a becomes
$$303·9874 a^2 - 26·1098(2a) - 96·7599 = 0,$$
which is satisfied by the values
$$a = +·6566 \text{ and } -·4848.$$

It is thus clear that no estimate of the relative potency of drug A compared with drug B exceeding ·6566, or rather less than two-thirds, is compatible with the data presented. The fact that the other value is negative shows that these data do not establish any positive soporific effect at all for drug A at the significance level used. It might, in fact, have exerted an antisoporific effect nearly one-half as potent as the soporific effect of drug B before the observed difference in efficacy between the two drugs would be significantly exceeded.

A method very similar in principle may be used to find limits for the value of the independent variate at which the regression function attains a given value, or the value at which two regression lines intersect.

If a_1 and a_2 are the true means and β_1, β_2 the true coefficients of regression of two dependent variates, then the point of intersection is the value of the unknown, X, satisfying the condition
$$a_1 + (X - \bar{x}_1)\beta_1 = a_2 + (X - \bar{x}_2)\beta_2.$$
Now the sampling variance of
$$a_1 + (X - \bar{x}_1)b_1 - a_2 - (X - \bar{x}_2)b_2$$
is
$$s^2 \left\{ \frac{1}{N_1} + \frac{1}{N_2} + \frac{(X - \bar{x}_1)^2}{S_1} + \frac{(X - \bar{x}_2)^2}{S_2} \right\},$$

§ 26·2] SIGNIFICANCE OF *MEANS*, ETC.   145

where S_1 and S_2 are the sums of squares of deviations of the independent variate for the two samples, and s^2 is the mean square deviation of the dependent variates from the fitted regression lines. Hence if we equate

to
$$\{a_1-a_2-b_1\bar{x}_1+b_2\bar{x}_2+X(b_1-b_2)\}^2$$

$$s^2t^2\left\{\frac{1}{N_1}+\frac{1}{N_2}+\frac{\bar{x}_1^2}{S_1}+\frac{\bar{x}_2^2}{S_2}-2X\left(\frac{\bar{x}_1}{S_1}+\frac{\bar{x}_2}{S_2}\right)+X^2\left(\frac{1}{S_1}+\frac{1}{S_2}\right)\right\},$$

we shall have a quadratic equation for X of which the roots are the limiting values possible at the level of significance represented by the value t.

Ex. 23·2. *The age at which girls become taller than boys.*—Karn (1934) gives values derived from measurements of 4007 school children in the borough of Croydon.

	Number N	Mean age $\bar{x}$ (years)	Mean Height a (inches)	Regression b (ins./yr.)	$S(x-\bar{x})^2$ (yr.)^2
Boys	1946	12·2016	56·004	1·60	337·894
Girls	2061	12·1300	56·550	2·45	382·835

The mean square deviation from the fitted regression lines, s^2, is

8·17915,

based on 3991 degrees of freedom. For limits at the 5 per cent. point we may therefore take

$$t = 1·96$$

and
$$s^2t^2 = 31·421.$$

From the remaining data we have, taking x as the excess of the age over 12 years,

$a_2 - a_1 - b_2 \bar{x}_2 + b_1 \bar{x}_1$	$\cdot 55016$
$(b_2 - b_1)X$	$\cdot 85 X$
$\dfrac{1}{N_1} + \dfrac{1}{N_2} + \dfrac{\bar{x}_1^2}{S_1} + \dfrac{\bar{x}_2^2}{S_2}$	$\cdot 001163582$
$-\left(\dfrac{\bar{x}_1}{S_1} + \dfrac{\bar{x}_2}{S_2}\right)2X$	$-\cdot 000936405(2X)$
$\left(\dfrac{1}{S_1} + \dfrac{1}{S_2}\right)X^2$	$\cdot 00557160(X^2)$.

The quadratic equation for X is therefore

$$(\cdot 547435)X^2 + (\cdot 497064)2X + \cdot 266122 = 0,$$

of which the roots are

$$-1 \cdot 490, \quad -\cdot 326$$

corresponding with ages

$$10 \cdot 510 \text{ and } 11 \cdot 674 \text{ years.}$$

The estimate derived from the means and regressions given is 11·353 years, much nearer to the upper than to the lower limit. The children were nearly all measured in their 11th and 12th years, and the precision of the comparison falls considerably at the lower ages, with the consequence that the lower limit differs widely from the direct estimate. With this number of children a much higher accuracy would have been obtained had they been measured a year earlier, or over a wider age-range.

27. The Fitting of Curved Regression Lines

But slight use has been made of the theory of the fitting of curved regression lines, save in the limited but most important case when the variability of the dependent variate is the same for all values of the independent variate, and is normal for each such value. When this is the case a technique has been fully worked out for fitting by successive stages any line of the form

$$Y = a+bx+cx^2+dx^3+ \ldots;$$

we shall give details of the case where the successive values of x are at equal intervals. The more general case, when varying numbers of observations occur at different values of x, is best treated by the method of Section 29·2 ; when the intervals also are unequal, the general method of Section 29 is available, using the powers of x as independent variates.

As it stands the form given would be inconvenient in practice, in that the fitting could not be carried through in successive stages. What is required is to obtain successively the mean of y, an equation linear in x, an equation quadratic in x, and so on, each equation being obtained from the last by adding a new term ; this being calculated by carrying a single process of computation through a new stage. In order to do this we take

$$Y = A+B\xi_1+C\xi_2+D\xi_3+ \ldots,$$

where ξ_1, ξ_2, ξ_3 shall be functions of x of the 1st, 2nd, and 3rd degrees, out of which the regression formula may be built.

These functions of x may be regarded as the coefficients of the corresponding observations in certain comparisons, or components of variation among them. Thus ξ_1 is always chosen to be $x-\bar{x}$; e.g. if there were 7 observations the values of ξ_1 would be -3, -2, -1, 0, 1, 2, 3; so that the comparison corresponding with the 1st degree in x is

(i) $\quad -3y_1-2y_2-y_3+0y_4+y_5+2y_6+3y_7.$

Again, ξ_2 might be taken as the coefficients in the comparison

(ii) $\quad 5y_1+0y_2-3y_3-4y_4-3y_5+0y_6+5y_7.$

Here the coefficients are expressible as a quadratic in x, namely

$$(x-\bar{x})^2-4 = \xi_1^2-4 = \xi_2,$$

and it is to be noticed that the sum of the coefficients, and the sum of their products with those of ξ_1, are both zero.

For the 3rd, 4th and 5th degrees we may use in turn

(iii) $\quad -y_1+y_2+y_3-y_5-y_6+y_7 \qquad \xi_1(\xi_1^2-7)/6$
(iv) $\quad 3y_1-7y_2+y_3+6y_4+y_5-7y_6+3y_7 \quad (7\xi_1^4-67\xi_1^2+72)/12$
(v) $\quad -y_1+4y_2-5y_3+5y_5-4y_6+y_7; \; (21\xi_1^5-245\xi_1^3+524\xi_1)/60.$

Note that the sum of the coefficients is zero in each case, so that each expression is properly a comparison among the values of y; moreover, the sum of the products of corresponding coefficients in any two expressions is zero, so that the comparisons made are properly independent.

In fitting a curve, the expressions in y are evaluated, each divided by the sum of the squares of its coefficients, and are then used as multipliers of the corresponding functions of x in the fitted curve. Thus the sums of squares of the five expressions above are 28, 84, 6,

§ 27]   SIGNIFICANCE OF *MEANS*, ETC.   149

154, 84. Consequently, the successive terms of the fitted curve are :—

$(-3y_1-2y_2-y_3+y_5+2y_6+3y_7)\xi_1/28$
$(5y_1-3y_3-4y_4-3y_5+5y_7)(\xi_1{}^2-4)/84$
$(-y_1+y_2+y_3-y_5-y_6+y_7)(\xi_1{}^3-7\xi_1)/36$
$(3y_1-7y_2+y_3+6y_4+y_5-7y_6+3y_7)(7\xi_1{}^4-67\xi_1{}^2+72)/1848$
$(-y_1+4y_2-5y_3+5y_5-4y_6+y_7)(21\xi_1{}^5-245\xi_1{}^3+524\xi_1)/5040$

The first of these expressions gives the best fitting straight line. By using the first two terms we have the best fitting parabola, the 3rd terms adjust it to the best fitting cubic, and so on.

Based on these orthogonal polynomials, independent comparisons convenient for fitting series to the 5th degree are given so far as $n' = 75$ in *Statistical Tables*. The general formulæ are given in editions 3 to 6 of this book, but for higher degrees and longer series it is best to use the arithmetical approach illustrated in the following sections.

The components are also expressible in terms of the successive differences of the series y_n'. Thus those given above might be written

(i) $\Delta(3y_1+5y_2+6y_3+6y_4+5y_5+3y_6)$
(ii) $\Delta^2(5y_1+10y_2+12y_3+10y_4+5y_5)$
(iii) $\Delta^3(y_1+2y_2+2y_3+y_4)$
(iv) $\Delta^4(3y_1+5y_2+3y_3)$
(v) $\Delta^5(y_1+y_2)$

where Δy_1 stands for y_2-y_1, and so on. In place of the sum of the squares of the coefficients of the explicit formulæ, we should then use the square of the sum of the coefficients of the differences of the appropriate degree, divided by

$$\frac{n'(n'^2-1)\ldots(n'^2-r^2)}{(2.6)(6.10)\ldots\{(4r-2)(4r+2)\}}$$

for the term of degree r. This device of using differences sometimes saves an immense amount of labour, since the differences are often smaller numbers than those from which they are derived, and fewer of them are to be used. The sign of the coefficients also is always positive. Coefficients of such expansions in differences may be found for any degree by starting with unity, and multiplying successively by

$$\frac{(r+1)(n'-r-1)}{1(n'-1)}, \frac{(r+2)(n'-r-2)}{2(n'-2)}, \ldots$$

a method which may be simply illustrated by constructing in this way the formula given above for $n' = 7, r = 4$; thence by differencing four times construct the actual coefficients of the fourth component.

Although, for arithmetical purposes, it is convenient to leave these expressions indeterminate in respect of constant factors, so that on removing any common factor, or clearing fractions, the expression may be used in its simplest form, algebraically, it is convenient to introduce the convention that in the polynomials the coefficient of the leading term is unity. Thus ξ_3 above is taken to be $\xi_1{}^3 - 7\xi_1$, with values 6 times the coefficients of the expression used. With this convention, the sum of the squares of the coefficients is found to be

$$\frac{n'(n'^2-1)\ldots(n'^2-r^2)}{12.15\ldots(16-4/r^2)} = \frac{(n'+r)!}{(n'-r-1)!} \cdot \frac{r!^4}{(2r)!(2r+1)!}$$

so that the process of fitting may now be represented by the equations

$$A = \bar{y} = \frac{1}{n'} S(y),$$

$$B = \frac{12}{n'(n'^2-1)} S(y\xi_1),$$

$$C = \frac{180}{n'(n'^2-1)(n'^2-4)} S(y\xi_2),$$

§ 28] SIGNIFICANCE OF *MEANS*, ETC.

where, in general, the coefficient of the term of the rth degree is
$$\frac{(2r)!(2r+1)!}{(r!)^4 n'(n'^2-1)\ldots(n'^2-r^2)} S(y\xi_r).$$

As each term is fitted the regression line approaches more nearly to the observed values, and the sum of the squares of the deviations
$$S(y-Y)^2$$
is diminished. It is desirable to be able to calculate this quantity, without evaluating the actual values of Y at each point of the series; this can be done by subtracting from $S(y^2)$ the successive quantities
$$n'A^2, \quad \frac{n'(n'^2-1)}{12}B^2, \quad \frac{n'(n'^2-1)(n'^2-4)}{180}C^2,$$
or more simply
$$AS(y), \quad BS(y\xi_1), \quad CS(y\xi_2),$$
and so on. These quantities represent the reduction which the sum of the squares of the residuals suffers each time the regression curve is fitted to a higher degree; and enable its value to be calculated at any stage by a mere extension of the process already used in the preceding examples. To obtain an estimate, s^2, of the residual variance, we divide by n, the number of degrees of freedom left after fitting, which is found from n' by subtracting from it the number of constants in the regression formula. Thus, if a straight line has been fitted, $n = n'-2$; while if a curve of the 5th degree has been fitted, $n = n'-6$.

28. The Arithmetical Procedure of Fitting

The main arithmetical labour of fitting curved regression lines to data of this type may be reduced to a repetition of the process of summation illustrated in

Ex. 23. We shall assume that the values of y are written down in a column in order of increasing values of x, and that at each stage the summation is commenced at the top of the column (not at the bottom, as in that example). The sums of the successive columns will be denoted by S_1, S_2, ... When these values have been obtained, each is divided by an appropriate divisor, which depends only on n', giving us a new series of quantities a, b, c, ... according to the following equations

$$a = \frac{1}{n'} S_1 = \frac{1}{n'} S(y) = \bar{y},$$

$$b = \frac{1.2}{n'(n'+1)} S_2,$$

$$c = \frac{1.2.3}{n'(n'+1)(n'+2)} S_3,$$

and so on.

From these a third series of quantities a', b', c'. ... is obtained by equations independent of n', of which we give below the first six, which are enough to carry the process of fitting up to the 5th degree:

$$a' = a,$$
$$b' = a-b,$$
$$c' = a-3b+2c,$$
$$d' = a-6b+10c-5d,$$
$$e' = a-10b+30c-35d+14e,$$
$$f' = a-15b+70c-140d+126e-42f.$$

The rule for the formation of the coefficients is to multiply successively by

$$\frac{r(r+1)}{1.2}, \quad \frac{(r-1)(r+2)}{2.3}, \quad \frac{(r-2)(r+3)}{3.4},$$

and so on till the series terminates.

§ 28] SIGNIFICANCE OF *MEANS*, ETC.

These new quantities are proportional to the required coefficients of the regression equation, and need only be divided by a second group of divisors to give the actual values. The equations are

$$A = a', \qquad B = \frac{6}{n'-1} b',$$

$$C = \frac{30}{(n'-1)(n'-2)} c', \qquad D = \frac{140}{(n'-1)(n'-2)(n'-3)} d',$$

$$E = \frac{630}{(n'-1)(n'-2)\ldots(n'-4)} e', \qquad F = \frac{2772}{(n'-1)\ldots(n'-5)} f',$$

the numerical part of the factor being

$$\frac{(2r+1)!}{(r!)^2}$$

for the term of degree r.

If an equation of degree r has been fitted, the estimate of the standard errors of the coefficients are all based upon the same value of s^2, *i.e.*

$$s^2 = \frac{1}{n'-r-1} \left\{ S(y^2) - n'A^2 - \frac{n'(n'^2-1)}{12} B^2 - \ldots \right\},$$

from which the estimated standard error of any coefficient, such as that of ξ_p, is obtained by dividing by

$$S(\xi_p^2) = \frac{(p!)^4}{(2p)!(2p+1)!} n'(n'^2-1) \ldots (n'^2-p^2)$$

and taking out the square root. The number of degrees of freedom upon which the estimate is based is $(n'-r-1)$, and this must be equated to n in using the Table of t.

A suitable example for using this method may be obtained by fitting the values of Ex. 22 (p. 136) with a curve of the 2nd or 3rd degree.

28·1. The Calculation of the Polynomial Values

The methods of the preceding sections provide an analysis of a series into the components which can be represented by polynomial terms of any required degree, and the remainder which cannot be so represented. For much work of this kind it is desirable to carry out this analysis without the labour of calculating the polynomial values, Y, at each point of the series. Sometimes, however, it is desirable to have these values, either to construct a graph, to examine the deviations in regions of special interest, or because doing so provides a completely satisfactory check upon the results calculated.

The very tedious procedure of calculating the individual values of ξ, and from them, and the calculated coefficients, forming the individual values of the polynomial, may be avoided by building up the whole series, by a continuous process, from its differences. The process is obvious when a straight line is fitted. For the terminal value, and the constant difference between successive values, we take

$$Y_1 = a' + 3b',$$
$$\Delta Y_1 = -\frac{6}{n'-1} b',$$

and build up all the other values of Y by continuous addition of the constant difference. The method is, however, applicable to polynomials of high order, and in such cases appears to save more than three-quarters of the labour of calculation. For curves of the 2nd degree the equations are:

$$Y_1 = a' + 3b' + 5c',$$
$$\Delta Y_1 = -\frac{6}{n'-1}(b' + 5c'),$$
$$\Delta^2 Y_1 = \frac{60}{(n'-1)(n'-2)} c'.$$

§ 28·1]   SIGNIFICANCE OF *MEANS*, ETC.   155

Starting with the terminal value ΔY_1, the series of first differences is built up by successive addition of the constant second difference $\Delta^2 Y_1$; then starting from Y_1, and adding successively the first differences, the series of values of Y is built up in turn.

The formulæ for any degree are constructed using the factors, with alternate positive and negative signs,

$$1, \quad \frac{-2.3}{n'-1}, \quad \frac{3.4.5}{(n'-1)(n'-2)}, \quad \frac{-4.5.6.7}{(n'-1)(n'-2)(n'-3)}, \ldots$$

together with expressions in a', b', c', ... with the same coefficients, as given in Table 30·2, whatever the degree of the curve.

The arithmetical procedure, which consists almost entirely of successive addition, may be illustrated on the series of Ex. 22. Table 30·1 shows on the left the

TABLE 30·1

Observed Values.	1st Sum.	2nd Sum.	3rd Sum.	Poly-nomial Values.	1st Difference.	2nd Difference.	3rd Difference.
...	...	...	...	...	...	...	
−0·25	117·88	960·77	4440·58	5·86	·739	−·1280	
+9·31	127·19	1087·96	5528·54	4·99	·871	−·1320	
−2·94	124·25	1212·21	6740·75	3·98	1·008	−·1361	
+7·07	131·32	1343·53	8084·28	2·84	1·148	−·1402	
+2·69	134·01	1477·54	9561·82	1·544	1·2919	−·14423	·004061
134·01	1477·54	9561·82	39167·21	134·00			
4·467000	3·177505	1·927786	0·957165				
4·467000	1·289495	−1·209943	−·105995				

last five lines of the summations needed to fit a curve of the 3rd degree, and on the right the first five lines of the summations by which the polynomial values are built up.

Below the first four columns are shown the values of a, ..., d derived directly from the totals, and of

a', ... d' derived from them. If we want the values of Y to two decimal places, it will be as well to calculate Y_1 to three places, and each difference to one more place than the last, discarding one place for the subsequent differences of each series. With this in view six decimal places will be sufficient for a, ..., d. Any further degree of accuracy required may be obtained merely by retaining additional digits. The sum of the column of polynomial values, which must tally with that of those observed, provides an excellent check of the latter parts of the procedure, but not of the correctness of the initial summations.

TABLE 30·2
COEFFICIENTS OF a', b', c', ... IN THE TERMINAL VALUES OF Y AND ITS DIFFERENCES

1	3	5	7	9	11	13	15	17	19	21
	1	5	14	30	55	91	140	204	285	385
		1	7	27	77	182	378	714	1254	2079
			1	9	44	156	450	1122	2508	5148
				1	11	65	275	935	2717	7007
					1	13	90	442	1729	5733
						1	15	119	665	2940
							1	17	152	952
								1	19	189
									1	21
										1

The coefficients used in this method in the expression for Y_1, ΔY_1, $\Delta^2 Y_1$, ... in terms of a', b', c', ... are given in Table 30·2 up to the 10th degree.

29. Regression with several Independent Variates

It frequently happens that the data enable us to express the average value of the dependent variate y, in terms of a number of different independent variates x_1, x_2, ... x_p. For example, the rainfall at any point within a district may be recorded at a number of stations for which the longitude, latitude, and altitude are all known. If all of these three variates influence the rainfall, it may be required to ascertain the average effect of each separately. In speaking of longitude, latitude, and altitude as independent variates, all that is implied is that it is in terms of them that the average rainfall is to be expressed; it is not implied that these variates vary independently, in the sense that they are uncorrelated. On the contrary, it may well happen that the more southerly stations lie on the whole more to the west than do the more northerly stations, so that for the stations available longitude measured to the west may be negatively correlated with latitude measured to the north If, then, rainfall increased to the west but was independent of latitude, we should obtain, merely by comparing the rainfall recorded at different latitudes, a fictitious regression indicating that rain decreased towards the north. What we require is an equation, taking account of all three variates at each station, and agreeing as nearly as possible with the values recorded; this is called a **partial** regression equation and its coefficients are known as partial regression coefficients.

To simplify the algebra we shall suppose that y, x_1, x_2, x_3, are all measured from their mean values, and that we are seeking a formula of the form

$$Y = b_1 x_1 + b_2 x_2 + b_3 x_3.$$

If S stands for summation over all the sets of observations we construct the three equations

$$b_1 S(x_1^2) + b_2 S(x_1 x_2) + b_3 S(x_1 x_3) = S(x_1 y),$$
$$b_1 S(x_1 x_2) + b_2 S(x_2^2) + b_3 S(x_2 x_3) = S(x_2 y),$$
$$b_1 S(x_1 x_3) + b_2 S(x_2 x_3) + b_3 S(x_3^2) = S(x_3 y),$$

of which the nine coefficients are obtained from the data either by direct multiplication and addition, or, if the data are numerous, by constructing correlation tables for each of the six pairs of variates. The three simultaneous equations for b_1, b_2, and b_3 may be solved in the ordinary way: first b_3 is eliminated from the first and third, and from the second and third equations, leaving two equations for b_1 and b_2; eliminating b_2 from these, b_1 is found, and thence by substitution, b_2 and b_3.

It frequently happens that, for the same set of values of the independent variates, it is desired to examine the regressions for more than one set of values of the dependent variate; as, for example, if for the same set of rainfall stations we had data for several different months or years. In such cases it is preferable to avoid solving the simultaneous equations afresh on each occasion, but to obtain a simpler formula which may be applied to each new case.

This may be done by solving once and for all the three sets, each consisting of three simultaneous equations:

$$b_1 S(x_1^2) + b_2 S(x_1 x_2) + b_3 S(x_1 x_3) = 1, \quad 0, \quad 0,$$
$$b_1 S(x_1 x_2) + b_2 S(x_2^2) + b_3 S(x_2 x_3) = 0, \quad 1, \quad 0,$$
$$b_1 S(x_1 x_3) + b_2 S(x_2 x_3) + b_3 S(x_3^2) = 0, \quad 0, \quad 1;$$

the three solutions of these three sets of equations may be written

$$b_1 = c_{11}, \ c_{12}, \ c_{13},$$
$$b_2 = c_{12}, \ c_{22}, \ c_{23},$$
$$b_3 = c_{13}, \ c_{23}, \ c_{33}.$$

Once the six values of c are known, then the partial regression coefficients may be obtained in any particular case merely by calculating $S(x_1 y)$, $S(x_2 y)$, $S(x_3 y)$ and substituting in the formulæ,

$$b_1 = c_{11}S(x_1 y) + c_{12}S(x_2 y) + c_{13}S(x_3 y),$$
$$b_2 = c_{12}S(x_1 y) + c_{22}S(x_2 y) + c_{23}S(x_3 y),$$
$$b_3 = c_{13}S(x_1 y) + c_{23}S(x_2 y) + c_{33}S(x_3 y).$$

The c-values, which are known as the covariance matrix, also serve to determine the precision of the regression co-efficients, so that this indirect method of obtaining them is generally to be recommended.

The method of partial regression is of very wide application. It is worth noting that the different independent variates may be related in any way; for example, if we desired to express the rainfall as a linear function of the latitude and longitude, and as a quadratic function of the altitude, the square of the altitude would be introduced as a fourth independent variate, without in any way disturbing the process outlined above, save in such points as that $S(x_3 x_4) = S(x_3^3)$ would be calculated directly from the distribution of altitude.

The analysis of sequences, exhibited in Sections 27 and 28 by means of orthogonal polynomials, could therefore alternatively have been carried out by the multiple regression method. In the case specially treated, in which we have a simple sequence of observations of a dependent variate, one for each of a series of equally spaced values of the independent variate, as in annual returns of economic and sociological data, the use of orthogonal polynomials presents manifest advantages. When, however, the number of observations is variable, or the intervals are not equally spaced, the method of orthogonal polynomials, which can be generalised to cover such cases, is artificial, and less direct than the treatment of the data by multiple regression. The equations of multiple regression are moreover equally applicable to regression equations involving not merely powers, but other functions such as logarithms, exponentials or trigonometric functions of the independent variate.

In estimating the sampling errors of partial regression coefficients we require to know how nearly our calculated value, Y, has reproduced the observed values of y; as in previous cases, the sum of the squares of $(y-Y)$ may be calculated by differences, for, with three variates,

$$S(y-Y)^2 = S(y^2) - b_1 S(x_1 y) - b_2 S(x_2 y) - b_3 S(x_3 y).$$

If we had n' sets of observations, and p independent variates, we should therefore first calculate

$$s^2 = \frac{1}{n'-p-1} S(y-Y)^2,$$

and to test if b_1 differed significantly from any hypothetical value, β_1, we should calculate

$$t = \frac{b_1 - \beta_1}{s\sqrt{c_{11}}},$$

entering the Table of t with $n = n'-p-1$.

In the practical use of a number of variates it is convenient to use cards, on each of which is entered the values of the several variates which may be required. By sorting these cards in suitable grouping units with respect to any two variates the corresponding correlation table may be constructed with little risk of error, and thence the necessary sums of squares and products obtained.

Ex. 24. *Dependence of rainfall on position and altitude.*—The situations of 57 rainfall stations in

§ 29] SIGNIFICANCE OF *MEANS*, ETC

Hertfordshire have a mean longitude $12'\cdot4$ W., a mean latitude $51°\ 48'\cdot5$ N., and a mean altitude 302 feet. Taking as units two minutes of longitude, one minute of latitude, and twenty feet of altitude, the following values of the sums of squares and products of deviations from the mean were obtained:

$S(x_1^2) = 1934\cdot1,$ $S(x_2x_3) = +119\cdot6,$
$S(x_2^2) = 2889\cdot5,$ $S(x_3x_1) = +924\cdot1,$
$S(x_3^2) = 1750\cdot8,$ $S(x_1x_2) = -772\cdot2.$

To find the multipliers suitable for any particular set of weather data from these stations, first solve the equations

$1934\cdot1\ c_{11} - 772\cdot2\ c_{12} + 924\cdot1\ c_{13} = 1$
$-772\cdot2\ c_{11} + 2889\cdot5\ c_{12} + 119\cdot6\ c_{13} = 0$
$924\cdot1\ c_{11} + 119\cdot6\ c_{12} + 1750\cdot8\ c_{13} = 0;$

using the last equation to eliminate c_{13} from the first two, we have

$2532\cdot3\ c_{11} - 1462\cdot5\ c_{12} = 1\cdot7508$
$-1462\cdot5\ c_{11} + 5044\cdot6\ c_{12} = 0;$

from these eliminate c_{12}, obtaining

$10,635\cdot5\ c_{11} = 8\cdot8321;$

whence

$c_{11} = \cdot00083044,\quad c_{12} = \cdot00024075,\quad c_{13} = -\cdot00045477,$

the last two being obtained successively by substitution.

Since the corresponding equations for c_{12}, c_{22}, c_{23} differ only in changes in the right-hand member, we can at once write down

$-1462\cdot5\ c_{12} + 5044\cdot6\ c_{22} = 1\cdot7508;$

whence, substituting for c_{12} the value already obtained,

$c_{22} = \cdot00041686,\quad c_{23} = -\cdot00015555;$

finally, to obtain c_{33} we have only to substitute in the equation

$$924\cdot 1\ c_{13} + 119\cdot 6\ c_{23} + 1750\cdot 8\ c_{33} = 1,$$

giving

$$c_{33} = \cdot 00082183.$$

It is usually worth while, to facilitate the detection of small errors by checking, to retain, as above, one more decimal place than the data warrant.

The partial regression of any particular weather data on these three variates can now be found with little labour. In January 1922 the mean rainfall recorded at these stations was 3·87 inches, and the sums of products of deviations with those of the three independent variates were (taking 0·1 inch as the unit for rain)

$$S(x_1 y) = +1137\cdot 4, \quad S(x_2 y) = -592\cdot 9, \quad S(x_3 y) = +891\cdot 8;$$

multiplying these first by c_{11}, c_{12}, c_{13} and adding, we have for the partial regression on longitude

$$b_1 = \cdot 39624;$$

similarly, using the multipliers c_{12}, c_{22}, c_{23} we obtain for the partial regression on latitude

$$b_2 = -\cdot 11204;$$

and finally, by using c_{13}, c_{23}, c_{33},

$$b_3 = \cdot 30788$$

gives the partial regression on altitude.

Remembering now the units employed, it appears that in the month in question rainfall increased by ·0198 of an inch for each minute of longitude westwards, it decreased by ·0112 of an inch for each minute of latitude northwards, and increased by ·00154 of an inch for each foot of altitude.

Let us calculate to what extent the regression on altitude is affected by sampling errors. For the 57 recorded deviations of the rainfall from its mean value, in the units previously used

$$S(y^2) = 1786\cdot 6\,;$$

whence, knowing the values of b_1, b_2, and b_3 we obtain by subtraction

$$S(y-Y)^2 = 994\cdot 9.$$

To find s^2, we must divide this by the number of degrees of freedom remaining after fitting a formula involving three variates—that is, by 53—so that

$$s^2 = 18\cdot 772\,;$$

multiplying this by c_{33} and taking the square root,

$$s\sqrt{c_{33}} = \cdot 12421.$$

Since n is as high as 53 we shall not be far wrong in taking the regression of rainfall on altitude to be in working units ·308, with a standard error ·124; or in inches of rain per 100 feet as ·154, with a standard error ·062.

The importance of the procedure developed in Ex. 24 lies in the generality of its applications, and in the fact that the same process is used to give in succession (*a*) the best regression equation of a given form, and (*b*) the materials for studying the residual variation, and the precision of the coefficients of our equation.

We have illustrated and used the fact that the sampling variance of any coefficient, such as b_1, is given by multiplying the estimated residual variance, s^2, by the factor c_{11} derived wholly from the independent variates. In many applications the calculation of the multipliers c is of further value owing to the fact that the sampling covariance of any two coefficients, such as b_1 and b_2, is given by multiplying the same estimated variance by c_{12}. We may, therefore, without repeating the primary calculations, review the results from a variety of different points of view. Although it would be of little interest in the meteorological problem, it will in other cases be frequently important to compare the magnitude of two different coefficients, *e.g.* to ask if b_1 is significantly greater than b_2. We need to compare the difference $b_1 - b_2$ with its estimated standard error, and this will be the square root of

$$s^2(c_{11} - 2c_{12} + c_{22}),$$

since the variance of the difference of any two quantities must be the sum of their variances, less twice their covariance, as is apparent from the algebraic identity

$$(x-y)^2 = x^2 - 2xy + y^2.$$

By the use of the c multipliers, we are thus able to test the significance of the sum or difference, or indeed any linear function, of two or more regression coefficients, by calculating its standard error, and recognising the ratio it bears to its standard error as t, having degrees of freedom appropriate to the estimation of the residual variance.

Although such single comparisons, chosen to answer questions of particular interest, are of the greatest practical use, yet a very comprehensive theoretical question also, is resolved by these relationships; namely, that of the *simultaneous* distribution, in the light of the data available, of the errors of random sampling of the regression coefficients

$$b_1 - \beta_1, b_2 - \beta_2, \ldots\ldots\ldots\ldots\ldots, b_p - \beta_p.$$

Since the values of b_1, b_2, ..., b_p are known with

exactitude and without uncertainty, the simultaneous distribution of these quantities is equally the simultaneous distribution of the unknown parameters $\beta_1, \beta_2, \ldots, \beta_p$. During the period of the earlier editions of this book, violent objection was often taken, on quasi-philosophical grounds, to the identification of parameters with random variables, (although such had been the 18th and 19th century practice), owing to the idea that this would imply that a parameter could have simultaneously more than one value. It is now generally recognised, save in a very restricted circle, that no such absurdity is implied, but that recognition is given to the undoubted fact that there is in the neighbourhood of any estimate a zone of uncertainty, any value within which might, on our data, be the true value. The exact nature of this uncertainty is accurately and comprehensively stated in terms of a probability distribution, using the word probability in its strict mathematical sense.

If S_{ij} stands for

$$S(x_i - \bar{x}_i)(x_j - \bar{x}_j),$$

with $|S|$ for the determinant of these, and Q^2 for the quadratic expression

$$S(b_i - \beta_i)(b_j - \beta_j)S_{ij}/s^2,$$

where both i and j take in the summation all integral values from 1 to p, then the simultaneous distribution inferred from the equations of multiple regression is

$$\frac{\Gamma\left(\frac{n+p}{2}\right)|S|^{1/2}\,d\beta_1\,d\beta_2\ldots\ldots\ldots\,d\beta_p}{(\pi n)^{\frac{1}{2}p}\,\Gamma\left(\frac{1}{2}n\right)s^p\quad(1+Q^2/n)^{\frac{1}{2}(n+p)}},$$

and the probability, in the light of the observations,

of the simultaneous values of $\beta_1, \beta_2, \ldots, \beta_p$ lying within any defined region of a p-fold Euclidean space, is simply the multiple integral of the expression above, over that region.

Regions over which it is easy to integrate are delimited (*a*) by linear functions of the β_i giving the t-test first developed and (*b*) regions inside or outside the closed boundary on which Q^2 is constant, equivalent to a z-test with $n_1 = p$ and $n_2 = n$ the number of degrees of freedom on which the estimate s^2 is based. The variance ratio is then equal to Q^2/p (*i.e.* $e^{2z} = Q^2/p$). The simultaneous distribution of the β_i (given the b_i) or of the b_i (given the β_i) is a generalised t-distribution, the properties of which have been examined by E. A. Cornish in a series of publications (*vide* Sources used for Data and Methods, p. 341 for references).

29.1. The Omission of an Independent Variate

It may happen that after a regression equation has been worked out, it appears that one of the independent variates used is of little interest, and that it would have been preferable to have omitted it, and to have calculated the regression on the others. This could be done by solving anew the set of equations involving only the squares and products of the remaining variates, but this labour may be avoided. The omission of a single variate will always increase the number of residual degrees of freedom by unity, and correspondingly will increase the sum of squares of deviations from the regression formula by a quantity corresponding to this 1 degree of freedom. If x_3 stands for the variate to be omitted, we may recall that the variance of the corresponding coefficient b_3 was given by the expression $\sigma^2 c_{33}$. The

variance of $b_3/\sqrt{c_{33}}$ will therefore be σ^2, and
$$b_3^2/c_{33}$$
must be the increment added to the sum of squares by the omission of the variate x_3.

Equally, if, in the regression formula, we had wished to replace b_3, not by zero, but by a theoretical value β_3, the increment would have been
$$(b_3-\beta_3)^2/c_{33}.$$

We may also wish to adjust the coefficients of the remaining variates, which have been already calculated, to what they would have been if any particular variate, such as x_3, had been omitted. This is easily done by subtracting from b_1 the quantity
$$\frac{c_{13}}{c_{33}}b_3,$$
and applying a similar adjustment of the other coefficients.

I owe to Professor H. Schultz of Chicago a more comprehensive application of this method than was given in the fifth edition. This is to recalculate the c-matrix from formulæ of the form
$$c'_{12} = c_{12} - \frac{c_{13}\, c_{23}}{c_{33}}.$$

The values c' supply the c-matrix which would have been obtained had variate (3) been omitted. These give the variances and covariances of the adjusted coefficients, and also the means of making the further adjustments needed should it be desired to omit a second variate, or indeed more, in succession.

Thus, if the regression of a dependent variate be worked out on a considerable group of six or more variates, which are regarded as possibly influential, it is always possible, with very little labour, if any one of them is found to be really unimportant, to obtain from our formula the result which would have been obtained had this one been omitted from the original calculations. More laboriously a succession of unwanted variates may be discarded in turn.

29.2. Polynomial Fitting when the Frequencies are Unequal

The advantages of the arithmetical procedure of Sections 28 and 28·1 may still be obtained when it is desired to fit a polynomial regression curve of any specified degree to a set of observations of the dependent variate, the frequencies of which at different values of the independent variate are unequal. Here we shall not be concerned to obtain a sequence of polynomials of different degrees, but only to obtain a single formula, the coefficients of which will not require separate tests of significance. We shall illustrate the process in detail for fitting a cubic curve to the times taken to run 100 yards by 988 boys at various ages from 9·25 to 19·25 years (H. Gray's data).

The addition process is applied separately to the frequencies and to the *totals* of sprinting time. Table 30·3 shows the frequencies in 21 half-year classes. To fit a cubic, these are summed seven times (numbered from 0 to 6), though the last summation need not be written out. Much labour is saved by choosing a "working zero," which we have placed at 14·25 years. Only the frequencies for age groups younger than this are summed forward. The frequencies for the older age groups are summed backward. The first backward summation (number 0) includes the working zero; the others each stop one step short of the summation before. For the columns of even number the forward and backward totals are added, while for those of odd number the forward is

TABLE 30·3

ABBREVIATED SUMMATION PROCESS FOR FREQUENCIES

Age.	Frequency.	Summation.						
		0.	1.	2.	3.	4.	5.	6.
9·25	6	6	6	6	6	6	6	
9·75	8	14	20	26	32	38	44	
10·25	10	24	44	70	102	140	184	
10·75	28	52	96	166	268	408	592	
11·25	29	81	177	343	611	1019	1611	
11·75	46	127	304	647	1258	2277	3888	
12·25	40	167	471	1118	2376	4653	8541	
12·75	53	220	691	1809	4185	8838	17379	
13·25	54	274	965	2774	6959	15797	33176	
13·75	66	340	1305	4079	11038	26835	60011	
14·25	87	648						
14·75	71	561	2296					
15·25	98	490	1735	4975				
15·75	84	392	1245	3240	7398			
16·25	85	308	853	1995	4158	7961		
16·75	67	223	545	1142	2163	3803	6309	
17·25	65	156	322	597	1021	1640	2506	
17·75	44	91	166	275	424	619	866	
18·25	25	47	75	109	149	195	247	
18·75	16	22	28	34	40	46	52	
19·25	6	6	6	6	6	6	6	
S		988	991	9054	−3640	34796	−53702	129109

subtracted from the backward total. The resulting sums are represented by S_0, S_1, ... S_6.

A similar process with only four summations (0 to 3) is then applied to the total sprinting times, as in Table 30·4, using the same working zero.

In the previous sections, where our regression

function is built up of polynomials of specially chosen simplicity, the coefficients were obtained by the solution of simple equations. In this, as in the last section, the equations are simultaneous. Four will

TABLE 30·4

SUMMATION OF TOTAL TIMES

Age	Total Times.	0.	1.	2.	3.
9·25	101·4	101·4	101·4	101·4	
9·75	127·2	228·6	330·0	431·4	
10·25	167·0	395·6	725·6	1157·0	
10·75	445·2	840·8	1566·4	2723·4	
11·25	475·6	1316·4	2882·8	5606·2	
11·75	713·0	2029·4	4912·2	10518·4	
12·25	612·0	2641·4	7553·6	18072·0	
12·75	800·3	3441·7	10995·3	29067·3	
13·25	810·0	4251·7	15247·0	44314·3	
13·75	943·8	5195·5	20442·5	64756·8	
14·25	1209·3	8354·9			
14·75	958·5	7145·6	28656·9		
15·25	1303·4	6187·1	21511·3	61169·6	
15·75	1075·2	4883·7	15324·2	39658·3	
16·25	1088·0	3808·5	10440·5	24334·1	
16·75	830·8	2720·5	6632·0	13893·6	
17·25	780·0	1889·7	3911·5	7261·6	
17·75	541·2	1109·7	2021·8	3350·1	
18·25	297·5	568·5	912·1	1328·3	
18·75	198·4	271·0	343·6	416·2	
19·25	72·6	72·6	72·6	72·6	
		13550·4	8214·4	125926·4	−86433·4

be needed for the coefficients of a cubic, and we take the four sums obtained above from the total times as the right-hand members of them. The coefficients of the unknowns on the left-hand sides are obtained

§ 29·2] SIGNIFICANCE OF *MEANS*, ETC. 171

from the totals $S_0 \ldots S_6$, derived from the frequencies, according to the following scheme:

S_0	S_1	S_2	S_3
S_1	$2S_2+S_1$	$3S_3+2S_2$	$4S_4+3S_3$
S_2	$3S_3+2S_2$	$6S_4+6S_3+S_2$	$10S_5+12S_4+3S_3$
S_3	$4S_4+3S_3$	$10S_5+12S_4+3S_3$	$20S_6+30S_5+12S_4+S_3$

This table is not changed, but if necessary extended, when curves are fitted of degrees other than three. It is a good intelligence test to write down the next two or three rows and columns from those given for a cubic curve. We are brought therefore to the equations

					Right-hand Values.	Check Column.
$988A +$	$991B +$	$9054C -$	$3640D =$		$13550·4$	$20943·4$
$991A +$	$19099B +$	$7188C +$	$128264D =$		$8214·4$	$163756·4$
$9054A +$	$7188B +$	$195990C -$	$130388D =$		$125926·4$	$207770·4$
$-3640A +$	$128264B -$	$130388C +$	$1385032D =$		$-86433·4$	$1292834·6$

where the unknowns A, B, C and D are the polynomial value at the working zero, and its first three advancing differences. The process of solution is shown in full below. Since the coefficients on the left form a symmetrical matrix, duplicate values may be omitted. The work in this example is also arranged to exhibit the use of a check column, which is merely the sum of the numbers in the same row, irrespective of which side of the equation they belong to. The numbers in this column are treated just as are those in the adjacent column at each stage of the solution of the equations, and afford a check for each row of figures as it is completed. The arithmetical details are given in Table 30·5 as arranged for machine calculation. When the number of equations has been reduced to one, the value of A is calculated; B is then found by substitution in the second equation, and a new value for A from the first of the pair of equations at the penultimate stage. In the same way C, B and A are calculated from the trio of equations, and D, C, B and A from the original equations by substitution for each

TABLE 30·5

STEPS IN THE DIRECT SOLUTION OF FOUR EQUATIONS

Coefficients of Unknowns.				Right-hand Side.	Check Column.
988	991	9054	−3640	13550·4	20943·4
	19099	7188	128264	8214·4	163756·4
		195990	−130388	125926·4	207770·4
			1385032	−86433·4	1292834·6
1·355162	1·839448	12·06547		18·45312	33·71320
	10·00107	26·67970		22·46350	60·98372
		254·4514		163·1422	456·3388
1·992473	1·461470			27·27035	30·72429
	18·32980			13·63284	33·42411
34·385737					

unknown always in its appropriate equation. Such a complete system of checking obviates all arithmetical errors, and from the extent of the variations observed in the solutions gives an idea of the extent to which the limited accuracy of the process of solution can affect the results.

To obtain the fitted polynomial values to 2 decimal places, we may retain 3, 4, 5 and 6 places in A, B, C, D

TABLE 30·6

SOLUTIONS CHECKED BY EACH EQUATION

A.	B.	C.	D.
13·95742	−·3690990	·01802630	·01015438
42	90	49	
42	83		
42			

TABLE 30·7

DEVELOPMENT OF POLYNOMIAL VALUES FROM THE SOLUTIONS OF THE EQUATIONS

Observed Mean Times.	Fitted Polynomial Values.	Differences.		
		1st.	2nd.	3rd.
16·9	16·40			
		+·009	−·0835	
15·9	16·41			
		−·074	−·0734	
16·7	16·34			
		−·148	−·0632	
15·9	16·19			
		−·211	−·0530	
16·4	15·98			
		−·264	−·0429	
15·5	15·72			
		−·307	−·0327	
15·3	15·41			
		−·340	−·0226	
15·1	15·07			
		−·362	−·0124	
15·0	14·71			
		−·375	−·0023	
14·3	14·33			
		−·377	+·0079	
13·9	13·957			
		−·3691	·01803	
13·5	13·59			·010154
		−·351	·0282	
13·3	13·24			
		−·323	·0383	
12·8	12·91			
		−·285	·0485	
12·8	12·63			
		−·236	·0586	
12·4	12·39			
		−·178	·0688	
12·0	12·21			
		−·109	·0790	
12·3	12·11			
		−·030	·0891	
11·9	12·08			
		+·059	·0993	
12·4	12·14			
		·159		
12·1	12·29			

and build up the polynomial by successive addition as in Table 30·7. It will be understood that in forming the second differences on a machine, 6 places are visible at each stage, although only 4 need be written down, using the nearest integer in the 4th place. For the rest, the table explains itself.

The sum of the squares of the polynomial values, multiplied by their appropriate frequencies, is found as usual by multiplying the solution of the regression equations by the right-hand values. Since in this case the regression equations contain an absolute term, A, this will not give the sum of squares of deviations from the mean, but from zero. To reduce to the mean we must deduct $(13550·4)^2 \div 988$, leaving for 3 degrees of freedom the value 1645·58. Deducting this from the 20 degrees of freedom for differences among classes, there remains 31·24 representing residual deviations from the function fitted.

	Degrees of Freedom.	Sum of Squares.	Mean Square.
Regression . . .	3	1645·58	548·53
Residual differences . .	17	31·24	1·838
Within age groups . .	967	1620·27	1·676
Total . . .	987	3297·09	

The adequacy of the form of curve chosen for representing the sequence of means observed may be judged by comparing the mean square derived from the deviations with that within age classes. The average sum of squares within age groups, derived from the standard deviations at each age given by

Gray, is 1620·27. The whole variation among the 988 times recorded has thus been analysed into three portions (see preceding Table).

Since the mean square for residuals approximates closely to that observed among runners of the same age, it is evident that no curve could fit the data appreciably better. In applying this test we have anticipated the method explained in Section 44.

TABLE IV—TABLE OF t

n	P=·9	·8	·7	·6	·5	·4	·3	·2	·1	·05	·02	·01
1	·158	·325	·510	·727	1·000	1·376	1·963	3·078	6·314	12·706	31·821	63·657
2	·142	·289	·445	·617	·816	1·061	1·386	1·886	2·920	4·303	6·965	9·925
3	·137	·277	·424	·584	·765	·978	1·250	1·638	2·353	3·182	4·541	5·841
4	·134	·271	·414	·569	·741	·941	1·190	1·533	2·132	2·776	3·747	4·604
5	·132	·267	·408	·559	·727	·920	1·156	1·476	2·015	2·571	3·365	4·032
6	·131	·265	·404	·553	·718	·906	1·134	1·440	1·943	2·447	3·143	3·707
7	·130	·263	·402	·549	·711	·896	1·119	1·415	1·895	2·365	2·998	3·499
8	·130	·262	·399	·546	·706	·889	1·108	1·397	1·860	2·306	2·896	3·355
9	·129	·261	·398	·543	·703	·883	1·100	1·383	1·833	2·262	2·821	3·250
10	·129	·260	·397	·542	·700	·879	1·093	1·372	1·812	2·228	2·764	3·169
11	·129	·260	·396	·540	·697	·876	1·088	1·363	1·796	2·201	2·718	3·106
12	·128	·259	·395	·539	·695	·873	1·083	1·356	1·782	2·179	2·681	3·055
13	·128	·259	·394	·538	·694	·870	1·079	1·350	1·771	2·160	2·650	3·012
14	·128	·258	·393	·537	·692	·868	1·076	1·345	1·761	2·145	2·624	2·977
15	·128	·258	·393	·536	·691	·866	1·074	1·341	1·753	2·131	2·602	2·947
16	·128	·258	·392	·535	·690	·865	1·071	1·337	1·746	2·120	2·583	2·921
17	·128	·257	·392	·534	·689	·863	1·069	1·333	1·740	2·110	2·567	2·898
18	·127	·257	·392	·534	·688	·862	1·067	1·330	1·734	2·101	2·552	2·878
19	·127	·257	·391	·533	·688	·861	1·066	1·328	1·729	2·093	2·539	2·861
20	·127	·257	·391	·533	·687	·860	1·064	1·325	1·725	2·086	2·528	2·845
21	·127	·257	·391	·532	·686	·859	1·063	1·323	1·721	2·080	2·518	2·831
22	·127	·256	·390	·532	·686	·858	1·061	1·321	1·717	2·074	2·508	2·819
23	·127	·256	·390	·532	·685	·858	1·060	1·319	1·714	2·069	2·500	2·807
24	·127	·256	·390	·531	·685	·857	1·059	1·318	1·711	2·064	2·492	2·797
25	·127	·256	·390	·531	·684	·856	1·058	1·316	1·708	2·060	2·485	2·787
26	·127	·256	·390	·531	·684	·856	1·058	1·315	1·706	2·056	2·479	2·779
27	·127	·256	·389	·531	·684	·855	1·057	1·314	1·703	2·052	2·473	2·771
28	·127	·256	·389	·530	·683	·855	1·056	1·313	1·701	2·048	2·467	2·763
29	·127	·256	·389	·530	·683	·854	1·055	1·311	1·699	2·045	2·462	2·756
30	·127	·256	·389	·530	·683	·854	1·055	1·310	1·697	2·042	2·457	2·750

VI

THE CORRELATION COEFFICIENT

30. No quantity has been more characteristic of biometrical work than the correlation coefficient, and no method has been applied to such various data as the method of correlation. Observational data in particular, in cases where we can observe the occurrence of various possible contributory causes of a phenomenon, but cannot control them, have been given by its means an altogether new importance. In experimental work proper its position is much less central; it will be found useful in the exploratory stages of an inquiry, as when two factors which had been thought independent appear to be associated in their occurrence; but it is seldom, with controlled experimental conditions, that it is desired to express our conclusion in the form of a correlation coefficient.

One of the earliest and most striking successes of the method of correlation was in the biometrical study of inheritance. At a time when nothing was known of the mechanism of inheritance, or of the structure of the germinal material, it was possible by this method to demonstrate the existence of inheritance, and to "measure its intensity"; and this in an organism in which experimental breeding could not be practised, namely, Man. By comparison of the results obtained from the physical measurements in man with those obtained from other organisms, it was established that man's nature is not less governed by heredity than that of the rest of the animate world. The scope of the analogy was further widened by demonstrating that correlation coefficients of the same magnitude were obtained for the mental and moral qualities in man as for the physical measurements.

These results are still of fundamental importance, for not only is inheritance in man still incapable of experimental study, and existing methods of mental testing are still unable to analyse the mental disposition, but even with organisms suitable for experiment and measurement, it is only in the most favourable cases that the several factors causing fluctuating variability can be resolved, and their effects studied, by Mendelian methods. Such fluctuating variability, with an approximately normal distribution, is characteristic of the majority of the useful qualities of domestic plants and animals; and although there is strong reason to think that inheritance in such cases is ultimately Mendelian, the biometrical method of study is at present alone capable of holding out hopes of immediate progress.

That this method was once centred on the correlation coefficient gives to this statistic a certain importance, even to those who prefer to develop their analysis in other terms.

We give in Table 31 an example of a correlation table. It consists of a record in compact form of the stature of 1376 fathers and daughters. (Pearson and Lee's data.) The measurements are grouped in inches, and those whose measurement was recorded as an integral number of inches have been split; thus a father recorded as of 67 inches would appear as $\frac{1}{2}$ under 66·5 and $\frac{1}{2}$ under 67·5. Similarly with the daughters; in consequence, when both measurements are whole

§ 30] THE CORRELATION COEFFICIENT 179

numbers the case appears in four quarters. This gives the table a confusing appearance, since the majority of entries are fractional, although they represent frequencies. The practice of splitting observations is not to be deliberately imitated. A little care in the choice of group limits will avoid all ambiguity. When many items are split, Sheppard's corrections are no longer accurate.

The most obvious feature of the table is that cases do not occur in which the father is very tall and the daughter very short, and *vice versa*; the upper right-hand and lower left-hand corners of the table are blank, so that we may conclude that such occurrences are too rare to occur in a sample of about 1400 cases. The observations recorded lie in a roughly elliptical figure lying diagonally across the table. If we mark out the region in which the frequencies exceed 10 it appears that this region, apart from natural irregularities, is similar, and similarly situated. The frequency of occurrence increases from all sides to the central region of the table, where a few frequencies over 30 may be seen. The lines of equal frequency are roughly similar and similarly situated ellipses. In the outer zone observations occur only occasionally, and therefore irregularly; beyond this we could only explore by taking a much larger sample.

The table has been divided into four quadrants by marking out central values of the two variates; these values, 67·5 inches for the fathers and 63·5 inches for the daughters, are near the means. When the table is so divided it is obvious that the lower right-hand and upper left-hand quadrants are distinctly more populous than the other two; not only are more squares occupied, but the frequencies are higher. It

TABLE

		____	____	____	Height of	____	____	____	____
		58·5	59·5	60·5	61·5	62·5	63·5	64·5	65·5	66·5

		58·5	59·5	60·5	61·5	62·5	63·5	64·5	65·5	66·5
Height of Daughters in Inches.	52·5	...	...	...	...	·25	·25	...	...	...
	53·5	...	...	...	...	·25	·25	...	...	...
	54·5	...	...	...	...	...	...	...	...	...
	55·5	...	...	...	...	...	...	...	1	...
	56·5	·25	·25	...	·25	1·25	·5	...	1	·5
	57·5	·25	·25	·5	1·5	4·5	1	1·5	1·5	2·5
	58·5	·25	·75	·5	·75	·75	1	1·75	1·25	5
	59·5	·5	1	2	...	6	4·75	5	6·25	11·75
	60·5	·75	·75	...	2·5	8	6·25	12·5	18·25	20·25
	61·5	...	·5	1·75	2	9·75	11·5	13	23·75	23·75
	62·5	...	1	2·25	2	4·5	12	22·75	26	33
	63·5	...	...	·25	2	6	8·25	11	27·25	35·75
	64·5	...	...	·25	2·5	1·75	3·25	9·25	23	18·75
	65·5	...	...	...	·5	1	·5	11	12·25	9·25
	66·5	...	...	...	·5	·5	1·5	3·25	7·25	8·75
	67·5	...	...	...	...	...	...	1	5·75	7
	68·5	...	...	...	...	·25	·25	·25	·25	1·5
	69·5	...	...	...	...	·25	·25	·25	·25	·25
	70·5	...	...	...	...	...	...	...	...	...
	71·5	...	...	...	...	...	...	...	...	..
	72·5	...	...	...	...	...	...	...	...	...
	Total	2	4·5	7·5	14·5	45	51·5	92·5	155	178

Fathers in Inches.

67·5	68·5	69·5	70·5	71·5	72·5	73·5	74·5	75·5	Total.
...	...	...	...	...	...	...	...	...	·5
...	...	...	...	...	...	...	...	...	·5
...	...	...	...	...	...	...	...	...	...
...	...	...	...	...	...	...	...	...	1
·5	...	...	...	...	...	...	...	...	4·5
...	·5	·5	...	...	...	...	...	...	14·5
2·75	·5	·25	...	...	...	...	...	...	15·5
3·5	3·5	2	1·75	·5	...	...	...	...	48·5
11	9	4·75	2·5	1·25	1·25	...	...	...	99
20·25	16·5	10·25	4·25	3	1·25	...	...	...	141·5
28·25	24·75	14·25	13·75	4·75	·75	·5	...	...	190·5
37·25	31·5	26·25	16·25	7·75	1·5	·75	·25	...	212
28·5	33	34·25	24·5	11·75	5·5	1	·25	1	198·5
19·75	30	26·5	22·25	15	4·75	3·75	2	1	159·5
16	26·25	26·75	20·5	18·5	7·75	4·25	·25	·5	142·5
4	14·25	13·25	12	11·25	4·5	3·75	·75	...	77·5
3	5·5	4·25	5·75	5·25	3·75	2·5	1·5	2	36
·25	1	2·5	6·5	2·25	2·75	2	1	...	19·5
...	1·75	·25	4·5	·75	1·25	·75	·25	...	9·5
...	·5	...	·5	·5	1·5	·75	·25	...	4
...	1	...	..	...	...	...	...	...	1
175	199·5	166	135	82·5	36·5	20	6·5	4·5	1376

is apparent that tall men have tall daughters more frequently than the short men, and *vice versa*. The method of correlation aims at measuring the degree to which this association exists.

The marginal totals show the frequency distributions of the fathers and the daughters respectively. These are both approximately normal distributions, as is frequently the case with biometrical data collected without selection. This marks a frequent difference between biometrical and experimental data. An experimenter would perhaps have bred from two contrasted groups of fathers of, for example, 63 and 72 inches in height; all his fathers would then belong to these two classes, and the correlation coefficient, if used, would be almost meaningless. Such an experiment would serve to ascertain the regression of daughter's height on father's height and so to determine the effect on the daughters of selection applied to the fathers, but it would not give us the correlation coefficient, which is a descriptive observational feature of the population as it is, and may be wholly vitiated by selection.

Just as normal variation with one variate may be specified by a frequency formula in which the logarithm of the frequency is a quadratic function of the variate, so with two variates the frequency may be expressible in terms of a quadratic function of the values of the two variates. We then have a normal correlation surface, for which the frequency may conveniently be written in the form

$$df = \frac{1}{2\pi\sigma_1\sigma_2\sqrt{1-\rho^2}} e^{-\frac{1}{2(1-\rho^2)}\left\{\frac{x^2}{\sigma_1^2} - \frac{2\rho xy}{\sigma_1\sigma_2} + \frac{y^2}{\sigma_2^2}\right\}} dx dy.$$

In this expression x and y are the deviations of

the two variates from their means, σ_1 and σ_2 are the two standard deviations, and ρ is the *correlation* between x and y. The correlation in the above expression may be positive or negative, but cannot exceed unity in magnitude; it is a pure number without physical dimensions. If $\rho = 0$, the expression for the frequency degenerates into the product of the two factors

$$\frac{1}{\sigma_1\sqrt{2\pi}} e^{-\frac{x^2}{2\sigma_1^2}} dx \cdot \frac{1}{\sigma_2\sqrt{2\pi}} e^{-\frac{y^2}{2\sigma_2^2}} dy,$$

showing that the limit of the normal correlation surface, when the correlation vanishes, is merely that of two normally distributed variates varying in complete independence. At the other extreme, when ρ is $+1$ or -1, the variation of the two variates is in strict proportion, so that the value of either may be calculated accurately from that of the other. In other words, we cease strictly to have two variates, but merely two measures of the same variable quantity.

If we pick out the cases in which one variate has an assigned value, we have what is termed an array; the columns and rows of the table may, except as regards variation within the group limits, be regarded as arrays. With normal correlation the variation within an array may be obtained from the general formula, by giving x a constant value, (say) a, and dividing by the total frequency with which this value occurs; then we have

$$df = \frac{1}{\sigma_2\sqrt{2\pi}\sqrt{1-\rho^2}} \cdot e^{-\frac{1}{2(1-\rho^2)\sigma_2^2}\left(y - \rho\frac{a\sigma_2}{\sigma_1}\right)^2} dy,$$

showing (i) that the variation of y within the array is normal; (ii) that the mean value of y for that array is

$\rho a \sigma_2 / \sigma_1$, so that the regression of y on x is linear, with regression coefficient

$$\rho \frac{\sigma_2}{\sigma_1},$$

and (iii) that the variance of y within the array is $\sigma_2^2(1-\rho^2)$, and is the same within each array. We may express this by saying that of the total variance of y the fraction $(1-\rho^2)$ is independent of x, while the remaining fraction ρ^2, is determined by, or calculable from, the value of x.

These relations are reciprocal; the regression of x on y is linear, with regression coefficient $\rho \sigma_1 / \sigma_2$; the correlation ρ is thus the geometric mean of the two regressions. The two regression lines representing the mean value of x for given y, and the mean value of y for given x, cannot coincide unless $\rho = \pm 1$. The variation of x within an array in which y is fixed is normal with variance equal to $\sigma_1^2(1-\rho^2)$, so that we may say that of the variance of x the fraction $(1-\rho^2)$ is independent of y, and the remaining fraction, ρ^2, is determined by, or calculable from, the value of y.

Such are the formal mathematical consequences of normal correlation. Much biometric material certainly shows a general agreement with the features to be expected on this assumption; though I am not aware that the question has been subjected to any sufficiently critical inquiry. Approximate agreement is perhaps all that is needed to justify the use of the correlation as a quantity descriptive of the population; its efficacy in this respect is undoubted, and it is not improbable that in some cases it affords, in conjunction with the means and variances, a complete description of the simultaneous variation of the variates.

31. The Statistical Estimation of the Correlation

Just as the variance of a normal population in one variate may be most satisfactorily estimated from the sum of the squares of deviations from the mean of the observed distribution, so, as we have seen, the only satisfactory estimate of the covariance, when the variates are normally correlated, is found from the sum of the products. The estimate used for the correlation is the ratio of the covariance to the geometric mean of the two variances. If x and y represent the deviations of the two variates from their means, we calculate the three statistics s_1, s_2, r by the three equations

$$ns_1^2 = \mathrm{S}(x^2), \quad ns_2^2 = \mathrm{S}(y^2), \quad nrs_1s_2 = \mathrm{S}(xy);$$

then s_1 and s_2 are estimates of the standard deviations σ_1 and σ_2, and r is an estimate of the correlation ρ. Such an estimate is called the *correlation coefficient*, or the *product moment correlation*, the latter term referring to the summation of the product terms, xy, in the last equation. The value used for n should properly be the number of degrees of freedom, or one less than the number of pairs of observations in the sample. So far as the value obtained for r is concerned, however, the value used for n is indifferent, and it is usually convenient to base the calculation directly on the sums of squares and products without dividing by n.

The method of calculation might have been derived from the consideration that the correlation of the population is the geometric mean of the two regression coefficients; for our estimates of these two regressions would be
$$\frac{\mathrm{S}(xy)}{\mathrm{S}(x^2)} \text{ and } \frac{\mathrm{S}(xy)}{\mathrm{S}(y^2)},$$

so that it is in accordance with these estimates to take as our estimate of ρ

$$r = \frac{S(xy)}{\sqrt{S(x^2) \cdot S(y^2)}},$$

which is in fact the product moment correlation.

Ex. 25. *Parental correlation in stature.*—The numerical work required to calculate the correlation coefficient is shown in Table 32.

The first eight columns require no explanation, since they merely repeat the usual process of finding the mean and variance of the two marginal distributions. It is not necessary actually to find the mean, by dividing the total of the 3rd column, 480·5, by 1376, since we may work all through with the undivided totals. The correction for the fact that our working mean is not the true mean is performed by subtracting $(480·5)^2 \div 1376$ in the 4th column; a similar correction appears at the foot of the 8th column, and at the foot of the last column. The correction for the sum of products is performed by subtracting $480·5 \times 260·5 \div 1376$. This correction of the product term may be positive or negative; if the total deviations of the two variates are of opposite sign, the correction must be added. The sum of squares, with and without Sheppard's adjustment ($1376 \div 12$), are shown separately; there is no corresponding adjustment to be made to the product term.

The 9th column shows the total deviations of the daughter's height for each of the 18 columns in which Table 31 is divided. When the numbers are small, these may usually be written down by inspection of the table. In the present case, where the numbers are large, and the entries are complicated by quartering, more care is required. The total of column 9

§ 31] THE CORRELATION COEFFICIENT 187

	Daughters.				Fathers.				Product.
Deviation.	Frequency.			Deviation.	Frequency.				
−11	·5	5·5	60·5	−9	2	18	162		+78·75
−10	·5	5	50	−8	4·5	36	288		+122
−9			—	−7	7·5	52·5	367·5		+133
−8	1	8	64	−6	14·5	87	522		+138
−7	4·5	31·5	220·5	−5	45	225	1125		+543·75
−6	14·5	87	522	−4	51·5	206	824		+324
−5	15·5	77·5	387·5	−3	92·5	277·5	832·5		+228·75
−4	48·5	194	776	−2	155	310	620		+177
−3	99	297	891	−1	178	178	178		+131·25
−2	141·5	283	566	0	175				...
−1	190·5	190·5	190·5	1	199·5	199·5	190·5		+183·25
0	212		...	2	166	332	664		+394·5
1	198·5	198·5	198·5	3	135	405	1215		+735
2	159·5	319	638	4	82·5	330	1320		+699
3	142·5	427·5	1282·5	5	36·5	182·5	912·5		+526·25
4	77·5	310	1240	6	20	120	720		+429
5	36	180	900	7	6·5	45·5	318·5		+176·75
6	19·5	117	702	8	4·5	36	288		+116
7	9·5	66·5	465·5						
8	4	32	256						
9	1	9	81						
	1376	+1659·5 −1179	9491·5 −167·8		1376	+1650·5 −1390	10556·5 −49·3	Total for Daughters.	+5136·25 −90·97
	Total Correction for mean	+480·5	9323·7		Total Correction for mean	+260·5	10507·2	480·5	+5045·28
	Sheppard's adjustment	.	114·7		Sheppard's adjustment	.	114·7	Total Correction for mean	
		.	9209·0			.	10392·5		

checks with that of the 3rd column. In order that it shall do so, the central entry $+15\cdot 5$, which does not contribute to the products, has to be included. Each entry in the 9th column is multiplied by the paternal deviation to give the 10th column. In the present case all the entries in column 10 are positive; frequently both positive and negative entries occur, and it is then convenient to form a separate column for each. A useful check is afforded by repeating the work of the last two columns, interchanging the variates; we should then find the total deviation of the fathers for each array of daughters, and multiply by the daughters' deviation. The uncorrected totals, $5136\cdot 25$, should then agree. This check is especially useful with small tables, in which the work of the last two columns, carried out rapidly, is liable to error.

The value of the correlation coefficient, using Sheppard's adjustment, is found by dividing $5045\cdot 28$ by the geometric mean of $9209\cdot 0$ and $10,392\cdot 5$; its value is $+\cdot 5157$. If Sheppard's adjustment had not been used, we should have obtained $+\cdot 5097$. The difference is in this case not large compared to the errors of random sampling, and the full effects on the distribution in random samples of using Sheppard's adjustment have never been fully examined, but there can be little doubt that Sheppard's adjustment should be used, and that its use gives generally an improved estimate of the correlation. On the other hand, the distribution in random samples of the uncorrected value is simpler and better understood, so that the uncorrected value should be used in tests of significance, in which the effect of correction need not, of course, be overlooked. For simplicity coarse grouping should be avoided where such tests are intended. The fact that with small samples the correlation obtained by the use of Sheppard's adjustment may exceed unity illustrates the disturbance introduced into the random sampling distribution.

32. Partial Correlations

A great extension of the utility of the idea of correlation lies in its application to groups of more than two variates. In such cases, where the correlation between each pair of three variates is known, it is possible to eliminate any one of them, and so find what the correlation of the other two would be in a population selected so that the third variate was constant.

When estimates of the three correlations are obtainable *from the same body of data* the process of elimination shown below will give an estimate of the partial correlation exactly comparable with a direct estimate.

Ex. 26. *Elimination of age in organic correlations with growing children.*—For example, it was found (Mumford and Young's data) in a group of boys of different ages, that the correlation of *standing height* with *chest girth* was $+ \cdot 836$. One might expect that part of this association was due to general growth with increasing age. It would be more desirable for many purposes to know the correlation between the variates for boys of a given age; but in fact only a few of the boys will be exactly of the same age, and even if we make age groups as broad as a year, we shall have in each group many fewer than the total number measured. In order to utilise the whole material, we only need to know the correlations of *standing height* with *age*, and of *chest girth* with *age*. These are given as ·714 and ·708.

The fundamental formula in calculating partial correlation coefficients may be written

$$r_{12\cdot 3} = \frac{r_{12}-r_{13}r_{23}}{\sqrt{(1-r_{13}^2)(1-r_{23}^2)}}.$$

Here the three variates are numbered 1, 2, and 3, and we wish to find the correlation between 1 and 2, when 3 is eliminated; this is called the "partial" correlation between 1 and 2, and is designated by $r_{12\cdot 3}$, to show that variate 3 has been eliminated. The symbols r_{12}, r_{13}, r_{23} indicate the correlations found directly between each pair of variates, these correlations being distinguished as "total" correlations.

Inserting the numerical values in the formula given we find $r_{12\cdot 3} = $ ·668, showing that when age is eliminated the correlation, though still considerable, has been markedly reduced. The mean value stated by the above-mentioned authors for the correlations found by grouping the boys by years, is ·653, not a greatly different value. In a similar manner, two or more variates may be eliminated in succession; thus with four variates, we may first eliminate variate 4, by thrice applying the formula to find $r_{12\cdot 4}$, $r_{13\cdot 4}$, and $r_{23\cdot 4}$. Then applying the same formula again, to these three new values, we have

$$r_{12\cdot 34} = \frac{r_{12\cdot 4}-r_{13\cdot 4}r_{23\cdot 4}}{\sqrt{(1-r_{13\cdot 4}^2)(1-r_{23\cdot 4}^2)}}.$$

The labour increases rapidly with the number of variates to be eliminated. To eliminate s variates, the number of operations involved, each one application of the same formula, is $\frac{1}{6}s(s+1)(s+2)$; for values of s from 1 to 6 this gives 1, 4, 10, 20, 35, 56 operations. Much of this labour may be saved by using tables of $\sqrt{1-r^2}$ such as that published by J. R. Miner.

Like the independent variates in regression, the variates eliminated in correlation analysis need not be distributed even approximately in normal distributions. Equally, and this is most frequently overlooked, random errors in them introduce systematic errors in the results. For example, if the partial correlation of variates (1) and (2) were really zero, so that r_{12} were equal to $r_{13}\,r_{23}$, random errors in the measurement or evaluation of variate (3) would tend to reduce both r_{13} and r_{23} numerically, so that their product must be numerically less than r_{12}. An apparent partial correlation between the first two variates will therefore be produced by random errors in the third.

The meaning of the correlation coefficient should be borne clearly in mind. The original aim to measure the "strength of heredity" by this method was based clearly on the supposition that the whole class of factors which tend to make relatives alike, in contrast to the unlikeness of unrelated persons, may be grouped together as heredity. That this is so for all practical purposes is, I believe, admitted, but the correlation does not tell us that this is so; it merely tells us the degree of resemblance in the actual population studied, between father and daughter. It tells us to what extent the height of the father is relevant information respecting the height of the daughter, or, otherwise interpreted, it tells us the relative importance of the factors which act alike upon the heights of father and daughter, compared to the totality of factors at work. If we know that B is caused by A, together with other factors, independent of A, and that B has no influence on A, then the correlation between A and B does tell us how important, in relation to the other causes at work, is the influence of A. If we have not such knowledge, the correlation does not tell us whether A causes B, or B causes A, or whether both influences are at work, with or without the effects of common causes.

This is true equally of partial correlations. If we know that a phenomenon A is not itself influential in determining certain other phenomena B, C, D, . . ., but on the contrary is probably directly influenced by them, then the calculation of the partial correlations A with B, C, D, . . ., in each case eliminating the remaining values, will form a most valuable analysis of the causation of A. If on the contrary we choose a group of social phenomena with no antecedent knowledge of the causation or absence of causation among them, then the calculation of correlation coefficients, total or partial, will not advance us a step towards evaluating the importance of the causes at work.

The correlation between A and B measures, on a conventional scale, the importance of the factors which (on a balance of like and unlike action) act alike in both A and B, as against the remaining factors which affect A and B independently. If we eliminate a third variate C, we are removing from the comparison all those factors which become inoperative when C is fixed. If these are only those which affect A and B independently, then the correlation between A and B, whether positive or negative, will be numerically increased. We shall have eliminated irrelevant disturbing factors, and obtained, as it were, a better controlled experiment. We may also require to eliminate C if these factors act alike, or oppositely on the two variates correlated; in such a case the variability of C actually masks the effect we wish to investigate. Thirdly, C may be one of the chain of events by the mediation of which A affects B, or *vice versa*. The extent to which C is the channel through which the influence passes may be estimated by eliminating C; as one may demonstrate the small effect of latent factors in human heredity by finding the correlation of grandparent and grandchild, eliminating the intermediate parent. In no case, however, can we judge whether or not it is profitable to eliminate a certain variate unless we know, or are willing to assume, a qualitative scheme of causation. For the purely descriptive purpose of specifying a population in respect of a number of variates, either partial or total correlations are effective, and correlations of either type may be of interest.

As an illustration we may consider in what sense the coefficient of correlation does measure the " strength of heredity," assuming that heredity only is concerned in causing the resemblance between relatives; that is, that any environmental effects are distributed at haphazard. In the first place, we may note that if such environmental effects are increased in magnitude, the correlations would be reduced; thus the same population, genetically speaking, would show higher correlations if reared under relatively uniform nutritional conditions, than they would if the nutritional conditions had been very diverse, although the genetical processes in the two cases were identical. Secondly, if environmental effects were at all influential (as in the population studied seems not to be indeed the case), we should obtain higher correlations from a mixed population of genetically very diverse strains than we should from a more uniform population. Thirdly, although the influence of father on daughter is in a certain sense direct, in that the father contributes to the germinal composition of his daughter, we must not assume that this fact is necessarily the cause of the whole of the correlation; for it has been shown that husband and wife also show considerable resemblance in stature, and consequently taller fathers tend to have taller daughters partly because they choose, or are chosen by, taller wives. For this reason, for example, we should expect to find a noticeable positive correlation between stepfathers and stepdaughters; also that, when the stature of the wife is eliminated, the partial correlation between father and daughter will be found to be lower than the total correlation. These considerations serve to some extent to define the sense in which the somewhat vague phrase "strength of heredity" must be interpreted, in speaking of the correlation coefficient. It will readily be understood that, in less well understood cases, analogous considerations may be of some importance, and should be critically considered with all possible care.

33. Accuracy of the Correlation Coefficient

With large samples, and moderate or small correlations, the correlation obtained from a sample of n pairs of values is distributed normally about the true value ρ, with variance,

$$\frac{(1-\rho^2)^2}{n-1};$$

it is therefore usual to attach to an observed value r, a standard error $(1-r^2)/\sqrt{n-1}$, or $(1-r^2)/\sqrt{n}$. This procedure is only valid under the restrictions stated above; with small samples the value of r is often very different from the true value, ρ, and the factor $1-r^2$, correspondingly in error; in addition, the distribution of r is far from normal, so that tests of significance based on the large-sample formula are often very deceptive. Since it is with small samples, less than 100, that the practical research worker ordinarily wishes to use the correlation coefficient, we shall give an account of more accurate methods of handling the results.

In all cases the procedure is alike for total and for partial correlations. Exact account may be taken of the differences in the distributions in the two cases, by deducting unity from the sample number for each variate eliminated; thus a partial correlation found by eliminating three variates, and based on data giving 13 values for each variate, is distributed exactly as is a total correlation based on 10 pairs of values.

34. The Significance of an Observed Correlation

In testing the significance of an observed correlation we require to calculate the probability that such a correlation should arise, by random sampling, from an uncorrelated population. If the probability is low we regard the correlation as significant. The Table of t given in the preceding chapter (p. 176) may be utilised to make an exact test. If n' be the number of pairs of observations on which the correlation is based, and r the correlation obtained, without using Sheppard's adjustment, then we take

$$t = \frac{r}{\sqrt{1-r^2}} \cdot \sqrt{n'-2},$$
$$n = n'-2,$$

and it may be demonstrated that the distribution of t so calculated, will agree with that given in the table.

It should be observed that this test, as is obviously necessary, is identical with that given in the last chapter for testing whether or not the linear regression coefficient differs significantly from zero.

Table V.A. (p. 211) allows this test to be applied directly from the value of r, for samples up to 100 pairs of observations. Taking the four definite levels of significance, represented by P = ·10, ·05, ·02, and ·01, the table shows for each value of n, from 1 to 20, and thence by larger steps to 100, the corresponding values of r.

Ex. 27. *Significance of a correlation coefficient between autumn rainfall and wheat crop.*—For the twenty years 1885-1904, the mean wheat yield of Eastern England was found to be correlated with the autumn rainfall; the correlation found was −·629. Is this value significant? We obtain in succession

$$1-r^2 = ·6044,$$
$$\sqrt{1-r^2} = ·7774,$$
$$r/\sqrt{1-r^2} = -·8091,$$
$$t = -3·433.$$

For $n = 18$, this shows that P is less than ·01, and the correlation is definitely significant. The same conclusion may be read off at once from Table V.A. entered with $n = 18$.

If we had applied the standard error,

$$\sigma_r = \frac{1-r^2}{\sqrt{n'-1}},$$

we should have

$$t = \frac{r}{\sigma_r} = \frac{r}{1-r^2}\sqrt{n'-1} = -4·536,$$

§ 34]   THE CORRELATION COEFFICIENT   197

a much greater value than the true one, very much exaggerating the significance. In addition, assuming that r was normally distributed ($n = \infty$), the significance of the result would be even further exaggerated. This illustration will suffice to show how deceptive, in small samples, is the use of the standard error of the correlation coefficient, on the assumption that it will be normally distributed. Without this assumption the standard error is without utility. The misleading character of the formula is increased if n' is substituted for $n'-1$, as is often done. Judging from the normal deviate 4·536, we should suppose that the correlation obtained would be exceeded in random samples from uncorrelated material only 6 times in a million trials. Actually it would be exceeded about 3000 times in a million trials, or with 500 times the frequency supposed.

It is necessary to warn the student emphatically against the misleading character of the standard error of the correlation coefficient deduced from a small sample, because the principal utility of the correlation coefficient lies in its application to subjects of which little is known, and upon which the data are relatively scanty. With extensive material appropriate for biometrical investigations there is little danger of false conclusions being drawn, whereas with the comparatively few cases to which the experimenter must often look for guidance, the uncritical application of methods standardised in biometry must be so frequently misleading as to endanger the credit of this most valuable weapon of research. It is not true, as the example above shows, that valid conclusions cannot be drawn from small samples; if accurate methods are used in calculating the probability, we thereby make full allowance for the size of the sample, and should be influenced in our judgment only by the value of the probability indicated. The great increase of certainty which accrues from increasing data is reflected in the value of P, if accurate methods are used.

Ex. 28. *Significance of a partial correlation coefficient.*—In a group of 32 poor law relief unions, Yule found that the percentage change from 1881 to 1891 in the percentage of the population in receipt of relief was correlated with the corresponding change in the ratio of the numbers given outdoor relief to the numbers relieved in the workhouse, when two other variates had been eliminated, namely, the corresponding changes in the percentage of the population over 65, and in the population itself.

The correlation found by Yule after eliminating the two variates was $+\cdot457$; such a correlation is termed a partial correlation of the second order. Test its significance.

It has been demonstrated that the distribution in random samples of partial correlation coefficients may be derived from that of total correlation coefficients merely by deducting from the number of the sample the number of variates eliminated. Deducting 2 from the 32 unions used, we have 30 as the effective number of the sample; hence

$$n = 28.$$

Calculating t from r as before, we find

$$t = 2\cdot719,$$

whence it appears from the table that P lies between ·02 and ·01. The correlation is therefore significant. This, of course, as in other cases, is on the assumption that the variates correlated (but not necessarily those eliminated) are normally distributed; economic variates seldom themselves give normal distributions, but the fact that we are here dealing with rates of change makes the assumption of normal distribution much more plausible. The values given in Table V.A. for $n = 25$, and $n = 30$, give a sufficient indication of the level of significance attained by this observation.

35. Transformed Correlations

In addition to testing the significance of a correlation, to ascertain if there is any substantial evidence of association at all, it is also frequently required to perform one or more of the following operations, for each of which the standard error would be used in the case of a normally distributed quantity. With correlations derived from large samples the standard error may, therefore, be so used, except when the correlation approaches ± 1; but with small samples such as frequently occur in practice, special methods must be applied to obtain reliable results.

(i) To test if an observed correlation differs significantly from a given theoretical value.
(ii) To test if two observed correlations are significantly different.
(iii) If a number of independent estimates of a correlation are available, to combine them into an improved estimate.
(iv) To perform tests (i) and (ii) with such average values.

Problems of these kinds may be solved by a method analogous to that by which we have solved the problem of testing the significance of an observed correlation.

In that case we were able from the given value r to calculate a quantity t which is distributed in a known manner, for which tables were available. The transformation led exactly to a distribution which had already been studied. The transformation which we shall now employ leads approximately to the normal distribution in which all the above tests may be carried out without difficulty. Let

$$z = \tfrac{1}{2}\{\log_e(1+r)-\log_e(1-r)\},$$
$$= r+\tfrac{1}{3}r^3+\tfrac{1}{5}r^5+ \ldots,$$

then as r changes from 0 to 1, z will pass from 0 to ∞. For small values of r, z is nearly equal to r, but as r approaches unity, z increases without limit. For negative values of r, z is negative. The advantage of this transformation of r into z lies in the distribution of these two quantities in random samples. The standard deviation of r depends on the true value of the correlation, ρ, as is seen from the formula

$$\sigma_r = \frac{1-\rho^2}{\sqrt{n'-1}}.$$

Since ρ is unknown, we have to substitute for it the observed value r, and this value will not, in small samples, be a very accurate estimate of ρ. The standard error of z is simpler in form, approximately

$$\sigma_z = \frac{1}{\sqrt{n'-3}},$$

and is practically independent of the value of the correlation in the population from which the sample is drawn.

In the second place, the distribution of r is not normal in small samples, and even for large samples it remains far from normal for high correlations. The distribution of z is not strictly normal, but it tends to normality rapidly as the sample is increased, whatever may be the value of the correlation. We shall give examples to test the effect of the departure of the z distribution from normality.

Finally, the distribution of r changes its form rapidly as ρ is changed; consequently no attempt can be made, with reasonable hope of success, to allow for the skewness of the distribution. On the contrary, the distribution of z is nearly constant in form, and the accuracy of tests may be improved by small corrections for departure from normality; such corrections are, however, too small to be of practical importance, and we shall not deal with them. The simple assumption that z is normally distributed will in all ordinary cases be sufficiently accurate.

These three advantages of the transformation from r to z may be seen by comparing Figs. 7 and 8. In Fig. 7 are shown the actual distributions of r, for 8 pairs of observations, from populations having correlations 0 and 0·8; Fig. 8 shows the corresponding distribution curves for z. The two curves in Fig. 7 are widely different in their modal heights; both are distinctly non-normal curves; in form also they are strongly contrasted, the one being symmetrical, the other highly unsymmetrical. On the contrary, in Fig. 8 the two curves do not differ greatly in height; although not exactly normal in form, they come so close to it, even for a small sample of 8 pairs of observations, that the eye cannot detect the difference; and this approximate normality holds up to the extreme limits $\rho = \pm 1$. One additional feature is brought out by Fig. 8; in the distribution for $\rho = $ 0·8, although the curve itself is as symmetrical as the eye can judge of, yet the ordinate of zero error is not centrally placed.

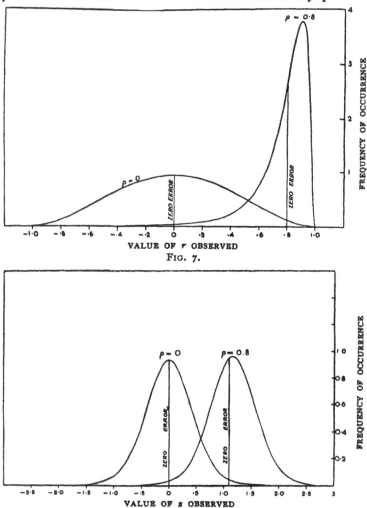

FIG. 7.

FIG. 8.

The figure, in fact, reveals the small bias which is introduced into the estimate of the correlation coefficient as ordinarily calculated: we shall treat further of this bias in the next section, and in the following chapter shall deal with a similar bias introduced in the calculation of intraclass correlations.

To facilitate the transformation we give in Table V.B. (p. 212) the values of r corresponding to values of z, proceeding by intervals of ·01, from 0 to 3. In the earlier part of this table it will be seen that the values of r and z do not differ greatly; but with higher correlations small changes in r correspond to relatively large changes in z. In fact, measured on the z-scale, a correlation of ·99 differs from a correlation ·95 by more than a correlation ·6 exceeds zero. The values of z give a truer picture of the relative importance of correlations of different sizes than do the values of r.

To find the value of z corresponding to a given value of r, say ·6, the entries in the table lying on either side of ·6 are first found, whence we see at once that z lies between ·69 and ·70; the interval between these entries is then divided proportionately to find the fraction to be added to 69. In this case we have 20/64, or ·31, so that $z = $ ·6931. Similarly, in finding the value of r corresponding to any value of z, say ·9218, we see at once that it lies between ·7259 and ·7306; the difference is 47, and 18 per cent. of this gives 8 to be added to the former value, giving us finally $r = $ ·7267. The same table may thus be used to transform r into z, and to reverse the process.

Ex. 29. *Test of the approximate normality of the distribution of z.*—In order to illustrate the kind of accuracy obtainable by the use of z, let us take the case that has already been treated by an exact method in Ex. 27. A correlation of $-·629$ has been obtained from 20 pairs of observations; test its significance.

For $r = -·629$ we have, using either a table of natural logarithms, or the special table for z, $z = -·7398$. To divide this by its standard error is equivalent to multiplying it by $\sqrt{17}$. This gives $-3·050$, which we interpret as a normal deviate. From the table of normal deviates it appears that this value will be exceeded about 23 times in 10,000 trials. The true frequency, as we have seen, is about 30 times in 10,000 trials. The error tends only slightly to exaggerate the significance of the result.

Ex. 30. *Further test of the normality of the distribution of z.*—A partial correlation $+·457$ was obtained from a sample of 32, after eliminating two variates. Does this differ significantly from zero? Here $z = ·4935$; deducting the two eliminated variates the effective size of the sample is 30, and the standard error of z is $1/\sqrt{27}$; multiplying z by $\sqrt{27}$, we have as a normal variate $2·564$. Table I (or the bottom line of Table IV) shows, as before, that P is just over ·01. There is a slight exaggeration of significance, but it is even slighter than in the previous example.

These examples indicate that the z transformation will give a variate which, for most practical purposes, may be taken to be normally distributed. In the case of simple tests of significance the use of the Table of t is to be preferred; in the following examples this method is not available, and the only method available which is both tolerably accurate and sufficiently rapid for practical use lies in the use of z.

Ex. 31. *Significance of deviation from expectation of an observed correlation coefficient.*—In a sample of 25 pairs of parent and child the correlation was found to be ·60. Is this value consistent with the view that the true correlation in that character was ·46?

The first step is to find the difference of the corresponding values of z. This is shown in Table 33.

To obtain the normal deviate we multiply by $\sqrt{22}$, and obtain ·918. The deviation is less than the standard deviation, and the value obtained is therefore quite in accordance with the hypothesis.

TABLE 33

	r.	z.
Sample value . .	·60	·6931
Population value . .	·46	·4973
Difference . .		·1958

Ex. 32. *Significance of difference between two observed correlations.*—Of two samples the first, of 20 pairs, gives a correlation ·6, the second, of 25 pairs, gives a correlation ·8 : are these values significantly different?

In this case we require not only the difference of the values of z, but the standard error of the difference. The variance of the difference is the sum of the reciprocals of 17 and 22 ; the work is shown below.

TABLE 34

	r.	z.	$n'-3$.	Reciprocal.
1st sample .	·60	·6931	17	·05882
2nd sample .	·80	1·0986	22	·04545
Difference .		·4055±·3230	Sum .	·10427

The standard error which is appended to the difference of the values of z is the square root of the variance found on the same line. The difference does not exceed twice the standard error, and cannot therefore be judged significant. There is thus no sufficient evidence to conclude that the two samples are not drawn from equally correlated populations.

Ex. 33. *Combination of values from small samples.*—Assuming that the two samples in the last example were drawn from equally correlated populations, estimate the value of the correlation.

The two values of z must be given weight inversely proportional to their variance. We therefore multiply the first by 17, the second by 22 and add, dividing the total by 39. This gives an estimated value of z for the population, and the corresponding value of r may be found from the table.

TABLE 35

	r	z	$n'-3$	$(n'-3)z$
1st sample	·60	·6931	17	11·7827
2nd sample	·80	1·0986	22	24·1692
	·7267	·9218	39	35·9519

The weighted average value of z is ·9218, to which corresponds the value $r = ·7267$; the value of z so obtained may be regarded as subject to normally distributed errors of random sampling with variance equal to 1/39. The accuracy is therefore equivalent to that of a single value obtained from 42 pairs of observations. Tests of significance may thus be applied to such averaged values of z, as to individual values.

36. Systematic Errors

In connexion with the averaging of correlations obtained from small samples it is worth while to consider the effects of two classes of systematic errors, which, although of little or no importance when single values only are available, become of increasing importance as larger numbers of samples are averaged.

The value of z obtained from any sample is an estimate of a true value, ζ, belonging to the sampled population, just as the value of r obtained from a sample is an estimate of a population value, ρ. If the method of obtaining the correlation were free from bias, the values of z would be normally distributed about a mean $\bar{z}$, which would agree in value with ζ. Actually there is a small bias which makes the mean value of z somewhat greater numerically than ζ; thus the correlation, whether positive or negative, is slightly exaggerated. This bias may effectively be corrected by subtracting from the value of z the correction

$$\frac{\rho}{2(n'-1)}.$$

For single samples this correction is unimportant, being small compared to the standard error of z. For example, if $n' = 10$, the standard error of z is ·378, while the correction is $\rho/18$ and cannot exceed ·056. If, however, $\bar{z}$ were the mean of 1000 such values of z, derived from samples of 10, the standard error of $\bar{z}$ is only ·012, and the correction, which is unaltered by taking the mean, may well be of great importance.

The second type of systematic error is that introduced by neglecting Sheppard's adjustment. In calculating the value z, we must always take the value of

r found without using Sheppard's adjustment, since the latter complicates the distribution.

But the omission of Sheppard's adjustment introduces a systematic error, in the opposite direction to that mentioned above; and which, though normally very small, appears in large as well as in small samples. In the case of averaging the correlations from a number of coarsely grouped small samples, the average z should be obtained from values of r found without Sheppard's adjustment, and to the result a correction, representing the average effect of Sheppard's adjustment, may be applied.

37. Correlation between Series

The extremely useful case in which it is required to find the correlation between two series of quantities, such as annual figures, arranged in order at equal intervals of time, may be regarded as a case of partial correlation, although it may be treated more directly by the method of fitting curved regression lines given in Section 27 (p. 147).

If, for example, we had a record of the number of deaths from a certain disease for successive years, and wished to study if this mortality were associated with meteorological conditions, or the incidence of some other disease, or the mortality of some other age group, the outstanding difficulty in the direct application of the correlation coefficient is that the number of deaths considered probably exhibits a progressive change during the period available. Such changes may be due to changes in the population among which the deaths occur, whether it be the total population of a district, or that of a particular age group, or to changes in the sanitary conditions in which the population lives, or in the skill and availability of medical assistance, or to changes in the racial or genetic composition of the population. In any case, it is usually found that the changes are still apparent when the number of deaths is converted into a death-rate on the existing population in each year, by which means one of the direct effects of changing population is eliminated.

If the progressive change could be represented effectively by a straight line it would be sufficient to consider the *time* as a third variate, and to eliminate it by calculating the corresponding partial correlation coefficient. Usually, however, the change is not so simple, and would need an expression involving the square and higher powers of the time adequately to represent it. The partial correlation required is one found by eliminating not only t, but t^2, t^3, t^4, . . ., regarding these as separate variates; for if we have eliminated all of these up to (say) the 4th degree, we have incidentally eliminated from the correlation any function of the time of the 4th degree, including that by which the progressive change is best represented.

This partial correlation may be calculated directly from the coefficients of the regression function obtained as in Section 28 (p. 151). If y and y' are the two quantities to be correlated, we obtained for y the coefficients A, B, C, . . ., and for y' the corresponding coefficients A', B', C', . . .; the sums of the squares of the deviations of the variates from the curved regression lines are obtained as before, from the equations

$$S(y-Y)^2 = S(y^2) - n'A^2 - \frac{n'(n'^2-1)}{12}B^2 - \ldots,$$

$$S(y'-Y')^2 = S(y'^2) - n'A'^2 - \frac{n'(n'^2-1)}{12}B'^2 - \ldots;$$

while the sum of the products may be obtained from the similar equation

$$S\{(y-Y)(y'-Y')\} = S(yy') - n'AA' - \frac{n'(n'^2-1)}{12}BB' - \ldots;$$

the required partial correlation being, then,

$$r = \frac{S\{(y-Y)(y'-Y')\}}{\sqrt{S(y-Y)^2 \cdot S(y'-Y')^2}}.$$

In this process the number of variates eliminated is equal to the degree of t to which the fitting has been carried; it will be understood that both variates must be fitted to the same degree, even if one of them is capable of adequate representation by a curve of lower degree than is the other.

TABLE]

§ 37] THE CORRELATION COEFFICIENT

TABLE V.A.—VALUES OF THE CORRELATION COEFFICIENT FOR DIFFERENT LEVELS OF SIGNIFICANCE

n.	$P = \cdot 1$.	·05.	·02.	·01.
1	·98769	·996917	·9995066	·9998766
2	·90000	·95000	·98000	·990000
3	·8054	·8783	·93433	·95873
4	·7293	·8114	·8822	·91720
5	·6694	·7545	·8329	·8745
6	·6215	·7067	·7887	·8343
7	·5822	·6664	·7498	·7977
8	·5494	·6319	·7155	·7646
9	·5214	·6021	·6851	·7348
10	·4973	·5760	·6581	·7079
11	·4762	·5529	·6339	·6835
12	·4575	·5324	·6120	·6614
13	·4409	·5139	·5923	·6411
14	·4259	·4973	·5742	·6226
15	·4124	·4821	·5577	·6055
16	·4000	·4683	·5425	·5897
17	·3887	·4555	·5285	·5751
18	·3783	·4438	·5155	·5614
19	·3687	·4329	·5034	·5487
20	·3598	·4227	·4921	·5368
25	·3233	·3809	·4451	·4869
30	·2960	·3494	·4093	·4487
35	·2746	·3246	·3810	·4182
40	·2573	·3044	·3578	·3932
45	·2428	·2875	·3384	·3721
50	·2306	·2732	·3218	·3541
60	·2108	·2500	·2948	·3248
70	·1954	·2319	·2737	·3017
80	·1829	·2172	·2565	·2830
90	·1726	·2050	·2422	·2673
100	·1638	·1946	·2301	·2540

For a total correlation, n is 2 less than the number of pairs in the sample; for a partial correlation, the number of eliminated variates also should be subtracted.

TABLE V.B.—TABLE OF r, FOR VALUES OF z FROM 0 TO 3

z.	.01	.02	.03	.04	.05	.06	.07	.08	.09	.10
.0	0100	0200	0300	0400	0500	0599	0699	0798	0898	0997
.1	1096	1194	1293	1391	1489	1586	1684	1781	1877	1974
.2	2070	2165	2260	2355	2449	2543	2636	2729	2821	2913
.3	3004	3095	3185	3275	3364	3452	3540	3627	3714	3800
.4	3885	3969	4053	4136	4219	4301	4382	4462	4542	4621
.5	4699	4777	4854	4930	5005	5080	5154	5227	5299	5370
.6	5441	5511	5580	5649	5717	5784	5850	5915	5980	6044
.7	6107	6169	6231	6291	6351	6411	6469	6527	6584	6640
.8	6696	6751	6805	6858	6911	6963	7014	7064	7114	7163
.9	7211	7259	7306	7352	7398	7443	7487	7531	7574	7616
1.0	7658	7699	7739	7779	7818	7857	7895	7932	7969	8005
1.1	8041	8076	8110	8144	8178	8210	8243	8275	8306	8337
1.2	8367	8397	8426	8455	8483	8511	8538	8565	8591	8617
1.3	8643	8668	8692	8717	8741	8764	8787	8810	8832	8854
1.4	8875	8896	8917	8937	8957	8977	8996	9015	9033	9051
1.5	9069	9087	9104	9121	9138	9154	9170	9186	9201	9217
1.6	9232	9246	9261	9275	9289	9302	9316	9329	9341	9354
1.7	9366	9379	9391	9402	9414	9425	9436	9447	9458	9468
1.8	94783	94884	94983	95080	95175	95268	95359	95449	95537	95624
1.9	95709	95792	95873	95953	96032	96109	96185	96259	96331	96403
2.0	96473	96541	96609	96675	96739	96803	96865	96926	96986	97045
2.1	97103	97159	97215	97269	97323	97375	97426	97477	97526	97574
2.2	97622	97668	97714	97759	97803	97846	97888	97929	97970	98010
2.3	98049	98087	98124	98161	98197	98233	98267	98301	98335	98367
2.4	98399	98431	98462	98492	98522	98551	98579	98607	98635	98661
2.5	98688	98714	98739	98764	98788	98812	98835	98858	98881	98903
2.6	98924	98945	98966	98987	99007	99026	99045	99064	99083	99101
2.7	99118	99136	99153	99170	99186	99202	99218	99233	99248	99263
2.8	99278	99292	99306	99320	99333	99346	99359	99372	99384	99396
2.9	99408	99420	99431	99443	99454	99464	99475	99485	99495	99505

For greater accuracy, and for values beyond the table,

$r = (e^{2z} - 1) \div (e^{2z} + 1)$;

$z = \frac{1}{2}\{\log(1+r) - \log(1-r)\}$.

VII

INTRACLASS CORRELATIONS AND THE ANALYSIS OF VARIANCE

38. A type of data, which is of very common occurrence, may be treated by methods closely analogous to that of the correlation table, while at the same time it may be more usefully and accurately treated by the analysis of variance, that is by the separation of the variance ascribable to one group of causes from the variance ascribable to other groups. We shall in this chapter treat first of those cases, arising in biometry, in which the analogy with the correlations treated in the last chapter may most usefully be indicated, and then pass to more general cases, prevalent in experimental results, in which the treatment by correlation appears artificial, and in which the analysis of variance appears to throw a real light on the problems before us. A comparison of the two methods of treatment illustrates the general principle, so often lost sight of, that tests of significance, in so far as they are accurately carried out, are bound to agree, whatever process of statistical reduction may be employed.

If we have measurements of n' pairs of brothers, we may ascertain the correlation between brothers in two slightly different ways. In the first place we may divide the brothers into two classes, as for instance elder brother and younger brother, and find the correlation between these two classes exactly as we do with parent and child. If we proceed in this manner we shall find the mean of the measurements of the elder brothers, and separately that of the younger brothers. Equally the standard deviations about the mean are found separately for the two classes. The correlation so obtained, being that between two classes of measurements, is termed for distinctness an **interclass** correlation. Such a procedure would be imperative if the quantities to be correlated were, for example, the *ages*, or some characteristic sensibly dependent upon age, at a fixed date. On the other hand, we may not know, in each case, which measurement belongs to the elder and which to the younger brother, or, such a distinction may be quite irrelevant to our purpose; in these cases it is usual to use a common mean derived from all the measurements, and a common standard deviation about that mean. If $x_1, x'_1; x_2, x'_2; \ldots ; x_{n'}, x'_{n'}$ are the pairs of measurements given, we calculate

$$\bar{x} = \frac{1}{2n'} S(x+x'),$$

$$s^2 = \frac{1}{2n'} \{S(x-\bar{x})^2 + S(x'-\bar{x})^2\},$$

$$r = \frac{1}{ns^2} S\{(x-\bar{x})(x'-\bar{x})\}.$$

When this is done, r is distinguished as an **intraclass** correlation, since we have treated all the brothers as belonging to the same class, and having the same mean and standard deviation. The intraclass correlation, when its use is justified by the irrelevance of any such distinction as age, may be expected to give a more accurate estimate of the true value than does any of the possible interclass correlations derived from the same material, for we have used estimates of the mean and standard deviation founded on $2n'$ instead of on n' values. This is in fact found to be the case; the intraclass correlation is not an estimate equivalent to an interclass correlation, but is somewhat more accurate. The error distribution is, however, as we shall see, affected also in other ways, which require the intraclass correlation to be treated separately.

The analogy of this treatment with that of interclass correlations may be further illustrated by the construction of what is called a symmetrical table. Instead of entering each pair of observations once in such a correlation table, it is entered twice, the co-ordinates of the two entries being, for instance, (x_1, x'_1) and (x'_1, x_1). The total entries in the table will then be $2n'$, and the two marginal distributions will be identical, each representing the distribution of the whole $2n'$ observations. The equations given, for calculating the intraclass correlation, bear the same relation to the symmetrical table as the equations for the interclass correlation bear to the corresponding unsymmetrical table with n' entries. Although the intraclass correlation is somewhat the more accurate, it is by no means so accurate as is an interclass correlation with $2n'$ independent pairs of observations.

The contrast between the two types of correlation becomes more obvious when we have to deal not with pairs, but with sets of three or more measurements; for example, if three brothers in each family have been measured. In such cases also a symmetrical table can be constructed. Each trio of brothers will provide three pairs, each of which gives two entries, so that each trio provides 6 entries in the table. To calculate the correlation from such a table is equivalent to the following equations:

$$\bar{x} = \frac{1}{3n}\, S(x+x'+x''),$$

$$s^2 = \frac{1}{3n}\, S\{(x-\bar{x})^2+(x'-\bar{x})^2+(x''-\bar{x})^2\},$$

$$r = \frac{1}{3ns^2}\, S\{(x'-\bar{x})(x''-\bar{x})+(x''-\bar{x})(x-\bar{x})+(x-\bar{x})(x'-\bar{x})\}.$$

In many instances of the use of intraclass correlations the number of observations in the same "family" is large, as when the resemblance between leaves on the same tree was studied by picking 26 leaves from a number of different trees, or when 100 pods were taken from each tree in another group of correlation studies. If k is the number in each family, then each set of k values will provide $k(k-1)$ values for the symmetrical table, which thus may contain an enormous number of entries, and be very laborious to construct. To obviate this difficulty Harris introduced an abbreviated method of calculation by which the value of the correlation given by the symmetrical table may be obtained directly from two distributions: (i) the distribution of the whole group of kn' observations, from which we obtain, as above, the values of $\bar{x}$ and s; (ii) the distribution of the n' means of families. If $\bar{x}_1, \bar{x}_2, \ldots, \bar{x}_{n'}$, represent these means each derived from k values, then

$$kS(\bar{x}_p - \bar{x})^2 = ns^2\{1 + (k-1)r\}$$

is an equation from which can be calculated the value of r, the intraclass correlation derived from the symmetrical table. It is instructive to verify this fact, for the case $k = 3$, by deriving from it the full formula for r given above for that case.

One salient fact appears from the above relation: the sum of a number of squares, and therefore the left hand of this equation, is necessarily positive. Consequently r cannot have a negative value less than $-1/(k-1)$. There is no such limitation to positive values, all values up to $+1$ being possible. Further, if k, the number in any family, is not necessarily less than some fixed value, the correlation in the population cannot be negative at all. For example, in card games, where the number of suits is limited to four, the correlation between the number of cards in different suits in the same hand may have negative values down to $-\frac{1}{3}$; but there is probably nothing in the production of a leaf or a child which necessitates that the number in such a family should be less than any number however great, and in the absence of such a necessary restriction we cannot expect to find negative correlations within such families. This is in the sharpest contrast to the unrestricted occurrence of negative values among interclass correlations, and it is obvious, since the extreme limits of variation are different in the two cases, that the distribution of values in random samples must be correspondingly modified.

39. Sampling Errors of Intraclass Correlations

The case $k = 2$, which is closely analogous to an interclass correlation, may be treated by the transformation previously employed, namely

$$z = \tfrac{1}{2}\{\log_e(1+r) - \log_e(1-r)\};$$

z is then distributed very nearly in a normal distribution, the distribution is wholly independent of the value of the correlation ρ in the population from which the sample is drawn, and the variance of z consequently depends only on the size of the sample, being given by the formula

$$\sigma_z^2 = \frac{1}{n'-3/2}.$$

The transformation has, therefore, the same advantages in this case as for interclass correlations. It will be observed that the slightly greater accuracy of the intraclass correlation, compared to an interclass correlation based on the same number of pairs, is indicated by the use of $n' - 3/2$ in place of $n' - 3$. The advantage is, therefore, equivalent to $1\frac{1}{2}$ additional pairs of observations. A second difference lies in the bias to which such estimates are subject. For interclass correlations the value found in samples, whether positive or negative, is exaggerated to the extent of requiring a correction,

$$-\frac{\rho}{2(n'-1)},$$

to be applied to the average value of z. With intraclass correlations the bias is always in the negative direction, and is independent of ρ; the correction necessary in these cases being $+\frac{1}{2}\log.\frac{n'}{n'-1}$, or, approximately, $+\frac{1}{2n'-1}$. This bias is characteristic of intraclass correlations for all values of k, and arises from the fact that the symmetrical table does not provide us with quite the best estimate of the correlation.

The effect of the transformation upon the error curves may be seen by comparing Figs. 9 and 10. Fig. 9 shows the actual error curves of r derived from a symmetrical table formed from 8 pairs of observations, drawn from populations having correlations 0 and 0·8. Fig. 10 shows the corresponding error curves for the distribution of z. The three chief advantages noted in Figs. 7 and 8 are equally visible in the comparison of Figs. 9 and 10. Curves of very unequal variance are replaced by curves of equal variance, skew curves by approximately normal curves, curves of dissimilar form by curves of similar form. In one respect the effect of the transformation is more perfect for the intraclass than it is for the interclass correlations, for, although in both cases the curves are not precisely

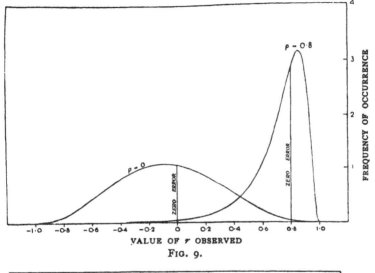

Fig. 9.

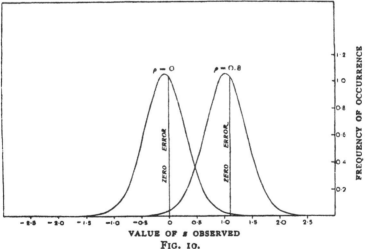

Fig. 10.

normal, with the intraclass correlations they are entirely constant in variance and form, whereas with interclass correlations there is a slight variation in both respects, as the correlation in the population is varied. Fig. 10 shows clearly the effect of the bias introduced in estimating the correlation from the symmetrical table; the bias, like the other features of these curves, is absolutely constant in the scale of z.

Ex. 34. *Accuracy of an observed intraclass correlation.*—An intraclass correlation ·6000 is derived from 13 pairs of observations: estimate the correlation in the population from which it was drawn, and find the limits within which it probably lies.

Placing the values of r and z in parallel columns, we have

TABLE 36

	r.	z.
Calculated value	+·6000	+ ·6931
Correction	...	+ ·0400
Estimate	+·6250	+ ·7331
Standard error	...	± ·2949
Upper limit	+·8675	+1·3229
Lower limit	+·1423	+ ·1433

The calculation is carried through in the z column, and the corresponding values of r found as required from the Table V.B. (p. 212). The value of r is obtained from the symmetrical table, and the corresponding value of z calculated. These values suffer from a small negative bias, and this is removed by adding to z the correction; the unbiased estimate of z is therefore ·7331, and the corresponding value of r, ·6250, is an unbiased estimate, based upon the sample, of the correlation in the population from which the sample was drawn. To find the limits within which this correlation may be expected to lie, the standard error of z is calculated, and twice this value is added and

§ 39] INTRACLASS CORRELATIONS 221

subtracted from the estimated value to obtain the values of z at the upper and lower limits. From these we obtain the corresponding values of r. The observed correlation must in this case be judged significant, since the lower limit is positive; we shall seldom be wrong in concluding that it exceeds ·14 and is less than ·87.

The sampling errors for the cases in which k exceeds 2 may be more satisfactorily treated from the standpoint of the analysis of variance; but whenever it is preferred to think in terms of correlation, it is possible to use an analogous transformation suitable for all values of k. Let

$$z = \tfrac{1}{2} \log_e \frac{1+(k-1)r}{1-r},$$

a transformation, which reduces to the form previously used when $k = 2$. Then, in random samples of sets of k observations the distribution of errors in z is independent of the true value, and approaches normality as n' is increased, though not so rapidly as when $k = 2$. The variance of z may be taken, when n' is sufficiently large, to be approximately

$$\frac{k}{2(k-1)(n'-2)}.$$

To find r for a given value of z in this transformation, Table V.B. may still be utilised, as in the following example.

Ex. 35. *Extended use of Table V.B.*—Find the value of r corresponding to $z = +1\cdot0605$, when $k = 100$.

First deduct from the given value of z half the natural logarithm of $(k-1)$; enter the difference as "z" in the table and multiply the corresponding value of "r" by k; add $k-2$ and divide by $2(k-1)$. The numerical work is shown below:

TABLE 37

z	$+1\cdot0605$
$\frac{1}{2}\log_e(k-1) = \frac{1}{2}\log 99$	$2\cdot2975$
"z"	$-1\cdot2370$
"r"	$-\cdot8446$
k"r" $= 100$"r"	$-84\cdot46$
$k-2$	98
$2r(k-1) = 198r$	$13\cdot54$
r	$+\cdot0684$

Ex. 36. *Significance of intraclass correlation from large samples.*—A correlation $+\cdot0684$ was found between the "ovules failing" in the different pods from the same tree of *Cercis Canadensis*. 100 pods were taken from each of 60 trees (Harris's data). Is this a significant correlation?

As the last example shows, $z = 1\cdot0605$; the standard error of z is $\cdot0933$. The value of z exceeds its standard error over 11 times, and the correlation is undoubtedly significant.

When n' is sufficiently large we have seen that, subject to somewhat severe limitations, it is possible to assume that the interclass correlation is normally distributed in random samples with standard error

$$\frac{1-\rho^2}{\sqrt{n'-1}}.$$

The corresponding formula for intraclass correlations, using k in a class, is

$$\frac{(1-\rho)\{1+(k-1)\rho\}}{\sqrt{\frac{1}{2}k(k-1)n'}}.$$

The utility of this formula is subject to even more drastic limitations than is that for the interclass correlation, for n' is more often small. In addition, the regions for which the formula is inapplicable, even when n' is large, are now not in the neighbourhood of ± 1, but in the neighbourhood of $+1$ and $-\dfrac{1}{k-1}$. When k is large the latter approaches zero, so that an extremely skew distribution for r is found not only with high correlations but also with very low ones. It is therefore not usually an accurate formula to use in testing significance. This abnormality in the neighbourhood of zero is particularly to be noticed, since it is only in this neighbourhood that much is to be gained by taking high values of k. Near zero, as the formula above shows, the accuracy of an intraclass correlation is with large samples equivalent to that of $\tfrac{1}{2}k(k-1)n'$ independent pairs of observations ; which gives to high values of k an enormous advantage in accuracy. For correlations near ·5, however great k be made, the accuracy is no higher than that obtainable from $9n'/2$ pairs ; while near $+1$ it tends to be no more accurate than would be n' pairs.

40. Intraclass Correlation as an Example of the Analysis of Variance

A very great simplification is introduced into questions involving intraclass correlation when we recognise that in such cases the correlation merely measures the relative importance of two groups of factors causing variation. We have seen that in the practical calculation of the intraclass correlation we merely obtain the two necessary quantities kns^2 and

$ns^2\{1+(k-1)r\}$, by equating them to the two quantities

$$\overset{kn'}{\underset{1}{S}}(x-\bar{x})^2, \quad k\overset{n'}{\underset{1}{S}}(\bar{x}_p-\bar{x})^2,$$

of which the first is the sum of the squares (kn' in number) of the deviations of all the observations from their general mean, and the second is k times the sum of the squares of the n' deviations of the mean of each family from the general mean. Now it may easily be shown that

$$\overset{kn'}{\underset{1}{S}}(x-\bar{x})^2 = k\overset{n'}{\underset{1}{S}}(\bar{x}_p-\bar{x})^2 + \overset{kn'}{\underset{1}{S}}(x-\bar{x}_p)^2,$$

in which the last term is the sum of the squares of the deviations of each individual measurement from the mean of the family to which it belongs. The following table summarises these relations by showing the number of degrees of freedom involved in each case, and, in the last column, the interpretation put upon each expression in the calculation of an intraclass correlation from a symmetrical table.

TABLE 38

	Degrees of Freedom.	Sum of Squares.	
Within families	$n'(k-1)$	$\overset{kn'}{\underset{1}{S}}(x-\bar{x}_p)^2$	$ns^2(k-1)(1-r)$
Between families	$n'-1$	$k\overset{n'}{\underset{1}{S}}(\bar{x}_p-\bar{x})^2$	$ns^2\{1+(k-1)r\}$
Total	$n'k-1$	$\overset{kn'}{\underset{1}{S}}(x-\bar{x})^2$	ns^2k

It will now be observed that z of the preceding section is, apart from a constant, half the difference of the logarithms of the two parts into which the sum of squares has been analysed. The fact that the form of the distribution of z in random samples is independent of the correlation of the population sampled, is thus a consequence of the fact that deviations of the individual observations from the means of their families are *independent* of the deviations of those means from the general mean. The data provide us with independent estimates of two variances; if these variances are equal the correlation is zero; if our estimates do not differ significantly the correlation is insignificant. If, however, they are significantly different, we may if we choose express the fact in terms of a correlation.

The interpretation of such an inequality of variance in terms of a correlation may be made clear as follows, by a method which also serves to show that the interpretation made by the use of the symmetrical table is slightly defective. Let a quantity be made up of two parts, each normally and independently distributed; let the variance of the first part be A, and that of the second part B; then it is easy to see that the variance of the total quantity is $A+B$. Consider a sample of n' values of the first part, and to each of these add a sample of k values of the second part, taking a fresh sample of k in each case. We then have n' families of values with k in each family. In the infinite population from which these are drawn the correlation between pairs of members of the same family will be

$$\rho = \frac{A}{A+B}.$$

From such a set of kn' values we may make estimates of the values of A and B, or in other words we may analyse the variance into the portions contributed by the two causes; the intraclass correlation will be merely the fraction of the total variance due to that cause which observations in the same family have in common. The value of B may be estimated directly, for variation within each family is due to this cause alone, consequently

$$\overset{kn'}{\underset{1}{S}}(x-\bar{x}_p)^2 = n'(k-1)B.$$

The mean of the observations in any family is made up of two parts, the first part with variance A, and a second part, which is the mean of k values of the second parts of the individual values, and has therefore a variance B/k; consequently from the observed variation of the means of the families, we have

$$k\overset{n'}{\underset{1}{S}}(\bar{x}_p-\bar{x})^2 = (n'-1)(kA+B).$$

Table 38 may therefore be rewritten, writing in the last column s^2 for $A+B$, and r for the unbiased estimate of the correlation.

TABLE 39

	Degrees of Freedom.	Sum of Squares.	
Within families	$n'(k-1)$	$\overset{kn'}{\underset{1}{S}}(x-\bar{x}_p)^2$	$n'(k-1)B = n's^2(k-1)(1-r)$
Between families	$n'-1$	$k\overset{n'}{\underset{1}{S}}(\bar{x}_p-\bar{x})^2$	$(n'-1)(kA+B) = (n'-1)s^2\{1+(k-1)r\}$
Total	$n'k-1$	$\overset{kn'}{\underset{1}{S}}(x-\bar{x})^2$	$(n'-1)kA+(n'k-1)B = s^2\{n'k-1-(k-1)r\}$

Comparing the last column with that of Table 38 it is apparent that the difference arises solely from putting n' for n in the first line and $n'-1$ for n in the second; the ratio between the sums of squares is altered in the ratio n' : $(n'-1)$, which precisely eliminates the negative bias observed in z derived by the previous method. The error of that method consisted in assuming that the total variance derived from n' sets of related individuals could be accurately estimated by equating the sum of squares of all the individuals from their mean, to ns^2k just as if they were all unrelated; this error is unimportant when n' is large, as it usually is when $k=2$, but with higher values of k, data may be of great value even when n' is very small, and in such cases serious discrepancies arise from the use of the uncorrected values.

The direct test of the significance of an intraclass correlation may be applied to such a table of the analysis of variance without actually calculating r. If there is no correlation, then A is not significantly different from zero; there is no difference between the several families which is not accounted for, as a random sampling effect of the differences within each family. In fact the whole group of observations is a homogeneous group with variance equal to B.

41. Test of Significance of Difference of Variance

The test of significance of intraclass correlations is thus simply an example of the much wider class of tests of significance which arise in the analysis of variance. These tests are all reducible to the single problem of testing whether one estimate of variance derived from n_1 degrees of freedom is significantly greater than a second such estimate derived from n_2 degrees of freedom. This problem is reduced to its simplest form by calculating z equal to half the difference of the *natural* logarithms of the estimates of the variance, or to the difference of the logarithms of the corresponding standard deviations. Then if P is the probability of exceeding this value by chance, it is possible to calculate the value of z corresponding to different values of P, n_1, and n_2.

A full table of this kind, involving three variables, would be very extensive; we therefore give tables for three especially important values of P, and for a number of combinations of n_1 and n_2, sufficient to indicate the values for other combinations (Table VI, pp. 242-249). We shall give various examples of the use of this table. When both n_1 and n_2 are large, and also for moderate values when they are equal or nearly equal, the distribution of z is sufficiently near normal for effective use to be made of its standard deviation, which may be written

$$\sqrt{\frac{1}{2}\left(\frac{1}{n_1}+\frac{1}{n_2}\right)}.$$

This includes the case of the intraclass correlation, when $k = 2$, for if we have n' pairs of values, the variation between classes is based on $n' - 1$ degrees of freedom, and that within classes is based on n' degrees of freedom, so that

$$n_1 = n'-1, \quad n_2 = n',$$

and for moderately large values of n' we may take z to be normally distributed as above explained. When k exceeds 2 we have

$$n_1 = n'-1, \quad n_2 = (k-1)n';$$

these may be very unequal, so that unless n' be quite large, the distribution of z will be perceptibly asymmetrical, and the standard deviation will not provide a satisfactory test of significance.

The values tabulated in Table VI were, even in the case of the small Table of the first edition (1925) calculated from the corresponding values of the

Variance Ratio, e^{2z}. Later several correspondents to whom small tables of natural logarithms were not readily accessible, suggested that logarithms to the base 10 would be more convenient. In all such cases I advised in preference the use of the original variance ratio, and this was tabulated by Mahalanobis (1932) with the symbol x, and by Snedecor (1934) with F. The wide use in the United States of Snedecor's symbol has led to the distribution being often referred to as the distribution of F. The values of the variance ratio are tabulated along with those of z in *Statistical Tables* (1938). The z values are more convenient for those who have not a computing machine at hand, and who possess and know how to use one of the available tables of natural logarithms. They are also more accurate for interpolation.

Ex. 37. *Sex difference in variance of stature.*—From 1164 measurements of males the sum of squares of the deviations was found to be 8590; while from 1456 measurements of females it was 9870 : is there a significant difference in absolute variability ?

TABLE 40

	Degrees of Freedom.	Sum of Squares.	Mean Square.	Log (Mean Square).	$1/n$.
Men	1163	8590	7·386	1·9996	·0008598
Women	1455	9870	6·783	1·9145	·0006873
			Difference ·0851		Sum ·0015471

The mean squares are calculated from the sum of squares by dividing by the degrees of freedom ; the difference of the logarithms is ·0851, so that z is ·0426. The variance of z is half the sum of the last column, so that the standard deviation of z is ·02781. The difference in variability, though suggestive of a real effect, cannot be judged significant on these data.

Ex. 38. *Homogeneity of small samples.*—In an experiment on the accuracy of counting soil bacteria, a soil sample was divided into four parallel samples, and from each of these after dilution seven plates were inoculated. The number of colonies on each plate is shown below. Do the results from the four samples agree within the limits of random sampling? In other words, is the whole set of 28 values homogeneous, or is there any perceptible intraclass correlation?

TABLE 41

Plate.	Sample.			
	I.	II.	III.	IV.
1	72	74	78	69
2	69	72	74	67
3	63	70	70	66
4	59	69	58	64
5	59	66	58	62
6	53	58	56	58
7	51	52	56	54
Total	426	461	450	440
Mean	60·86	65·86	64·28	62·86

From these values we obtain

TABLE 42

	Degrees of Freedom.	Sum of Squares.	Mean Square.	S.D.	Log S.D.
Within classes	24	1446	60·25	7·762	2·0493
Between classes	3	94·96	31·65	5·626	1·7274
Total	27	1540·96	57·07	7·55	—·3219
				(Difference)	=z

The variation within classes is actually the greater, so that if any correlation is indicated it must be negative. The numbers of degrees of freedom are small and unequal, so we shall use Table VI. This is entered with n_1 equal to the degrees of freedom corresponding to the larger variance, in this case 24; also, $n_2 = 3$. The table gives 1·0781 for the 5 per cent. point; so that the observed difference, ·3219, is really very moderate, and quite insignificant. The whole set of 28 values appears to be homogeneous with variance about 57·07.

It should be noticed that if only two samples had been present, the test of homogeneity would have been equivalent to testing the significance of t, as explained in Chapter V. In fact the values for $n_1 = 1$ in the table of z (p. 244) are nothing but the logarithms of the values, for P = ·05 and ·01, in the Table of t (p. 176). Similarly, the values for $n_2 = 1$ in Table VI are the logarithms of the reciprocals of the values, which would appear in Table IV under P = ·95 and ·99. The present method may be regarded as an extension of the method of Chapter V, appropriate when we wish to compare more than two means. Equally it may be regarded as an extension of the methods of Chapter IV, for if n_2 were infinite z would equal $\frac{1}{2}\log_e \frac{\chi^2}{n}$ of Table III for P = ·05 and ·01, and if n_1 were infinite it would equal $-\frac{1}{2}\log_e \frac{\chi^2}{n}$ for P = ·95 and ·99. Tests of goodness of fit, in which the sampling variance is not calculable *a priori*, but may be estimated from the data, may therefore be made by means of Table VI. (See Chap. VIII.)

Ex. 39. *Comparison of intraclass correlations.—*

The following correlations are given (Harris's data) for the number of ovules in different pods of the same tree, 100 pods being counted on each tree (*Cercis Canadensis*):

Meramec Highlands . . 60 trees $+\cdot 3527$
Lawrence, Kansas . . 22 trees $+\cdot 3999$

Is the correlation at Lawrence significantly greater than that in the Meramec Highlands?

First we find z in each case from the formula

$$z = \tfrac{1}{2}\{\log_e(1+99r) - \log_e(1-r)\}$$

(p. 221); this gives $z = 2\cdot 0081$ for Meramec and $2\cdot 1071$ for Lawrence; since these were obtained by the method of the symmetrical table we shall insert the small correction $1/(2n'-1)$ and obtain $2\cdot 0165$ for Meramec, and $2\cdot 1304$ for Lawrence, as the values which would have been obtained by the method of the analysis of variance.

To ascertain to what errors these determinations are subject, consider first the case of Lawrence, which being based on only 22 trees is subject to the larger errors. We have $n_1 = 21$, $n_2 = 22 \times 99 = 2178$. These values are not given in the table, but from the value for $n_1 = 24$, $n_2 = \infty$ it appears that positive errors exceeding $\cdot 2085$ will occur in rather more than 5 per cent. of samples. This fact alone settles the question of significance, for the value for Lawrence only exceeds that obtained for Meramec by $\cdot 1139$.

In other cases greater precision may be required. In the Table for z the five values 6, 8, 12, 24, ∞ are chosen for being in harmonic progression, and so facilitating interpolation, if we use $1/n$ as the variable. If we have to interpolate both for n_1 and n_2, we proceed in three steps. We find first the values of z for $n_1 = 12$, $n_2 = 2178$, and for $n_1 = 24$, $n_2 = 2178$, and from these obtain the required value for $n_1 = 21$, $n_2 = 2178$.

To find the value for $n_1 = 12$, $n_2 = 2178$, observe that

$$\frac{60}{2178} = \cdot 0275,$$

for $n_2 = \infty$ we have $\cdot 2804$, and for $n_2 = 60$ a value higher by $\cdot 0450$, so that $\cdot 2804 + \cdot 0275 \times \cdot 0450 = \cdot 2816$ gives the approximate value for $n_2 = 2178$.

Similarly for $n_1 = 24$

$$\cdot 2085 + \cdot 0275 \times \cdot 0569 = \cdot 2101.$$

From these two values we must find the value for $n_1 = 21$; now

$$\frac{24}{21} = 1 + \frac{1}{7},$$

so that we must add to the value for $n_1 = 24$ one-seventh of its difference from the value for $n_1 = 12$; this gives

$$\cdot 2101 + \frac{\cdot 0715}{7} = \cdot 2203,$$

which is approximately the positive deviation which would be exceeded by chance in 5 per cent. of random samples.

Just as we have found the 5 per cent. point for positive deviations, so the 5 per cent. point for negative deviations may be found by interchanging n_1 and n_2; this turns out to be $\cdot 2978$. If we assume that our observed value does not transgress the 5 per cent. point in either deviation, that is to say that it lies in the central nine-tenths of its frequency distribution, we may say that the value of z for Lawrence, Kansas, lies between $1 \cdot 9101$ and $2 \cdot 4282$; these fiducial limits being found respectively by subtracting the positive deviation and adding the negative deviation to the observed value.

The fact that the two deviations are distinctly unequal, as is generally the case when n_1 and n_2 are unequal and not both large, shows that such a case cannot be treated accurately by means of a probable error.

Somewhat more accurate values than the above may be obtained by improved methods of interpolation; the method given will, however, suffice for all ordinary requirements, except in the corner of the table where n_1 exceeds 24 and n_2 exceeds 30. For cases which fall into this region, the following formula gives the 5 per cent. point within one-hundredth of its value. If h is the harmonic mean of n_1 and n_2, so that

$$\frac{2}{h} = \frac{1}{n_1} + \frac{1}{n_2}$$

then
$$z = \frac{1 \cdot 6449}{\sqrt{h-1}} - \cdot 7843 \left(\frac{1}{n_1} - \frac{1}{n_2}\right).$$

Similarly, the 1 per cent. point is given approximately by the formula

$$z = \frac{2 \cdot 3263}{\sqrt{h-1 \cdot 4}} - 1 \cdot 235 \left(\frac{1}{n_1} - \frac{1}{n_2}\right).$$

For the 0·1% point we may use

$$z = \frac{3 \cdot 0902}{\sqrt{h-2 \cdot 1}} - 1 \cdot 925 \left(\frac{1}{n_1} - \frac{1}{n_2}\right).$$

The modification of the $\sqrt{h-1}$, which is good for the 5% point, used for the higher levels of significance, is due to W. G. Cochran. For a fuller examination of approximations of this sort, see Cornish and Fisher, 1937.

Let us apply this formula to find the 5 per cent. points for the Meramec Highlands, $n_1 = 59, n_2 = 5940$; the calculation is as follows:

TABLE 43

$1/n_1$	·01695	$\sqrt{h-1}$	10·76		
$1/n_2$	·00017	$1/\sqrt{h-1}$	·09294	First term	·15288
$2/h$	·01712	$1/n_1 - 1/n_2$	·01678	Second term	·01316
$1/h$	·00856			Difference	·1397
h	116·8			Sum	·1660

The 5 per cent. point for positive deviations is therefore ·1397, and for negative deviations ·1660; with the same standards as before, therefore, we may say that the value for Meramec lies between 1·8768 and 2·1825 with a fiducial probability of 90 per cent.; the large overlap of this range with that of Lawrence shows that the correlations found in the two districts are not significantly different.

42. Analysis of Variance into more than Two Portions

It is often necessary to divide the total variance into more than two portions; it sometimes happens both in experimental and in observational data that the observations may be grouped into classes in more than one way; each observation belongs to one class of type A and to a different class of type B. In such a case we can find separately the variance between classes of type A and between classes of type B; the balance of the total variance may represent only the variance within each subclass, or there may be in addition an interaction of causes so that a change in class of type A does not have the same effect in all B classes. If the observations do not occur singly in the subclasses, the variance within the subclasses may be determined independently, and the presence or absence of interaction verified. Sometimes also, for example, if the observations are frequencies, it is possible to calculate the variance to be expected in the subclasses.

Ex. 40. *Diurnal and annual variation of rain*

TABLE 44

Hour.	Jan.	Feb.	Mar.	Apr.	May.	June.	July.	Aug.	Sept.	Oct.	Nov.	Dec.	Total
1	19	16	34	17	21	26	18	18	11	27	31	28	266
2	27	16	32	21	18	25	17	24	14	37	27	29	287
3	27	19	27	22	21	31	15	32	16	36	27	33	306
4	19	20	23	25	26	28	15	31	20	39	28	34	308
5	23	18	23	19	28	31	23	23	16	42	28	31	305
6	20	23	33	19	22	29	15	22	17	42	24	26	292
7	20	22	35	19	19	33	15	18	20	36	19	23	285
8	22	22	28	25	24	23	15	18	20	32	21	31	282
9	22	20	25	26	24	28	11	22	16	35	24	28	276
10	21	19	19	21	18	22	16	19	16	31	20	24	240
11	17	19	23	15	26	25	17	13	14	33	22	21	248
12	20	21	31	18	25	27	24	25	20	36	31	25	304
13	16	18	35	19	23	32	24	25	18	36	29	28	305
14	21	20	35	28	23	32	20	27	15	29	24	28	301
15	17	18	39	28	22	27	25	30	20	28	31	32	321
16	18	31	38	32	31	32	31	28	15	28	27	34	360
17	26	29	37	38	22	32	24	31	20	37	24	38	351
18	21	25	41	30	24	28	25	27	21	38	26	33	332
19	26	18	35	24	23	30	25	27	20	37	26	31	322
20	24	18	30	25	28	25	28	18	16	36	30	25	297
21	23	20	27	21	18	21	25	20	14	34	27	39	292
22	25	16	33	23	29	23	22	22	14	35	26	36	290
23	22	15	33	22	24	23	21	19	11	22	28	28	273
24	22	18	30	19	26	27	18	22	14	22	28	27	273
Total	518	481	746	555	565	660	489	561	398	803	628	712	7116

§ 42] INTRACLASS CORRELATIONS 237

frequency.—The frequencies of rain at different hours in different months (Table 44) were observed at Richmond during 10 years (quoted from Shaw, with two corrections in the totals).

The variance may be analysed as follows :

TABLE 45

	Degrees of Freedom.	Sum of Squares.	Mean Square.
Months	11	6,568·58	597·144
Hours	23	1,539·33	66·928
Remainder	253	3,819·58	15·097
Total	287	11,927·50	

The mean of the 288 values given in the table is 24·7, and if the original data had represented independent sampling chances, we should expect the mean square residue to be nearly as great as this, or greater, if the rain distribution during the day differs in different months. Clearly the residual variance is subnormal, and the reason for this is obvious when we consider that the probability that it should be raining in the 2nd hour is not independent of whether it is raining or not in the 1st hour of the *same day*. Each shower will thus often have been entered twice or more often, and the values for neighbouring hours in the same month will be positively correlated. Much of the random variation has thus been included in that ascribed to the months, and probably accounts for the very irregular sequence of the monthly totals. The variance between the 24 hours is, however, quite significantly greater than the residual variance, and this shows that the rainy hours have been on the whole similar in the different months, so that the figures clearly indicate the influence of time of day. From the data it is not possible to estimate the influence of time of year, or to discuss whether the effect of time of day is the same in all months.

Ex. 41. *Analysis of variation in experimental field trials.*—The table on the following page gives the yield in lb. per plant in an experiment with potatoes (Rothamsted data). A plot of land, the whole of which had received a dressing of dung, was divided into 36 patches, on which 12 varieties were grown, each variety having 3 patches scattered over the area. Each patch was divided into three lines, one of which received, in addition to dung, a basal dressing only, containing no potash, while the other two received additional dressings of sulphate and chloride of potash respectively.

From data of this sort a variety of information may be derived. The total yields of the 36 patches give us 35 degrees of freedom, of which 11 represent differences among the 12 varieties, and 24 represent the differences between different patches growing the same variety. By comparing the variance in these two classes we may test the significance of the varietal differences in yield for the soil and climate of the experiment. The 72 additional degrees of freedom given by the yields of the separate rows consist of 2 due to manurial treatment, which we can subdivide into one representing the differences due to a potash dressing as against the basal dressing, and a second representing the manurial difference between the sulphate and the chloride ; and 70 more representing the differences observed in manurial response in the different patches. These latter may in turn be divided into 22 representing the difference in manurial response of the different varieties, and 48 representing the differences in manurial response in different patches

TABLE 46

Variety.	Sulphate Row.	Sulphate Row.	Sulphate Row.	Chloride Row.	Chloride Row.	Chloride Row.	Basal Row.	Basal Row.	Basal Row.
Ajax	3·20	4·00	3·86	2·55	3·04	4·13	2·82	1·75	4·71
Arran Comrade	2·25	2·56	2·58	1·96	2·15	2·10	2·42	2·17	2·17
British Queen	3·21	2·82	3·82	2·71	2·68	4·17	2·75	2·75	3·32
Duke of York	1·11	1·25	2·25	1·57	2·00	1·75	1·61	2·00	2·46
Epicure	2·36	1·64	2·29	2·11	1·93	2·64	1·43	2·25	2·79
Great Scot	3·38	3·07	3·89	2·79	3·54	4·14	3·07	3·25	3·50
Iron Duke	3·43	3·00	3·96	3·33	3·08	3·32	3·50	2·32	3·29
K. of K.	3·71	4·07	4·21	3·39	4·63	4·21	2·89	4·20	4·32
Kerr's Pink	3·04	3·57	3·82	2·96	3·18	4·32	2·00	3·00	3·88
Nithsdale	2·57	2·21	3·58	2·04	2·93	3·71	1·96	2·86	3·56
Tinwald Perfection	3·46	3·11	2·50	2·83	2·96	3·21	2·55	3·39	3·36
Up-to-Date	4·29	2·93	4·25	3·39	3·68	4·07	4·21	3·64	4·11

TABLE 47

Variety.	Manuring.			Total.	Plot.		
	Sulphate.	Chloride.	Basal.		I.	II.	III.
Ajax	11·06	9·72	9·28	30·06	8·57	8·79	12·70
Arran Comrade .	7·39	6·21	6·76	20·36	6·63	6·88	6·85
British Queen .	9·85	9·56	8·82	28·23	8·67	8·25	11·31
Duke of York .	4·61	5·32	6·07	16·00	4·29	5·25	6·46
Epicure . . .	6·29	6·68	6·47	19·44	5·90	5·82	7·72
Great Scot . .	10·34	10·47	9·82	30·63	9·24	9·86	11·53
Iron Duke . .	10·39	9·73	9·11	29·23	10·26	8·40	10·57
K. of K. . .	11·99	12·23	11·41	35·63	9·99	12·90	12·74
Kerr's Pink . .	10·43	10·46	8·88	29·77	8·00	9·75	12·02
Nithsdale . .	8·36	8·68	8·38	25·42	6·57	8·00	10·85
Tinwald Perfection	9·07	9·00	9·30	27·37	8·84	9·46	9·07
Up-to-Date . .	11·47	11·14	11·96	34·57	11·89	10·25	12·43
Total	111·25	109·20	106·26	326·71	...	...	...

growing the same variety. To test the significance of the manurial effects, we may compare the variance in each of the two manurial degrees of freedom with that in the remaining 48; to test the significance of the differences in varietal response to manure, we compare the variance in the 22 degrees of freedom with that in the 48; while to test the significance of the difference in yield of the same variety in different patches, we compare the 24 degrees of freedom representing the differences in the yields of different patches growing the same variety with the 48 degrees representing the differences of manurial response on different patches growing the same variety.

For each variety we shall require the total yield for the whole of each patch, the total yield for the 3 patches and the total yield for each manure; we shall also need the total yield for each manure for the aggregate of the 12 varieties; these values are given on page 240 (Table 47).

The sum of the squares of the deviations of all the 108 values from their mean is 71·699; divided, according to patches, in 36 classes of 3, the value for the 36 patches is 61·078; dividing this again according to varieties into 12 classes of 3, the value for the 12 varieties is 43·638. We may express the facts so far as follows:

TABLE 48

Variance.	Degrees of Freedom.	Sum of Squares.	Mean Square.	Log (S.D.)
Between varieties	11	43·6384	3·967	·6890
Between patches for same variety	24	17·4401	·727	−·1594
Within patches	72	10·6204	...	...
Total	107	71·6989		

The value of z, found as the difference of the logarithms in the last column, is ·8484, the corresponding 1 per cent. value being about ·564; the effect of variety is therefore very significant.

Of the variation within the patches the portion ascribable to the two differences of manurial treatment may be derived from the totals for the three manurial treatments. The sum of the squares of the three deviations, divided by 36, is ·3495; of this the square of the difference of the totals for the two potash dressings, divided by 72, contributes ·0584, while the square of the difference between their mean and the total for the basal dressing, divided by 54, gives the remainder, ·2911. It is possible, however, that the whole effect of the dressings may not appear in these figures, for if the different varieties had responded in different ways, or to different extents, to the dressings, the whole effect would not appear in the totals. The 70 remaining degrees of freedom would not be homogeneous. The 36 values, giving the totals for each manuring and for each variety, give us 35 degrees of freedom, of which 11 represent the differences of variety, 2 the differences of manuring, and the remaining 22 show the differences in manurial response of the different varieties. The analysis of this group is shown below:

TABLE 49

Variance due to	Degrees of Freedom.	Sum of Squares.	Mean Square.
Potash dressing	1	·2911	·2911
Sulphate v. chloride	1	·0584	·0584
Differential response of varieties	22	2·1911	·0996
Differential response in patches with same variety	48	8·0798	·1683
Total	72	10·6204	

To test the significance of the variation observed in the yield of patches bearing the same variety, we may compare the value ·727 found above from 24 degrees of freedom, with ·1683 just found from 48 degrees. The value of z, half the difference of the logarithms, is ·7316, while the 1 per cent. point is about ·394. The evidence for unequal fertility of the different patches is therefore unmistakable. As is always found in careful field trials, local irregularities in the nature or depth of the soil materially affect the yields. In this case the soil irregularity was perhaps combined with unequal quality or quantity of the dung supplied.

There is no sign of differential response among the varieties; indeed, the difference between patches with different varieties is less than that found for patches with the same variety. The difference between the values is not significant; $z = $ ·2623, while the 5 per cent. point is about ·33.

Finally, the effect of the manurial dressings tested is small; the difference due to potash is indeed greater than the value for the differential effects, which we may now call random fluctuations, but z is only ·3427, and would require to be about ·7 to be significant. With no total response, it is of course to be expected, though not as a necessary consequence, that the differential effects should be insignificant. Evidently the plants with the basal dressing had all the potash necessary, and in addition no apparent effect on the yield was produced by the difference between chloride and sulphate ions.

[TABLE

TABLE

5 PER CENT. POINTS OF

		Values			
		1.	2.	3.	4.
Values of n_2	1	2·5421	2·6479	2·6870	2·7071
	2	1·4592	1·4722	1·4765	1·4787
	3	1·1577	1·1284	1·1137	1·1051
	4	1·0212	·9690	·9429	·9272
	5	·9441	·8777	·8441	·8236
	6	·8948	·8188	·7798	·7558
	7	·8606	·7777	·7347	·7080
	8	·8355	·7475	·7014	·6725
	9	·8163	·7242	·6757	·6450
	10	·8012	·7058	·6553	·6232
	11	·7889	·6909	·6387	·6055
	12	·7788	·6786	·6250	·5907
	13	·7703	·6682	·6134	·5783
	14	·7630	·6594	·6036	·5677
	15	·7568	·6518	·5950	·5585
	16	·7514	·6451	·5876	·5505
	17	·7466	·6393	·5811	·5434
	18	·7424	·6341	·5753	·5371
	19	·7386	·6295	·5701	·5315
	20	·7352	·6254	·5654	·5265
	21	·7322	·6216	·5612	·5219
	22	·7294	·6182	·5574	·5178
	23	·7269	·6151	·5540	·5140
	24	·7246	·6123	·5508	·5106
	25	·7225	·6097	·5478	·5074
	26	·7205	·6073	·5451	·5045
	27	·7187	·6051	·5427	·5017
	28	·7171	·6030	·5403	·4992
	29	·7155	·6011	·5382	·4969
	30	·7141	·5994	·5362	·4947
	60	·6933	·5738	·5073	·4632
	∞	·6729	·5486	·4787	·4319

VI

THE DISTRIBUTION OF z

of n_1.

5.	6.	8.	12.	24.	∞.
2·7194	2·7276	2·7380	2·7484	2·7588	2·7693
1·4800	1·4808	1·4819	1·4830	1·4840	1·4851
1·0994	1·0953	1·0899	1·0842	1·0781	1·0716
·9168	·9093	·8993	·8885	·8767	·8639
·8097	·7997	·7862	·7714	·7550	·7368
·7394	·7274	·7112	·6931	·6729	·6499
·6896	·6761	·6576	·6369	·6134	·5862
·6525	·6378	·6175	·5945	·5682	·5371
·6238	·6080	·5862	·5613	·5324	·4979
·6009	·5843	·5611	·5346	·5035	·4657
·5822	·5648	·5406	·5126	·4795	·4387
·5666	·5487	·5234	·4941	·4592	·4156
·5535	·5350	·5089	·4785	·4419	·3957
·5423	·5233	·4964	·4649	·4269	·3782
·5326	·5131	·4855	·4532	·4138	·3628
·5241	·5042	·4760	·4428	·4022	·3490
·5166	·4964	·4676	·4337	·3919	·3366
·5099	·4894	·4602	·4255	·3827	·3253
·5040	·4832	·4535	·4182	·3743	·3151
·4986	·4776	·4474	·4116	·3668	·3057
·4938	·4725	·4420	·4055	·3599	·2971
·4894	·4679	·4370	·4001	·3536	·2892
·4854	·4636	·4325	·3950	·3478	·2818
·4817	·4598	·4283	·3904	·3425	·2749
·4783	·4562	·4244	·3862	·3376	·2685
·4752	·4529	·4209	·3823	·3330	·2625
·4723	·4499	·4176	·3786	·3287	·2569
·4696	·4471	·4146	·3752	·3248	·2516
·4671	·4444	·4117	·3720	·3211	·2466
·4648	·4420	·4090	·3691	·3176	·2419
·4311	·4064	·3702	·3255	·2654	·1644
·3974	·3706	·3309	·2804	·2085	0

TABLE

1 PER CENT. POINTS OF

		Values			
		1.	2.	3.	4.
Values of n_z	1	4·1535	4·2585	4·2974	4·3175
	2	2·2950	2·2976	2·2984	2·2988
	3	1·7649	1·7140	1·6915	1·6786
	4	1·5270	1·4452	1·4075	1·3856
	5	1·3943	1·2929	1·2449	1·2164
	6	1·3103	1·1955	1·1401	1·1068
	7	1·2526	1·1281	1·0672	1·0300
	8	1·2106	1·0787	1·0135	·9734
	9	1·1786	1·0411	·9724	·9299
	10	1·1535	1·0114	·9399	·8954
	11	1·1333	·9874	·9136	·8674
	12	1·1166	·9677	·8919	·8443
	13	1·1027	·9511	·8737	·8248
	14	1·0909	·9370	·8581	·8082
	15	1·0807	·9249	·8448	·7939
	16	1·0719	·9144	·8331	·7814
	17	1·0641	·9051	·8229	·7705
	18	1·0572	·8970	·8138	·7607
	19	1·0511	·8897	·8057	·7521
	20	1·0457	·8831	·7985	·7443
	21	1·0408	·8772	·7920	·7372
	22	1·0363	·8719	·7860	·7309
	23	1·0322	·8670	·7806	·7251
	24	1·0285	·8626	·7757	·7197
	25	1·0251	·8585	·7712	·7148
	26	1·0220	·8548	·7670	·7103
	27	1·0191	·8513	·7631	·7062
	28	1·0164	·8481	·7595	·7023
	29	1·0139	·8451	·7562	·6987
	30	1·0116	·8423	·7531	·6954
	60	·9784	·8025	·7086	·6472
	∞	·9462	·7636	·6651	·5999

§42] INTRACLASS CORRELATIONS 247

VI.—Continued

THE DISTRIBUTION OF z

of n_1.

5.	6.	8.	12	24.	∞.
4·3297	4·3379	4·3482	4·3585	4·3689	4·3794
2·2991	2·2992	2·2994	2·2997	2·2999	2·3001
1·6703	1·6645	1·6569	1·6489	1·6404	1·6314
1·3711	1·3609	1·3473	1·3327	1·3170	1·3000
1·1974	1·1838	1·1656	1·1457	1·1239	1·0997
1·0843	1·0680	1·0460	1·0218	·9948	·9643
1·0048	·9864	·9614	·9335	·9020	·8658
·9459	·9259	·8983	·8673	·8319	·7904
·9006	·8791	·8494	·8157	·7769	·7305
·8646	·8419	·8104	·7744	·7324	·6816
·8354	·8116	·7785	·7405	·6958	·6408
·8111	·7864	·7520	·7122	·6649	·6061
·7907	·7652	·7295	·6882	·6386	·5761
·7732	·7471	·7103	·6675	·6159	·5500
·7582	·7314	·6937	·6496	·5961	·5269
·7450	·7177	·6791	·6339	·5786	·5064
·7335	·7057	·6663	·6199	·5630	·4879
·7232	·6950	·6549	·6075	·5491	·4712
·7140	·6854	·6447	·5964	·5366	·4560
·7058	·6768	·6355	·5864	·5253	·4421
·6984	·6690	·6272	·5773	·5150	·4294
·6916	·6620	·6196	·5691	·5056	·4176
·6855	·6555	·6127	·5615	·4969	·4068
·6799	·6496	·6064	·5545	·4890	·3967
·6747	·6442	·6006	·5481	·4816	·3872
·6699	·6392	·5952	·5422	·4748	·3784
·6655	·6346	·5902	·5367	·4685	·3701
·6614	·6303	·5856	·5316	·4626	·3624
·6576	·6263	·5813	·5269	·4570	·3550
·6540	·6226	·5773	·5224	·4519	·3481
·6028	·5687	·5189	·4574	·3746	·2352
·5522	·5152	·4604	·3908	·2913	0

TABLE

0·1 PER CENT. POINTS OF

		Values		
	1.	2.	3.	4.
1	6·4577	6·5612	6·5966	6·6201
2	3·4531	3·4534	3·4535	3·4535
3	2·5604	2·5003	2·4748	2·4603
4	2·1529	2·0574	2·0143	1·9892
5	1·9255	1·8002	1·7513	1·7184
6	1·7849	1·6479	1·5828	1·5433
7	1·6874	1·5384	1·4662	1·4221
8	1·6177	1·4587	1·3809	1·3332
9	1·5646	1·3982	1·3160	1·2653
10	1·5232	1·3509	1·2650	1·2116
11	1·4900	1·3128	1·2238	1·1683
12	1·4627	1·2814	1·1900	1·1326
13	1·4400	1·2553	1·1616	1·1026
14	1·4208	1·2332	1·1376	1·0772
15	1·4043	1·2141	1·1169	1·0553
16	1·3900	1·1976	1·0989	1·0362
17	1·3775	1·1832	1·0832	1·0195
18	1·3665	1·1704	1·0693	1·0047
19	1·3567	1·1591	1·0569	·9915
20	1·3480	1·1489	1·0458	·9798
21	1·3401	1·1398	1·0358	·9691
22	1·3329	1·1315	1·0268	·9595
23	1·3264	1·1240	1·0186	·9507
24	1·3205	1·1171	1·0111	·9427
25	1·3151	1·1108	1·0041	·9354
26	1·3101	1·1050	·9978	·9286
27	1·3055	1·0997	·9920	·9223
28	1·3013	1·0947	·9866	·9165
29	1·2973	1·0903	·9815	·9112
30	1·2936	1·0859	·9768	·9061
60	1·2413	1·0248	·9100	·8345
∞	1·1910	·9663	·8453	·7648

Values of n_2.

The author is indebted to Dr Deming

§ 42] INTRACLASS CORRELATIONS 249

VI.—*Continued*

THE DISTRIBUTION OF z

of n_1.

5.	6.	8.	12.	24.	∞.
6·6323	6·6405	6·6508	6·6611	6·6715	6·6819
3·4535	3·4535	3·4536	3·4537	3·4536	3·4536
2·4511	2·4446	2·4361	2·4272	2·4179	2·4081
1·9728	1·9612	1·9459	1·9294	1·9118	1·8927
1·6964	1·6808	1·6596	1·6370	1·6123	1·5845
1·5177	1·4986	1·4730	1·4449	1·4134	1·3783
1·3927	1·3711	1·3417	1·3090	1·2721	1·2296
1·3008	1·2770	1·2443	1·2077	1·1662	1·1169
1·2304	1·2047	1·1694	1·1293	1·0830	1·0279
1·1748	1·1475	1·1098	1·0668	1·0165	·9557
1·1297	1·1012	1·0614	1·0157	·9619	·8957
1·0926	1·0628	1·0213	·9733	·9162	·8450
1·0614	1·0306	·9875	·9374	·8774	·8014
1·0348	1·0031	·9586	·9066	·8439	·7635
1·0119	·9795	·9336	·8800	·8147	·7301
·9920	·9588	·9119	·8567	·7891	·7005
·9745	·9407	·8927	·8361	·7664	·6740
·9590	·9246	·8757	·8178	·7462	·6502
·9442	·9103	·8605	·8014	·7277	·6285
·9329	·8974	·8469	·7867	·7115	·6086
·9217	·8858	·8346	·7735	·6964	·5904
·9116	·8753	·8234	·7612	·6828	·5738
·9024	·8657	·8132	·7501	·6704	·5583
·8939	·8569	·8038	·7400	·6589	·5440
·8862	·8489	·7953	·7306	·6483	·5307
·8791	·8415	·7873	·7220	·6385	·5183
·8725	·8346	·7800	·7140	·6294	·5066
·8664	·8282	·7732	·7066	·6209	·4957
·8607	·8223	·7679	·6997	·6129	·4853
·8554	·8168	·7610	·6932	·6056	·4756
·7798	·7377	·6760	·5992	·4955	·3198
·7059	·6599	·5917	·5044	·3786	0

for this section of the Table of z.

VIII

FURTHER APPLICATIONS OF THE ANALYSIS OF VARIANCE

43. We shall in this chapter give examples of the further applications of the method of the analysis of variance developed in the last chapter in connexion with the theory of intraclass correlations. It is impossible in a short space to give examples of all the different applications which may be made of this method; we shall therefore limit ourselves to those of the most immediate practical importance, paying especial attention to those cases where erroneous methods have been largely used, or where no alternative method of attack has hitherto been put forward.

44. Fitness of Regression Formulæ

There is no more pressing need in connexion with the examination of experimental results than to test whether a given body of data is or is not in agreement with any suggested hypothesis. The previous chapters have largely been concerned with such tests appropriate to hypotheses involving frequency of occurrence, such as the Mendelian hypothesis of segregating genes, or the hypothesis of linear arrangement in linkage groups, or the more general hypotheses of the independence or correlation of variates. More frequently, however, it is desired to test hypotheses involving, in statistical language, the form of regression lines. We may wish to test, for example, if the growth of an animal, plant or population follows an assigned law, if for example it increases with time in arithmetic or geometric progression, or according to the so-called "autocatalytic," or "logistic," law of increase; we may wish to test if with increasing applications of manure, plant growth increases in accordance with the laws which have been put forward, or whether in fact the data in hand are inconsistent with such a supposition. Such questions arise not only in crucial tests of widely recognised laws, but in every case where a relation, however empirical, is believed to be descriptive of the data, and are of value not only in the final stage of establishing the laws of nature, but in the early stages of testing the efficiency of a technique. The methods we shall put forward for testing the Goodness of Fit of regression lines are aimed both at simplifying the calculations by reducing them to a standard form, and so making *accurate* tests possible, and at so displaying the whole process that it may be apparent exactly what questions can be answered by such a statistical examination of the data.

If for each of a number of selected values of the independent variate x a number of observations of the dependent variate y is made, let the number of values of x available be a; then a is the number of arrays in our data. Designating any particular array by means of the suffix p, the number of observations in any array will be denoted by n_p, and the mean of their values by $\bar{y}_p$; $\bar{y}$ being the general mean of all the values of y. Then whatever be the nature of the data, the purely algebraic identity

$$S(y-\bar{y})^2 = S\{n_p(\bar{y}_p-\bar{y})^2\} + SS(y-\bar{y}_p)^2$$

expresses the fact that the sum of the squares of the deviations of all the values of y from their general mean may be broken up into two parts, one representing the sum of the squares of the deviations of the means of the arrays from the general mean, each multiplied by the number in the array, while the second is the sum of the squares of the deviations of each observation from the mean of the array in which it occurs. This resembles the analysis used for intraclass correlations, save that now the number of observations may be different in each array. The deviations of the observations from the means of the arrays are due to causes of variation, including errors of grouping, errors of observation, and so on, which are not dependent upon the value of x; the standard deviation due to these causes thus provides a basis for comparison by which we can test whether the deviations of the means of the arrays from the values expected by hypothesis are or are not significant.

Let Y_p represent in any array the mean value expected on the hypothesis to be tested, then

$$S\{n_p(\bar{y}_p - Y_p)^2\}$$

will measure the discrepancy between the data and the hypothesis. In comparing this with the variation within the arrays, we must of course consider how many degrees of freedom are available, in which the observations may differ from the hypothesis. In some cases, which are relatively rare, the hypothesis specifies the actual mean value to be expected in each array; in such cases a degrees of freedom are available, a being the number of the arrays. More frequently, the hypothesis specifies only the form of the regression line, having one or more parameters to be determined

THE ANALYSIS OF VARIANCE

from the observations, as when we wish to test if the regression can be represented by a straight line, so that our hypothesis is justified if any straight line fits the data. In such cases to find the number of degrees of freedom we must deduct from a the number of parameters obtained from the data.

Ex. 42. *Test of straightness of regression line.*—The following data are taken from a paper by A. H. Hersh on the influence of temperature on the number of eye facets in *Drosophila melanogaster*, in various homozygous and heterozygous phases of the " bar " factor. They represent females heterozygous for " full " and " double-bar," the facet number being measured in factorial units, effectively a logarithmic scale. Can the influence of temperature on facet number be represented by a straight line, in these units ?

[TABLE

TABLE 50

Temperature °C.	15°.	17°.	19°.	21°.	23°.	25°.	27°.	29°.	31°.	Total.
+ 8·07	3	1	1	...	...	...	...	...	...	5
+ 7·07	5	2	5	1	...	...	...	...	...	13
+ 6·07	13	7	3	...	...	...	...	...	...	23
+ 5·07	25	9	2	1	...	...	...	...	...	37
+ 4·07	22	10	16	...	...	2	...	...	...	50
+ 3·07	12	10	12	6	1	3	...	...	...	44
+ 2·07	7	5	14	16	2	2	...	...	...	46
+ 1·07	3	4	14	21	8	9	...	...	...	59
+ ·07	...	3	7	26	7	19	1	...	...	63
− ·93	...	1	7	12	11	24	3	1	...	59
− 1·93	...	...	1	9	14	22	8	6	...	60
− 2·93	...	2	1	5	12	15	15	4	...	54
− 3·93	...	...	...	2	19	18	44	10	1	94
− 4·93	...	...	...	1	4	4	26	6	6	47
− 5·93	...	...	...	...	2	2	19	14	13	50
− 6·93	...	...	...	...	2	...	11	28	9	50
− 7·93	...	...	...	...	3	1	8	8	8	28
− 8·93	...	...	...	...	1	...	2	5	5	13
− 9·93	...	...	...	...	...	...	...	4	4	8
−10·93	...	...	...	...	...	...	...	10	2	12
−11·93	...	...	...	...	...	1	...	1	2	4
−12·93	...	...	...	...	...	...	...	·5	1·5	2
−13·93	...	...	...	...	...	...	...	·5	·5	1
−14·93	...	...	...	...	...	...	...	...	...	...
−15·93	...	...	...	...	...	...	...	...	1	1
Total	90	54	83	100	86	122	137	98	53	823

There are 9 arrays representing 9 different temperatures. Taking a working mean at −1·93 we calculate the total and average excess over the working mean from each array, and for the aggregate of all 9. Each average is found by dividing the total excess by the number in the array; three decimal places are sufficient save in the aggregate, where four are needed. We have

[TABLE

THE ANALYSIS OF VARIANCE

TABLE 51

Array.	15.	17.	19.	21.	23.	25.	27.	29.	31.	Aggregate
Total excess	583	294	367	225	−43	+37	−369	−463·5	−306·5	+324
Mean excess	6·478	5·444	4·422	2·250	−·500	+·303	−2·693	−4·730	−5·783	+·3937

The sum of the products of these nine pairs of numbers, less the product of the final pair, gives the value of

$$S\{n_p(\bar{y}_p - \bar{y})^2\} = 12{,}370,$$

while from the distribution of the aggregate of all the values of y we have

$$S(y - \bar{y})^2 = 16{,}202,$$

whence is deduced the following table:

TABLE 52

Variance.	Degrees of Freedom.	Sum of Squares.	Mean Square.
Between arrays .	8	12,370	...
Within arrays .	814	3,832	4·708
Total . .	822	16,202	

The variance within the arrays is thus only about 4·7; the variance between the arrays will be made up of a part which can be represented by a linear regression, and of a part which represents the deviations of the observed means of arrays from a straight line.

To find the part represented by a linear regression, calculate
$$S(x-\bar{x})^2 = 4742{\cdot}21$$
and
$$S(x-\bar{x})(y-\bar{y}) = -7535{\cdot}38,$$
which latter can be obtained by multiplying the above total excess values by $x-\bar{x}$; then since
$$\frac{(7535{\cdot}38)^2}{4742{\cdot}21} = 11{,}974$$
we may complete the analysis as follows:

TABLE 53

Variance between Arrays due to	Degrees of Freedom.	Sum of Squares.	Mean Square.
Linear regression	1	11,974	...
Deviations from regression	7	396	56·6
Total	8	12,370	

It is useful to check the figure, 396, found by differences, by calculating the actual value of Y for the regression formula and evaluating
$$S\{n_p(\bar{y}_p - Y_p)^2\};$$
such a check has the advantage that it shows to which arrays in particular the bulk of the discrepancy is due, in this case to the observations at 23 and 25° C.

The deviations from linear regression are evidently larger than would be expected, if the regression were really linear, from the variations within the arrays. For the value of z, we have

THE ANALYSIS OF VARIANCE

TABLE 54

Degrees of Freedom.	Mean Square.	Natural Log.	½ Log$_e$.
7	56·6	4·0360	2·0180
814	4·708	1·5493	·7746
	Difference (z)	...	1·2434

while the 1 per cent. point is about ·488. There can therefore be no question of the statistical significance of the deviations from the straight line, although the latter accounts for the greater part of the variation.

Note that Sheppard's adjustment is not to be applied in making this test; a certain proportion both of the variation within arrays and of the deviations from the regression line is ascribable to errors of grouping, but to deduct from each the average error due to this cause would be unduly to accentuate their inequality, and so to render inaccurate the test of significance.

The example of regression worked out in Section 29·2 supplies a further illustration, to which the test given in this section is equally applicable.

45. The "Correlation Ratio" η

We have seen how, from the sum of the squares of the deviations of all observations from the general mean, a portion may be separated representing the differences between different arrays. The ratio which this bears to the whole is often denoted by the symbol η^2, so that

$$\eta^2 = S\{n_p(\bar{y}_p - \bar{y})^2\} \div S(y - \bar{y})^2,$$

and the square root of this ratio, η, is called the correlation ratio of y on x. Similarly, if Y is the hypothetical regression function, we may define R, so that

$$R^2 = S\{n_x(Y-\bar{y})^2\} \div S(y-\bar{y})^2,$$

then R will be the correlation coefficient between y and Y, and if the regression is linear, $R^2 = r^2$, where r is the correlation coefficient between x and y. From these relations it is obvious that η exceeds R, and thus that η provides an upper limit, such that no regression function can be found, the correlation of which with y is higher than η.

As a descriptive statistic the utility of the correlation ratio is extremely limited. It will be noticed that the number of degrees of freedom in the numerator of η^2 depends on the number of the arrays, so that, for instance in Example 42, the value of η obtained will depend, not only on the range of temperatures explored, but on the number of temperatures employed within a given range.

To test if an observed value of the correlation ratio is significant is to test if the variation between arrays is significantly greater than is to be expected, in the absence of differentiation, from the variation within the arrays; and this can be done from the analysis of variance (Table 52) by means of the Table of z. Attempts have been made to test the significance of the correlation ratio by calculating for it a standard error, but such attempts overlook the fact that, even with indefinitely large samples, the distribution of η for undifferentiated arrays does not tend to normality, unless the number of arrays also is increased without limit. On the contrary, with very large samples, when N is the total number of observations, $N\eta^2$ tends to be distributed as is χ^2 when n, the number of degrees of freedom, is equal to $(a-1)$, that is, to one less than the number of arrays.

46. Blakeman's Criterion

In the same sense that η^2 measures the difference between different arrays, so $(\eta^2 - R^2)/(1 - R^2)$ measures the aggregate deviation of the means of the arrays from the hypothetical regression line. The attempt to obtain a criterion of linearity of regression by comparing this quantity to its standard error results in the test known as Blakeman's criterion. In this test, also, no account is taken of the number of the arrays, and in consequence it does not provide even a first approximation in estimating what values of $\eta^2 - r^2$ are permissible. Similarly with η^2 with zero regression, so with $\eta^2 - r^2$, the regression being linear, if the number of observations is increased without limit, the distribution does not tend to normality, but that of $N(\eta^2 - r^2)/(1 - r^2)$ tends to be distributed as is χ^2 when $n = a - 2$. Its mean value is then $(a-2)$, and to ignore the value of a is to disregard the main feature of its sampling distribution.

In Example 42 we have seen that with 9 arrays the departure from linearity was very markedly significant; it is easy to see that had there been 90 arrays, with the same values of η^2 and r^2, the departure from linearity would have been even less than the expectation based on the variation within each array. Using Blakeman's criterion, however, these two opposite conditions are indistinguishable.

As in other cases of testing goodness of fit, so in testing regression lines it is essential that if any parameters have to be fitted to the observations, this process of fitting shall be efficiently carried out.

Some account of efficient methods has been given in Chapter V. In general, save in the more complicated cases, of which this book does not treat, the necessary condition may be fulfilled by the procedure known as the Method of Least Squares, by which the measure of deviation

$$S\{n_p(\bar{y}_p - Y_p)^2\}$$

is reduced to a minimum subject to the hypothetical conditions which govern the form of Y.

In the cases to which it is appropriate this method is a special application of the Method of Maximum Likelihood, from which it may be derived, and which will be more fully discussed in Chapter IX.

47. Significance of the Multiple Correlation Coefficient

If, as in Section 29 (p. 156), the regression of a dependent variate y on a number of independent variates x_1, x_2, x_3 is expressed in the form

$$Y = b_1 x_1 + b_2 x_2 + b_3 x_3,$$

then the correlation between y and Y is greater than the correlation of y with any other linear function of the independent variates, and thus measures, in a sense, the extent to which the value of y depends upon, or is related to, the combined variation of these variates. The value of the correlation so obtained, denoted by R, may be calculated from the formula

$$R^2 = \{b_1 S(x_1 y) + b_2 S(x_2 y) + b_3 S(x_3 y)\} \div S(y^2).$$

The multiple correlation, R, differs from the correlation obtained with a single independent variate in that it is always positive ; moreover, it has been recognised in the case of the multiple correlation that its random sampling distribution must depend on the number of independent variates employed. The exact treatment is in fact strictly parallel to that developed above (Section 45) for the correlation ratio, with a similar analysis of variance.

In the section referred to we made use of the fact that

$$S(y^2) = S(y-Y)^2 + \{b_1 S(x_1 y) + b_2 S(x_2 y) + b_3 S(x_3 y)\};$$

if n' is the number of observations of y, and p the number of independent variates, these three terms will represent respectively $n'-1$, $n'-p-1$, and p degrees of freedom. Consequently the analysis of variance takes the form :

TABLE 55

Variance due to	Degrees of Freedom.	Sum of Squares.
Regression function . .	p	$b_1 S(x_1 y) + \ldots$
Deviations from the regression function	$n'-p-1$	$S(y-Y)^2$
Total . .	$n'-1$	$S(y^2)$

it being assumed that y is measured from its mean value.

If in reality there is no connexion between the independent variates and the dependent variate y, the values in the column headed " sum of squares " will be divided approximately in proportion to the number of degrees of freedom ; whereas if a significant connexion exists, then the p degrees of freedom in the regression function will obtain distinctly more than their share. The test, whether R is or is not significant, is in fact exactly the test whether the mean square ascribable to the regression function is or is not significantly greater than the mean square of deviations from the regression function, and may be carried out, as in all such cases, by means of the Table of z.

Ex. 43. *Significance of a multiple correlation.*—To illustrate the process we may perform the test whether the rainfall data of Example 24 was significantly related to the longitude, latitude, and altitude of the recording stations. From the values found in that example, the following table may be immediately constructed:

TABLE 56

Variance due to	Degrees of Freedom.	Sum of Squares.	Mean Square.	½ Log.
Regression formula	3	791·7	263·9	2·7878
Deviations . .	53	994·9	18·77	1·4661
Total .	56	1786·6		

The value of z is thus 1·3217 while the 1 per cent. point is about ·714, showing that the multiple correlation is clearly significant. The actual value of the multiple correlation may easily be calculated from the above table, for

$$R^2 = 791\cdot7 \div 1786\cdot6 = \cdot4431,$$
$$R = \cdot6657;$$

but this step is not necessary in testing the significance

48. Technique of Plot Experimentation

The statistical procedure of the analysis of variance is essential to an understanding of the principles underlying modern methods of arranging field experiments. This section and the two following illustrate its application to these methods. Since they were written the cognate subject of experimental design has developed rapidly, and a much fuller account of the principles and logic of experimentation will be found in *The Design of Experiments*.

The first requirement which governs all well-planned experiments is that the experiment should yield not only a comparison of different manures, treatments, varieties, etc., but also a means of testing the significance of such differences as are observed. Consequently all treatments must at least be duplicated, and preferably further replicated, in order that a comparison of replicates may be used as a standard with which to compare the observed differences. This is a requirement common to most types of experimentation; the peculiarity of agricultural field experiments lies in the fact, verified in all careful uniformity trials, that the area of ground chosen for the experimental plots may be assumed to be markedly heterogeneous, in that its fertility varies in a systematic, and often a complicated manner from point to point. For our test of significance to be valid the differences in fertility between plots chosen as parallels must be truly representative of the differences between plots with different treatment; and we cannot assume that this is the case if our plots have been chosen in any way according to a prearranged system; for the systematic arrangement of our plots may have, and tests with the results of uniformity trials show that it often does have, features in common with the systematic variation of fertility, and thus the test of significance is wholly vitiated.

Ex. 44. *Accuracy attained by random arrangement.*—The direct way of overcoming this difficulty is to arrange the plots wholly at random. For example, if 20 strips of land were to be used to test 5 different treatments each in quadruplicate, we might take such an arrangement as the following, found by shuffling 20 cards thoroughly and setting them out in order :

TABLE 57

B	C	A	C	E	E	E	A	D	A
3504	3430	3376	3334	3253	3314	3287	3361	3404	3366
B	C	B	D	D	B	A	D	C	E
3416	3291	3244	3210	3168	3195	3330	3118	3029	3085

The letters represent 5 different treatments; beneath each is shown the weight of mangold roots obtained by Mercer and Hall in a uniformity trial with 20 such strips.

The deviations in the total yield of each treatment are

A	B	C	D	E
+290	+216	−59	−243	−204;

in the analysis of variance the sum of squares corresponding to " treatment " will be a quarter of the sum of the squares of these deviations. Since the sum of the squares of the 20 deviations from the general mean is 289,766, we have the following analysis :

TABLE 58

Variance due to	Degrees of Freedom.	Sum of Squares.	Mean Square.	Standard Deviation.
Treatment	4	58,726	14,681	121·1
Experimental error	15	231,040	15,403	124·1
Total	19	289,766	15,251	123·5

It will be seen that the standard error of a single plot estimated from such an arrangement is 124·1, whereas, in this case, we know its true value to be 123·5; this is an exceedingly close agreement, and illustrates the manner in which a purely random arrangement of plots ensures that the experimental error calculated shall be an unbiased estimate of the errors actually present.

Ex. 45. *Restrictions upon random arrangement.*—While adhering to the essential condition that the errors by which the observed values are affected shall be a random sample of the errors which contribute to our estimate of experimental error, it is still possible to eliminate much of the effect of soil heterogeneity, and so increase the accuracy of our observations, by laying restrictions on the order in which the strips are arranged. As an illustration of a method which is widely applicable, we may divide the 20 strips into 4 blocks, and impose the condition that each treatment shall occur once in each block; we shall then be able to separate the variance into three parts representing (i) local differences between blocks, (ii) differences due to treatment, (iii) experimental errors; and if the 5 treatments are arranged at random within each block, our estimate of experimental error will be an unbiased estimate of the actual errors in the differences due to treatment. As an example of a random arrangement subject to this restriction, the following was obtained :

AECDB | CBEDA | ADEBC | CEBAD.

Analysing out, with the same data as before, the contributions of local differences between blocks, and of treatment, we find

TABLE 59

Variance due to	Degrees of Freedom.	Sum of Squares.	Mean Square.	Standard Deviation.
Local differences	3	154,483	51,494	...
Treatment	4	40,859	10,215	...
Experimental error	12	94,424	7,869	88·7
Treatment+error	16	135,283	8,455	92·0

The local differences between the blocks are very significant, so that the accuracy of our comparisons is much improved, in fact the remaining variance is reduced almost to 55 per cent. of its previous value. The arrangement arrived at by chance has happened to be a slightly unfavourable one, the errors in the treatment values being a little more than usual, while the estimate of the standard error is 88·7 against a true value 92·0. Such variation is to be expected, and indeed upon it is our calculation of significance based.

It might have been thought preferable to arrange the experiment in a systematic order, such as

ABCDE | EDCBA | ABCDE | EDCBA.

and, as a matter of fact, owing to the marked fertility gradient exhibited by the yields in the present example, such an arrangement would have produced smaller errors in the totals of the 5 treatments. With such an arrangement, however, we have no guarantee that an estimate of the standard error derived from the discrepancies between parallel plots is really representative of the differences produced between the different treatments, consequently no such estimate of the standard error can be trusted, and no valid test of significance is possible.

That part of the fertility gradient which is not included in the differences between blocks may, however, be eliminated by regarding position within the blocks, *i.e.*, the ordinal numbers, 1, 2, 3, 4, 5, or more simply $-2, -1, 0, 1, 2$, as an independent variate for each plot, from which the yield as dependent variate may be predicted by the regression. An analysis of covariance of the ordinal numbers (x) and the yields (y), as explained in Section 49·1, gives the following results:—

TABLE 59·1
ANALYSIS OF COVARIANCE OF YIELD (y), AND ORDER WITHIN BLOCK (x)

	Degrees of Freedom.	x^2.	xy.	y^2.	y'^2_x.	Mean Square.
Blocks	3	0	0	154483		
Treatments	4	5·5	$-268·25$	40859	28678	7169·5
Error	12	34·5	$-1206·75$	94424	52214	4746·7
Treatments and Error	16	40·0	$-1475·00$	135283	80892	5392·8

It will be seen that the precision of the experiment has been increased. The mean square for treatments plus error is now 5393; in contrast, using blocks only, it was 8455, while disregarding blocks it was 15251.

If we take as having unit value an experiment

giving comparable yields subject to a standard error of 10 per cent., the value of such an experiment as this, in quadruplicate, may be found by squaring one-tenth of the mean yield, multiplying by four (giving 431816), and dividing by the mean square obtained by each method of procedure. For randomisation without blocks we have then

$$\frac{431816}{15250 \cdot 8} = 28 \cdot 18$$

units of information. Using randomised blocks we have 51·07, while adjustment for ordinal position within the block raises the value to 80·07 units.

In this case, as in many others, the lower mean square is obtained at the expense of some reduction of the number of degrees of freedom on which the estimate of error is based. This makes the tests of significance somewhat less stringent. If n is the number of degrees of freedom for error, the loss of information due to this cause is found (*Design of Experiments*, xi.) to be the fraction $2/(n+3)$, so that, taking this factor into consideration, we may summarise the results as follows :—

TABLE 59·2
AMOUNTS OF INFORMATION ELICITED BY DIFFERENT METHODS

	Degrees of Freedom for Error.	Units of Information.	
		Crude.	Adjusted.
Randomisation of 20 plots .	15	28·18	25·05
Randomisation in 4 blocks .	12	51·07	44·26
Eliminating order in block .	11	80·07	68·63

§ 49]   THE ANALYSIS OF VARIANCE   269

Even when allowance is thus made for the degrees of freedom absorbed, it is clear that in this case both the use of blocks, and that of order within the block, have been exceedingly profitable. The latter, of course, is due to the exceptionally regular gradient of fertility which these data exhibit.

It should be stated that the adjustment used in Table 59·2 is correct and exact, since more complicated and inexact adjustments have been proposed by later writers who evidently have not understood the problem.

49. The Latin Square

The method of laying restrictions on the distribution of the plots and eliminating the corresponding degrees of freedom from the variance is, however, capable of some extension in suitably planned experiments. In a block of 25 plots arranged in 5 rows and 5 columns, to be used for testing 5 treatments, we can arrange that each treatment occurs once in each row, and also once in each column, while allowing free scope to chance in the distribution subject to these restrictions. Then out of the 24 degrees of freedom, 4 will represent treatment; 8, representing soil differences between different rows or columns, may be eliminated; and 12 will remain for the estimation of error. These 12 will provide an unbiased estimate of the errors in the comparison of treatments, provided that every pair of plots, not in the same row or column, belong equally frequently to the same treatment.

Ex. 46. *Doubly restricted arrangements.* — The following root weights for mangolds were found by Mercer and Hall in 25 plots; we have distributed letters representing 5 different treatments at random in such a way that each appears once in each row and column.

TABLE 60

						Total of Row.
	D 376	E 371	C 355	B 356	A 335	1793
	B 316	D 338	E 336	A 356	C 332	1678
	C 326	A 326	B 335	D 343	E 330	1660
	E 317	B 343	A 330	C 327	D 336	1653
	A 321	C 332	D 317	E 318	B 306	1594
Total	1656	1710	1673	1700	1639	8378

Analysing out the contributions of rows, columns, and treatments we have

TABLE 61

Differences between	Degrees of Freedom.	Sum of Squares.	Mean Square.	S.D.
Rows	4	4240·24	...	...
Columns	4	701·84	...	...
Treatments	4	330·24 ⎫	130·3	11·41
Remainder	12	1754·32 ⎭		
Total	24	7026·64	292·8	17·11

By eliminating the soil differences between different rows and columns the mean square has been reduced to less than half, and the value of the experiment as a means of detecting differences due to treatment is therefore more than doubled. This method of equalising the rows and columns may with advantage be combined with that of equalising the distribution over different blocks of land, so that very accurate results may be obtained by using a number of blocks each arranged in, for example, 5 rows and columns. In this way the method may be applied even to cases with only 3 treatments to be compared. Further, since the method is suitable whatever may be the differences in actual fertility of the soil, the same statistical method of reduction may be used when, for instance, the plots are 25 strips lying side by side. Treating each block of 5 strips in turn as though they were successive columns in the former arrangement, we may eliminate, not only the difference between the blocks, but such differences as those due to a fertility gradient, which affect the yield according to the order of the strips in the block. When, therefore, the number of strips employed is the square of the number of treatments, each treatment can be not only *balanced* but completely *equalised* in respect to order in the block, and we may rely upon the (usually) reduced value of the standard error obtained by eliminating the corresponding degrees of freedom. Such a double elimination may be especially fruitful if the blocks of strips coincide with some physical feature of the field such as the ploughman's " lands," which often produce a characteristic periodicity in fertility due to variations in depth of soil, drainage, and such factors.

To sum up : systematic arrangements of plots in field trials should be avoided, since with these it is usually possible to estimate the experimental error in several different ways, giving widely different results, each way depending on some one set of assumptions as to the distribution of natural fertility, which may or may not be justified. With unrestricted random arrangement of plots the experimental error, though accurately estimated, will usually be unnecessarily large. In a well-planned experiment certain restrictions may be imposed upon the random arrangement of the plots in such a way that the experimental error may still be accurately estimated, while the greater part of the influence of soil heterogeneity may be eliminated.

It must be emphasised that when, by an improved method of arranging the plots, we can reduce the standard error to one-half, the value of the experiment is increased at least fourfold; for only by repeating the experiment four times in its original form could the same accuracy have been attained. This argument really under-estimates the preponderance in the scientific value of the more accurate experiments, for, in agricultural plot work, the experiment cannot in practice be repeated upon identical conditions of soil and climate.

49·1. The Analysis of Covariance

It has been shown that the precision of an experiment may be greatly increased by equalising, among the different treatments to be compared, certain potential sources of error. Thus in dividing the area available for an agricultural experiment into blocks, in each of which all treatments are equally represented, the differences of fertility between the different blocks of land, which without this precaution would be a source of experimental error, have been eliminated from the comparisons, and, by the analysis of variance, are eliminated equally from our estimate of error. In the Latin square any differences in fertility between entire rows, or between entire columns, have been eliminated from the comparisons, and from the estimates of error, so that the real and apparent precision of the comparison is the same as if the experiment had been performed on land in which the entire rows, and also the entire columns, were of equal fertility.

A strictly analogous equalisation is widely applied in all kinds of experimental work. Thus in nutritional experiments the growth rates of males and females may be distinctly different, while nevertheless both sexes may be equally capable of showing the advantage of one diet over another. The effect of sex, on the growth rates compared, will, therefore, be eliminated by assigning the same proportion of males to each experimental treatment, and, what is more often neglected, eliminating the average difference between the sexes from the estimate of error. Notably different reactions are often found also in different strains or breeds of animals, and for this reason each strain employed should be used equally for all treatments. The effect of strain will then be eliminated from the comparisons, and may be easily eliminated by the analysis of variance from the estimate of error. It is sometimes assumed that all the animals in the same experiment must be of the same strain, and adequate replication is in consequence believed to be impossible for lack of a sufficient quantity of homogeneous material. The examples already discussed show that this requirement is superfluous, and adds nothing to the precision of the comparisons actually attained. Indeed, while adding nothing to the precision, this course detracts definitely from the applicability of the results; for results obtained from a number of strains are evidently applicable to a wider range of material than results only established for a single strain; and, working from highly homogeneous material, there is a real danger of drawing inferences, which, had we had a wider inductive basis, would have been seen to be insecure.

There are, however, many factors relevant to the precision of our comparisons, which, while they cannot be equalised, can be measured, and for which we may reasonably attempt to make due allowance. Such are the age and weight of experimental animals, the initial weight being particularly relevant in experiments on the growth rate. In field experiments with roots the yield is often notably affected by the plant number, and if we have reason to be willing to ignore any effect our treatments may have on plant number, it would be preferable to make our comparisons on plots with an equal number of plants. Again, although we cannot equalise the fertility of the plots used for different treatments, the same land may be cropped in a previous year under uniform treatment, and the yields of this uniformity trial will clearly be relevant to the interpretation of our experimental yields. This principle is of particular importance with perennial crops, for there is here continuity, not only of the soil, but of the individual plants growing upon it; and the much more limited facilities for confirming results on a new, or unused, plantation make it especially important to increase the precision of such material as we have.

Ex. 46·1. *Covariance of tea yields in successive periods.*—T. Eden gives data for successive periods each of fourteen pluckings from sixteen plots of tea bushes intended for experimental use in Ceylon. The yields are given in per cent. of the average for each period, but the process to be exemplified would apply equally to actual yields. We give below (Tables 61·1, 61·2) data for his second and third periods, which

§ 49·1]   THE ANALYSIS OF VARIANCE   275

for our purpose may be regarded as preliminary and experimental yields respectively. The sixteen plots are arranged in a 4 × 4 square.

TABLE 61·1
PRELIMINARY YIELDS OF TEA PLOTS

88	102	91	88	369
94	110	109	118	431
109	105	115	94	423
88	102	91	96	377
379	419	406	396	1600

TABLE 61·2
EXPERIMENTAL YIELDS OF TEA PLOTS

90	93	85	81	349
93	106	114	121	434
114	106	111	93	424
92	107	92	102	393
389	412	402	397	1600

Let us suppose the area in the experimental period had been occupied by a Latin square in 4 treatments. Of the 15 degrees of freedom, 6 representing differences between rows and columns would then be eliminated, and the remaining 9 would be made up of 3 for differences between treatments, and 6 for the estimation of error. Since no actual treatment differences were applied, we shall use all 9 for the estimation of error. The experimental yields then give

[TABLE

TABLE 61·3
ANALYSIS OF EXPERIMENTAL YIELDS

	Degrees of Freedom.	Sum of Squares.	Mean Square.
Rows	3	1095·5	...
Columns	3	69·5	...
Error	9	875·0	97·22
Total	15	2040·0	136·00

Even after eliminating the large variance among rows, the residual variance is as high as 97·22; the standard error of a single plot is, therefore, about 9·86 per cent., and that for the total of four plots about 4·93 per cent.

It is, however, evident that a great part of this variance of yield in the experimental period has been foreshadowed in the yields of the preliminary period. A glance at the table will show that of the eight plots which were above the average in the experimental period, seven were above the average in the preliminary period. In fact, by choosing sets of plots which in the first period yielded nearly the same total for each set, and assigning these sets to treatments in the experimental period, we might have very materially reduced the experimental error of our treatment comparisons. The equalisation of the total preliminary yields has often been advocated, but seldom practised for reasons which will become apparent. The commonsense inference that sets of plots, giving equal total yields in the preliminary period, should under equal treatment give equal totals in the experimental period, implies that the expectation of subsequent yield of any plot is well represented in terms of the preliminary yield by a linear regression function. The important point is that the adjustments of the results of the experiment appropriate to any regression formula (of which the linear form is obviously the most important) may be made from the results of the experiment themselves without taking any notice, in the arrangement of the plots, of the previous yields. The method of regression also avoids two difficulties which are encountered in the equalisation of previous yields, namely, that the advantage of eliminating differences between rows and columns (or blocks) would often have to be sacrificed to equalisation, and that such equalisation as would be possible would always be inexact.

The adjustment to be made in the difference in yield between two plots, the previous yields of which are known, is evidently the difference to be expected in the subsequent yields, judged from the difference observed between plots treated alike. The appropriate coefficient of linear regression is given by the ratio of the covariance to the variance of the independent variate, which in this case is the variance in the preliminary yields ascribable, in our experimental arrangement, to error. To find this variance of the independent variate, the preliminary yields are analysed in exactly the same way as the experimental yields. A third table, this time an analysis of the covariance of the preliminary and the experimental yields, is constructed by using at every stage, products of the yields in these two periods in place of squares of yields at either one period.

TABLE 61·4

ANALYSIS OF PRELIMINARY YIELDS

	Degrees of Freedom.	Sum of Squares
Rows . . .	3	745·0
Columns . .	3	213·5
Error . . .	9	567·5
Total . .	15	1526·0

The exact similarity of the arithmetic in constructing these three tables may be illustrated by taking out in parallel the contributions of "columns" to each table. In Tables 61·1 and 61·2 the mean of the column totals is 400, the deviations in the first columns are -21 and -11; denoting these by x and y, the squares and products of these pairs of numbers are written in parallel below :—

TABLE 61·5

x^2	xy	y^2
441	+231	121
361	+228	144
36	+12	4
16	+12	9
854	+483	278

Dividing these totals each by 4 (the number of plots contributing to each), we have the corresponding entries in the triple table :—

§ 49·1]   THE ANALYSIS OF VARIANCE   279

TABLE 61·6
SUMS OF SQUARES AND PRODUCTS

	Degrees of Freedom.	x^2	xy	y^2
Rows . . .	3	745·0	837·0	1095·5
Columns . .	3	213·5	120·75	69·5
Error . . .	9	567·5	654·25	875·0
Total . .	15	1526·0	1612·00	2040·0

in which the variances of the two variates, and their covariance are analysed in parallel columns.

Relationships expressed either by regression or by correlation, between the two variates, may now be determined independently for the different rows of the table. In particular we need the ratio 654·25/567·5 representing the regression of y on x, for plots treated alike, after eliminating the differences between rows and columns. This is evidently the correct allowance to be deducted from any experimental yield y, for each unit by which the corresponding x is in excess of the average.

The correction, being linear, may be applied to individual plots, or to the composite totals represented by rows, columns or treatments. More comprehensively, the result of applying the correction and analysing the variance of the adjusted yields, may be derived directly from the analysis of sums and products already presented. For, if b stand for the regression coefficient, comparisons of adjusted yields will be in fact comparisons of quantities $(y-bx)$. Now

$$(y-bx)^2 = b^2x^2 - 2bxy + y^2;$$

so that, to obtain the sum of squares for the adjusted yields in any line, we need only multiply the entries in the table already constructed by b^2, $-2b$ and unity, and add the products.

In the present example $b = 1\cdot 1529$, $b^2 = 1\cdot 3291$, giving :—

TABLE 61·7

ANALYSIS OF ADJUSTED YIELDS

	Degrees of Freedom.	Sum of Squares.	Mean Square.
Rows	3	155·8	51·93
Columns	3	74·8	24·93
Error	8	120·7	15·09
Total	14	351·3	25·09

It will be noticed that the total number of degrees of freedom has been diminished from 15 to 14, to allow for the one adjustable constant in the regression formula, and that this 1 degree has been subtracted from the particular line from which the numerical value of the regression has been estimated. In this line, in fact, b has been chosen so that

$$bS(x^2) = S(xy)$$

and consequently, so that

$$S(y-bx)^2 = S(y^2) - S^2(xy)/S(x^2);$$

showing that the entry in this line is always diminished by the contribution of 1 degree of freedom. In the other lines the entry may be either increased or diminished by the adjustment.

§ 49·1] THE ANALYSIS OF VARIANCE

The value of b used in obtaining the adjusted yields is a statistical estimate subject to errors of random sampling. In consequence, although the quantities $y - bx$ are appropriate estimates of the corrected yields, they are of varying precision, as shown in Section 26; the sums of their squares in the lines of the table from which b has not been calculated do not therefore supply exact material for testing the homogeneity of deviations from the simple regression formula. This test we should wish to make if real differences of treatment had been given to our plots, in which case for each variate we should have 3 degrees of freedom assigned to treatments, and only 6 left for error, from which 6 the value of b would be calculated. In our example there are no real treatments, and we shall illustrate the test of significance by applying it to the rows, the significance of which is in reality of no consequence to the result of the experiment.

Taking those parts of Table 61·6 which refer to rows and error only, we obtain the reduced values of the sum of squares of the dependent variate y, respectively for the error and for the total, by deducting in each case from $S(y^2)$, the quantity

$$S^2(xy)/S(x^2)$$

derived from the same line. This gives for the error, the reduced value of the sum of squares of y, 120·7, as in Table 61·7; for the total we have $1970·5 - 1694·3 = 276·2$, corresponding to 11 degrees of freedom. Subtracting the first from the second, we find the reduced sum of squares ascribable to the 3 degrees of freedom for rows to be 155·5, which is the value to be compared with the reduced sum of squares for error, in making an exact test. The whole process is shown in Table 61·71. (See also Table 59·1.) In this case it is obvious that the sum of squares of $(y-bx)$ would have provided an excellent approximation. As such, however, it is, always to some extent, and sometimes greatly, inflated by the sampling errors of b; and there is no difficulty in applying the exact test, which makes proper allowance for these sampling errors.

TABLE 61·71

TEST OF SIGNIFICANCE WITH REDUCED VARIANCE

	Degrees of Freedom.	x^2	xy	y^2	Degrees of Freedom.	Reduced y^2	Mean Square
Rows	3	745·0	837·0	1095·5	3	155·5	51·83
Error	9	567·5	654·25	875·0	8	120·7	15·09
Total	12	1312·5	1491·25	1970·5	11	276·2	

Comparing the analysis of the adjusted yields with that obtained without using the preliminary pluckings, the most striking change is the reduction of the mean square error per plot from 97·22 to 15·09, in spite of the reduction in the degrees of freedom, showing that the precision of the comparison has been increased over six-fold. A second point should also be noticed. The large difference in yield between different rows, which appears in the original analysis, has fallen to about one-seventh of its original value. It appears therefore that the greater part of this element of heterogeneity may be eliminated in favourable cases by the use of preliminary yields; but this does not diminish the importance, when such preliminary yields are available, of eliminating from the comparisons differences between the larger areas of land, blocks, rows, columns, etc. In fact, the elimination of rows and columns is more important in the adjusted yields, where it reduces the mean square from 24·08 to 15·09, than in the unadjusted yields, where it reduced it from 136 to 97·2. If, for example, we take an experiment with 10 per cent. error in the means of treatments, to have unit value, the elimination of rows and columns in the unadjusted yields only increased the value from 2·94 to 4·12, a net gain of 1·18 units; while the same elimination in the adjusted yields increases the value from 16·61 to 26·51, a net gain of 9·90 units, or nearly nine times as much. In practice, however, especially when the numbers of degrees of freedom are small, it is desirable to base such comparisons on the quantity of information realised, making due allowance for the number of degrees of freedom available in each of the cases to be compared as in Table 59·2.

An examination of the process exemplified in the foregoing example shows that it combines the advantages and reconciles the requirements of the two very widely applicable procedures known as regression and analysis of variance. Once the simple procedure of building up the covariance tables is recognised, there will be found no difficulty in applying the analysis to three or more variates and the complete set of their covariances, and so making allowance simultaneously for two or more measurable but uncontrolled concomitants of our observations. These observations are treated as the dependent variate, the variability of which may be partially accounted for in terms of concomitant observations, by the method of multiple regression. Thus, if we were concerned to study the effects of agricultural treatments upon the purity index of the sugar extracted from sugar-beet, a variate which might be much affected by concomitant variations in (*a*) sugar-percentage, and (*b*) root weight, an analysis of covariance applied to the three variates, purity, sugar percentage and root weight, for the different plots of the experiment, would enable us to make a study of the effects of experimental treatments on purity alone ; *i.e.*, after allowance for any effect they may have on root weight or concentration, without our needing to have observed in fact any two plots agreeing exactly in both root weight and sugar percentage.

In such a research it would again be open to the investigator to eliminate not merely the mean root weight of the plots, but, if he judged it profitable, also its square, so using a regression non-linear in root weight. Again, if he possessed not merely the mean root weight for the different plots, but the individual values of which the mean is the average, he could eliminate simultaneously mean root weight and mean square root weight, or, in other words, make his purity comparisons with corrections appropriate to equalising both the means and the variances of the roots from the different plots.

In considering, in respect to any given body of data, what particular adjustments are worth making, it is sufficient for our immediate guidance to note their effect upon the residual error. If, in Example 46·1, we compare Tables 61·3 and 61·7, it is apparent that we may divide the 9 degrees of freedom for error of unadjusted yields into two parts, one of which comprises the 1 degree of freedom eliminated by the regression equation, and the other the 8 degrees of freedom remaining after this equation has been used for adjusting the yields. This analysis of error is shown below in Table 61·8.

TABLE 61·8

ANALYSIS OF RESIDUAL ERROR

	Degrees of Freedom.	Sum of Squares.	Mean Square.
Regression	1	754·3	754·3
Error of adjusted yields . .	8	120·7	15·09
Error of unadjusted yields .	9	875·0	...

The great advantage of making due allowance for the preliminary yields is evidently due to the very large share of the residual error which is contained in the 1 degree of freedom specified by our regression formula. We need not test the significance of a regression before using it, but any advantage it may confer will be slight unless it is in fact significant.

The chief advantage of the analysis of covariance lies, however, not in its power of getting the most out of an existing body of data, but in the guidance it is capable of giving in the design of an observational programme, and in the choice of which of many possible concomitant observations shall in fact be recorded. The example of the tea yields shows that in that case the value for experimental purposes of a plantation was increased six-fold by the comparatively trifling additional labour of recording separately the yields from different plots for a period prior to the experiment. With annual agricultural crops, to crop the experimental area in the previous year is nearly to double the labour of the experiment. What is often more serious, a year's delay is incurred before the result is made available. Analysis of covariance on successive yields on uniformly treated land shows that the value of the experiment is usually increased, but seldom by more than about 60 per cent., by a knowledge of the yields of the previous year. It seems therefore to be always more profitable to lay down an adequately replicated experiment on untried land than to expend the time and labour available in exploring the irregularities of its fertility.

In most kinds of experimentation, however, the possibilities of obtaining greatly increased precision from comparatively simple supplementary observations are almost entirely unexplored, and, indeed, in many fields the possibility of making a critically valid use of such observations is scarcely recognised. The probability that methods of experimentation can be greatly improved, either by a great increase of precision, or by a proportionate decrease in the labour required, is naturally greatest in these fields.

An analysis of covariance always involves the primary classification of the analysis, in addition to the relation between a dependent and an independent variate. Sometimes the classification may be complex, as is a hierarchical classification in three or more stages; also there may be more than one dependent variate, and possibly a number of independent variates may need to be eliminated. An example involving these complications, and with the working procedure exhibited in detail, is referred to in the bibliography (with B. Day, 1937).

49.2. The Discrimination of Groups by Means of Multiple Measurements; Appropriate Scores

A valuable application of the technique of calculation used in multiple regression consists in finding which of all possible linear compounds of a set of measurements will best discriminate between two different groups. For example, a human mandible or jaw bone may be found in circumstances in which, apart from the evidence provided by its form, the sex of its possessor is unknown. The anthropologist desires, so far as is possible, to assign the right sex to such finds. If he has a number of mandibles of known sex, measurements of these may provide a clue. Some measurements, in fact, show significant differences, but, as these are likely to be highly correlated, the evidence they provide cannot be treated as independent. For the same reason other measurements, which by themselves provide no means of discrimination, may in conjunction with the rest aid considerably. Only when that particular linear function is determined which, better than any other, discriminates mandibles of the two sexes, can we recognise that some measurements are useless, while others are of real evidential value.

To illustrate the formal equivalence with multiple regression let us suppose we have N_1 male and N_2 female mandibles, on each of which measurements $x_1, \ldots, x_p$ can be made. The mean differences (male—female) will be represented by $d_1, \ldots d_p$; further, we represent the sums of squares and products of the measurements, ignoring sex, by

$$S_{ij} = S(x_i - \bar{x}_i)(x_j - \bar{x}_j)$$

Then it has been shown that the solutions $b_1, \ldots, b_p$ of the equations

$$S_{11}b_1 + S_{12}b_2 + \ldots + S_{1p}b_p = d_1$$
$$S_{12}b_1 + S_{22}b_2 + \ldots + S_{2p}b_p = d_2$$

and so on, will be proportional to the coefficients of that linear function,

$$X = b_1 x_1 + b_2 x_2 + \ldots + b_p x_p,$$

which, as judged from the data, will most successfully discriminate mandibles of unknown sex.

If we had introduced a formal variate y, equal to $N_2/(N_1+N_2)$ for all males and to $-N_1/(N_1+N_2)$ for all females, the equations for the coefficients of multiple regression of y on $x_1, \ldots, x_p$ would in fact only differ from those written above by a factor $N_1N_2/(N_1+N_2)$ on the right. The value of the coefficient of multiple correlation of y with $x_1, \ldots, x_p$ is therefore given by

$$R^2 = \frac{N_1 N_2}{N_1 + N_2} (b_1 d_1 + \ldots + b_p d_p).$$

Hotelling (1931) has shown that, if the variates x are normally distributed within groups, the significance of the correlation can be tested, in an analysis of variance test, with p, and $n-p+1$ degrees of freedom, where n is the number of degrees of freedom within groups. So that

$$e^{2z} = \frac{n-p+1}{p} \cdot \frac{R^2}{1-R^2}.$$

This, of course, is the basic test as to whether any significant discrimination has been achieved. We may also wish to test whether any proposed discriminant function,

$$X' = \beta_1 x_1 + \ldots + \beta_p x_p,$$

specifying the ratios of the coefficients, but not their absolute values, is compatible with the observational facts. It has been shown that this can be easily done, merely by finding the correlation coefficient, within groups, between X and X'. If this is r, the value of R^2 in Hotelling's test may be multiplied by $(1-r^2)$, and used as before with 1 less degree of freedom to test the special form of discriminant proposed.

The value of z may now be obtained from

$$e^{2z} = \frac{n-p+1}{p-1} \cdot \frac{R'^2}{1-R'^2},$$

when $R'^2 = R^2(1-r^2)$ and the degrees of freedom are $n_1 = p-1$ and $n_2 = n-p+1$. The test thus rejects any proposed formula having r so small that the value of z given above is significant.

Instead of the differences between the means of the variates from two samples, the method may be applied equally to the regressions of the means of several samples on any variate characteristic of these samples. Thus Barnard has used the regressions of the means of certain measurements of Egyptian skulls on the approximate date of burial, to ascertain what linear function of the cranial measurements obtainable shows the most distinct change with time. An important application to plant selection has been made by Fairfield Smith to determine how the different observable characters of plant progenies should be combined in selecting for any particular end.

If, in a replicated variety trial, observables x_1, ..., x_p are recorded from each plot, we may obtain sums of squares and products, first, for varieties, which we shall denote t_{ij} and next for errors e_{ij}. Subtracting the second from the first we obtain unbiased estimates of the varietal effects $g_{ij} = t_{ij} - e_{ij}$. If, now, the value of a variety, for which $x_1, \ldots, x_p$ were exactly known, is judged to be correctly assessed by the formula

$$a_1 x_1 + a_2 x_2 + \ldots + a_p x_p,$$

where the coefficients a may be positive or negative, we may at once calculate

$$A_i = a_1 g_{1i} + a_2 g_{2i} + \ldots + a_p g_{pi}$$

for each value of i. The appropriate scores, $b_1, \ldots, b_p$ for rating the selective value of any variety will then be found from the simultaneous equations

$$b_1 t_{11} + \ldots + b_p t_{1p} = A_1,$$
$$b_2 t_{12} + \ldots + b_p t_{2p} = A_2,$$

and so on.

On solving these we compare the values of the compound score

$$X = b_1 \bar{x}_1 + b_2 \bar{x}_2 + \ldots + b_p \bar{x}_p$$

for each variety, X being the function of the observables most highly correlated with the true value of the variety.

The foregoing examples all illustrate the general principle that we may determine a set of adjustable coefficients in such a way as to maximise the ratio of the square of one chosen component to the sum of squares of a set of other components in an analysis of variance. The same principle may be applied to maximise the ratio which the sum of squares for n_1 degrees of freedom bears to that of a residue of n_2 degrees of freedom. After making the adjustment to obtain the maximal ratio, involving p adjustable constants, we shall, as the best available approximation, test the significance of $n_1 + p$ compared with $n_2 - p$ degrees of freedom.

§ 49·2]   THE ANALYSIS OF VARIANCE   291

When only a single component is to be maximised relative to the rest, the equations are linear, and the procedure of multiple regression may be used. Other cases may lead to equations of higher degree. Thus, given a two-way table of non-numerical observations we may ask what values, or scores, shall be assigned to them in order that the observations shall be as additive as possible.

Ex. 46·2. *The derivation of an additive scoring system from serological readings.*—Twelve samples of human blood tested with twelve different sera gave reactions represented by the five symbols −, ?, w, (+), and +, according to Table 61·9 on next page (G. L. Taylor, Galton Laboratory).

[TABLE

TABLE 61·9
NON-NUMERICAL TWO-WAY TABLE OF SEROLOGICAL READINGS

Sera \ Cells	1	2	3	4	5	6	7	8	9	10	11	12
12	W	?	W	?	W	W	W	W	W	?	W	W
11	W	W	W	W	W	W	(+)	(+)	W	W	W	+
10	(+)	W	W	W	(+)	(+)	(+)	(+)	(+)	W	W	(+)
9	W	W	W	W	W	W	W	W	W	W	W	W
8	W	W	W	W	(+)	W	(+)	(+)	(+)	W	W	(+)
7	?	?	W	—	?	?	W	W	W	?	?	W
6	(+)	W	W	W	W	W	+	(+)	(+)	W	W	(+)
5	W	W	W	W	W	W	+	W	W	W	W	(+)
4	(+)	W	W	W	(+)	(+)	(+)	(+)	(+)	W	W	(+)
3	W	?	W	W	W	(+)	(+)	(+)	(+)	W	(+)	+
2	W	W	W	W	(+)	W	(+)	+	(+)	?	W	(+)
1	W	?	W	W	W	W	(+)	W	W	?	W	W

§ 49·2]   THE ANALYSIS OF VARIANCE   293

If we arbitrarily assign the value o to the symbol —, and the value 1 to the symbol +, the values corresponding to the symbols ?, w, and (+) may be given the algebraic values x, y and z. Then by counting the numbers of the different kinds of symbol in each row and column, we find the sum of squares corresponding to rows and columns to be :—

TABLE 61·901
MATRIX FOR ROWS AND COLUMNS

	x	y	z	1
x	718	2	−672	−106
y	2	1630	−1416	−218
z	−672	−1416	1944	216
1	−106	−218	216	118

where it is convenient to write the quadratic expression as a symmetrical 4×4 matrix. Thus the coefficient of x^2 is 718, while those of xy and yx are both 2, making together the term $4xy$. The whole has been multiplied by 144 to avoid fractions. Similarly, the total sum of squares for 143 degrees of freedom is found to be :—

TABLE 61·902
MATRIX FOR TOTAL

	x	y	z	1
x	1703	−1157	−468	−65
y	−1157	4895	−3204	−445
z	−468	−3204	3888	−180
1	−65	−445	−180	695

To find the values of x, y and z which will make the ratio of the first of these expressions to the second as large as possible, it is necessary to solve an equation of the 4th degree. If from each element of the first matrix a multiple (θ) of the corresponding element of the second matrix is subtracted, the determinant of the sixteen values so found when equated to zero gives the equation. What is wanted is this equation's largest solution.

It is not necessary to calculate the coefficients of the equation. It is usually more convenient to evaluate the determinant exactly for chosen values of θ, and to apply the method of divided differences to calculate the required solution. The following table shows the values obtained at six chosen values of θ, simplified by dividing by 3456:

TABLE 61·91

TRIAL VALUES OF A DETERMINANT, AND THEIR DIVIDED DIFFERENCES

θ	Determinant.	First Divided Difference.	Second Divided Difference.	Third Divided Difference.	Fourth Divided Difference.
0	429106				
0·2	49982·376	−1,895618·12			
0·4	−598370·560	−3,241764·68	−3,365366·4		
0·6	−1,536668·072	−4,691487·56	−3,624307·2	−431568	
0·8	4,123941·552	28,303048·12	82,486339·2	143,517744	179,936640
1·0	30,181877	130,289677·24	254,966572·8	287,467056	179,936640

The second column is found by dividing the successive differences of the first column by 0·2, the interval between successive values of θ; the third column is likewise found from the second, the divisor in this case being the difference between values of θ separated by two steps, which in this table is constantly 0·4. Since, for any expression of the 4th degree, the fourth divided difference is constant, the exactitude of the values is checked in the last column, if enough values of θ are used.

It is apparent that the value required lies between 0·6 and 0·8. Since the fourth difference is constant whether the intervals are equal or unequal, positive or negative, the equation may be solved by choosing successive values of θ to continue the table so as to make the determinant approximate to zero. Thus in calculating the value for 0·7 a new line is added in which the third divided difference is increased by the fourth difference multiplied by 0·3, the multiplier being simply the new value less the value in the table four steps back. The new third difference is then multiplied by 0·1, the difference in θ taken three steps back, and added to the second difference. The factor by which the new second difference is multiplied is −0·1, since the new value of θ is ·1 less than that used two steps back. Finally, the new first difference is multiplied by −0·3 and added to the value of the determinant at 1·0 to find its value at 0·7. In the table on p. 296 (Table 61·92) this line has been filled in with exact values. In the subsequent lines sufficient figures have been retained for a very accurate determination. Notice that the value chosen for the third line is too high by 3 units in the fifth place of decimals, but that this circumstance does not interrupt the straightforward course of the work. For lower accuracy fewer figures would be needed in each column, and the process would be terminated in fewer steps. For machine calculation, however, the work shown is not heavy, and completely avoids the algebraic manipulation of the determinant.

TABLE 61·92
Steps in the Solution of an Algebraic Equation by Divided Differences; Fourth Difference 179,936640 throughout

θ	Determinant.	1st.	2nd.	3rd.
1·0	30,181877	130,289677·24	254,966572·8	287,467056
0·7	−231684·844	101,378539·48	289,111377·6	341,448048
0·708	− 18463·0046	26,652729·92	255,910306·7	360,881205
0·70869	+860·1817	28,004617·84	155,568230·5	344,451190
0·7086593	−2·7600	28,108851·19	158,096991·4	292,028323
0·7086593982	−·0002	28,104007·21	158,290581·8	293,586466

The value of θ so obtained is actually the fraction of the total sum of squares ascribable to rows and columns, when this fraction is maximised. To obtain the corresponding score values, x, y and z, the matrix for total sum of squares is multiplied by this value of θ, and subtracted from that for rows and columns, to give the following equations:—

$$-488 \cdot 8470x + 821 \cdot 9189y - 340 \cdot 3474z = 59 \cdot 9371$$
$$821 \cdot 9189x - 1838 \cdot 8878y + 854 \cdot 5447z = -97 \cdot 3534$$
$$-340 \cdot 3474x + 854 \cdot 5447y - 811 \cdot 2677z = -343 \cdot 5587,$$

of which the solution is

$$x = \cdot 192959,$$
$$y = \cdot 584453,$$
$$z = \cdot 958163,$$

the values appropriate to the symbols ?, w, and (+) if zero is assigned to −, and unity to +. It will be observed that the numerical values, of which only the first two figures need be used, lie between 0 and 1 in the proper order for increasing reaction. This is not a consequence of the procedure by which they have been obtained, but a property of the data examined.

Without evaluating the scores we may test the significance of rows and columns directly from the value of θ, for only the ratio of the sums of squares is needed. An approximate test is supplied by adding 3 *degrees of freedom* for the 3 unknown adjusted, to the 22 for rows and columns, and subtracting 3 from the remainder. Thus we have :—

TABLE 61·93
ANALYSIS OF VARIANCE OF A NON-NUMERICAL TABLE

	Degrees of Freedom.	Sums of Square.	Mean Square.	½ Log$_e$
Rows and columns	25	·70866	·028346	1·6723
Remainder	118	·29134	·002469	·4519
Total	143	1·00000		$z = 1·2204$

The differences between different rows and columns are thus very highly significant. We may infer that large differences exist in the strengths of the sera, or in the sensitivities of the different cells used. This is important, since it is only on this condition that the scores are worth anything.

49·3. The Precision of Estimated Scores

The numerical values obtained for the scores are, of course, subject to sampling errors. The notion of a standard error is not, however, very simply applicable to such scores, which cannot be used except in conjunction with the other scores of the system, including the two which have been assigned arbitrary values. This difficulty may be overcome comprehensively by developing a test whether the data differ significantly from expectation based on any given system of scores. Thus, retaining the score zero for a negative reading, we might have given to the readings ?, w, (+) and + the scores ·25, ·50, ·75 and 1·00, or equally 1, 2, 3 and 4. Then a test of significance exactly analogous to that made above will show whether such a system is sufficient to explain the whole of the apparent differentiation of rows and columns.

To perform the test, which is indeed of a kind for which extensive data, rather than a single table, should be used, we may denote the new variate by ξ. Then in Tables 61·901 and 61·902 (p. 293) we may make a new column by multiplying the four columns by 1, 2, 3 and 4 and adding; this gives the two sets of values :—

Rows and Columns.	Totals.
−1718	−2275
−1858	−2759
3192	4068
578	1285

If we multiply the four rows by 1, 2, 3 and 4 and add we shall obtain an analysis of variance for ξ; equally, if we multiply by the system of scores we have derived from the data, we shall have an analysis of covariance for X and ξ, where X stands for the system of scores previously derived. Similarly, we may find the analysis of variance for X, giving :—

TABLE 61·94

ANALYSIS OF COVARIANCE FOR ARBITRARY AND EMPIRICAL SCORES

	$S(\xi^2)$	$S(\xi X)$	$S(X^2)$
Rows and columns . .	6454	2219·039	770·496
Remainder . . .	3097	912·280	316·762
Total . .	9551	3131·319	1087·258

§ 49·3]   THE ANALYSIS OF VARIANCE   299

If now we eliminate ξ according to the general procedure, by deducting from $S(X^2)$ the square of $S(\xi X)$ divided by $S(\xi^2)$, using the lines for remainder and total, and obtaining that for rows and columns by subtraction, we find :—

TABLE 61·95
ANALYSIS OF VARIANCE OF EMPIRICAL SCORES, ELIMINATING ARBITRARY SCORES

	Degrees of Freedom.	Sum of Squares.	Mean Square.	½ Log$_e$
Rows and columns .	24	12·615	·5256	·8297
Remainder . .	118	48·033	·4071	·6982
Total . .	142	60·648	z	·1315

The degrees of freedom have been reduced for rows and columns, since, after eliminating ξ, there are only two values adjustable ; the value of z exceeds the 20 per cent. point, but falls far short of the 5 per cent. point. The table of data examined, with its very few − and + entries, is thus not sufficient to show that the linear series of scores is inadequate.

In this, as in Table 61·93, the z test is only approximate, though in both cases it is sufficient to answer the question at issue. In Table 61·93 the distribution of the fraction of sums of squares ·70866 depends on the three parameters 22 and 121 for original degrees of freedom, and 4, the degree of the equation solved, which is one more than the number of adjustable scores. In Table 61·95 the corresponding ratio, ·2080, likewise depends on the numbers 22, 120 and 3. The general solution of this problem of distribution has been found, but no exact tables are yet available.

The comprehensive method outlined in this section is applicable to a great variety of practical problems. It often happens that the statistician is provided with data on aggregates which it is required to allocate to different items. Thus, we may have data on the total consumption of different households, without knowing how this consumption is allocated between a man and his wife, or among children of different ages. If the composition of each household is known, the relative importance of each class of consumer may be obtained by minimising the deviation between the consumption recorded, and that expected, on assigned scores, from the composition of the family. Where continuous variables, such as age, are involved, it is preferable not to assign a separate unknown score to each age recorded, but to introduce the age, its square and possibly its cube, or higher powers, as independent variates, as in fitting curved regressions. Thus Day of the U.S. Forest Service has succeeded in allocating the cost of hauling logs of different diameters, from data giving only the composition by diameter of seventy different loads, each load involving the same haulage cost. An equation, quadratic in the diameter, was found sufficient to represent the curve of true cost.

IX

THE PRINCIPLES OF STATISTICAL ESTIMATION

50. The practical importance of using satisfactory methods of statistical estimation, and the widespread use in statistical literature of inefficient statistics, in the sense explained in Section 3, makes it necessary for the research worker, in interpreting his own results, or studying those reported by others, to discriminate between those conclusions which flow from the nature of the observations themselves, and those which are due solely to faulty methods of estimation.

Ex. 47.—As an example which brings out the main principles of the theory, and which does not involve data so voluminous that we cannot easily try out a variety of methods, we shall choose the estimation of linkage from the progeny of self-fertilised heterozygotes. Thus for two factors in maize, Starchy *v.* Sugary and Green *v.* White base leaf we may have (W. A. Carver's data) such observations as the following seedling counts :—

TABLE 62

Starchy.		Sugary.		Total.
Green.	White.	Green.	White.	
1997	906	904	32	3839

51. The Significance of Evidence for Linkage

It is a useful preliminary before making a statistical estimate, such as one of the intensity of linkage, to test if there is anything to justify estimation at all. We therefore test the possibility that the two factors are inherited independently. If such were the case the two factors, each segregating in a 3 : 1 ratio, would give the four combinations in the ratio 9 : 3 : 3 : 1, or with expectations, and corresponding contributions to χ^2, shown in Table 63.

TABLE 63

Expectation (m) .	2159·4	719·8	719·8	239·9	
Difference (d) .	−162·4	+186·2	+184·2	−207·9	
d^2/m . .	12·21	48·17	47·14	180·17	287·69

Since for 3 degrees of freedom the 1 per cent. point is only 11·34, the observed values are clearly in contradiction to the expectations. Such a result would, however, be produced either by linkage or by a departure from the 3 : 1 ratios; the test may be made specific by analysing χ^2 into its components as in Section 22. For this purpose, designating the four observed frequencies by a, b, c, d, and their total by n, the deviations from expectation in the ratio of starchy and sugary will be measured by

$$x = (a+b) - 3(c+d) = +95,$$

that of the other factor by

$$y = (a+c) - 3(b+d) = +87,$$

while to complete the analysis we need

$$z = a - 3b - 3c + 9d = -3145.$$

§ 52] STATISTICAL ESTIMATION 303

Then dividing the square of each discrepancy by its sampling variance, namely $3n$ for x and y, and $9n$ for z, we have the components

$$x^2 \div 3n \quad . \quad . \quad . \quad \cdot 784$$
$$y^2 \div 3n \quad . \quad . \quad . \quad \cdot 657$$
$$z^2 \div 9n \quad . \quad . \quad . \quad 286 \cdot 273$$

$$\text{Total} \quad . \quad . \quad 287 \cdot 714$$

agreeing with the former total as nearly as its limited accuracy will allow. The conclusion is evident that neither of the single factor ratios is abnormal, and that all but an insignificant fraction of the discrepancy is ascribable to linkage. The principles on which the deviations x, y, and z are constructed will be made more clear in Section 55.

52. The Specification of the Progeny Population for Linked Factors

When, as in the present case, the results are to be interpreted in terms of a definite theory, the specification of the population consists merely in following out the logical consequences of that theory. The theory we have to consider is that in both male and female gametogenesis, while each gamete has an equal chance of bearing the starchy or the sugary gene, and again of bearing the gene for green or white base leaf, yet the parental combinations Starchy White and Sugary Green are produced more frequently than the recombination classes Starchy Green and Sugary White. If the probability of the two latter classes is p in female gametogenesis and p' in male gametogenesis, the probability of the four types of ovules and of pollen will be

[TABLE 64

TABLE 64

	Starchy.		Sugary.	
	Green.	White.	Green.	White.
Ovules	$\tfrac{1}{2}p$	$\tfrac{1}{2}(1-p)$	$\tfrac{1}{2}(1-p)$	$\tfrac{1}{2}p$
Pollen	$\tfrac{1}{2}p'$	$\tfrac{1}{2}(1-p')$	$\tfrac{1}{2}(1-p')$	$\tfrac{1}{2}p'$

The theory further asserts that each grain of pollen will with equal probability fertilise each ovule, and that the seeds and seedlings produced will be equally viable. Then the probability that a seedling will be the double recessive Sugary White, which can only happen if both pollen and ovule carry these characters, will be $\tfrac{1}{4}pp'$. The probability of each of the other three classes of seedlings may be deduced at once, for the total probability of the two Sugary classes is $\tfrac{1}{4}$ irrespective of linkage, which leaves $\tfrac{1}{4}(1-pp')$ for the Sugary Green class. Similarly, the probability of the Starchy White class is $\tfrac{1}{4}(1-pp')$, leaving $\tfrac{1}{4}(2+pp')$ for Starchy Green.

Since these probabilities involve only the quantity pp', it is only of this and not of the separate values of p and p' that the data can provide an estimate. We shall therefore illustrate the problem of estimating the unknown quantity pp', which we may designate by θ. If p and p' were equal, then $\sqrt{\theta}$ would give the recombination fraction in both sexes, and if these are unequal it will still give their geometric mean. The data before us, however, throw direct light only on the value of θ. It is to be observed that in the case of coupling, when both dominant genes are received from the same grandparent, exactly the same specification is used, only it is $1-\sqrt{\theta}$ instead of $\sqrt{\theta}$ which is to be interpreted as the recombination fraction.

The statistical problem now takes the definite form: the probabilities of four events are
$$\tfrac{1}{4}(2+\theta),\ \tfrac{1}{4}(1-\theta),\ \tfrac{1}{4}(1-\theta),\ \tfrac{1}{4}\theta;$$
estimate the value of the parameter θ from the observed frequencies a, b, c, d.

53. The Multiplicity of Consistent Statistics

Nothing is easier than to invent methods of estimation. It is the chief purpose of this chapter to explain how satisfactory methods may be distinguished from unsatisfactory ones. The late development of this branch of the subject seems to be chiefly due to the lack of recognition of the number and variety of the plausible statistics which present themselves. We shall consider five of these.

In our example we may observe that the probability of the first and fourth class increases, and that of the two other classes diminishes as θ is increased. The expression
$$a-b-c+d$$
will therefore afford a convenient estimate of θ. To make a consistent estimate on these lines, we substitute the expected values
$$\frac{n}{4}(2+\theta,\ 1-\theta,\ 1-\theta,\ \theta),$$
for a, b, c, and d, and finding the result to be $n\theta$, we define our first estimate, T_1, by the equation
$$nT_1 = a-b-c+d.$$

Since the definition of consistency here used, was first put forward in 1921 (*Mathematical Foundations*), and was illustrated since the second edition (1928) in this work, it is a measure of the isolation of some departments purporting to teach statistics, that when restated in a new book (*Scientific Inference*) in 1957, it was reported by reviewers as *new*.

Alternatively, we might take the expression for z in Section 51, which appears there as a measure of linkage for the purpose of testing its significance; substituting the expected values, as before, we obtain $n(4\theta-1)$, and may define a new estimate, T_2, by the equation
$$n(4T_2-1) = a-3b-3c+9d,$$
or
$$4nT_2 = 2a-2b-2c+10d.$$
Obviously any number of similar estimates may be formed by the same method.

Instead of considering the sum of the extreme frequencies a and d we might have considered their product. The ratio of the product ad to the product bc clearly increases with θ; on substitution we have an equation for a third estimate in the form
$$\frac{\theta(2+\theta)}{(1-\theta)^2} = \frac{ad}{bc},$$
a quadratic equation of which T_3 is taken to be the positive solution.

As a fourth statistic we shall choose that given by the method of maximum likelihood. This method consists in multiplying the logarithm of the number expected in each class by the number observed, summing for all classes and finding the value of θ for which the sum is a maximum.

Now,
$$a \log(2+\theta) + b \log(1-\theta) + c \log(1-\theta) + d \log \theta$$
may be seen, by differentiating with respect to θ, to be a maximum if
$$\frac{a}{2+\theta} + \frac{d}{\theta} = \frac{b+c}{1-\theta},$$
leading to the quadratic equation
$$n\theta^2 - (a-2b-2c-d)\theta - 2d = 0,$$
of which the positive solution, T_4, satisfies the condition of maximum likelihood.

Finally, for any value adopted for θ, we shall be able to make a comparison of observed with expected frequencies, and to calculate the discrepancy, χ^2, between them. In fact χ^2 can be expressed in the form

$$\chi^2 = \frac{4}{n}\left(\frac{a^2}{2+\theta} + \frac{b^2}{1-\theta} + \frac{c^2}{1-\theta} + \frac{d^2}{\theta}\right) - n,$$

and the value for which this is a minimum will be the positive solution of the equation of the 4th degree

$$\frac{a^2}{(2+\theta)^2} + \frac{d^2}{\theta^2} = \frac{b^2+c^2}{(1-\theta)^2},$$

a statistic which we shall designate by T_5.

54. The Comparison of Statistics by means of the Test of Goodness of Fit

All the statistics mentioned, except the last, are easily calculated. The reader should calculate the first four, and verify that the value of the fifth given below approximately satisfies its equation. For each statistic we may calculate the numbers expected in the four classes of seedlings, and compare them with those observed. This is done in Table 65, where also the values of χ^2 derived from this comparison are given.

TABLE 65

COMPARISON OF FIVE STATISTICAL ESTIMATES OF LINKAGE

Method.	1.	2.	3.	4.	5.	
T . . .	·057046	·045194	·035645	·035712	·035785	
Recombination per cent. .	23·88	21·26	18·880	18·898	18·917	Observed
Numbers expected	1974·25	1962·875	1953·711	1953·775	1953·845	1997
	905·00	916·375	925·539	925·475	925·405	906
	905·00	916·375	925·539	925·475	925·405	904
	54·75	43·375	34·211	34·275	34·345	32
χ^2 . . .	9·717	3·860	2·0158	2·0154	2·0153	...

In the actual values of the estimates the first three methods differ considerably, but the last three are closely alike; so closely that the expectations of methods (3) and (5) differ from those of (4) by only about one-fifteenth of a seedling in each class. In the comparisons between the numbers expected and those observed, the most important discrepancies are in the fourth class, where method (2) gives a large and method (1) a very large discrepancy. The contrast between the first three methods in the values of χ^2 is very striking. For 2 degrees of freedom—not 3 because on fitting a linkage value 1 degree should be eliminated—a value above 9·21 should only occur once in a hundred trials. The value given by method (2) is not in itself significant, but since its value is nearly double that of methods (3), (4), and (5) we may be sure that the test of goodness of fit, if correct for the latter, must be highly erroneous for method (2), as well as for method (1). The general theorem which this illustrates is that the test of goodness of fit is only valid when efficient statistics are used in fitting a hypothesis to the data; in this case, as will be seen in the next section, methods (3), (4), and (5) are efficient, while methods (1) and (2) are not.

55. The Sampling Variance of Statistics

A more searching examination of the merits of various statistics may be made by calculating the sampling variance of each. Since the subject of sampling variance is usually treated by somewhat elaborate mathematical methods, it will be as well to give a number of simple formulæ by which the majority of ordinary cases may be treated.

First, if x is a linear function of the observed frequencies, such as

$$k_1 a + k_2 b + k_3 c + k_4 d,$$

then, designating the theoretical probability of any class by p, the mean value of x will be

$$n\mathrm{S}(pk).$$

The random sampling variance of x is given by the formula

$$\frac{1}{n} \mathrm{V}(x) = \mathrm{S}(pk^2) - \mathrm{S}^2(pk), \qquad . \qquad . \quad (\mathrm{A})$$

and if the mean value of x is zero, the variance of x becomes simply

$$n\mathrm{S}(pk^2).$$

Further, if a second linear function of the frequencies, y, is specified by coefficients, k', then the covariance of x and y is

$$n\mathrm{S}(pkk').$$

In view of this theorem the choice of the linear functions used for analysing χ^2 in Section 51 will no longer appear arbitrary, and the values taken for their sampling variance will be apparent. For the values of p are

$$\frac{1}{16}(9, 3, 3, 1),$$

and for x the values of k are

$$1, 1, -3, -3,$$

giving

$$\mathrm{S}(pk) = 0, \quad \mathrm{S}(pk^2) = 3,$$

so that the variance is $3n$, the value adopted. For y we evidently have the same values, with the additional

fact that the mean value of xy is zero. For z again
$$S(pk) = 0, \quad S(pk^2) = 9,$$
while the mean values of xz and yz are each zero. In analysing χ^2 into its components we always use linear functions of the frequencies, the mean value of each being zero, and such that all the covariances shall vanish.

It should be noted that the mean of xy is only zero in the absence of linkage. When linkage is present the values of p are $\tfrac{1}{4}(2+\theta,\ 1-\theta,\ 1-\theta,\ \theta)$,
giving for the covariance of x and y,
$$nS(pkk') = n(4\theta-1),$$
and for the correlation between them,
$$\tfrac{1}{3}(4\theta-1).$$

A statistic used for estimation will not be a linear function of the frequencies, for it must tend to a finite value as the sample is increased indefinitely; it will, however, often be of the form
$$T = \frac{1}{n}(k_1 a + k_2 b + k_3 c + k_4 d),$$
as in our example are T_1 and T_2.

For such cases a convenient formula is
$$nV(T) = S(pk^2) - \theta^2 \quad . \quad . \quad . \quad (B)$$
the statistic being supposed to be consistent. Now for T_1, k is always $+1$ or -1, and we have at once
$$V(T_1) = \frac{1-\theta^2}{n},$$
while for T_2, with $k = \tfrac{1}{2},\ -\tfrac{1}{2},\ -\tfrac{1}{2},\ 2\tfrac{1}{2}$, and $p = \tfrac{1}{4}(2+\theta,\ 1-\theta,\ 1-\theta,\ \theta)$ it is easy to find
$$V(T_2) = \frac{1+6\theta-4\theta^2}{4n}$$

These two sampling variances are very different; if θ is small (close linkage in repulsion), the variance of T_2 is only a quarter of that of T_1, and we may say that T_2 utilises four times as much of the available information as does T_1. This advantage diminishes, but persists over the whole range of repulsion linkages, for at $\theta = \frac{1}{4}$ the ratio of the variances is as five to three. The variances become equal at $\theta = \frac{1}{2}$, at which value the coupling recombination, $1 - \sqrt{\theta}$, is about ·29, and for closer linkage than this, in the coupling phase, T_1 is the better statistic.

The standard error to which either estimate, T, is subject is, of course, found by taking the square root of the variance; it will be of more practical interest to find the standard error of the recombination fraction, $\sqrt{\theta}$. For this purpose the above variances are divided by 4θ, before taking the square root. Putting $\theta = \cdot 0357$, in the variances, we then have the two estimates of the recombination percentage,

$$23 \cdot 88 \pm 4 \cdot 268 \text{ and } 21 \cdot 26 \pm 2 \cdot 348,$$

from the first of which we might judge roughly that the recombination per cent. lay between 15·3 and 32·4, while the second indicates the much closer limits 16·6 to 26·0.

For any function of the frequencies, whether the sample number n appears explicitly or not, we can obtain the approximation to the sampling variance appropriate to the theory of large samples in the form

$$\frac{1}{n} V(T) = S\left\{ p\left(\frac{\partial T}{\partial a}\right)^2 \right\} - \left(\frac{\partial T}{\partial n}\right)^2, \quad . \quad . \quad (C)$$

a formula which involves the differential coefficients of the function in question with respect to each observed frequency, and to the total, n. After differentiation the expectation pn is substituted for each frequency a. If we apply formula (C) to the function

$$F = \log(ad) - \log(bc) = \log\{T_3(2+T_3)\} - 2\log(1-T_3),$$

the values of $\partial F/\partial a$ are

$$\frac{1}{a}, \quad -\frac{1}{b}, \quad -\frac{1}{c}, \quad \frac{1}{d},$$

while, since n does not appear explicitly, $\partial F/\partial n = 0$. Hence, substituting pn for a, and the known values of p in terms of θ, we have

$$\frac{n}{4}V(F) = \frac{1}{2+\theta} + \frac{2}{1-\theta} + \frac{1}{\theta} = \frac{2(1+2\theta)}{\theta(1-\theta)(2+\theta)}.$$

To obtain the variance of T_3 we must divide this by the square of dF/dT_3, putting T_3 equal to θ after differentiation; but

$$\frac{dF}{dT_3} = \frac{1}{2+T_3} + \frac{2}{1-T_3} + \frac{1}{T_3},$$

hence

$$nV(T_3) = \frac{2\theta(1-\theta)(2+\theta)}{1+2\theta}.$$

For the variance of the statistic which satisfies the conditions of maximum likelihood a very simple and direct general method is available. The expression obtained by direct differentiation, and which, equated to zero, gave the equation for T_4 in Section 53, was

$$\frac{a}{2+\theta} - \frac{b+c}{1-\theta} + \frac{d}{\theta}.$$

If this is differentiated again with respect to θ, and the expected values substituted for a, b, c, and d, we obtain

$$-\frac{n}{4}\left(\frac{1}{2+\theta} + \frac{2}{1-\theta} + \frac{1}{\theta}\right);$$

and this is simply equated to $-1/V(T_4)$, giving

$$nV(T_4) = \frac{2\theta(1-\theta)(2+\theta)}{1+2\theta},$$

the same expression as we have obtained for the sampling variance of T_3. This expression is of great importance for our problem, for it has been proved that no statistic can have a smaller sampling variance, in the theory of large samples, than has the solution of the equation of maximum likelihood. This group of statistics (to which the minimum χ^2 solution also always belongs), which agree in their sampling variance with the maximum likelihood solution, are therefore of particular value, and are designated *efficient* statistics, on the ground that for large samples they may be said to make use of the whole of the relevant information available, whereas less efficient statistics such as T_1 and T_2 utilise only a portion of it.

The expression for the minimum variance

$$\frac{2\theta(1-\theta)(2+\theta)}{(1+2\theta)n}$$

represents, therefore, an intrinsic property of the data, irrespective of the methods of estimation actually used. For large samples we may interpret its reciprocal

$$I = \frac{(1+2\theta)n}{2\theta(1-\theta)(2+\theta)}$$

as a numerical measure of the total amount of information, relevant to the value of θ, which the sample contains; and it is evident that each seedling observed contributes a definite amount of information, measured by

$$\frac{1+2\theta}{2\theta(1-\theta)(2+\theta)}$$

relevant to the estimation of the value of θ. This consideration affords a basis for the exact treatment of sampling problems even for small samples, for once we know how to calculate the amount of information in the data, the amount extracted by any proposed method of analysis may be evaluated likewise, though this may be difficult, and a comparison of the two quantities gives an objective measure of the efficiency of the method proposed in conserving the relevant information available.

The actual fraction of the information utilised by inefficient statistics in large samples is obtained by expressing the random sampling variance of efficient statistics as a fraction of that of the statistic in question. Thus for T_1 and T_2 we have the fractions,

$$E(T_1) = V(T_4) \div V(T_1) = \frac{2\theta(2+\theta)}{(1+2\theta)(1+\theta)},$$

which rises to unity at $\theta = 1$, but is less at all other values; and

$$E(T_2) = V(T_4) \div V(T_2) = \frac{8\theta(1-\theta)(2+\theta)}{(1+2\theta)(1+6\theta-4\theta^2)},$$

which rises to unity at $\theta = \frac{1}{4}$, falling to zero if $\theta = 0$, or $\theta = 1$.

Fig. 11 shows the course of these fractions expressed as a percentage, for all values of the recombination percentage, $\sqrt{\theta}$ for repulsion, and $1-\sqrt{\theta}$ for coupling. It will be seen that for our actual value of about 19 per cent. in repulsion, the efficiency of T_1 is about 13 per cent., while that of T_2 is about 44 per cent. The use of T_1 wastes about seven-eighths of the information utilised by T_3, T_4, and T_5, while the use of T_2 wastes more than half of it. In other words, T_1 is only as good an estimate as should be obtained from a count of 503 seedlings, while T_2 is as good as should be obtained from 1661 out of the 3839 actually counted.

The standard error of the efficient estimates of recombination value is 1·545 per cent., giving probable limits of 15·8 to 22·0 for the true value. The use of inefficient statistics is therefore liable to give not merely inferior estimates of the value sought, but

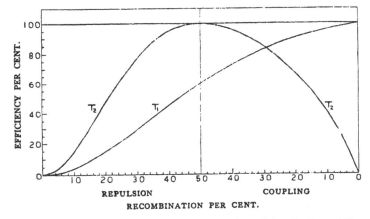

Fig. 11.—Efficiency of T_1 and T_2 for all values of θ. T_3, T_4, and T_5 having 100 per cent. efficiency throughout the range, are represented by the upper line.

estimates which are distinctly contradicted by the data from which they are derived. The value 23·88 per cent. obtained for T_1 differs from the better estimates by more than three times the standard error of the latter. It is highly misleading to derive such an estimate from data which themselves prove it to be erroneous.

The second respect in which the use of inefficient statistics is liable to be misleading is in the use of the χ^2 test of goodness of fit. Using T_1, we should naturally be led to conclude that the simple hypothesis of linked factors was in ill accord with the observations and that the results must be complicated by some such additional factor as differential viability. Finding only 32 double recessives against an expectation of 55, it would be natural to draw the conclusion that this genotype suffered from a low viability; whereas the data rightly interpreted give no significant indication of this sort. In the second place, whether the discrepancy were ascribed to differential viability or not, its existence would provide a very good reason for distrusting the linkage value obtained from such data; if, on the contrary, satisfactory methods of estimation are used, the grounds for this distrust are seen to fall away.

56. Comparison of Efficient Statistics

It has been seen that the three efficient statistics tested give closely similar results. This is in accordance with a general theorem that the correlation between any two efficient statistics tends to $+1$, as the sample is indefinitely increased. The conclusions drawn from their use will therefore ordinarily be equivalent. It appears from Fig. 11 that, for special values of θ, T_1 and T_2 also rank as efficient.

T_2 is efficient when θ is $\frac{1}{4}$, or in the absence of linkage. This accords with the use of s in Section 51 for testing the significance of linkage, for we are then testing the hypothesis that the factors are unlinked, and the test may be applied simply by seeing whether or not z^2 exceeds (say) $36n$. Any test based upon an efficient estimate of linkage compared to its standard error must agree with this. It is by no means uncommon to find statistics such as T_2 which provide excellent tests of significance, yet which become highly inefficient in estimating the magnitude of a significant effect. An outstanding example of this is the use of the third and fourth moments to measure the departure from normality of a frequency curve. The third and fourth moments provide excellent tests of the significance of the departure from normality, but when the distribution is one of the Pearsonian types differing considerably from the normal, the third and fourth moments are very inefficient statistics to use in estimating the form of the curve. This is the more noteworthy as the method of moments is ordinarily used for this purpose. The fact is that the efficiency of each of these statistics rises to 100 per cent. only for the normal form, just as that of T_2 reaches 100 per cent. only for zero linkage; but that the efficiency depends on the form of the curve, just as that of T_2 depends on the value of θ, and falls rapidly away as we leave the special region of high efficiency.

The statistic, T_1, is fully efficient when $\theta = 1$, that is, for very high linkage in the coupling phase; and therefore, in the theory of large samples, should give an estimate equivalent to T_3, T_4, and T_5. This extreme case, $\theta = 1$, is interesting in bringing out a limitation of the theory of large samples, which it is sometimes important to bear in mind; for the theory is valid only if none of the numbers counted, a, b, c, and d, is very small. Now for high linkage in coupling the recombination types, b and c, may be very scarce. It is true that for any proportion of crossing-over, however small, it is possible theoretically to take a sample so big that b and c will be large enough numbers; and in such cases the theory of large samples is justified. But it is also true for a sample of any given size, that linkage may be so high that seedlings of types b and c will be few; then, it is easy to see that some of the efficient statistics will fail. If, for example, either b or c is zero, T_3 will necessarily be unity, indicating complete linkage, whereas two or three seedlings in the other recombination class will show that crossing-over has really taken place. In the same way T_5 also fails, for it makes the recombination fraction proportional to $\sqrt{b^2+c^2}$, while T_1 and T_4 make it proportional to $b+c$. In general, the equation for minimising χ^2 is never satisfactory when some of the classes are thinly occupied, as one might expect from the nature of χ^2; the method therefore fails whenever the number of classes possible is infinite, as it usually is when we are concerned with the distributions of continuous variates. The two remaining efficient statistics T_1 and T_4 give equivalent estimates

$$\frac{b+c}{n}$$

for the recombination fraction, when the linkage is very high. Of course, as shown by Fig. 11, for any incomplete linkage the efficiency of T_1 is slightly below 100 per cent., so that the exact value of T_4 is slightly preferable. T_1, however, does provide a distinctly better estimate than T_3 or T_5 if b and c are small.

57. The Interpretation of the Discrepancy χ^2

The statistic obtained by the method of maximum likelihood stands in a peculiar relation to the measure of discrepancy, χ^2, and an examination of this relation will serve to illuminate the method, using degrees of freedom, which we have adopted in Chapter IV, and throughout the book. It has been stated that although in the distribution of a given number of individuals among four classes there are 3 degrees of freedom, yet if, as in the present problem, the expected numbers have been calculated from those observed by means of an adjustable parameter (θ), then only 2 degrees of freedom remain in which observation can differ from hypothesis. Consequently the value of χ^2 calculated in such a case is to be compared with the values characteristic of its distribution for 2 degrees of freedom. This principle has been disputed, but the common-sense considerations upon which it was based have since received complete theoretical verification. In the present instance we can in fact identify the 2 degrees of freedom concerned. For the observed numbers in each class will be entirely specified if we know :

(i) The number in the sample ;
(ii) The ratio of starchy to sugary plants ;
(iii) The ratio of green to white base leaf ;
(iv) The intensity of linkage.

Now if the expected series agrees in items (i) and (iv), it can only differ in items (ii) and (iii) and these will be completely given by the two quantities x and y defined by
$$x = a+b-3c-3d,$$
$$y = a-3b+c-3d,$$
specifying the ratios by linear functions of the frequencies.

The mean values of x and y are zero, and the random sampling variance of each is $3n$. In the absence of linkage their deviations will be independent, but if linkage is present the mean value of xy has

been found to be
$$-3n\frac{1-4\theta}{3},$$
and the correlation between x and y to be
$$\rho = -\frac{1-4\theta}{3}.$$

The simultaneous deviation of x and y from zero will therefore be measured (compare Section 30) by

$$Q^2 = \frac{1}{1-\rho^2}\left\{\frac{x^2-2\rho xy+y^2}{3n}\right\}$$
$$= \frac{3}{8n(1-\theta)(1+2\theta)}\left\{x^2+y^2+\frac{2}{3}(1-4\theta)xy\right\}.$$

This expression, which of course depends upon θ, is a quadratic function of the frequencies; in this it resembles χ^2, and on comparing term by term the two expressions it appears that

$$\chi^2 = Q^2 + \frac{1}{I}\left\{\frac{a}{2+\theta} - \frac{b+c}{1-\theta} + \frac{d}{\theta}\right\}^2,$$

where I is the quantity of information contained in the data as defined in Section 55.

This identity has two important consequences: first, that $\chi^2 = Q^2$ for the particular value of θ given by the equation of maximum likelihood, and for no other value. At this point, then, even for finite samples, the deviations between observation and expectation represent precisely the deviations in the two single factor ratios.

The second point is, that for any value of θ, χ^2 is the sum of two positive parts of which one is Q^2, while the other measures the deviation of the value of θ considered from the maximum likelihood solution; this latter part is the contribution to χ^2 of errors of estimation, while the discrepancy of observation from hypothesis, allowing any value of θ, is measured by Q^2 only.

Fig. 12 shows the values of χ^2 and Q^2 over the region covering the three efficient solutions.

The contact of the graphs at the maximum likelihood solution makes it evident why the solution based

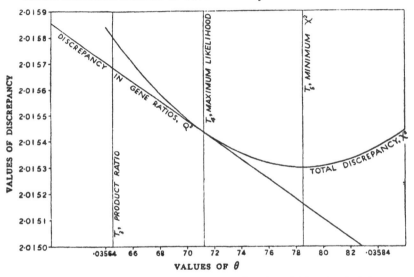

FIG. 12.—Graphs of χ^2 and Q^2 for varying θ in the neighbourhood of the efficient estimates

on minimum χ^2 should be of no special interest, although χ^2 is a valid measure of discrepancy between observation and hypothesis. As the hypothetical value, θ, is changed the value of Q^2 changes, and, although this change is very minute, it gives the line a sufficient slope to make an appreciable shift in the point of contact.

If we set aside the portion ascribable to errors of estimation, which satisfactory methods of estimation will always reduce to a trifling amount, it is apparent that the measure of discrepancy, χ^2, in our chosen problem, merely measures the deviation from expectation of the two single factor ratios, and its significance must therefore be judged by comparison with expectation for 2 degrees of freedom. Such a comparison will give an objective test dependent only on the data, and independent of our methods of treating it, if and only if the error of estimation measured from the maximum likelihood solution is sufficiently small. This, of course, where the theory of large samples is applicable, will be true if any efficient statistic is used ; it will always be true for the method of maximum likelihood.

57·1. Fragmentary Data

It very frequently happens, in a statistical enumeration, that only a portion of the whole sample is completely classified, the remaining members showing various degrees of incompleteness in their classification. Since the treatment of such data appears extremely troublesome, it is proper to lay great stress upon completeness of classification, whenever this is possible. In many cases, however, some degree of incompleteness is unavoidable, and the problem of framing an adequate statistical treatment, which shall utilise the whole of the information actually available, should be fairly faced. It will be shown that if approached in the right manner, and on the basis of a comprehensive theory of estimation, such problems offer no insuperable difficulties. We may again find a good example in the estimation of linkage, remembering that the type of difficulty to be discussed occurs in statistical work of all kinds.

Ex. 48. Tedin, working with two linked factors in *Pisum*, **Ar** and **Oh**, obtained, by selfing the double heterozygote, a progeny of 216 plants which could be classified as 99 **OhAr** 71 **ohAr** and 46 **ar**. The factor **Oh** could not be discriminated in the last group of plants, and, as is inevitable with moderate numbers and high linkage in repulsion, the proportions of this progeny give little information as to linkage value. From 63 of the **OhAr** group progenies were raised by self-fertilisation, which enabled their parents to be classified; 3 were homozygous for **Ar** but not for **Oh**, 8 for **Oh** and not for **Ar**, while 52 were heterozygous for both factors. Further, all of these 52 showed repulsion. Finally, of 47 plants of the **ohAr** group the progenies raised showed only 3 to be heterozygous for **Ar**, the remaining 44 being homozygous.

We may now set out the distribution of those 110 plants which in the end were completely classified alongside a table showing the relative frequencies with which plants completely classified should fall into the several classes, the recombination proportion being represented by p.

TABLE 66

	Ar Ar	Ar ar		ar ar
OhOh	0	8		—
Ohoh	3	0	52	—
ohoh	44	3		—

[TABLE 67

TABLE 67

	Ar Ar	Ar ar	ar ar
OhOh	p^2	$2p(1-p)$	$(1-p)^2$
Ohoh	$2p(1-p)$	$2p^2 \quad 2(1-p)^2$	$2p(1-p)$
ohoh	$(1-p)^2$	$2p(1-p)$	p^2

Next we have 60 plants, from which progenies were not grown, but which could be classified by their appearance as follows :—

TABLE 68

	Ar Ar Ar ar	ar ar
OhOh ⎫ Ohoh ⎭	36	0
ohoh	24	0

TABLE 69

	Ar Ar Ar ar	ar ar
OhOh ⎫ Ohoh ⎭	$2+p^2$	$1-p^2$
ohoh	$1-p^2$	p^2

and finally 46 plants, of which the classification is still less complete :—

TABLE 70

	Ar Ar	Ar ar	ar ar
OhOh Ohoh ohoh		0	46

TABLE 71

	Ar Ar	Ar ar	ar ar
OhOh Ohoh ohoh		3	1

If now it may be assumed that those plants, which, within any class, are incompletely specified, are a random sample of the members of that class, we may apply the method of maximum likelihood, as in Section 53, by multiplying the logarithm of the expectation in any class by the number recorded in that class, and adding all classes together, *irrespective of the completeness of classification*. When the expectations of any two classes are the same, the numbers in such classes may therefore be pooled, and we obtain

$$(8+3+3) \log \{2p(1-p)\} + 52 \log \{2(1-p)^2\} + 44 \log (1-p)^2$$
$$+ 36 \log (2+p^2) + 24 \log (1-p^2)$$
$$+ 46 \log (1)$$

as the logarithm of the likelihood, which is to be maximised. Any constant factor, such as 2, in the expectations makes a constant contribution to this quantity, independent of p, and may therefore be ignored. In particular the expectation in the **arar** class being entirely independent of p, the number in that class makes no contribution whatever to our knowledge of the linkage, and the whole class must be ignored. With these simplifications, and using the fact that the logarithm of a product is the sum of the logarithms of its factors, the expression to be maximised is reduced to

$$14 \log p + 206 \log (1-p) + 36 \log (2+p^2) + 24 \log (1-p^2).$$

By differentiating this expression with respect to p, we obtain the equation of maximum likelihood in the form

$$\frac{14}{p} - \frac{206}{1-p} + \frac{72p}{2+p^2} - \frac{48p}{1-p^2} = 0;$$

the first two terms are due to plants completely classified, and may be expected to contain the bulk of the desired information, the latter pair including the supplementary information due to the 60 **Ar** plants less completely classified. From the former only we should judge that p was nearly $14 \div 220$, or between 6 and 7 per cent. The exact estimate of the method of maximum likelihood may be most rapidly approached by substituting likely values for p and interpolating. Thus putting p equal to ·06 and ·07 we obtain :—

§57.2]   STATISTICAL ESTIMATION   327

TABLE 72

	$p = \cdot 06$.	$p = \cdot 07$.	$p = \cdot 0638$.
$14/p$	233·33	200·00	219·436
$-206/(1-p)$	−219·15	−221·51	−220·038
$72\,p/(2+p^2)$	2·16	2·51	2·292
$-48\,p/(1-p^2)$	−2·89	−3·38	−3·075
Total $\dfrac{\partial L}{\partial p}$	+13·45	−22·38	−1·385

The result of substituting ·06 being 13·45, while with ·07 we obtain −22·38, the true value which gives zero must be near to ·06+·01 (13·45÷35·83), or ·0638. The effect of substituting this value is shown in the final column, which serves both as a check to the previous work, and as a basis, if such were needed, for a more accurate solution. The improved value is ·06345, from which as an exercise in the method the student may rapidly obtain a still more accurate value.

57.2. The Amount of Information: Design and Precision

The standard error to be attached to such an estimate is derived directly from the amount of information in the data. In cases in which the data are fragmentary, we proceed as usual in differentiating the left-hand side of the equation of maximum likelihood, and in changing the sign of the terms, but in substituting the expected for the observed frequencies note should be taken of the basis on which these are expected, as well as of the expectation in the classes which do not appear in our sample. Thus in the classification of the first year the expectations from 216 plants are $54(2+p^2)$ OhAr and $54(1-p^2)$ ohAr;

these will make contributions to the information available of

$$54(2+p^2)\left(\frac{2p}{2+p^2}\right)^2 + 54(1-p^2)\left(\frac{-2p}{1-p^2}\right)^2$$

or

$$216p^2\left\{\frac{1}{2+p^2} + \frac{1}{1-p^2}\right\}$$

$$= 216\frac{3p^2}{(2+p^2)(1-p^2)} \quad . \quad . \quad . \quad . \quad (A)$$

and this, a very trifling amount numerically, is the amount of information available from the first year's classification.

If we now consider the 47 **ohAr** plants from which progenies were grown, we have expectations $47(1-p)^2 \div (1-p^2)$ **ArAr** and $47 \times 2p(1-p) \div (1-p^2)$ **Arar**. The *additional* information which these will contribute will be

$$47\frac{1-p}{1+p}\left(\frac{-1}{1-p}\right)^2 + 94\frac{p}{1+p}\left(\frac{1}{p}\right)^2 - 47\left(\frac{1}{1+p}\right)^2$$

the expected frequency in each portion being multiplied by the square of its logarithmic differential, and a like term deducted for the total; this gives

$$47\left\{\frac{1}{1-p^2} + \frac{2}{p(1+p)} - \frac{1}{(1+p)^2}\right\}$$

$$= 47\frac{2}{p(1-p)(1+p)^2}.$$

The additional information per plant of this group is therefore

$$\frac{2(1-p)}{p(1-p^2)^2} \quad . \quad . \quad . \quad . \quad (B)$$

Finally, the observed distribution of the 63 **ArOh** plants into 52 **ohAr/Ohar**, 11 **OhOh/Arar** or **Ohoh/**

ArAr, and o **OhOh/ArAr** or **OhAr/ohar** must be replaced by the expectations

$$\frac{63}{(2+p^2)}\left\{2(1-p)^2,\ 4p(1-p),\ 3p^2\right\}.$$

The additional information per plant in this group is therefore

$$\frac{1}{2+p^2}\left\{2(1-p)^2\left(\frac{-2}{1-p}\right)^2 + 4p(1-p)\left(\frac{1}{p}-\frac{1}{1-p}\right)^2 + 3p^2\left(\frac{2}{p}\right)^2\right\} - \left(\frac{2p}{2+p^2}\right)^2,$$

or $\quad \dfrac{1}{2+p^2}\left\{8 + \dfrac{4(1-2p)^2}{p(1-p)} + 12\right\} - \dfrac{4p^2}{(2+p^2)^2}$,

which may be reduced to

$$\frac{4(2+2p-p^2)}{p(1-p)(2+p^2)^2} \qquad . \qquad . \qquad . \quad (C)$$

At 6·345 per cent. recombination the numerical contribution per plant under (A), (B) and (C) are ·006051, 29·76 and 35·58. The second year's classifications thus give nearly 5000 and 6000 times as much information per plant as the first year's classification. On the actual numbers available the total information is 3642. The reciprocal of this, ·0002746 is the variance of the recombination fraction; whence 2·746 is the variance of the recombination percentage, and 1·657 per cent. is the standard error.

The advantage of examining the amount of information gained at each stage of the experiment lies in the fact that the precision attainable in the majority of experiments is limited by the amount of land, labour and supervision available, and much guidance may be gained as to how these resources should best be allocated, by considering the quantity of information to be anticipated. In the experiment in question, for example, it appears that progenies from **OhAr** plants are somewhat more profitable than those from **ohAr** plants.

If, on the contrary, our object is merely to assign a standard error to a particular result, we may estimate the amount of information available directly by differentiating the expression for $\partial L/\partial p$ in the equation of maximum likelihood, using the actual numbers recorded in the classes observed. We should then obtain

$$\frac{14}{p^2} + \frac{206}{(1-p)^2} - \frac{72(2-p^2)}{(2+p^2)^2} + \frac{48(1+p^2)}{(1-p^2)^2};$$

this gives 3725 as the total amount of information upon which our estimate has been based, and 1·638 as the standard error of the estimate of the recombination percentage. It should be noted that an estimate obtained thus is in no way inferior to one obtained from the theoretical expectations; only that it gives no guidance as to the improvement of the conduct of the experiment. It might be said that owing to chance the experiment has given a somewhat higher amount of information than should be expected from the numbers classified.

The difference between the amount of information actually supplied by the data and the average value to be expected from an assigned set of observations is of theoretical interest, and being often small requires the rather exact calculations illustrated above. For the purpose of merely estimating the precision of the result attained a much briefer method may be indicated. The values obtained in Table 72 show that for a change of ·01 in p, the value of $\partial L/\partial p$ falls by 35·83; from this the amount of information may be estimated at once to be 3583 units, and the standard error to be 1·67 per cent., a sufficiently good estimate for most purposes.

In some cases this very crude approximation will not be good enough. It really estimates the amount of information appropriate to a value about 6·5 per cent., half-way between the two trial values. We want its value at 6·345 per cent. the actual value obtained from our estimate. An improved value may easily be obtained where three trial values have been used. From $p = ·06$ and $p = ·0638$, we have

$$\frac{13·45 + 1·385}{0·0038} = 3904$$

at $p = ·0619$.

From $p = ·0638$ and $p = ·07$

$$\frac{-1·385 + 22·38}{0·0062} = 3386$$

at $p = ·0669$.

Whence for $p = ·06345$ we should take

$$\frac{·00155 \times 3386 + ·00345 \times 3904}{·005} = 3743,$$

corresponding to a standard error 1·635 per cent., a result of amply sufficient accuracy, obtained without the evaluation of the algebraical expressions for quantity of information.

57·3. Test of Homogeneity of Evidence used in Estimation

When diverse data throw light on what is theoretically the same quantity, the evidence from different sources may be combined, as in the last Section, to provide a single estimate based on the whole of the evidence. The need for such methods can scarcely be overlooked. In practical research, however, it is often of equal or greater importance to test whether the different sources of information fully concur in the estimate towards which they lead, or whether, on the contrary, this is a compromise between bodies of evidence which are significantly discrepant. We shall now show how a χ^2 test of homogeneity may be applied, making use of the same computational procedure as that employed in finding the combined estimate.

In tetrasomic inheritance each chromosome is capable of pairing, not with a single mate, but with any other of the set of four homologous chromosomes to which it belongs. If different parts of it pair with different partners it is possible for the two homologous genes carried by a single gamete to have been in origin identical. The proportion of such gametes will be designated by a, in respect of any particular factor. It is thus possible for a plant containing one dominant gene, out of the four present, to transmit two such genes in the same gamete. The frequencies with which it transmits 0, 1 and 2 dominant genes being then $2+a$, $2-2a$, and a out of 4. The corresponding frequencies for a duplex plant (carrying two dominant genes) will be $1+2a$, $4-4a$, and $1+2a$, out of 6.

For a gene determining top-necrosis of potato plants grafted with a scion infected with virus X, Cadman gives data from four sources: the backcross and intercross progenies of simplex plants, and the backcross and intercross progenies from duplex plants. These are as shown in Table 73.

From these data we may estimate the magnitude of a, and test the homogeneity of the evidence. A standard form of calculation is shown in Table 74.

§ 57·3]   STATISTICAL ESTIMATION   333

Values of a, ·120 and ·122, are sufficiently closely approximate to give both an improved joint estimate

TABLE 73

		Necrotic.	Non-necrotic.	Total.
Simplex plants	Backcross	762	842	1604
	Intercross	122	41	163
Duplex plants	Backcross	144	38	182
	Intercross	122	10	132

TABLE 74

	$a = ·120$.	$a = ·122$.	I.	D^2/I	D^2/I.
Simplex backcross					
$842/(2+a)$	397·1698	396·7955			
$-762/(2-a)$	$-405·3191$	$-405·7508$			
D	$-8·1493$	$-8·9553$	403·0	·1990	
Simplex intercross—					·5131
$82/(2+a)$	38·6792	38·6428			
$-244(2+a)/\{16-(2+a)^2\}$	$-44·9590$	$-45·0346$			
D	$-6·2798$	$-6·3918$	56·0	·7296	
Duplex backcross—					
$76/(1+2a)$	61·2903	61·0932			
$-288/(5-2a)$	$-60·5042$	$-60·5551$			
D	$+·7861$	$+·5381$	124·0	·0023	
Duplex intercross—					1·0984
$40/(1+2a)$	32·2581	32·1543			
$-488(1+2a)/\{36-(1+2a)^2\}$	$-17·5588$	$-17·6206$			
D	$+14·6993$	$+14·5337$	82·8	2·5511	
Total	$+1·0563$	$-·2753$	665·8	$-·0001$	$-·0001$
			χ^2	3·4819	1·6114
			n	3	1

of a, 12·16 per cent., and the amount of information, I, provided by the several parts of the data. These are given sufficiently nearly by dividing the difference between the discrepancies found for these two estimates by ·002. (Table 74.)

The amount of information is estimated for $a = 12\cdot1$ per cent., near to the true value. At the true value $\chi^2 = D^2/I$, as shown in section 57; in this case we add the contributions from the separate parts of the data, subtracting that for the data as a whole, which is almost negligible. In the table the values of D used are for $a = \cdot122$; the reader may be interested to make the test using those for $a = \cdot120$.

It should be noted that the exact equivalence, demonstrated in Section 57, requires that I from each batch of data should be calculated as the amount of information expected from the total number of observations in each batch. *E.g.* for the simple backcross I would be $1604/(4-a)^2$. This process gives amounts of information slightly different from those used in Table 74, namely, for the four sections, 402·5, 56·71, 123·0, and 61·302, or 643·5 in all. The corresponding values of D^2/I are then ·1992, ·7204, ·0023, and 3·4457, with a total χ^2 of 4·3675. These last values check exactly with the contributions to χ^2 found by calculating the expected numbers in each of the eight classes enumerated. The discrepancy between the two methods of calculating χ^2 is due to errors of random sampling, and tends to zero as the size of the sample is increased; both methods therefore tend to give the theoretical χ^2 distribution for large samples of homogeneous material, and there appears to be no good reason for preferring one to the other. The method of this Section is available, however, when estimation is based on measurements and not on frequencies, so that no alternative value based on frequencies can be calculated.

The test is applied both for all three degrees of freedom among the four kinds of data, and for the one degree of freedom contrasting simplex with duplex parents. On both tests the homogeneity is satisfactory, though we should perhaps wish to repeat the test with a larger amount of information in all than the 665·8 units here available.

57.4. Compiling and summarizing data of use in estimation.

In the foregoing examples we have considered the problem of estimation under the aspect of a closed record. That is to say we have considered the needs of an investigator who wishes to know exactly what conclusions can properly be drawn from a body of data already completed and assembled. Many unusual determinations, however, both in biology and in the non-biological sciences are perpetually provisional. In obtaining a value for the frequency of double reduction at a particular locus, we wish indeed to make all possible use of the observations so far available, but we should at the same time anticipate that later workers using the same locus will obtain additional information, which will inevitably modify, and should improve our estimate.

The form of calculation set out in Table 74 supplies what in contradistinction may be called an open record, for the entries under the two trial values used for α, may be regarded as additive scores derived from different sections of Cadman's data, and to which data from later experiments may be added at will, using the same trial values. Thus the following table ;—

	Scores		Information
Frequency of double reduction	12.0%	12.2%	
Simplex backcross	−8.1493	−8.9553	403.0
Simplex intercross	−6.2798	−6.3918	56.0
Duplex backcross	+ .7861	+ .5381	124.0
Duplex intercross	+14.6993	+14.5337	82.8
	+ 1.0563	− .2753	665.8

or even the last line of it, giving the totals, can be carried over directly into any further compilation on the subject, with confidence that, the scores being efficient, the whole of the relevant information contained in the data will be made available, and that any estimate, or tests of homogeneity obtained after the addition of further observations, reduced to the same form, will have the same validity, and will lead to the same result as if the calculations had been undertaken afresh upon the whole compilation.

In such an open record, estimates of parameters and tests of homogeneity do not appear explicitly. It is a property of the record that they can easily be made as required at any stage. The trial values at which the scores are appropriate will not be varied, unless subsequent observations should shift the estimate, at which the total score is to vanish, so far as to make one or other of the two values adopted seem unlikely. Only in that event will any recalculations from the original data be necessary. Using a more elaborate record with three trial values as was seen in Table 72, the amount of information may also be adjusted to variations in the estimate, and the scores at other neighbouring trial values may be predicted with accuracy from the summary record.

In the general application of the method, the score assigned to each body of data is

$$S\left(a\,\frac{1}{m}\,\frac{\delta m}{\delta \theta}\right)$$

where a stands for the frequency observed in any distinguishable class and m for the frequency expected at each of the values of θ chosen. Since for each body of data the expectations in each observable class take account of the number of observations, of the frequency distribution and of what classes are or are not distinguishable in the circumstances of the experiment, all these considerations are taken into account by the scores, which are designed for combination by simple addition: whereas sums of frequencies from data of different types may be quite meaningless. In effect the method exploits the fact that the equations of maximal likelihood are linear in the observed frequencies.

58. Summary of Principles

In any problem of estimation innumerable methods may be invented arbitrarily, all of which will tend to give the correct results as the available data are increased indefinitely. Each of these methods supplies a formula from which a statistic, intended as an estimate of the unknown, can be calculated from the observed frequencies. These statistics are of very different value.

A test of five such statistics in a simple genetical problem has shown that a particular group of them give closely concordant results, while the estimates obtained by the remainder are discrepant. This discrepancy is particularly marked in the misleading values found for χ^2.

An examination of the sampling errors shows that the concordant group have in large samples a variance equal to that of the maximum likelihood solution, and therefore as small as possible. These are efficient statistics; the variances of the inefficient statistics are larger, and may be so large that their values are quite inconsistent with the data from which they are derived.

Efficient statistics give closely equivalent results if the samples are sufficiently large, but when the theory of large samples no longer holds, such statistics, other than that obtained by the method of maximum likelihood, may fail.

The measure of discrepancy, χ^2, may be divided into two parts, one measuring the real discrepancy between observation and hypothesis, while the other measures merely the discrepancy between the value adopted and that given by the method of maximum likelihood. Using this fact, the homogeneity of data drawn from various sources may be tested in the process of obtaining the estimate.

The amount of information supplied by the data is capable of exact measurement, and the fraction of the information available which is utilised by any inefficient statistic can thereby be calculated. The same method may, though more laboriously, be applied to compare efficient statistics when the sample of data is small.

The method of maximum likelihood is directly applicable to fragmentary data, of which part is less completely classified than the remainder. Each fraction then contributes to the total amount of information utilised, according to the completeness with which it is classified. The knowledge of the amount of information supplied by the different fractions may be profitably utilised in planning the allocation of labour, and other resources, to observations of different kinds.

It will be readily understood that the thorough investigation which we have given to three somewhat slight genetical examples is not all necessary to their practical treatment. Its purpose has been to elucidate principles which are applicable to all problems involving statistical estimation. In many cases one need do no more than solve, at least to a good approximation the equation of maximum likelihood, and calculate the sampling variance of the estimate so obtained.

SOURCES USED FOR DATA AND METHODS

A. C. AITKEN (1931). Note on the computation of determinants. Transactions of the Faculty of Actuaries, xiii. 12-15.

A. C. AITKEN (1932). On the evaluation of determinants, the formation of their adjugates and the general solution of simultaneous linear equations. Proceedings of the Edinburgh Mathematical Society, Ser. II, iii. 207-219.

F. E. ALLAN (1930). The general form of the orthogonal polynomials for simple series, with proofs of their simple properties. Proceedings of the Royal Society of Edinburgh, l. 310-320.

M. M. BARNARD (1935). The secular variations of skull characters in four series of Egyptian skulls. Annals of Eugenics, vi. 352-371.

T. BAYES (1763). An essay towards solving a problem in the doctrine of chances. Philosophical Transactions of the Royal Society of London, liii. 370-418.

W.-U. BEHRENS (1929). Ein Beitrag zur Fehlen-Berechnung bei wenigen Beobachtungen. Landwirtschaftliche Jahrbüchen, 68, 807-837.

J. W. BISPHAM (1923). An experimental determination of the distribution of the partial correlation coefficient in samples of thirty. Metron, ii. 684-696.

J. BLAKEMAN (1905). On tests for linearity of regression in frequency distributions. Biometrika, iv. 332.

C. I BLISS (1935). The calculation of the dosage-mortality curve. Annals of Applied Biology, xxii. 134-167, particularly Appendix, 164, by R. A. Fisher.

C. I. BLISS (1935). The comparison of dosage-mortality data. Annals of Applied Biology, xxii. 307-333.

M. BRISTOL-ROACH (1925). On the relation of certain soil algæ to some soluble organic compounds. Annals of Botany, xxxix. 149-201.

J. BURGESS (1895). On the definite integral, etc. Transactions of the Royal Society of Edinburgh, xxxix. 257-321.

C. H. CADMAN (1942). Autotetraploid inheritance in the potato : some new evidence. Journal of Genetics, xliv. 33-52.

W. A. CARVER (1927). A genetic study of certain chlorophyll deficiencies in maize. Genetics, xii. 415-440.

SOURCES OF DATA AND METHODS 341

W. G. COCHRAN (1940). Note on an approximate formula for the significance levels of z. Annals of Mathematical Statistics, xl. 93-96.

C. G. COLCORD and LOLA S. DEMING (1936). The one-tenth per cent level of z. Sankhya 2, 413-423.

E. A. CORNISH (1954). The multivariate t-distribution associated with a set of normal sample deviates. Australian Journal of Physics, vii. 531-542.

E. A. CORNISH (1955). The sampling distributions of statistics derived from the multivariate t-distribution. Australian Journal of Physics, viii. 193-199.

E. A. CORNISH (1960). Fiducial limits for parameters in compound hypotheses. Australian Journal of Statistics, ii. 32-40.

B. B. DAY (1937). A suggested method for allocating logging costs to log sizes. Journal of Forestry, xxxv. 69-71.

T. EDEN (1931). Studies in the yield of tea. I. The experimental errors of field experiments with tea. Journal of Agricultural Science, xxi. 547-573.

W. P. ELDERTON (1902). Tables for testing the goodness of fit of theory to observation. Biometrika, i. 155.

E. C. FIELLER (1940). The biological standardisation of insulin. Supplement to the Journal of the Royal Statistical Society, vii. 1-53.

R. A. FISHER. *See* Bibliography, p. 345.

J. G. H. FREW (1924). On *Chlorops Tæniopus* Meig. (The gout fly of barley.) Annals of Applied Biology, xi. 175-219.

A. GEISSLER (1889). Beiträge zur Frage des Geschlechts verhältnisses der Geborenen. Zeitschrift des K. Sachsischen Statistischen Bureaus.

J. W. L. GLAISHER (1871). On a class of definite integrals. Philosophical Magazine, Series IV, xlii. 421-436.

H. GRAY (1935). Athletic performance as a function of growth: speed in sprinting. Journal of Pediatrics, vi. 14-21.

M. GREENWOOD and G. U. YULE (1915). The statistics of antityphoid and anticholera inoculations, and the interpretation of such statistics in general. Proceedings of the Royal Society of Medicine; Section of Epidemiology and State Medicine, viii. 113.

R. P. GREGORY, D. DE WINTON, and W. BATESON (1923). Genetics of *Primula Sinensis*. Journal of Genetics, xiii. 219-253.

J. A. HARRIS (1913). On the calculation of intraclass and interclass coefficients of correlation from class moments when the number of possible combinations is large. Biometrika, ix. 446-472.

J. A. HARRIS (1916). A contribution to the problem of homotyposis. Biometrika, xi. 201-214.

F. R. HELMERT (1875). Ueber die Berechnung des wahrscheinlichen Fehlers aus einer endlichen Anzahl wahrer Beobachtungsfehler. Zeitschrift für Mathematik und Physik, xx. 300-303.

A. H. HERSH (1924). The effects of temperature upon the heterozygotes in the bar series of *Drosophila*. Journal of Experimental Zoology, xxxix. 55-71.

H. HOTELLING (1931). The generalisation of Student's ratio. Annals of Mathematical Statistics, ii. 360-378.

J. S. HUXLEY (1923). Further data on linkage in *Gammarus Chevreuxi* and its relation to cytology. British Journal of Experimental Biology, i. 79-96.

M. N. KARN (1934). An investigation of the records of height and weight taken in school medical inspections in the county borough of Croydon. Annals of Eugenics, vi. 83-107.

T. L. KELLEY (1923). Statistical method. Macmillan and Co.

J. LANGE (1931). Crime and destiny. Allen and Unwin. (Translated by C. Haldane.)

LAPLACE (1820). Théorie analytique des probabilités. Paris. 3rd Edition.

K. MATHER (1938). The measurement of linkage in heredity. Methuen & Co. Ltd., London.

P. C. MAHALANOBIS (1932). Auxiliary tables for Fisher's z-test in analysis of variance. Indian Journal of Agricultural Science, ii. 679-693.

M. MASUYAMA (1951). An improved binomial probability paper and its use with tables. Report of Statistical Application Research, Union of Japanese Scientists and Engineers, i. 15-22.

" MATHETES " (1924). Statistical study on the effect of manuring on infestation of barley by gout fly. Annals of Applied Biology, xi. 220-235.

W. B. MERCER and A. D. HALL (1911). The experimental error of field trials. Journal of Agricultural Science, iv. 107-132.

J. R. MINER (1922). Tables of $\sqrt{1-r^2}$ and $1-r^2$. Johns Hopkins Press, Baltimore.

SOURCES OF DATA AND METHODS

F. MOSTELLER and J. W. TUKEY (1949). The uses and usefulness of binomial probability paper. Journal of the American Statistical Association, xliv. 174-212.

A. A. MUMFORD and M. YOUNG (1923). The interrelationships of the physical measurements and the vital capacity. Biometrika, xv. 109-133.

K. PEARSON (1900). On the criterion that a given system of deviations from the probable in the case of a correlated system of variables is such that it can be reasonably supposed to have arisen from random sampling. Philosophical Magazine, Series V, l. 157-175.

K. PEARSON and A. LEE (1903). Inheritance of physical characters. Biometrika, ii. 357-462.

J. RASMUSSON (1934). Genetically changed linkage values in *Pisum*. Hereditas, xix. 323-340.

N. SHAW (1922). The air and its ways. Cambridge University Press.

W. F. SHEPPARD (1907). Table of deviates of the normal curve. Biometrika, v. 404-406.

W. F. SHEPPARD (1938). Tables of the Probability Integral, completed and edited by the British Association Committee for the calculation of Mathematical Tables. British Association Mathematical Tables, vii.

H. FAIRFIELD SMITH (1936). A discriminant function for plant selection. Annals of Eugenics, vii. 240-250.

G. W. SNEDECOR (1934). Analysis of variance and covariance. Collegiate Press, Inc., Ames, Iowa.

"Student" (1907). On the error of counting with a hæmacytometer. Biometrika, v. 351-360.

"Student" (1908). The probable error of a mean. Biometrika, vi. 1-25.

"Student" (1925). New tables for testing the significance of observations. Metron, V, No. 3, 105-120.

P. V. SUKHATMÉ (1938). On Fisher and Behrens' test of significance for the difference in means of two normal samples. Sankyha, iv. 39-48.

H. TEDIN and O. TEDIN (1928). Contributions to the Genetics of *Pisum*. V, Seed coat colour, linkage and free combination. Hereditas, xi. 11-62.

O. TEDIN (1931). The influence of systematic plot arrangement upon the estimate of error in field experiments. Journal of Agricultural Science, xxi. 191-208.

T. N. THIELE (1903). Theory of observations. C. & E. Layton, London, 143 pp.

J. F. TOCHER (1908). Pigmentation survey of school children in Scotland. Biometrika, vi. 129-235.

W. L. WACHTER (1927). Linkage studies in mice. Genetics, xii. 108-114.

H. WORKING and H. HOTELLING (1929). The application of the theory of error to the interpretation of trends. Journal of the American Statistical Association, xxiv. 73-85.

F. YATES (1934). Contingency tables involving small numbers and the χ^2 test. Supplement to Journal of the Royal Statistical Society, i. 217-235.

G. U. YULE (1917). An introduction to the theory of statistics. C. Griffen and Co., London.

G. U. YULE (1923). On the application of the χ^2 method to association and contingency tables, with experimental illustrations. Journal of the Royal Statistical Society, lxxxv. 95-104.

BIBLIOGRAPHY

The following list includes the statistical publications of the author, together with a few other mathematical publications.

1912
On an absolute criterion for fitting frequency curves. Messenger of Mathematics, xli. 155-160.

1915
Frequency distribution of the values of the correlation coefficient in samples from an indefinitely large population. Biometrika, x. 507-521.

1918
The correlation between relatives on the supposition of Mendelian inheritance. Transactions of the Royal Society of Edinburgh, lii. 399-433.

1919
The genesis of twins. Genetics, iv. 489-499.

1920
A mathematical examination of the methods of determining the accuracy of an observation by the mean error and by the mean square error. Monthly Notices of the Royal Astronomical Society, lxxx. 758-770.

1921
Some remarks on the methods formulated in a recent article on "the quantitative analysis of plant growth." Annals of Applied Biology, vii. 367-372.
On the mathematical foundations of theoretical statistics. Philosophical Transactions of the Royal Society of London, A, ccxxii. 309-368.
Studies in crop variation. I. An examination of the yield of dressed grain from Broadbalk. Journal of Agricultural Science, xi. 107-135.
On the "probable error" of a coefficient of correlation deduced from a small sample. Metron, i, pt. 4, 1-32.

1922
On the interpretation of χ^2 from contingency tables, and the

1922—(cont.)

calculation of P. Journal of the Royal Statistical Society, lxxxv. 87-94.

The goodness of fit of regression formulæ, and the distribution of regression coefficients. Journal of the Royal Statistical Society, lxxxv. 597-612.

The systematic location of genes by means of crossover ratios. American Naturalist, lvi. 406-411.

[*with* W. A. MACKENZIE.] The correlation of weekly rainfall. Quarterly Journal of the Royal Meteorological Society, xlviii. 234-245.

[*with* H. G. THORNTON and W. A. MACKENZIE.] The accuracy of the plating method of estimating the density of bacterial populations. Annals of Applied Biology, ix. 325-359.

On the dominance ratio. Proceedings of the Royal Society of Edinburgh, xlii. 321-341.

1923

[*with* W. A. MACKENZIE.] Studies in crop variation. II, The manurial response of different potato varieties. Journal of Agricultural Science, xiii. 311-320.

Statistical tests of agreement between observation and hypothesis. Economica, iii. 139-147.

Note on Dr Burnside's recent paper on errors of observation. Proceedings of the Cambridge Philosophical Society, xxi. 655-658.

1924

The distribution of the partial correlation coefficient. Metron, iii. 329-332.

[*with* SVEN ODÉN.] The theory of the mechanical analysis of sediments by means of the automatic balance. Proceedings of the Royal Society of Edinburgh, xliv. 98-115.

The influence of rainfall on the yield of wheat at Rothamsted. Philosophical Transactions of the Royal Society of London, B, ccxiii. 89-142.

On a distribution yielding the error functions of several well-known statistics. Proceedings of the International Mathematical Congress, Toronto, 1924, pp. 805-813.

The conditions under which χ^2 measures the discrepancy between observation and hypothesis. Journal of the Royal Statistical Society, lxxxvii. 442-449.

A method of scoring coincidences in tests with playing cards. Proceedings of the Society for Psychical Research, xxxiv. 181-185.

BIBLIOGRAPHY

1925

Sur la solution de l'équation intégrale de M. V. Romanovsky. Comptes Rendus de l'Académie des Sciences, clxxxi. 88-89.

[*with* P. R. ANSELL.] Note on the numerical evaluation of a Bessel function derivative. Proceedings of the London Mathematical Society, xxiv. 54-56.

The resemblance between twins, a statistical examination of Lauterbach's measurements. Genetics, x. 569-579.

Statistical methods for research workers. Oliver & Boyd, Edinburgh. (Editions 1925, 1928, 1930, 1932, 1934, 1936, 1938, 1941, 1944, 1946, 1950, 1954, 1958, 1970.)

Theory of statistical estimation. Proceedings of the Cambridge Philosophical Society, xxii. 700-725.

1926

On the capillary forces in an ideal soil; correction of formulæ given by W. B. Haines. Journal of Agricultural Science, xvi. 492-505.

Periodical health surveys. Journal of State Medicine, xxxiv. 446-449.

Applications of "Student's" distribution. Metron, v. pt. 3, 90-104.

Expansion of "Student's" integral in powers of n^{-1}. Metron, v. pt. 3, 109-112.

On the random sequence. Quarterly Journal of the Royal Meteorological Society, lii. 250.

The arrangement of field experiments. Journal of the Ministry of Agriculture, xxxiii. 503-513.

Bayes' theorem and the fourfold table. The Eugenics Review, xviii. 32-33.

1927

[*with* H. G. THORNTON.] On the existence of daily changes in the bacterial numbers in American soil. Soil Science, xxiii. 253-257.

[*as* SECRETARY.] Recommendations of the British Association Committee on Biological Measurements. British Association, Section D, Leeds, 1927, pp. 13.

[*with* T. EDEN.] Studies in crop variation. IV, The experimental determination of the value of top dressings with cereals. Journal of Agricultural Science, xvii. 548-562.

Triplet children in Great Britain and Ireland. Proceedings of the Royal Society of London, B, cii. 286-311.

1927—(*cont.*)

[*with* J. WISHART.] On the distribution of the error of an interpolated value, and on the construction of tables. Proceedings of the Cambridge Philosophical Society, xxiii. 912-921.

On some objections to mimicry theory: statistical and genetic. Transactions of the Entomological Society of London, lxxv. 269-278.

1928

The possible modification of the response of the wild type to recurrent mutations. American Naturalist, lxii. 115-126.

[*with* L. H. C. TIPPETT.] Limiting forms of the frequency distribution of the largest or smallest member of a sample. Proceedings of the Cambridge Philosophical Society, xxiv. 180-190.

Further note on the capillary forces in an ideal soil. Journal of Agricultural Science, xviii. 406-410.

[*with* BHAI BALMUKAND.] The estimation of linkage from the offspring of selfed heterozygotes. Journal of Genetics, xx. 79-92.

[*with* E. B. FORD.] The variability of species in the Lepidoptera, with reference to abundance and sex. Transactions of the Entomological Society of London, lxxvi. 367-384.

Two further notes on the origin of dominance. American Naturalist, lxii. 571-574.

The general sampling distribution of the multiple correlation coefficient. Proceedings of the Royal Society of London, A, cxxi. 654-673.

[*with* T. N. HOBLYN.] Maximum- and Minimum-correlation tables in comparative climatology. Geografiska Annaler, iii. 267-281.

On a property connecting the χ^2 measure of discrepancy with the method of maximum likelihood. Bologna. Atti del Congresso Internazionale dei Matematici, vi. 94-100.

1929

A preliminary note on the effect of sodium silicate in increasing the yield of barley. Journal of Agricultural Science, xix. 132-139.

[*with* T. EDEN.] Studies in crop variation. VI, Experiments on the response of the potato to potash and nitrogen. Journal of Agricultural Science, xix. 201-213.

The over-production of food. The Realist, i. pt. 4, 45-60.

Tests of significance in harmonic analysis. Proceedings of the Royal Society of London, A, cxxv. 54-59.

1929—(cont.)
The statistical method in psychical research. Proceedings of the Society for Psychical Research, xxxix. 189-192.

Moments and product moments of sampling distributions. Proceedings of the London Mathematical Society (Series 2), xxx. 199-238.

The evolution of dominance; reply to Professor Sewall Wright. American Naturalist, lxiii. 553-556.

The sieve of Eratosthenes. The Mathematical Gazette, xiv. 564-566.

1930
The distribution of gene ratios for rare mutations. Proceedings of the Royal Society of Edinburgh, l. 205-220.

The evolution of dominance in certain polymorphic species. The American Naturalist, lxiv. 385-406.

[*with* J. WISHART.] The arrangement of field experiments and the statistical reduction of the results. Imperial Bureau of Soil Science: Technical Communication No. 10. 24 pp.

Inverse probability. Proceedings of the Cambridge Philosophical Society, xxvi. 528-535.

The moments of the distribution for normal samples of measures of departure from normality. Proceedings of the Royal Society of London, A, cxxx. 16-28.

The genetical theory of natural selection. Oxford: at the Clarendon Press, 1930; Dover Publications, Inc., New York, 1958.

1931
The evolution of dominance. Biological Reviews, vi. 345-368.

[*with* J. WISHART.] The derivation of the pattern formulæ of two-way partitions from those of simpler patterns. Proceedings of the London Mathematical Society (Series 2), xxxiii. 195-208.

The sampling error of estimated deviates, together with other illustrations of the properties and applications of the integrals and derivatives of the normal error function. British Association: Mathematical Tables, vol. 1. xxvi-xxxv.

1932
[*with* F. R. IMMER and O. TEDIN.] The genetical interpretation of statistics of the third degree in the study of quantitative inheritance. Genetics, xvii. 107-124.

Inverse probability and the use of likelihood. Proceedings of the Cambridge Philosophical Society, xxviii. 257-261.

The bearing of genetics on theories of evolution. Science Progress, xxvii. 273-287.

1933

The concepts of inverse probability of fiducial probability referring to unknown parameters. Proceedings of the Royal Society of London, A, cxxxix. 343-348.

On the evidence against the chemical induction of melanism in Lepidoptera. Proceedings of the Royal Society of London, B, cxii. 407-416.

Selection in the production of the ever-sporting stocks. Annals of Botany, clxxxviii. 727-733.

Number of Mendelian factors in quantitative inheritance. Nature, cxxxi. 400.

The contribution of Rothamsted to the development of statistics. Rothamsted Experimental Station, Harpenden. Report for 1933, pp. 43-50.

1934

Two new properties of mathematical likelihood. Proceedings of the Royal Society of London, A, cxliv. 285-307.

[*with* C. DIVER.] Crossing-over in the land snail *Cepæa nemoralis* L. Nature, cxxxiii. 834.

Professor Wright on the theory of dominance. The American Naturalist, lxviii. 370-374.

Probability, likelihood and quantity of information in the logic of uncertain inference. Proceedings of the Royal Society of London, A, cxlvi. 1-8.

[*with* F. YATES.] The 6×6 Latin squares. Proceedings of the Cambridge Philosophical Society, xxx. 492-507.

Randomization, and an old enigma of card play. Mathematical Gazette, xviii. 294-297.

The effect of methods of ascertainment upon the estimation of frequencies. Annals of Eugenics, vi. 13-25.

The amount of information supplied by records of families as a function of the linkage in the population sampled. Annals of Eugenics, vi. 66-70.

The use of simultaneous estimation in the evaluation of linkage. Annals of Eugenics, vi. 71-76.

1935

The logic of inductive inference. Journal of the Royal Statistical Society, xcviii. 39-82.

On the selective consequences of East's (1927) theory of heterostylism in *Lythrum*. Journal of Genetics, xxx. 369-382.

Some results of an experiment on dominance in poultry, with special reference to polydactyly. Proceedings of the Linnean Society of London, Session 147, pt. 3, 71-88.

1935—(*cont.*)

The detection of linkage with " Dominant " abnormalities. Annals of Eugenics, vi. 187-201.

Dominance in poultry. Philosophical Transactions of the Royal Society of London, B, ccxxv. 195-226.

The mathematical distributions used in the common tests of significance. Econometrica, iii. 353-365.

The sheltering of lethals. American Naturalist, lxix. 446-455.

The detection of linkage with recessive abnormalities. Annals of Eugenics, vi. 339-351.

The fiducial argument in statistical inference. Annals of Eugenics, vi. 391-398.

The design of experiments. Oliver & Boyd, Edinburgh. (Editions 1935, 1937, 1942, 1946, 1947, 1949, 1951, 1966).

1936

Has Mendel's work been rediscovered? Annals of Science, i. 115-137.

Heterogeneity of Linkage data for Friedreich's Ataxia and the spontaneous antigens. Annals of Eugenics, vii. 17-21.

Tests of significance applied to Haldane's data on partial sex linkage. Annals of Eugenics, vii. 87-104.

The use of multiple measurements in taxonomic problems. Annals of Eugenics, vii. 179-188.

[*with* S. BARBACKI.] A test of the supposed precision of systematic arrangements. Annals of Eugenics, vii. 189-193.

The coefficient of racial likeness. Journal of the Royal Anthropological Institute, lxvi. 57-63.

Uncertain Inference. Proceedings of the American Academy of Arts and Sciences, lxxi. 245-258.

[*with* K. MATHER.] A linkage test with mice. Annals of Eugenics, vii. 303-318.

1937

The relation between variability and abundance shown by the measurements of the eggs of British-nesting birds. Proceedings of the Royal Society of London, B, cxxii. 1-26.

Professor Karl Pearson and the Method of Moments. Annals of Eugenics, vii. 303-318.

On a point raised by M. S. Bartlett on fiducial probability. Annals of Eugenics, vii. 370-375.

[*with* B. DAY.] The comparison of variability in populations having unequal means. An example of the analysis of covariance with multiple dependent and independent variates Annals of Eugenics, vii. 333-348.

1937—(cont.)

The wave of advance of advantageous genes. Annals of Eugenics, vii. 355-369.

[*with* H. GRAY.] Inheritance in man: Boas's data studied by the method of analysis of variance. Annals of Eugenics, viii. 74-93.

[*with* E. A. CORNISH.] Moments and cumulants in the specification of distributions. Revue de l'Institut International de Statistique, 1937, v. 307-322.

1938

Dominance in Poultry: Feathered Feet, Rose Comb, Internal Pigment and Pile. Proceedings of the Royal Society, B, cxxv. 25-48.

[*with* F. YATES.] Statistical Tables. Oliver & Boyd, Edinburgh. (Editions 1938, 1943, 1948, 1953, 1957, 1963).

The mathematics of experimentation. Nature, 142, 442.

Quelques remarques sur l'estimation statistique. Biotypologie, vi. 153-159.

On the statistical treatment of the relation between sea-level characteristics and high-altitude acclimatization. Proceedings of Royal Society, B, cxxvi. 25-29.

The statistical utilization of multiple measurements. Annals of Eugenics, viii. 376-386.

Statistical Theory of Estimation. Calcutta University Readership Lectures. Published by the University of Calcutta.

1939

" Student." Annals of Eugenics, ix. 1-9.

The precision of the product formula for the estimation of linkage. Annals of Eugenics, ix. 50-54.

Presidential Address. Proceedings of the Indian Statistical Conference, Calcutta, 1938. Statistical Publishing Society, Calcutta.

Selective forces in wild populations of *Paratettix texanus*. Annals of Eugenics, ix. 109-122.

The comparison of samples with possibly unequal variances. Annals of Eugenics, ix. 174-180.

[*with* J. HUXLEY and E. B. FORD.] Taste-testing the anthropoid apes. Nature, cxliv. 750.

The sampling distribution of some statistics obtained from non-linear equations. Annals of Eugenics, ix. 238-249.

Stage of development as a factor influencing the variance in the number of offspring, frequency of mutants and related quantities. Annals of Eugenics, ix. 406-408.

1940

Scandinavian influence in Scottish ethnology. Nature, cxlv. 500.

On the similarity of the distributions found for the test of significance in harmonic analysis, and in Stevens's problem in geometrical probability. Annals of Eugenics, x. 14-17.

An examination of the different possible solutions of a problem in incomplete blocks. Annals of Eugenics, x. 52-75.

[*with* W. H. DOWDESWELL and E. B. FORD.] The quantitative study of populations in the Lepidoptera. 1. *Polyommatus icarus* Rott. Annals of Eugenics, x. 123-135.

The estimation of the proportion of recessives from tests carried out on a sample not wholly unrelated. Annals of Eugenics, x. 160-170.

A note on fiducial inference. Annals of Mathematical Statistics, x. 383-388.

The precision of discriminant functions. Annals of Eugenics, x. 422-429.

[*with* K. MATHER.] Non-lethality of the mid factor in *Lythrum salicaria*. Nature, cxlvi. 521.

1941

The theoretical consequences of polyploid inheritance for the mid style form of *Lythrum salicaria*. Annals of Eugenics, xi. 31-38.

Average excess and average effect of a gene substitution. Annals of Eugenics, xi. 53-63.

The asymptotic approach to Behrens' integral with further tables for the d test of significance. Annals of Eugenics, xi. 141-172.

The negative binomial distribution. Annals of Eugenics, xi. 182-187.

1942

The likelihood solution of a problem in compounded probabilities. Annals of Eugenics, xi. 306-307.

The theory of confounding in factorial experiments in relation to the theory of groups. Annals of Eugenics, xi. 341-353.

1943

Some combinational theorems and enumerations connected with the numbers of diagonal types of a Latin Square. Annals of Eugenics, xi. 395-401.

[*with* A. S. CORBET and C. B. WILLIAMS.] The relation between the number of species and the number of individuals in a random sample of an animal population. Journal of Animal Ecology, xii. 42-58.

1943—(cont.)

[with K. MATHER.] The inheritance of style-length in *Lythrum salicaria*. Annals of Eugenics, xii. 1-23.

1944

[with S. B. HOLT.] The experimental modification of dominance in Danforth's short-tailed mutant mice. Annals of Eugenics, xii. 102-120.

Allowance for double reduction in the calculation of genotype frequencies with polysomic inheritance. Annals of Eugenics xii. 169-171.

1945

A system of confounding for factors with more than two alternatives, giving completely orthogonal cubes and higher powers. Annals of Eugenics, xii. 283-290.

The logical inversion of the notion of the random variable. Sankhyā, vii. 129-132.

1946

A system of scoring linkage data, with special reference to the pied factors in mice. Amer. Nat., lxxx. 497-592.

The fitting of gene frequencies to data on Rhesus reactions. Annals of Eugenics, xiii. 150-155, and addendum, Note on the calculation of the frequencies of Rhesus allelomorphs. Annals of Eugenics, xiii. 223-224.

1947

The Rhesus factor. A study in scientific method. Amer. Scientist, xxxv. 95-103.

The theory of linkage in polysomic inheritance. Phil. Trans. Roy. Soc. B., no. 594, ccxxxiii. 55-87.

The analysis of covariance method for the relation between a part and the whole. Biometrics, iii. 65-68.

[with V. C. MARTIN.] Spontaneous occurrence in *Lythrum salicaria* of plants duplex for the short-style gene. Nature, clx. 541.

[with E. B. FORD.] The spread of a gene in natural conditions in a colony of the moth *Panaxia dominula* L. Heredity, i. 143-174.

Number of self-sterility alleles. Nature, clx. 797.

[with M. F. LYON and A. R. G. OWEN.] The sex chromosome in the house mouse. Heredity, i. 355-365.

1948

Conclusions fiduciaires. Annales de l'Institut Henri Poincare, x. 191-213.

1948—(cont.)

[with DANIEL DUGUÉ.] Un résultat assez inattendu d'arithmetique des lois de probabilite. Comptes rendus des séances de l'Academie des Sciences, ccxxvii. 1205-1206.

A quantitative theory of genetic recombination and chiasma formation. Biometrics, iv. 1-13.

1949

The linkage problem in a tetrasomic wild plant, *Lythrum salicaria*. Proceedings of the Eighth International Congress of Genetics (Hereditas Suppl. Vol. 1949).

[with W. H. DOWDESWELL and E. B. FORD.] The quantitative study of populations in the *Lepidoptera*. 2. *Maniola jurtina* L. Heredity, 3, 67-84.

A preliminary linkage test with *agouti* and *undulated* mice. Heredity, 3, 229-241.

Note on the test of significance for differential viability in frequency data from a complete three-point test. Heredity, 3, 215-219.

A theoretical system of selection for homostyle *Primula*. Sankhyā, 9, 325-342.

A biological assay of tuberculins. Biometrics, 5, 300-316.

The Theory of Inbreeding. Oliver and Boyd, Edinburgh. (Editions 1949, 1965).

1950

A class of enumerations of importance in genetics. Proceedings of the Royal Society, B, 136, 509-520.

Polydactyly in mice. Nature, 165, 407.

The significance of deviations from expectation in a Poisson series. Biometrics, 6, 17-24.

Gene frequencies in a cline determined by selection and diffusion. Biometrics, 6, 353-361.

1951

A combinatorial formulation of multiple linkage tests. Nature, 167, 520.

Standard calculations for evaluating a blood-group system. Heredity, 5, 95-102.

[with L. Martin.] The hereditary and familial aspects of toxic nodular goitre (secondary thyrotoxicosis). Quarterly Journal of Medicine, New Series, xx, 293-297.

1952

Statistical methods in genetics. (Bateson Lecture for 1951.) Heredity, 6, 1-12.

1953

The expansion of statistics. (Presidential Address for 1952.) Journal of the Royal Statistical Society, A, cxvi, 1-6.

Dispersion on a sphere. Proceedings of the Royal Society, A, 217, 295-305.

The variation in strength of the human blood group P. Heredity, 7, 81-89.

The linkage of *polydactyly* with *leaden* in the house-mouse. Heredity, 7, 91-95.

[*with* W. Landauer.] Sex differences of crossing-over in close linkage. American Naturalist, lxxxvii. 116.

Note on the efficient fitting of the negative binomial. Biometrics, 9, 197-200.

Population genetics. (Croonian Lecture.) Proceedings of the Royal Society, B, 141, 510-523.

1954

The analysis of variance with various binomial transformations Biometrics, 10, 130-139.

A fuller theory of "Junctions" in inbreeding. Heredity, 8, 187-197.

1955

Statistical methods and scientific induction. Journal of the Royal Statistical Society, B, 17. 69-78.

1956

On a test of significance in Pearson's *Biometrika Tables* (No. 11). Journal of the Royal Statistical Society, B. 18. 56-60.

New tables of Behrens' test of significance. Journal of the Royal Statistical Society, B, 18, 212-216.

Statistical Methods and Scientific Inference, Oliver and Boyd, Edinburgh. (Editions 1956, 1959).

1957

The underworld of probability. Sankhyā, 18, 201-210.

Blood groups and population genetics. Acta Genetica et Statistica Medica, 6, 507-509.

Comment on the notes by Neyman, Bartlett and Welch in the Journal of the Royal Statistical Society, B, 18 (1956). Journal of the Royal Statistical Society, B, 19, 179.

Methods in Human Genetics. Acta Genetica et Statistica Medica, 7, 7-10.

1958

Polymorphism and Natural Selection. Bulletin de L'Institut International de Statistique, 36, 284-293 (reprinted in Journal of Ecology, 46, 289-293).

1958—(cont.)

Mathematical Probability in the Natural Sciences. Presidential Address at Symposium 111 of the Congrès International des Sciences Pharmaceutiques, Brussels, (reprinted in Technometrics, 1, 21-29, 1959; Metrika, 2, 1-10, 1959; Ente Nazionale Idrocarburi, La Scuola in Azione, estratto dal notiziario, 20, 5-19, 1960, in Italian).

1959

An algebraically exact examination of junction formation and transmission in parent-offspring inbreeding. Heredity, 13, 179-186.

1960

[*with* E. A. Cornish]. The percentile points of distributions having known cumulants. Technometrics, 2, 209-226.

On some extensions of Bayesian inference proposed by Mr Lindley. Journal of the Royal Statistical Society, B, 22, 299-301.

1961

Possible differentiation in the wild population of *Oenothera organensis*. Australian Journal of Biological Sciences, 14, 76-78.

Sampling the reference set. Sankhyā, A, 23, 3-8.

The weighted mean of two normal samples with unknown variance ratio. Sankhyā, A, 23, 103-114.

A model for the generation of self-sterility alleles. Journal of Theoretical Biology, 1, 411-414.

1962

Confidence limits for a cross-product ratio. Australian Journal of Statistics, 4, 41.

The simultaneous distribution of correlation coefficients, Sankhyā, A, 24, 1-8.

Enumeration and classification in polysomic inheritance. Journal of Theoretical Biology, 2, 309-311.

Some examples of Bayes' Method of the Experimental Determination of Probabilities *a priori*. Journal of the Royal Statistical Society, B, 24, 118-124.

Letter to the Editor (*vide* Lewis, D. Journal of Theoretical Biology, 2, 69). Journal of Theoretical Biology, 3, 146-147.

The detection of a sex difference in recombination values using double heterozygotes. Journal of Theoretical Biology, 3, 509-513.

1963

The place of the Design of Experiments in the Logic of Scientific Inference. Centre National de la recherche scientifique, Colloques internationaux, No. 110 Le Plan D'Experiences, 13-19 (reprinted in Contributions to Statistics, edited by C. R. Rao, Oxford and London, Pergamon Press, Calcutta, Statistical Publishing Society, 1965, and in Ente Nazionale Idrocarburi, La Scuola in Azione, estratto dal numero 9, 33-42, 1962, in Italian).

INDEX

Age, 189
Alga, 140
Altitude, 156 seq.
Analysis of variance, 213 seq.
Arithmetic mean, 15, 41
Array, 183, 251
Association, degree of, 89
Autocatalytic curve, 29

Baby, growth of, 25
Bacteria, 58
Barley, 68
Barnard, 289
Bateson, 82, 101
Bayes, 20
Bernoulli, 9, 63
Binomial distribution, 9, 42, 63 seq.
Bispham, 98
Blakeman, 259
Bortkewitch, 55
Brandt, 87
Bristol-Roach, 140
Broadbalk, 30, 136
Buttercups, 37

Cadman, 332
Cards, 33, 160
Carver, 301
Cercis Canadensis, 222, 232
χ^2, distribution, 16, 22 seq., 78 seq.
χ^2, minimum, 307
Chloride of potash, 238
Cochran, 234
Combination of tests of significance, 99
Consistent statistics, 12, 305, 310
Contingency tables, 85 seq.

Contingency 2×2 tables; exact treatment, 96
Continuity, correction for, 92
Cornish, 166
Correlation, 5
Correlation coefficient, 23, 177 seq.
Correlation coefficient, tables, 211
Correlation diagrams, 29
Correlation ratio, 17, 257 seq.
Correlation table, 30, 33, 178
Covariance, x, 132, 272 seq.
Covariation, 5
Criminals, 94
Cumulants, 21, 73, 75
Cushny, 121

Day, 286, 300
Death-rates, 33, 38
Deming, xi, 248
Dependent variate, 129
De Winton, 82, 101
Diagrams, 24 seq.
Dice records, 63
Dilution method, 57
Discontinuous distributions, 54 seq.
Discontinuous variates, 36
Discriminant function, 287
Dispersion, 80
Dispersion, index of, 16, 58, 69
Distribution, kinds of, 3
Distribution, problems of, 8
Distributions, 41 seq.
Dot diagram, 30
Drosophila melanogaster, 253

Eden, 274
Efficiency, 13, 314 seq.
Efficient statistics, 13, 313, 316

INDEX

Elderton, 79
Errors, theory of, 2
Errors of fitting, 14, 321
Errors of grouping, 51
Errors of random sampling, 14, 49, 309
Estimation, 8, 21, 301 *seq.*
Eye facets, 253

Fairfield Smith, 289
Fertility gradient, 267
Fragmentary data, 322 *seq.*
Frequency curve, 5, 36
Frequency diagrams, 34
Frequency distributions, 4 *seq.*, 41 *seq.*
Frew, 69
Fungal colonies, 57

Galton, 5
Gammarus, 91
Gases, kinetic theory, 2
Gauss, 21
Geissler, 66
Glaisher, 341
Goodness of fit, 8, 78 *seq.*, 251 *seq.*
Gout fly, 68
Gray, 168
Greenwood, 85
Grouping, 35, 46 *seq.*, 76
Growth rate, 24
g-statistics, 75

Hæmacytometer, 56
Hair colour, 87
Half-invariants, 72, 74
Hall, 264, 269
Harris, 216, 222, 232
Helmert, 16, 22
Heredity, 177 *seq.*
Hersh, 253
Hertfordshire, 161
Hierarchical subdivisions, xi, 106
Histogram, 35
Homogeneity, 78 *seq.*

Horse-kick, 55
Hotelling, 135, 289
Huxley, 91

Inconsistent statistics, 11
Independence, 85 *seq.*
Independent variate, 129
Index of dispersion, 16, 58, 69
Information, amount of, 314, 327 *seq.*
Information, relevance of, 6 *seq.*
Interclass correlation, 213
Intraclass correlation, 17, 213 *seq*

Karn, 145
Kinetic theory of gases, 2
k-statistics, 72

Laplace, 9, 21
Latin square, 269 *seq.*
Latitude, 157 *seq.*
Lawrence, Kansas, 232
Least squares, 22, 260
Lee, 34, 117, 178
Lethal, 89
Lexis, 80
Likelihood, 10, 14, 306 *seq.*
Linkage, 89, 105, 301 *seq.*
Logarithmic scale, 28, 253
Logistic, 251
Logs, 300
Longitude, 157 *seq.*

Maize, 301
Mangolds, 264
Mass action, theory of, 2
Mather, x
Mean, 15, 41, 114 *seq.*
Mean, error curve of, 3
Mendelian frequencies, 81, 89 *seq.*, 302
Meramec Highlands, 232
Mercer, 264
Method of least squares, 22, 260
Method of maximum likelihood, 14, 21, 260, 306 *seq.*
Mice, 88

INDEX

Miner, 191
Modal class, 34
Moments, 46, 70 *seq.*, 317
Motile organisms, 61
Multiple births, 68
Multiple correlation, 17, 250 *seq.*
Multiple measurements, 287 *seq.*
Mumford, 189

Natural selection, theory of, 2
Nitrogenous fertilisers, 136
Normal distribution, 9, 12, 43 *seq.*
Normal distribution, tables, 77
Normality, test of, 52 *seq.*, 317

Omission of variate, 166
Organisms, presence and absence, 61 *seq.*
Ovules, 222, 232

Parameters, 6, 305
Partial correlation, 189 *seq.*
Partition of χ^2, 101 *seq.*, 302, 309
Pearson, 5, 16, 22, 34, 37, 52, 73, 79, 86, 117, 178
Peebles, 121
Pisum, 323
Plant selection, 289
Plot experimentation, 263 *seq.*
Poisson series, 9, 15, 16, 42, 54 *seq.*
Polynomial, 131, 147 *seq.*
Poor law relief, 198
Populations, 1 *seq.*, 33, 41
Potatoes, 238, 332
Primula, 82
Probability, theory, 9
Probable error, 18, 45
Product moment, correlation, 185
Pure cultures, 62

Quartile, 45

Rain frequency, 235
Rainfall, 32, 52, 156 *seq.*, 196
Rasmusson, 106
Ratio, fiducial limits of, 147
Reduction of data, 1

Regression coefficients, 17, 129 *seq.*
Regression formulæ, 129 *seq.*, 250
Relative growth rate, 24, 140
Richmond, 237
Rothamsted, 30, 52, 238

Schultz, 167
Selection, theory, 2
Sera, 291
Series, correlation between, 208
Sex difference, 229
Sex ratio, 66
Shaw, 235
Sheppard's adjustment, 48, 76, 188 *seq.*, 208 *seq.*, 257
Significance, 41
Skew curves, 46, 201, 220, 317
Skulls, 289
Small samples, 57 *seq.*, 68 *seq.*, 119 *seq.*, 134 *seq.*, 199, 230 *seq.*
Snedecor, 87
Soporifics, 121
Specification, 8, 303
Square-root chart, 39
Standard deviation, 3, 43 *seq.*
Standard error, 45, 49, 315
Statistic, 1 *seq.*, 6 *seq.*, 41, 305
Stature, 34, 46, 116, 178, 186, 229
"Student," 16, 19, 23, 56, 118, 119, 126
Sufficient statistics, 15
Sugar refinery products, 58
Sukhatmé, 125
Sulphate of potash, 238
Summation, 138 *seq.*
Systematic errors, 207 *seq.*

t, distribution of, 17, 23, 120, 166
t, use of, Chapter V, 114 *seq.*
Tables, 77, 112, 176, 211, 244
Taylor, 291
Tea yields, 274
Tedin, 323
Temperature, 253
Tests of significance, meaning of, 41

Thiele, 21, 72
Thornton, 128
Tocher, 87
Transformed correlation, 199 *seq.*, 217 *seq.*
Twin births, 68
Twins, 94
Typhoid, 85

Variance, 12, 75
Variate, 4
Variation, 1 *seq*

Wachter, 88
Weldon, 64
Wheat, 32 *seq.*
Working, 135
Working mean, 46

Yates, 92
Yeast, 56
Young, 189
Yule, 79, 85, 198

z distribution, 17, 23, 244 *seq.*

The Design of Experiments

The Design of Experiments

By

Sir Ronald A. Fisher, Sc.D., F.R.S.

Honorary Research Fellow, Division of Mathematical Statistics, C.S.I.R.O., University of Adelaide; Foreign Associate, United States National Academy of Sciences, and Foreign Honorary Member, American Academy of Arts and Sciences; Foreign Member of the Swedish Royal Academy of Sciences, and the Royal Danish Academy of Sciences and Letters; Member of the Pontifical Academy; Member of the German Academy of Sciences (Leopoldina); formerly Galton Professor, University of London, and Arthur Balfour Professor of Genetics, University of Cambridge

HAFNER PUBLISHING COMPANY
New York
1971

Copyright © 1971, The University of Adelaide

First Published 1935
Second Edition 1937
Third Edition 1942
Fourth Edition 1947
Fifth Edition 1949
Sixth Edition 1951
Reprinted 1953
Seventh Edition 1960
Eighth Edition 1966

Reprinted by Arrangement, 1971

By
HAFNER PUBLISHING COMPANY, INC.
866 Third Avenue
New York, N. Y. 10022

All Rights Reserved

PREFACE TO FIRST EDITION

IN 1925 the author wrote a book (*Statistical Methods for Research Workers*) with the object of supplying practical experimenters and, incidentally, teachers of mathematical statistics, with a connected account of the applications in laboratory work of some of the more recent advances in statistical theory. Some of the new methods, such as the analysis of variance, were found to be so intimately related with problems of experimental design that a considerable part of the eighth chapter was devoted to the technique of agricultural experimentation, and these sections have been progressively enlarged with subsequent editions, in response to frequent requests for a fuller treatment of the subject. The design of experiments is, however, too large a subject, and of too great importance to the general body of scientific workers, for any incidental treatment to be adequate. A clear grasp of simple and standardised statistical procedures will, as the reader may satisfy himself, go far to elucidate the principles of experimentation; but these procedures are themselves only the means to a more important end. Their part is to satisfy the requirements of sound and intelligible experimental design, and to supply the machinery for unambiguous interpretation. To attain a clear grasp of these requirements we need to study designs which have been widely successful in many fields, and to examine their structure in relation to the requirements of valid inference.

The examples chosen in this book are aimed at

illustrating the principles of successful experimentation; first, in their simplest possible applications, and later, in regard to the more elaborate structures by which the different advantages sought may be combined. Statistical discussion has been reduced to a minimum, and all the processes required will be found more fully exemplified in the previous work. The reader is, however, advised that the detailed working of numerical examples is essential to a thorough grasp, not only of the technique, but of the principles by which an experimental procedure may be judged to be satisfactory and effective.

GALTON LABORATORY
July 1935

PREFACE TO SEVENTH EDITION

THE second edition differed little from the first, published a year earlier. Apart from numerical corrections the principal changes were the fuller treatment of completely orthogonal squares in Section 35, and the addition of examples in Section 47.1, representing some of the newly developed combinatorial arrangements, which had attracted considerable interest. In the third edition Sections 45.1 and 45.2 were added, giving a more comprehensive view of the possibilities of confounding with many factors, and introducing the method of double confounding. In the fourth edition, Section 62.1 has been added on the fiducial limits of a ratio. In the fifth edition, Section 35.01 on configurations in three or more dimensions has been added. In the sixth edition attention may be called to the addition which has been made to Section 65, "Comparisons with Interactions," with a view to clarifying the differences in logical status between different sorts of categories which may appear in a factorial analysis. The numbers of sections have not been changed.

In the seventh edition, 1960, Sections 12.1 and 21.1 are new, while smaller additions and clarifications are scattered throughout.

DEPARTMENT OF STATISTICS, CSIRO, ADELAIDE, AUSTRALIA
Oct. 1959

NOTE TO THIS REPRINT EDITION

THE EIGHTH EDITION, 1966, which is reprinted here, is the same as the seventh, except for small additions and clarifications (mostly in Chapter X), introduced from notes written for this purpose by Sir Ronald Fisher some time before his death on 29 July, 1962.

J.H.B.

CONTENTS

I. INTRODUCTION

		PAGE
1.	The Grounds on which Evidence is Disputed	1
2.	The Mathematical Attitude towards Induction	3
3.	The Rejection of Inverse Probability	6
4.	The Logic of the Laboratory	7

II. THE PRINCIPLES OF EXPERIMENTATION, ILLUSTRATED BY A PSYCHO-PHYSICAL EXPERIMENT

5.	Statement of Experiment	11
6.	Interpretation and its Reasoned Basis	12
7.	The Test of Significance	13
8.	The Null Hypothesis	15
9.	Randomisation; the Physical Basis of the Validity of the Test	17
10.	The Effectiveness of Randomisation	19
11.	The Sensitiveness of an Experiment. Effects of Enlargement and Repetition	21
12.	Qualitative Methods of increasing Sensitiveness	22
12·1.	Scientific Inference and Acceptance Procedures	25

III. A HISTORICAL EXPERIMENT ON GROWTH RATE

13.		27
14.	Darwin's Discussion of the Data	27
15.	Galton's Method of Interpretation	29
16.	Pairing and Grouping	32
17.	"Student's" t Test	34
18.	Fallacious Use of Statistics	38
19.	Manipulation of the Data	40
20.	Validity and Randomisation	41

CONTENTS

	PAGE
21. Test of a Wider Hypothesis	44
21·1. "Non-parametric" tests	48

IV. AN AGRICULTURAL EXPERIMENT IN RANDOMISED BLOCKS

22. Description of the Experiment	50
23. Statistical Analysis of the Observations	52
24. Precision of the Comparisons	58
25. The Purposes of Replication	60
26 Validity of the Estimation of Error	62
27. Bias of Systematic Arrangements	64
28. Partial Elimination of Error	65
29. Shape of Blocks and Plots	66
30. Practical Example	67

V. THE LATIN SQUARE

31. Randomisation subject to Double Restriction	70
32. The Estimation of Error	73
33. Faulty Treatment of Square Designs	74
34. Systematic Squares	76
35 Græco-Latin and Higher Squares	80
35·01. Configurations in Three or more Dimensions	85
35·1. An Exceptional Design	88
36. Practical Exercises	90

VI. THE FACTORIAL DESIGN IN EXPERIMENTATION

37. The Single Factor	93
38. A Simple Factorial Scheme	95
39. The Basis of Inductive Inference	101
40. Inclusion of Subsidiary Factors	102
41. Experiments without Replication	106

VII. CONFOUNDING

42. The Problem of Controlling Heterogeneity	109
43. Example with 8 Treatments, Notation	111
44. Design suited to Confounding the Triple Interaction	113

CONTENTS

	PAGE
45. Effect on Analysis of Variance	114
45·1. General Systems of Confounding in Powers of 2	116
45·2. Double Confounding	121
46. Example with 27 Treatments	122
47. Partial Confounding	129
47·1. Practical Exercises	135

VIII. SPECIAL CASES OF PARTIAL CONFOUNDING

48.	137
49. Dummy Comparisons	137
50. Interaction of Quantity and Quality	139
51. Resolution of Three Comparisons among Four Materials	141
52. An Early Example	142
53. Interpretation of Results	152
54. An Experiment with 81 Plots	154

IX. THE INCREASE OF PRECISION BY CONCOMITANT MEASUREMENTS. STATISTICAL CONTROL

55. Occasions suitable for Concomitant Measurements	163
56. Arbitrary Corrections	168
57. Calculation of the Adjustment	171
58. The Test of Significance	175
58·1. Missing Values	177
59. Practical Examples	180

X. THE GENERALISATION OF NULL HYPOTHESES. FIDUCIAL PROBABILITY

60. Precision regarded as Amount of Information	184
61. Multiplicity of Tests of the same Hypothesis	187
62. Extension of the t Test	191
62·1. Fiducial Limits of a Ratio	194
63. The χ^2 Test	195
64. Wider Tests based on the Analysis of Variance	198
65. Comparisons with Interactions	207

XI. THE MEASUREMENT OF AMOUNT OF INFORMATION IN GENERAL

66. Estimation in General	214
67. Frequencies of Two Alternatives	216

		PAGE
68.	Functional Relationships among Parameters	218
69.	The Frequency Ratio in Biological Assay	224
70.	Linkage Values inferred from Frequency Ratios	226
71.	Linkage Values inferred from the Progeny of Self-fertilised or Intercrossed Heterozygotes	231
72.	Information as to Linkage derived from Human Families	236
73.	The Information elicited by Different Methods of Estimation	239
74.	The Information lost in the Estimation of Error	242
	INDEX	247

I AM very sorry, *Pyrophilus*, that to the many (elsewhere enumerated) difficulties which you may meet with, and must therefore surmount, in the serious and effectual prosecution of experimental philosophy I must add one discouragement more, which will perhaps as much surprise as dishearten you; and it is, that besides that you will find (as we elsewhere mention) many of the experiments published by authors, or related to you by the persons you converse with, false and unsuccessful (besides this, I say), you will meet with several observations and experiments which, though communicated for true by candid authors or undistrusted eye-witnesses, or perhaps recommended by your own experience, may, upon further trial, disappoint your expectation, either not at all succeeding constantly, or at least varying much from what you expected.

ROBERT BOYLE, 1673, *Concerning the Unsuccessfulness of Experiments.*

LE seul moyen de prévenir ces écarts, consiste à supprimer, ou au moins à simplifier, autant qu'il est possible, le raisonnement qui est de nous, & qui peut seul nous égarer, à la mettre continuellement a l'épreuve de l'expérience; à ne conserver que les faits qui sont des vérites données par la nature, & qui ne peuvent nous tromper; à ne chercher la verité que dans l'enchaînement des expériences & des observations, sur-tout dans l'ordre dans lequel elles sont présentées, de la même manière que les mathématiciens parviennent à la solution d'un problême par le simple arrangement des données, & en réduisant le raisonnement à des opérations si simples, à des jugemens si courts, qu'ils ne perdent jamais de vue l'évidence qui leur sert de guide.

Methode de Nomenclature chimique,
A. L. LAVOISIER, 1787.

THE DESIGN OF EXPERIMENTS

I

INTRODUCTION

1. The Grounds on which Evidence is Disputed

WHEN any scientific conclusion is supposed to be proved on experimental evidence, critics who still refuse to accept the conclusion are accustomed to take one of two lines of attack. They may claim that the *interpretation* of the experiment is faulty, that the results reported are not in fact those which should have been expected had the conclusion drawn been justified, or that they might equally well have arisen had the conclusion drawn been false. Such criticisms of interpretation are usually treated as falling within the domain of *statistics*. They are often made by professed statisticians against the work of others whom they regard as ignorant of or incompetent in statistical technique; and, since the interpretation of any considerable body of data is likely to involve computations, it is natural enough that questions involving the logical implications of the results of the arithmetical processes employed, should be relegated to the statistician. At least I make no complaint of this convention. The statistician cannot evade the responsibility for understanding the processes he applies or recommends. My immediate point is that the questions involved can be dissociated from all that is strictly technical in the statistician's craft, and, *when so detached*, are questions only of the right use of human reasoning powers, with which all intelligent people, who hope to be intelligible, are equally concerned, and on which the statistician, as such, speaks with no special authority. The statistician cannot excuse himself from the duty of getting his head clear on the principles of scientific inference, but equally no other thinking man can avoid a like obligation.

The other type of criticism to which experimental results are exposed is that the experiment itself was ill designed, or, of course, badly executed. If we suppose that the experimenter did what he intended to do, both of these points come down to the question of the *design*, or the *logical structure* of the experiment. This type of criticism is usually made by what I might call a heavyweight *authority*. Prolonged experience, or at least the long possession of a scientific reputation, is almost a pre-requisite for developing successfully this line of attack. Technical details are seldom in evidence. The authoritative assertion " His *controls* are *totally* inadequate " must have temporarily discredited many a promising line of work ; and such an authoritarian method of judgment must surely continue, human nature being what it is, so long as theoretical notions of the principles of experimental design are lacking—notions just as clear and explicit as we are accustomed to apply to technical details.

Now the essential point is that the two sorts of criticism I have mentioned are aimed only at different aspects of the same whole, although they are usually delivered by different sorts of people and in very different language. If the design of an experiment is faulty, any method of interpretation which makes it out to be decisive must be faulty too. It is true that there are a great many experimental procedures which are well designed in that they *may* lead to decisive conclusions, but on other occasions may fail to do so; in such cases, if decisive conclusions are in fact drawn when they are unjustified, we may say that the fault is wholly in the interpretation, not in the design. But the fault of interpretation, even in these cases, lies in overlooking the characteristic features of the design which lead to the result being sometimes inconclusive, or conclusive on some questions but not on all. To understand correctly the one aspect of the problem is to understand the other. Statistical procedure and experimental design are only two different aspects of the same whole, and that whole comprises all the logical requirements of the complete process of adding to natural knowledge by experimentation.

2. The Mathematical Attitude towards Induction

In the foregoing paragraphs the subject-matter of this book has been regarded from the point of view of an experimenter, who wishes to carry out his work competently, and having done so wishes to safeguard his results, so far as they are validly established, from ignorant criticism by different sorts of superior persons. I have assumed, as the experimenter always does assume, that it *is* possible to draw valid inferences from the results of experimentation; that it is possible to argue from consequences to causes, from observations to hypotheses; as a statistician would say, from a sample to the population from which the sample was drawn, or, as a logician might put it, from the particular to the general. It is, however, certain that many mathematicians, if pressed on the point, would say that it is not possible rigorously to argue from the particular to the general; that all such arguments must involve some sort of guesswork, which they might admit to be plausible guesswork, but the rationale of which, they would be unwilling, as mathematicians, to discuss. We may at once admit that any inference from the particular to the general must be attended with some degree of uncertainty, but this is not the same as to admit that such inference cannot be absolutely rigorous, for the nature and degree of the uncertainty may itself be capable of rigorous expression. In the theory of probability, as developed in its application to games of chance, we have the classic example proving this possibility. If the gamblers' apparatus are really *true* or unbiased, the probabilities of the different possible events, or combinations of events, can be inferred by a rigorous deductive argument, although the outcome of any particular game is recognised to be uncertain. The mere fact that inductive inferences are uncertain cannot, therefore, be accepted as precluding perfectly rigorous and unequivocal inference.

Naturally, writers on probability have made determined efforts to include the problem of inductive inference within the ambit of the theory of mathematical probability, developed in discussing deductive problems arising in games of chance. To illustrate how much was at one time thought to have been achieved in this way, I may quote a very lucid statement by Augustus de Morgan, published in 1838, in the preface to his essay on probabilities in *The Cabinet Cyclopædia*. At this period confidence in the theory of inverse probability, as it was called, had reached, under the influence of Laplace, its highest point. Boole's criticisms had not yet been made, nor the more decided rejection of the theory by Venn, Chrystal, and later writers. De Morgan is speaking of the advances in the theory which were leading to its wider application to practical problems.

" There was also another circumstance which stood in the way of the first investigators, namely, the not having considered, or, at least, not having discovered the method of reasoning from the happening of an event to the probability of one or another cause. The questions treated in the third chapter of this work could not therefore be attempted by them. Given an hypothesis presenting the necessity of one or another out of a certain, and not very large, number of consequences, they could determine the chance that any given one or other of those consequences should arrive; but given an event as having happened, and which might have been the consequence of either of several different causes, or explicable by either of several different hypotheses, they could not infer the probability with which the happening of the event should cause the different hypotheses to be viewed. But, just as in natural philosophy the selection of an hypothesis by means of observed facts is always preliminary to any attempt at deductive discovery; so in the application of the notion of probability to the actual affairs of life, the process of reasoning from observed events to their most probable antecedents must go before the direct use of any such antecedent, cause, hypothesis, or whatever it may be correctly termed. These two obstacles, therefore, the mathematical difficulty, and the want of an inverse method, prevented the science from extending its views beyond problems of that simple nature which games of chance present."

Referring to the inverse method, he later adds: "This was first used by the Rev. T. Bayes, and the author, though now almost forgotten, deserves the most honourable remembrance from all who treat the history of this science."

3. The Rejection of Inverse Probability

Whatever may have been true in 1838, it is certainly not true to-day that Thomas Bayes is almost forgotten. That he seems to have been the first man in Europe to have seen the importance of developing an exact and quantitative theory of inductive reasoning, of arguing from observational facts to the theories which might explain them, is surely a sufficient claim to a place in the history of science. But he deserves honourable remembrance for one fact, also, in addition to those mentioned by de Morgan. Having perceived the problem and devised an axiom which, if its truth were granted, would bring inverse inferences within the scope of the theory of mathematical probability, he was sufficiently critical of its validity to try to avoid the axiomatic approach, and, perhaps for the same reason, to withhold his entire treatise from publication until his doubts should have been satisfied. In the event, the work was published after his death by his friend, Price, and we cannot say what views he ultimately held on the subject.

The discrepancy of opinion among historical writers on probability is so great that to mention the subject is unavoidable. It would, however, be out of place here to argue the point in detail. I will only state three considerations which will explain why, in the practical applications of the subject, I shall not assume the truth of Bayes' axiom. Two of these reasons would, I think, be generally admitted, but the first, I can well imagine, might be indignantly repudiated in some quarters. The first is this: The axiom leads to apparent mathematical contradictions. In explaining these contradictions away, advocates of inverse probability seem forced to regard mathematical probability, not as an objective quantity measured by observable frequencies, but as measuring merely psychological tendencies, theorems respecting which are useless for scientific purposes.

My second reason is that it is the nature of an axiom that its truth should be apparent to any rational mind which fully apprehends its meaning. The axiom of Bayes has certainly been fully apprehended by a good many rational minds, including that of its author, without carrying this conviction of necessary truth. This, alone, shows that it cannot be accepted as the axiomatic basis of a rigorous argument.

My third reason is that inverse probability has been only very rarely used in the justification of conclusions from experimental facts, although the theory has been widely taught, and is widespread in the literature of probability. Whatever the reasons are which give experimenters confidence that they can draw valid conclusions from their results, they seem to act just as powerfully whether the experimenter has heard of the theory of inverse probability or not.

4. The Logic of the Laboratory

In fact, in the course of this book, I propose to consider a number of different types of experimentation, with especial reference to their logical structure, and to show that when the appropriate precautions are taken to make this structure complete, entirely valid inferences may be drawn from them, without using the disputed axiom. *If* this can be done, we shall, in the course of studies having directly practical aims, have overcome the theoretical difficulty of inductive inferences.

Inductive inference is the only process known to us by which essentially new knowledge comes into the world. To make clear the authentic conditions of its validity is the kind of contribution to the intellectual development of mankind which we should expect experimental science would ultimately supply. Men have always been capable of some mental processes of the kind we call "learning by experience." Doubtless this experience was often a very imperfect basis, and the reasoning processes used in interpreting it were very insecure; but there must have been in these processes a sort of embryology of knowledge, by which new knowledge was gradually produced. Experimental observations are only experience carefully planned in advance, and designed to form a secure basis of new knowledge; that is, they are systematically related to the body of knowledge already acquired, and the results are deliberately observed, and put on record accurately. As the art of experimentation advances the principles should become clear by virtue of which this planning and designing achieve their purpose.

It is as well to remember in this connection that the principles and method of even *deductive* reasoning were probably unknown for several thousand years after the establishment of prosperous and cultured civilisations. We take a knowledge of these principles for granted, only because geometry is universally taught in schools. The method and material taught is essentially that of Euclid's text-book of the third century B.C., and no one can make any progress in that subject without thoroughly familiarising his mind with the requirements of a precise deductive argument. Assuming the axioms, the body of their logical consequences is built up systematically and without ambiguity. Yet it is certainly something of an accident historically that this particular discipline should have become fashionable in the Greek Universities, and later embodied in the curricula of secondary education. It would be difficult to overstate how much the liberty of human thought has owed to this fortunate circumstance. Since Euclid's time there have been very long periods during which the right of unfettered individual judgment has been successfully denied in legal, moral, and historical questions, but in which it has, none the less, survived, so far as purely deductive reasoning is concerned, within the shelter of apparently harmless mathematical studies.

The liberation of the human intellect must, however, remain incomplete so long as it is free only to work out the consequences of a prescribed body of dogmatic data, and is denied the access to unsuspected truths, which only direct observation can give. The development of experimental science has therefore done much more than to multiply the technical competence of mankind; and if, in these introductory lines, I have seemed to wander far from the immediate purpose of this book, it is only because the two topics with which we shall be concerned, the arts of experimental design and of the valid interpretation of experimental results, in so far as they can be technically perfected, must constitute the core of this claim to the exercise of full intellectual liberty.

The chapters which follow are designed to illustrate the principles which are common to all experimentation, by means of examples chosen for the simplicity with which these principles are brought out. Next, to exhibit the principal designs which have been found successful in that field of experimentation, namely agriculture, in which questions of design have been most thoroughly studied, and to illustrate their applicability to other fields of work. Many of the most useful designs are extremely simple, and these deserve the greatest attention, as showing in what ways, and on what occasions, greater elaboration may be advantageous. The careful reader should be able to satisfy himself not only, in detail, *why* some experiments have a complex structure, but also *how* a complex observational record may be handled with intelligibility and precision.

The subject is a new one, and in many ways the most that the author can hope is to suggest possible lines of attack on the problems with which others are confronted. Progress in recent years has been rapid, and the few sections devoted to the subject in the author's *Statistical Methods for Research Workers*, first published in 1925, have, with each succeeding edition, come to appear more and more inadequate. On purely statistical questions the reader must be referred to that book; on logic, and the analysis of meaning, to *Statistical Methods and Scientific Inference*. The present volume is an attempt to do more thorough justice to the problems of planning and foresight with which the experimenter is confronted.

REFERENCES AND OTHER READING

T. BAYES (1763). An essay towards solving a problem in the doctrine of chances. Philosophical Transactions of the Royal Society, liii. 370.

A. DE MORGAN (1838). An essay on probabilities and on their application to life contingencies and insurance offices. Preface, vi. Longman & Co.

R. A. FISHER (1930). Inverse probability. Proceedings of the Cambridge Philosophical Society, xxvi. 528-535.

R. A. FISHER (1932). Inverse probability and the use of likelihood. Proceedings of the Cambridge Philosophical Society, xxviii. 257-261.

R. A. FISHER (1935). The logic of inductive inference. Journal Royal Statistical Society, xcviii. 39-54.

R. A. FISHER (1936). Uncertain inference. Proceedings of the American Academy of Arts and Sciences, 71. 245-258.

R. A. FISHER (1925-1963). Statistical methods for research workers. Oliver and Boyd Ltd., Edinburgh.

R. A. FISHER (1956, 1959) Statistical methods and scientific inference. Oliver and Boyd Ltd., Edinburgh.

II

THE PRINCIPLES OF EXPERIMENTATION, ILLUSTRATED BY A PSYCHO-PHYSICAL EXPERIMENT

5. Statement of Experiment

A LADY declares that by tasting a cup of tea made with milk she can discriminate whether the milk or the tea infusion was first added to the cup. We will consider the problem of designing an experiment by means of which this assertion can be tested. For this purpose let us first lay down a simple form of experiment with a view to studying its limitations and its characteristics, both those which appear to be essential to the experimental method, when well developed, and those which are not essential but auxiliary.

Our experiment consists in mixing eight cups of tea, four in one way and four in the other, and presenting them to the subject for judgment in a random order. The subject has been told in advance of what the test will consist, namely that she will be asked to taste eight cups, that these shall be four of each kind, and that they shall be presented to her in a random order, that is in an order not determined arbitrarily by human choice, but by the actual manipulation of the physical apparatus used in games of chance, cards, dice, roulettes, etc., or, more expeditiously, from a published collection of random sampling numbers purporting to give the actual results of such manipulation. Her task is to divide the 8 cups into two sets of 4, agreeing, if possible, with the treatments received.

6. Interpretation and its Reasoned Basis

In considering the appropriateness of any proposed experimental design, it is always needful to forecast all possible results of the experiment, and to have decided without ambiguity what interpretation shall be placed upon each one of them. Further, we must know by what argument this interpretation is to be sustained. In the present instance we may argue as follows. There are 70 ways of choosing a group of 4 objects out of 8. This may be demonstrated by an argument familiar to students of " permutations and combinations," namely, that if we were to choose the 4 objects in succession we should have successively 8, 7, 6, 5 objects to choose from, and could make our succession of choices in $8 \times 7 \times 6 \times 5$, or 1680 ways. But in doing this we have not only chosen every possible set of 4, but every possible set in every possible order; and since 4 objects can be arranged in order in $4 \times 3 \times 2 \times 1$, or 24 ways, we may find the number of possible choices by dividing 1680 by 24. The result, 70, is essential to our interpretation of the experiment. At best the subject can judge rightly with every cup and, knowing that 4 are of each kind, this amounts to choosing, out of the 70 sets of 4 which might be chosen, that particular one which is correct. A subject without any faculty of discrimination would in fact divide the 8 cups correctly into two sets of 4 in one trial out of 70, or, more properly, with a frequency which would approach 1 in 70 more and more nearly the more often the test were repeated. Evidently this frequency, with which unfailing success would be achieved by a person lacking altogether the faculty under test, is calculable from the number of cups used. The odds could be made much higher by enlarging the experiment, while if the experiment were much smaller even the greatest possible success would give odds so low that the result might, with considerable probability, be ascribed to chance.

7. The Test of Significance

It is open to the experimenter to be more or less exacting in respect of the smallness of the probability he would require before he would be willing to admit that his observations have demonstrated a positive result. It is obvious that an experiment would be useless of which no possible result would satisfy him. Thus, if he wishes to ignore results having probabilities as high as 1 in 20—the probabilities being of course reckoned from the hypothesis that the phenomenon to be demonstrated is in fact absent—then it would be useless for him to experiment with only 3 cups of tea of each kind. For 3 objects can be chosen out of 6 in only 20 ways, and therefore complete success in the test would be achieved without sensory discrimination, *i.e.* by " pure chance," in an average of 5 trials out of 100. It is usual and convenient for experimenters to take 5 per cent. as a standard level of significance, in the sense that they are prepared to ignore all results which fail to reach this standard, and, by this means, to eliminate from further discussion the greater part of the fluctuations which chance causes have introduced into their experimental results. No such selection can eliminate the whole of the possible effects of chance coincidence, and if we accept this convenient convention, and agree that an event which would occur by chance only once in 70 trials is decidedly " significant," in the statistical sense, we thereby admit that no isolated experiment, however significant in itself, can suffice for the experimental demonstration of any natural phenomenon ; for the " one chance in a million " will undoubtedly occur, with no less and no more than its appropriate frequency, however surprised we may be that it should occur to *us*. In order to assert that a natural phenomenon is experimentally demonstrable we need, not an isolated record, but a reliable method of procedure. In relation to the test of significance, we may say that a phenomenon is experimentally demonstrable when we know how to conduct an experiment which will rarely fail to give us a statistically significant result.

Returning to the possible results of the psychophysical experiment, having decided that if every cup were rightly classified a significant positive result would be recorded, or, in other words, that we should admit that the lady had made good her claim, what should be our conclusion if, for each kind of cup, her judgments are 3 right and 1 wrong? We may take it, in the present discussion, that any error in one set of judgments will be compensated by an error in the other, since it is known to the subject that there are 4 cups of each kind. In enumerating the number of ways of choosing 4 things out of 8, such that 3 are right and 1 wrong, we may note that the 3 right may be chosen, out of the 4 available, in 4 ways and, independently of this choice, that the 1 wrong may be chosen, out of the 4 available, also in 4 ways. So that in all we could make a selection of the kind supposed in 16 different ways. A similar argument shows that, in each kind of judgment, 2 may be right and 2 wrong in 36 ways, 1 right and 3 wrong in 16 ways and none right and 4 wrong in 1 way only. It should be noted that the frequencies of these five possible results of the experiment make up together, as it is obvious they should, the 70 cases out of 70.

It is obvious, too, that 3 successes to 1 failure, although showing a bias, or deviation, in the right direction, could not be judged as statistically significant evidence of a real sensory discrimination. For its frequency of chance occurrence is 16 in 70, or more than 20 per cent. Moreover, it is not the best possible result, and in judging of its significance we must take account not only of its own frequency, but also of the frequency of any better result. In the present instance "3 right and 1 wrong" occurs 16 times, and "4 right" occurs once in 70 trials, making 17 cases out of 70 as good as or better than that observed. The reason for including cases better than that observed becomes obvious on considering what our conclusions would have been had the case of 3 right and 1 wrong only 1 chance, and the case of 4 right 16 chances of occurrence out of 70. The rare case of 3 right and 1 wrong could not be judged significant merely because it was rare, seeing that a higher degree of success would frequently have been scored by mere chance.

8. The Null Hypothesis

Our examination of the possible results of the experiment has therefore led us to a statistical test of significance, by which these results are divided into two classes with opposed interpretations. Tests of significance are of many different kinds, which need not be considered here. Here we are only concerned with the fact that the easy calculation in permutations which we encountered, and which gave us our test of significance, stands for something present in every possible experimental arrangement; or, at least, for something required in its interpretation. The two classes of results which are distinguished by our test of significance are, on the one hand, those which show a significant discrepancy from a certain hypothesis; namely, in this case, the hypothesis that the judgments given are in no way influenced by the order in which the ingredients have been added; and on the other hand, results which show no significant discrepancy from this hypothesis. This hypothesis, which may or may not be impugned by the result of an experiment, is again characteristic of all experimentation. Much confusion would often be avoided if it were explicitly formulated when the experiment is designed. In relation to any experiment we may speak of this hypothesis as the "null hypothesis," and it should be noted that the null hypothesis is never proved or established, but is possibly disproved, in the course of experimentation. Every experiment may be said to exist only in order to give the facts a chance of disproving the null hypothesis.

It might be argued that if an experiment can disprove the hypothesis that the subject possesses no sensory discrimination between two different sorts of object, it must therefore be able to prove the opposite hypothesis, that she can make some such discrimination. But this last hypothesis, however reasonable or true it may be, is ineligible as a null hypothesis to be tested by experiment, because it is inexact. If it were asserted that the subject would never be wrong in her judgments we should again have an exact hypothesis, and it is easy to see that this hypothesis could be disproved by a single failure, but could never be proved by any finite amount of experimentation. It is evident that the null hypothesis must be exact, that is free from vagueness and ambiguity, because it must supply the basis of the "problem of distribution," of which the test of significance is the solution. A null hypothesis may, indeed, contain arbitrary elements, and in more complicated cases often does so: as, for example, if it should assert that the death-rates of two groups of animals are equal, without specifying what these death-rates actually are. In such cases it is evidently the equality rather than any particular values of the death-rates that the experiment is designed to test, and possibly to disprove.

In cases involving statistical "estimation" these ideas may be extended to the simultaneous consideration of a series of hypothetical possibilities. The notion of an error of the so-called "second kind," due to accepting the null hypothesis "when it is false" may then be given a meaning in reference to the quantity to be estimated. It has no meaning with respect to simple tests of significance, in which the only available expectations are those which flow from the null hypothesis being true. Problems of the more elaborate type involving estimation are discussed in Chapter IX.

9. Randomisation; the Physical Basis of the Validity of the Test

We have spoken of the experiment as testing a certain null hypothesis, namely, in this case, that the subject possesses no sensory discrimination whatever of the kind claimed; we have, too, assigned as appropriate to this hypothesis a certain frequency distribution of occurrences, based on the equal frequency of the 70 possible ways of assigning 8 objects to two classes of 4 each; in other words, the frequency distribution appropriate to a classification by pure chance. We have now to examine the physical conditions of the experimental technique needed to justify the assumption that, if discrimination of the kind under test is absent, the result of the experiment will be wholly governed by the laws of chance. It is easy to see that it might well be otherwise. If all those cups made with the milk first had sugar added, while those made with the tea first had none, a very obvious difference in flavour would have been introduced which might well ensure that all those made with sugar should be classed alike. These groups might either be classified all right or all wrong, but in such a case the frequency of the critical event in which all cups are classified correctly would not be 1 in 70, but 35 in 70 trials, and the test of significance would be wholly vitiated. Errors equivalent in principle to this are very frequently incorporated in otherwise well-designed experiments.

It is no sufficient remedy to insist that " all the cups must be exactly alike " in every respect except that to be tested. For this is a totally impossible requirement in our example, and equally in all other forms of experimentation. In practice it is probable that the cups will differ perceptibly in the thickness or smoothness of their material, that the quantities of milk added to the different cups will not be exactly equal, that the strength of the infusion of tea may change between pouring the first and the last cup, and that the temperature also at which the tea is tasted will change during the course of the experiment. These are only examples of the differences probably present; it would be impossible to present an exhaustive list of such possible differences appropriate to any one kind of experiment, because the uncontrolled causes which may influence the result are always strictly innumerable. When any such cause is named, it is usually perceived that, by increased labour and expense, it could be largely eliminated. Too frequently it is assumed that such refinements constitute improvements to the experiment. Our view, which will be much more fully exemplified in later sections, is that it is an essential characteristic of experimentation that it is carried out with limited resources, and an essential part of the subject of experimental design to ascertain how these should be best applied; or, in particular, to which causes of disturbance care should be given, and which *ought* to be deliberately ignored. To ascertain, too, for those which are not to be ignored, to what *extent* it is worth while to take the trouble to diminish their magnitude. For our present purpose, however, it is only necessary to recognise that, whatever degree of care and experimental skill is expended in equalising the conditions, other than the one under test, which are liable to affect the result, this equalisation must always be to a greater or less extent incomplete, and in many important practical cases will certainly be grossly defective. We are concerned, therefore, that this inequality, whether it be great or small, shall not impugn the exactitude of the frequency distribution, on the basis of which the result of the experiment is to be appraised.

10. The Effectiveness of Randomisation

The element in the experimental procedure which contains the essential safeguard is that the two modifications of the test beverage are to be prepared " in random order." This, in fact, is the only point in the experimental procedure in which the laws of chance, which are to be in exclusive control of our frequency distribution, have been explicitly introduced. The phrase "random order" itself, however, must be regarded as an incomplete instruction, standing as a kind of shorthand symbol for the full procedure of randomisation, by which the validity of the test of significance may be guaranteed against corruption by the causes of disturbance which have not been eliminated. To demonstrate that, with satisfactory randomisation, its validity is, indeed, wholly unimpaired, let us imagine all causes of disturbance—the strength of the infusion, the quantity of milk, the temperature at which it is tasted, etc.—to be predetermined for each cup; then since these, on the null hypothesis, are the only causes influencing classification, we may say that the probabilities of each of the 70 possible choices or classifications which the subject can make are also predetermined. If, now, after the disturbing causes are fixed, we assign, strictly at random, 4 out of the 8 cups to each of our experimental treatments, then every set of 4, whatever its probability of being so classified, will certainly have a probability of exactly 1 in 70 of *being* the 4, for example, to which the milk is added first. However important the causes of disturbance may be, even if they were to make it certain that one particular set of 4 should receive this classification, the probability that the 4 so classified and the 4 which ought to have been so classified should be the same, must be rigorously in accordance with our test of significance.

It is apparent, therefore, that the random choice of the objects to be treated in different ways would be a complete guarantee of the validity of the test of significance, if these treatments were the last in time of the stages in the physical history of the objects which might affect their experimental reaction. The circumstance that the experimental treatments cannot always be applied last, and may come relatively early in their history, causes no practical inconvenience; for subsequent causes of differentiation, if under the experimenter's control, as, for example, the choice of different pipettes to be used with different flasks, can either be predetermined before the treatments have been randomised, or, if this has not been done, can be randomised on their own account; and other causes of differentiation will be either (*a*) consequences of differences already randomised, or (*b*) natural consequences of the difference in treatment to be tested, of which on the null hypothesis there will be none, by definition, or (*c*) effects supervening by chance independently from the treatments applied. Apart, therefore, from the avoidable error of the experimenter himself introducing with his test treatments, or subsequently, other differences in treatment, the effects of which the experiment is not intended to study, it may be said that the simple precaution of randomisation will suffice to guarantee the validity of the test of significance, by which the result of the experiment is to be judged.

11. The Sensitiveness of an Experiment. Effects of Enlargement and Repetition

A probable objection, which the subject might well make to the experiment so far described, is that only if every cup is classified correctly will she be judged successful. A single mistake will reduce her performance below the level of significance. Her claim, however, might be, not that she could draw the distinction with invariable certainty, but that, though sometimes mistaken, she would be right more often than not; and that the experiment should be enlarged sufficiently, or repeated sufficiently often, for her to be able to demonstrate the predominance of correct classifications in spite of occasional errors.

An extension of the calculation upon which the test of significance was based shows that an experiment with 12 cups, six of each kind, gives, on the null hypothesis, 1 chance in 924 for complete success, and 36 chances for 5 of each kind classified right and 1 wrong. As 37 is less than a twentieth of 924, such a test could be counted as significant, although a pair of cups have been wrongly classified; and it is easy to verify that, using larger numbers still, a significant result could be obtained with a still higher proportion of errors. By increasing the size of the experiment, we can render it more sensitive, meaning by this that it will allow of the detection of a lower degree of sensory discrimination, or, in other words, of a quantitatively smaller departure from the null hypothesis. Since in every case the experiment is capable of disproving, but never of proving this hypothesis, we may say that the value of the experiment is increased whenever it permits the null hypothesis to be more readily disproved.

The same result could be achieved by repeating the experiment, as originally designed, upon a number of different occasions, counting as a success all those occasions on which 8 cups are correctly classified. The chance of success on each occasion being 1 in 70, a simple application of the theory of probability shows that 2 or more successes in 10 trials would occur, by chance, with a frequency below the standard chosen for testing significance; so that the sensory discrimination would be demonstrated, although, in 8 attempts out of 10, the subject made one or more mistakes. This procedure may be regarded as merely a second way of enlarging the experiment and, thereby, increasing its sensitiveness, since in our final calculation we take account of the aggregate of the entire series of results, whether successful or unsuccessful. It would clearly be illegitimate, and would rob our calculation of its basis, if the unsuccessful results were not all brought into the account.

12. Qualitative Methods of increasing Sensitiveness

Instead of enlarging the experiment we may attempt to increase its sensitiveness by qualitative improvements; and these are, generally speaking, of two kinds: (*a*) the reorganisation of its structure, and (*b*) refinements of technique. To illustrate a change of structure we might consider that, instead of fixing in advance that 4 cups should be of each kind, determining by a random process how the subdivision should be effected, we might have allowed the treatment of each cup to be determined independently by chance, as by the toss of a coin, so that each treatment has an equal chance of being chosen. The chance of classifying correctly 8 cups randomised in this way, without the aid of sensory discrimination, is 1 in 2^8, or 1 in 256 chances, and there are only 8 chances of classifying 7 right and 1 wrong; consequently the sensitiveness of the experiment has been increased, while still using only 8 cups, and it is possible to score a significant success, even if one is classified wrongly. In many types of experiment, therefore, the suggested change in structure would be evidently advantageous. For the special requirements of a psycho-physical experiment, however, we should probably prefer to forego this advantage, since it would occasionally occur that all the cups would be treated alike, and this, besides bewildering the subject by an unexpected occurrence, would deny her the real advantage of judging by comparison.

Another possible alteration to the structure of the experiment, which would, however, decrease its sensitiveness, would be to present determined, but unequal, numbers of the two treatments. Thus we might arrange that 5 cups should be of the one kind and 3 of the other, choosing them properly by chance, and informing the subject how many of each to expect. But since the number of ways of choosing 3 things out of 8 is only 56, there is now, on the null hypothesis, a probability of a completely correct classification of 1 in 56. It appears in fact that we cannot by these means do better than by presenting the two treatments in equal numbers, and the choice of this equality is now seen to be justified by its giving to the experiment its maximal sensitiveness.

With respect to the refinements of technique, we have seen above that these contribute nothing to the validity of the experiment, and of the test of significance by which we determine its result. They may, however, be important, and even essential, in permitting the phenomenon under test to manifest itself. Though the test of significance remains valid, it may be that without special precautions even a definite sensory discrimination would have little chance of scoring a significant success. If some cups were made with India and some with China tea, even though the treatments were properly randomised, the subject might not be able to discriminate the relatively small difference in flavour under investigation, when it was confused with the greater differences between leaves of different origin. Obviously, a similar difficulty could be introduced by using in some cups raw milk and in others boiled, or even condensed milk, or by adding sugar in unequal quantities. The subject has a right to claim, and it is in the interests of the sensitiveness of the experiment, that gross differences of these kinds should be excluded, and that the cups should, not as far as *possible*, but as far as is practically convenient, be made alike in all respects except that under test.

How far such experimental refinements should be carried is entirely a matter of judgment, based on experience. The validity of the experiment is not affected by them. Their sole purpose is to increase its sensitiveness, and this object can usually be achieved in many other ways, and particularly by increasing the size of the experiment. If, therefore, it is decided that the sensitiveness of the experiment should be increased, the experimenter has the choice between different methods of obtaining equivalent results; and will be wise to choose whichever method is easiest to him, irrespective of the fact that previous experimenters may have tried, and recommended as very important, or even essential, various ingenious and troublesome precautions.

12·1. Scientific Inference and Acceptance Procedures

In "The Improvement of Natural Knowledge", that is, in learning by experience, or by planned chains of experimentation, conclusions are always provisional and in the nature of progress reports, interpreting and embodying the evidence so far accrued. Convenient as it is to note that a hypothesis is contradicted at some familiar level of significance such as 5% or 2% or 1% we do not, in Inductive Inference, ever need to lose sight of the exact strength which the evidence has in fact reached, or to ignore the fact that with further trial it might come to be stronger, or weaker. The situation is entirely different in the field of Acceptance Procedures, in which irreversible action may have to be taken, and in which, whichever decision is arrived at, it is quite immaterial whether it is arrived at on strong evidence or on weak. All that is needed is a Rule of Action which is to be taken automatically, and without thought devoted to the individual decision. The procedure as a whole is arrived at by minimising the losses due to wrong decisions, or to unnecessary testing, and to frame such a procedure successfully the cost of such faulty decisions must be assessed in advance; equally, also, prior knowledge is required of the expected distribution of the material in supply. In the field of pure research no assessment of the cost of wrong conclusions, or of delay in arriving at more correct conclusions can conceivably be more than a pretence, and in any case such an assessment would be inadmissible and irrelevant in judging the state of the scientific evidence; moreover, accurately assessable prior information is ordinarily known to be lacking. Such differences between the logical situations should be borne in mind whenever we see tests of significance spoken of as " Rules of Action ". A good deal of confusion has certainly been caused by the attempt to formalise the exposition of tests of significance in a logical framework different from that for which they were in fact first developed.

REFERENCES AND OTHER READING

R. A. Fisher (1925-1963). Statistical methods for research workers. Chap. III., §§ 15-19
R. A. Fisher (1926). The arrangement of field experiments. Journal of Ministry of Agriculture, xxxiii. 503-513.

III

A HISTORICAL EXPERIMENT ON GROWTH RATE

13. WE have illustrated a psycho-physical experiment, the result of which depends upon judgments, scored " right " or " wrong," and may be appropriately interpreted by the method of the classical theory of probability. This method rests on the enumeration of the frequencies with which different combinations of right or wrong judgments will occur, on the hypothesis to be tested. We may now illustrate an experiment in which the results are expressed in quantitative measures, and which is appropriately interpreted by means of the theory of errors.

In the introductory remarks to his book on " The effects of cross and self-fertilisation in the vegetable kingdom," Charles Darwin gives an account of the considerations which guided him in the design of his experiments and in the presentation of his data, which will serve well to illustrate the principles on which biological experiments may be made conclusive. The passage is of especial interest in illustrating the extremely crude and unsatisfactory statistical methods available at the time, and the manner in which careful attention to commonsense considerations led to the adoption of an experimental design, in itself greatly superior to these methods of interpretation.

14. Darwin's Discussion of the Data

" I long doubted whether it was worth while to give the measurements of each separate plant, but have decided to do so, in order that it may be seen that the superiority of the crossed plants over the self-fertilised does not commonly depend on the presence of two or three extra fine plants on the one side, or of a few very poor plants on the other side. Although several observers have insisted in general terms on the offspring from intercrossed varieties being superior to either parent-form, no precise measurements have been given; and I have met with no observations on the effects of crossing and self-fertilising the individuals of the same variety. Moreover, experiments of this kind require so much time—mine having been continued during eleven years—that they are not likely soon to be repeated.

"As only a moderate number of crossed and self-fertilised plants were measured, it was of great importance to me to learn how far the averages were trustworthy. I therefore asked Mr Galton, who has had much experience in statistical researches, to examine some of my tables of measurements, seven in number, namely those of *Ipomœa, Digitalis, Reseda lutea, Viola, Limnanthes, Petunia,* and *Zea.* I may premise that if we took by chance a dozen or score of men belonging to two nations and measured them, it would I presume be very rash to form any judgment from such small numbers on their average heights. But the case is somewhat different with my crossed and self-fertilised plants, as they were of exactly the same age, were subjected from first to last to the same conditions, and were descended from the same parents. When only from two to six pairs of plants were measured, the results are manifestly of little or no value, except in so far as they confirm and are confirmed by experiments made on a larger scale with other species. I will now give the report on the seven tables of measurements,

15. Galton's Method of Interpretation

which Mr Galton has had the great kindness to draw up for me."

"I have examined the measurements of the plants with care, and by many statistical methods, to find out how far the means of the several sets represent constant realities, such as would come out the same so long as the general conditions of growth remained unaltered. The principal methods that were adopted are easily explained by selecting one of the shorter series of plants, say of *Zea mays*, for an example.

"The observations as I received them are shown in columns II. and III., where they certainly have no *primâ facie* appearance of regularity. But as soon as we arrange them in the order of their magnitudes, as in columns IV. and V., the case is materially altered. We now see, with few exceptions, that the largest plant on the crossed side in each pot exceeds the largest plant on the self-fertilised side, that the second exceeds the second, the third the third, and so on. Out of the fifteen cases in the table, there are only two exceptions to this rule.* We may therefore confidently affirm that a crossed series will always be found to exceed a self-fertilised series, within the range of the conditions under which the present experiment has been made.

"Next as regards the numerical estimate of this excess. The mean values of the several groups are so discordant, as is shown in the table just given, that a fairly precise numerical estimate seems impossible. But the consideration arises, whether the difference between pot and pot may not be of much the same order of importance as that of the other conditions upon which the growth of the plants has been modified. If so, and only on that condition, it would follow that when all the measurements, either of the crossed or the self-fertilised plants, were combined into a single series, that series would be statistically regular. The experiment is tried in columns VII. and VIII., where the regularity is abundantly clear, and justifies us in considering its mean as perfectly reliable

* Galton evidently did not notice that this is true also before rearrangement.

TABLE 1
Zea mays (young plants)

Column I.	As recorded by Mr Darwin.		Arranged in Order of Magnitude.				
			In Separate Pots.		In a Single Series.		
	II.	III.	IV.	V.	VI.	VII.	VIII.
	Crossed.	Self-fert.	Crossed.	Self-fert.	Crossed.	Self-fert.	Difference.
	Inches.	Inches.	Inches.	Inches.	Inches.	Inches.	Inches.
Pot I. . .	23 4/8	17 3/8	23 4/8	20 3/8	23 4/8	20 3/8	−3 1/8
	12	20 3/8	21	20	23 2/8	20	−3 2/8
	21	20	12	17 3/8	23	20	−3
Pot II. . .	22	20	22	20	22 1/8	18 5/8	−3 4/8
	19 1/8	18 3/8	21 4/8	18 5/8	22 1/8	18 5/8	−3 4/8
	21 4/8	18 5/8	19 1/8	18 3/8	22	18 3/8	−3 5/8
Pot III. . .	22 1/8	18 5/8	23 2/8	18 5/8	21 5/8	18	−3 5/8
	20 3/8	15 1/8	22 1/8	18	21 4/8	18	−3 4/8
	18 2/8	16 4/8	21 5/8	16 4/8	21	18	−3
	21 5/8	18	20 3/8	16 2/8	21	17 3/8	−3
	23 2/8	16 2/8	18 2/8	15 1/8	20 3/8	16 4/8	−3 7/8
Pot IV. . .	21	18	23	18	19 1/8	16 2/8	−2 7/8
	22 1/8	12 6/8	22 1/8	18	18 2/8	15 4/8	−2 6/8
	23	15 4/8	21	15 4/8	12	15 1/8	+3 2/8
	12	18	12	12 6/8	12	12 6/8	+0 6/8

GALTON'S METHOD OF INTERPRETATION

I have protracted these measurements, and revised them in the usual way, by drawing a curve through them with a free hand, but the revision barely modifies the means derived from the original observations. In the present, and in nearly all the other cases, the difference between the original and revised means is under 2 per cent. of their value. It is a very remarkable coincidence that in the seven kinds of plants, whose measurements I have examined, the ratio between the heights of the crossed and of the self-fertilised ranges in five cases within very narrow limits. In *Zea mays* it is as 100 to 84, and in the others it ranges between 100 to 76 and 100 to 86.

TABLE 2

Pot.	Crossed.	Self-fert.	Difference.
I.	18 7/8	19 3/8	+0 4/8
II.	20 7/8	19	−1 7/8
III.	21 1/8	16 7/8	−4 2/8
IV.	19 5/8	16	−3 5/8

"The determination of the variability (measured by what is technically called the 'probable error') is a problem of more delicacy than that of determining the means, and I doubt, after making many trials, whether it is possible to derive useful conclusions from these few observations. We ought to have measurements of at least fifty plants in each case, in order to be in a position to deduce fair results. . . ."

"Mr Galton sent me at the same time graphical representations which he had made of the measurements, and they evidently form fairly regular curves. He appends the words 'very good' to those of *Zea* and *Limnanthes*. He also calculated the average height of the crossed and self-fertilised plants in the seven tables by a more correct method than that followed by me, namely by including the heights, as estimated in accordance with statistical rules, of a few plants which died before they were measured; whereas I merely added up the heights of the survivors, and divided the sum by their number. The difference in our results is in one way highly satisfactory, for the average heights of the self-fertilised plants, as deduced by Mr Galton, is less than mine in all the cases excepting one, in which our averages are the same; and this shows that I have by no means exaggerated the superiority of the crossed over the self-fertilised plants."

16. Pairing and Grouping

It is seen that the method of comparison adopted by Darwin is that of pitting each self-fertilised plant against a cross-fertilised one, in conditions made as equal as possible. The pairs so chosen for comparison had germinated at the same time, and the soil conditions in which they grew were largely equalised by planting in the same pot. Necessarily they were not of the same parentage, as it would be difficult in maize to self-fertilise two plants at the same time as raising a cross-fertilised progeny from the pair. However, the parents were presumably grown from the same batch of seed. The evident object of these precautions is to increase the sensitiveness of the experiment, by making such differences in growth rate as were to be observed as little as possible dependent from environmental circumstances, and as much as possible, therefore, from intrinsic differences due to their mode of origin.

The method of pairing, which is much used in modern biological work, illustrates well the way in which an appropriate experimental design is able to reconcile two desiderata, which sometimes appear to be in conflict. On the one hand we require the utmost uniformity in the biological material, which is the subject of experiment, in order to increase the sensitiveness of each individual observation; and, on the other, we require to multiply the observations so as to demonstrate so far as possible the reliability and consistency of the results. Thus an experimenter with field crops may desire to replicate his experiments upon a large number of plots, but be deterred by the consideration that his facilities allow him to sow only a limited area on the same day. An experimenter with small mammals may have only a limited supply of an inbred and highly uniform stock, which he believes to be particularly desirable for experimental purposes. Or, he may desire to carry out his experiments on members of the same litter, and feel that his experiment is limited by the size of the largest litter he can obtain. It has indeed frequently been argued that, beyond a certain moderate degree, further replication can give no further increase in precision, owing to the increasing heterogeneity with which, it is thought, it must be accompanied. In all these cases, however, and in the many analogous cases which constantly arise, there is no real dilemma. Uniformity is only requisite between the objects whose response is to be contrasted (that is, objects treated differently). It is not requisite that all the parallel plots under the same treatment shall be sown on the same day, but only that each such plot shall be sown so far as possible simultaneously with the differently treated plot or plots with which it is to be compared. If, therefore, only two kinds of treatments are under examination, pairs of plots may be chosen, one plot for each treatment; and the precision of the experiment will be given its highest value if the members of each pair are treated closely alike, but will gain nothing from similarity of treatment applied to different pairs, nor lose anything if the conditions in these are somewhat varied. In the same way, if the numbers of animals available from any inbred line are too few for adequate replication, the experimental contrasts in treatments may be applied to pairs of animals from different inbred lines, so long as each pair belongs to the same line. In these two cases it is evident that the principle of combining similarity between controls to be compared, with diversity between parallels, may be extended to cases where three or more treatments are under investigation. The requirement that animals to be contrasted must come from the same litter limits, not the amount of replication, but the number of different treatments that can be so tested. Thus we might test three, but not so easily four or five treatments, if it were necessary that each set of animals must be of the same sex and litter. Paucity of homogeneous material limits the number of different treatments in an experiment, not the number of replications. It may cramp the scope and comprehensiveness of an experimental enquiry, but sets no limit to its possible precision.

17. " Student's " t Test [*]

Owing to the historical accident that the theory of errors, by which quantitative data are to be interpreted, was developed without reference to experimental methods, the vital principle has often been overlooked that the actual and physical conduct of an experiment must govern the statistical procedure of its interpretation. In using the theory of errors we rely for our conclusion upon one or more estimates of error, derived from the data, and appropriate to the one or more sets

[*] A full account of this test in more varied applications, and the tables for its use, will be found in *Statistical Methods for Research Workers*. Its originator, who published anonymously under the pseudonym " Student," possesses the remarkable distinction that, without being a professed mathematician, but a research chemist, he made early in life this revolutionary refinement of the classical theory of errors.

of comparisons which we wish to make. Whether these estimates are valid, for the purpose for which we intend them, depends on what has been actually done. It is possible, and indeed it is all too frequent, for an experiment to be so conducted that no valid estimate of error is available. In such a case the experiment cannot be said, strictly, to be capable of proving anything. Perhaps it should not, in this case, be called an *experiment* at all, but be added merely to the body of *experience* on which, for lack of anything better, we may have to base our opinions. All that we need to emphasise immediately is that, if an experiment does allow us to calculate a valid estimate of error, its structure must completely determine the statistical procedure by which this estimate is to be calculated. If this were not so, no interpretation of the data could ever be unambiguous; for we could never be sure that some other equally valid method of interpretation would not lead to a different result.

The object of the experiment is to determine whether the difference in origin between inbred and cross-bred plants influences their growth rate, as measured by height at a given date; in other words, if the numbers of the two sorts of plants were to be increased indefinitely, our object is to determine whether the average heights, to which these two aggregates of plants will tend, are equal or unequal. The most general statement of our null hypothesis is, therefore, that the limits to which these two averages tend are equal. The theory of errors enables us to test a somewhat more limited hypothesis, which, by wide experience, has been found to be appropriate to the metrical characters of experimental material in biology. The disturbing causes which introduce discrepancies in the means of measurements of similar material are found to produce quantitative effects which conform satisfactorily to a theoretical distribution known as the normal law of frequency of error. It is this circumstance that makes it appropriate to choose, as the null hypothesis to be tested, one for which an exact statistical criterion is available, namely that the two groups of measurements are samples drawn from the same normal population. On the basis of this hypothesis we may proceed to compare the average difference in height, between the cross-fertilised and the self-fertilised plants, with such differences as might be expected between these averages, in view of the observed discrepancies between the heights of plants of like origin.

We must now see how the adoption of the method of pairing determines the details of the arithmetical procedure, so as to lead to an unequivocal interpretation. The pairing procedure, as indeed was its purpose, has equalised any differences in soil conditions, illumination, air-currents, etc., in which the several pairs of individuals may differ. Such differences having been eliminated from the experimental comparisons, and contributing nothing to the real errors of our experiment, must, for this reason, be eliminated likewise from our estimate of error, upon which we are to judge what differences between the means are compatible with the null hypothesis, and what differences are so great as to be incompatible with it. We are therefore not concerned with the differences in height among plants of like origin, but only with differences in height between members of the same pair, and with the discrepancies among these differences observed in different pairs. Our first step, therefore, will be to subtract from the height of each cross-fertilised plant the height of the self-fertilised plant belonging to the same pair. The differences are shown below in eighths of an inch.

With respect to these differences our null hypothesis asserts that they are normally distributed about a mean value at zero, and we have to test whether our 15 observed differences are compatible with the supposition that they are a sample from such a population.

TABLE 3

Differences in eighths of an inch between cross- and self-fertilised plants of the same pair

49	23	56
−67	28	24
8	41	75
16	14	60
6	29	−48

The calculations needed to make a rigorous test of the null hypothesis stated above involve no more than the sum, and the sum of the squares, of these numbers. The sum is 314, and, since there are 15 plants, the mean difference is $20\frac{14}{15}$ in favour of the cross-fertilised plants. The sum of the squares is 26,518, and from this is deducted the product of the total and the mean, or 6573, leaving 19,945 for the sum of squares of deviations from the mean, representing discrepancies among the differences observed in the 15 pairs. The algebraic fact here used is that

$$S(x-\bar{x})^2 = S(x^2) - \bar{x}S(x)$$

where S stands for summation over the sample, and $\bar{x}$ for the mean value of the observed differences, x.

We may make from this measure of the discrepancies an estimate of a quantity known as the *variance* of an individual difference, by dividing by 14, one less than the number of pairs observed. Equally, and what is more immediately required, we may make an estimate of the variance of the mean of 15 such pairs, by dividing again by 15, a process which yields 94·976 as the estimate.

The square root of the variance is known as the standard error, and it is by the ratio which our observed mean difference bears to its standard error that we shall judge of its significance. Dividing our difference, 20·933, by *its* standard error 9·746, we find this ratio (which is usually denoted by t) to be 2·148.

The object of these calculations has been to obtain from the data a quantity measuring the average difference in height between the cross-fertilised and the self-fertilised plants, in terms of the observed discrepancies among these differences; and which, moreover, shall be distributed in a known manner when the null hypothesis is true. The mathematical distribution for our present problem was discovered by "Student" in 1908, and depends only upon the number of independent comparisons (or the number of degrees of freedom) available for calculating the estimate of error. With 15 observed differences we have among them 14 independent discrepancies, and our degrees of freedom are 14. The available tables of the distribution of t show that for 14 degrees of freedom the value 2·145 is exceeded by chance, either in the positive or negative direction, in exactly 5 per cent. of random trials. The observed value of t, 2·148, thus just exceeds the 5 per cent. point, and the experimental result may be judged significant, though barely so.

18. Fallacious Use of Statistics

We may now see that Darwin's judgment was perfectly sound, in judging that it was of importance to learn how far the averages were trustworthy, and that this could be done by a statistical examination of the tables of measurements of individual plants, though not of their averages. The example chosen, in fact, falls just on the border-line between those results which can suffice by themselves to establish the point at issue, and those which are of little value except in so far as they confirm or are confirmed by other experiments of a like nature. In particular, it is to be noted that Darwin recognised that the reliability of the result must be judged by the consistency of the superiority of the crossed plants over the self-fertilised, and not only on the difference of the averages, which might depend, as he says, on the presence of two or three extra-fine plants on the one side, or of a few very poor plants on the other side; and that therefore the presentation of the experimental evidence depended essentially on giving the measurements of each independent plant, and could not be assessed from the mere averages.

It may be noted also that Galton's scepticism of the value of the probable error, deduced from only 15 pairs of observations, though, as it turned out, somewhat excessive, was undoubtedly right in principle. The standard error (of which the probable error is only a conventional fraction) can only be estimated with considerable uncertainty from so small a sample, and, prior to "Student's" solution of the problem, it was by no means clear to what extent this uncertainty would invalidate the test of significance. From "Student's" work it is now known that the cause for anxiety was not so great as it might have seemed. Had the standard error been known with certainty, or derived from an effectively infinite number of observations, the 5 per cent. value of t would have been 1·960. When our estimate is based upon only 15 differences, the 5 per cent. value, as we have seen, is 2·145, or less than 10 per cent. greater. Even using the inexact theory available at the time, a calculation of the probable error would have provided a valuable guide to the interpretation of the results.

19. Manipulation of the Data

A much more serious fallacy appears to be involved in Galton's assumption that the value of the data, for the purpose for which they were intended, could be increased by rearranging the comparisons. Modern statisticians are familiar with the notions that any finite body of data contains only a limited amount of information, on any point under examination; that this limit is set by the nature of the data themselves, and cannot be increased by any amount of ingenuity expended in their statistical examination: that the statistician's task, in fact, is limited to the extraction of the whole of the available information on any particular issue. If the results of an experiment, as obtained, are in fact irregular, this evidently detracts from their value; and the statistician is not elucidating but falsifying the facts, who rearranges them so as to give an artificial appearance of regularity.

In rearranging the results of Darwin's experiment it appears that Galton thought that Darwin's experiment would be equivalent to one in which the heights of pairs of contrasted plants had been those given in his columns headed VI. and VII., and that the reliability of Darwin's average difference of about $2\frac{5}{8}$ inches could be fairly judged from the constancy of the 15 differences shown in column VIII.

How great an effect this procedure, if legitimate, would have had on the significance of the result, may be seen by treating these artificial differences as we have treated the actual differences given by Darwin. Applying the same arithmetical procedure as before, we now find t equals 5·171, a value which would be exceeded by chance only about once or twice in 10,000 trials, and is far beyond the level of significance ordinarily required. The falsification, inherent in this mode of procedure, will be appreciated if we consider that the tallest plant, of either the crossed or the self-fertilised series, will have become the tallest by reason of a number of favourable circumstances, including among them those which produce the discrepancies between those pairs of plants, which were actually grown together. By taking the difference between these two favoured plants we have largely eliminated real causes of error which have affected the value of our observed mean. We have, in doing this, grossly violated the principle that the estimate of error must be based on the effects of the very same causes of variation as have produced the real errors in our experiment. Through this fallacy Galton is led to speak of the mean as perfectly reliable, when, from its standard error, it appears that a repetition of the experiment would often give a mean quite 50 per cent. greater or less than that observed in this case.

20. Validity and Randomisation

Having decided that, when the structure of the experiment consists in a number of independent comparisons between pairs, our estimate of the error of the average difference must be based upon the discrepancies between the differences actually observed, we must next enquire what precautions are needed in the practical conduct of the experiment to guarantee that such an estimate shall be a valid one; that is to say that the very same causes that produce our real error shall also contribute the materials for computing an estimate of it. The logical necessity of this requirement is readily apparent, for, if causes of variation which do not influence our real error are allowed to affect our estimate of it, or equally, if causes of variation affect the real error in such a way as to make no contribution to our estimate, this estimate will be vitiated, and will be incapable of providing a correct statement as to the frequency with which our real error will exceed any assigned quantity; and such a statement of frequency is the sole purpose for which the estimate is of any use. Nevertheless, though its logical necessity is easily apprehended, the question of the validity of the estimates of error used in tests of significance was for long ignored, and is still often overlooked in practice. One reason for this is that standardised methods of statistical analysis have been taken over ready-made from a mathematical theory, into which questions of experimental detail do not explicitly enter. In consequence the assumptions which enter implicitly into the bases of the theory have not been brought prominently under the notice of practical experimenters. A second reason is that it has not until recently been recognised that any simple precaution would supply an absolute guarantee of the validity of the calculations.

In the experiment under consideration, apart from chance differences in the selection of seeds, the sole source of the experimental error in the average of our fifteen differences lies in the differences in soil fertility, illumination, evaporation, etc., which make the site of each crossed plant more or less favourable to growth than the site assigned to the corresponding self-fertilised plant. It is for this reason that every precaution, such as mixing the soil, equalising the watering and orienting the pot so as to give equal illumination, may be expected to increase the precision of the experiment. If, now, when the fifteen pairs of sites have been chosen, and in so doing all the differences in environmental circumstances, to which the members of the different pairs will be exposed during the course of the experiment, have been predetermined, we then assign at random, as by tossing a coin, which site shall be occupied by the crossed and which by the self-fertilised plant, we shall be assigning by the same act whether this particular ingredient of error shall appear in our average with a positive or a negative sign. Since each particular error has thus an equal and independent chance of being positive or negative, the error of our average will necessarily be distributed in a sampling distribution, centred at zero, which will be symmetrical in the sense that to each possible positive error there corresponds an equal negative error, which, as our procedure guarantees, will in fact occur with equal probability.

Our estimate of error is easily seen to depend only on the same fifteen ingredients, and the arithmetical processes of summation, subtraction and division may be designed, and have in fact been designed, so as to provide the estimate appropriate to the system of chances which our method of choosing sites had imposed on the data. This is to say much more than merely that the experiment is unbiased, for we might still call the experiment unbiased if the whole of the cross-fertilised plants had been assigned to the west side of the pots, and the self-fertilised plants to the east side, by a single toss of the coin. That this would be insufficient to ensure the validity of our estimate may be easily seen ; for it might well be that some unknown circumstance, such as the incidence of different illumination at different times of the day, or the desiccating action of the air-currents prevalent in the greenhouse, might systematically favour all the plants on one side over those on the other. The effect of any such prevailing cause would then be confounded with the advantage, real or apparent, of cross-breeding over inbreeding, and would be eliminated from our estimate of error, which is based solely on the discrepancies between the differences shown by different pairs of plants. Randomisation properly carried out, in which each pair of plants are assigned their positions independently at random, ensures that the estimates of error will take proper care of all such causes of different growth rates, and relieves the experimenter from the anxiety of considering and estimating the magnitude of the innumerable causes by which his data may be disturbed. The one flaw in Darwin's procedure was the absence of randomisation.

Had the same measurements been obtained from pairs of plants properly randomised the experiment would, as we have shown, have fallen on the verge of significance. Galton was led greatly to overestimate its conclusiveness through the major error of attempting to estimate the reliability of the comparisons by rearranging the two series in order of magnitude. His discussion shows, in other respects, an over-confidence in the power of statistical methods to remedy the irregularities of the actual data. In particular, the attempt mentioned by Darwin to improve on the simple averages of the two series " by a more correct method . . . by including the heights, as estimated in accordance with statistical rules, of a few plants which died before they were measured," seems to go far beyond the limits of justifiable inference, and is one of many indications that the logic of statistical induction was in its infancy, even at a time when the technique of accurate experimentation had already been notably advanced.

21. Test of a Wider Hypothesis

It has been mentioned that " Student's " t test, in conformity with the classical theory of errors, is appropriate to the null hypothesis that the two groups of measurements are samples drawn from the same normally distributed population. This is the type of null hypothesis which experimenters, rightly in the author's opinion, usually consider it appropriate to test, for reasons not only of practical convenience, but because the unique properties of the normal distribution make it alone suitable for general application. There has, however, in recent years, been a tendency for theoretical statisticians, not closely in touch with the requirements of experimental data, to stress the element of normality, in the hypothesis tested, as though it were a serious limitation to the test applied. It is, indeed, demonstrable that, as a test of this hypothesis, the exactitude of " Student's " t test is absolute. It may, nevertheless, be legitimately asked whether we should obtain a materially different result were it possible to test the wider hypothesis which merely asserts that the two series are drawn from the same population, without specifying that this is normally distributed.

In these discussions it seems to have escaped recognition that the physical act of randomisation, which, as has been shown, is necessary for the validity of any test of significance, affords the means, in respect of any particular body of data, of examining the wider hypothesis in which no normality of distribution is implied. The arithmetical procedure of such an examination is tedious, and we shall only give the results of its application in order to show the possibility of an independent check on the more expeditious methods in common use.

On the hypothesis that the two series of seeds are random samples from identical populations, and that their sites have been assigned to members of each pair independently at random, the 15 differences of Table 3 would each have occurred with equal frequency with a positive or with a negative sign. Their sum, taking account of the two negative signs which have actually occurred, is 314, and we may ask how many of the 2^{15} numbers, which may be formed by giving each component alternatively a positive and a negative sign, exceed this value. Since *ex hypothesi* each of these 2^{15} combinations will occur by chance with equal frequency, a knowledge of how many of them are equal to or greater than the value actually observed affords a direct arithmetical test of the significance of this value.

It is easy to see that if there were no negative signs, or only one, every possible combination would exceed 314, while if the negative signs are 7 or more, every possible combination will fall short of this value. The distribution of the cases, when there are from 2 to 6 negative values, is shown in the following table:—

TABLE 4

Number of combinations of differences, positive or negative, which exceed or fall short of the total observed

Number of negative values.	>314	$=314$	<314	Total.
0	1	...	...	1
1	15	...	...	15
2	94	1	10	105
3	263	3	189	455
4	302	11	1,052	1,365
5	138	12	2,853	3,003
6	22	1	4,982	5,005
7 or more	...	...	22,819	22,819
Total .	835	28	31,905	32,768

In just 863 cases out of 32,768 the total deviation will have a positive value as great as or greater than that observed. In an equal number of cases it will have as great a negative value. The two groups together constitute 5·267 per cent. of the possibilities available, a result very nearly equivalent to that obtained using the t test with the hypothesis of a normally distributed population. Slight as it is, indeed, the difference between the tests of these two hypotheses is partly due to the continuity of the t distribution, which effectively counts only half of the 28 cases which give a total of exactly 314, as being as great as or greater than the observed value.

Both tests prove that, in about 5 per cent. of trials, samples from the same batch of seed would show differences just as great, and as regular, as those observed; so that the experimental evidence is scarcely sufficient to stand alone. In conjunction with other experiments, however, showing a consistent advantage of cross-fertilised seed, the experiment has considerable weight; since only once in 40 trials would a chance deviation have been observed both so large, and in the right direction.

How entirely appropriate to the present problem is the use of the distribution of t, based on the theory of errors, when accurately carried out, may be seen by inserting an adjustment, which effectively allows for the discontinuity of the measurements. This adjustment is not usually of practical importance, with the t test, and is only given here to show the close similarity of the results of testing the two hypotheses, in one of which the errors are distributed according to the normal law, whereas in the other they may be distributed in any conceivable manner. The adjustment * consists in calculating the value of t as though the total difference between the two sets of measurements were less than that actually observed by half a unit of grouping;

---

* This adjustment is an extension to the distribution of t of Yates' adjustment for continuity, which is of greater importance in the distribution of χ^2, for which it was developed.

i.e. as if it were 313 instead of 314, since the possible values advance by steps of 2. The value of *t* is then found to be 2·139 instead of 2·148. The following table shows the effect of the adjustment on the test of significance, and its relation to the test of the more general hypothesis.

TABLE 5

	t.	Probability of a Positive Difference exceeding that observed.
Normal hypothesis { unadjusted	2·148	2·485 per cent.
{ adjusted	2·139	2·529 ,,
General hypothesis		2·634 ,,

The difference between the two hypotheses is thus equivalent to little more than a probability of one in a thousand.

21·1. " Non-parametric " Tests

In recent years tests using the physical act of randomisation to supply (on the Null Hypothesis) a frequency distribution, have been largely advocated under the name of " Non-parametric " tests. Somewhat extravagant claims have often been made on their behalf. The example of this Chapter, published in 1935, was by many years the first of its class. The reader will realise that it was in no sense put forward to supersede the common and expeditious tests based on the Gaussian theory of errors. The utility of such non-parametric tests consists in their being able to supply confirmation whenever, rightly or, more often, wrongly, it is suspected that the simpler tests have been appreciably injured by departures from normality.

They assume less knowledge, or more ignorance, of the experimental material than do the standard tests, and this has been an attraction to some mathematicians who often discuss experimentation without personal knowledge of the material. In inductive logic, however, an erroneous assumption of ignorance is not innocuous; it often leads to manifest absurdities. Experimenters should remember that they and their colleagues usually know more about the kind of material they are dealing with than do the authors of text-books written without such personal experience, and that a more complex, or less intelligible, test is not likely to serve their purpose better, in any sense, than those of proved value in their own subject.

REFERENCES AND OTHER READING

C. DARWIN (1876). The effects of cross- and self-fertilisation in the vegetable kingdom. John Murray, London.

R. A. FISHER (1925). Applications of " Student's " distribution. Metron, v. 90-104.

R. A. FISHER (1925-1963). Statistical methods for research workers. Chap. V., §§ 23-24.

R. A. FISHER (1956, 1959). Statistical methods and scientific inference. Oliver and Boyd Ltd., Edinburgh.

" STUDENT " (1908). The probable error of a mean. Biometrika, vi. 1-25.

IV

AN AGRICULTURAL EXPERIMENT IN RANDOMISED BLOCKS

22. Description of the Experiment

IN pursuance of the principles indicated by the discussions in the previous chapters we may now take an example from agricultural experimentation, the branch of the subject in which these principles have so far been most explicitly developed, and in which the advantages and disadvantages of the different methods open to the experimenter may be most clearly discussed.

We will suppose that our experiment is designed to test the relative productivity, or yield, of five different varieties of a farm crop; and that a decision has already been arrived at as to what produce shall be regarded as yield. In the case of cereal crops, for example, we may decide to measure the yield as total grain, or as grain sufficiently large not to pass a specified sieve, or as grain and straw valued together at predetermined prices, or in whatever method may be deemed appropriate for the purposes of the experiment. Our object is to determine whether, on the soil or in the climatic conditions experienced by the test, any of the varieties tested yield more than others, and, if so, to evaluate the differences with a determinate degree of precision.

We shall suppose that the experimental area is divided into eight compact, or approximately square, blocks, and that each of these is divided into five plots running from end to end of the block, and lying side by side, making forty plots in all. Apart from the differences in variety to be used, the whole area is to have uniform agricultural treatment. At harvest, narrow edges about a foot in width for cereal crops, or the width of a single row for larger plants, such as roots and potatoes, are to be discarded from experimental yields; the central portions, cut to be of equal area, are to be harvested, and the produce weighed, or, if preferred, measured in some other manner.

In each block the five plots are assigned one to each of the five varieties under test, and this assignment is made at random. This does not mean that the experimenter writes down the names of the varieties, or letters standing for them, in any order that may occur to him, but that he carries out a physical experimental process of randomisation, using means which shall ensure that each variety has an equal chance of being tested on any particular plot of ground. A satisfactory method is to use a pack of cards numbered from 1 to 100, and to arrange them in random order by repeated shuffling. The varieties are then numbered from 1 to 5, and any card such as number 33, for example, is deemed to correspond to variety number 3, because on dividing by 5 this number is found as the remainder. Numbers divisible by 5 will correspond to variety number 5. The order of varieties in each block may then be quickly determined from the order of the cards in the pack, after thoroughly shuffling. The remainder corresponding to any variety is disregarded after its first occurrence in the block.

Since 5 is a divisor of a hundred, each variety will be represented by 20 cards, and the probabilities of each appearing in any particular place will be equal. If we had been randomising six varieties we should have used a number of cards divisible by 6, for example

96, and could, for this purpose, use the same pack as before, discarding the 4 cards numbers 97 to 100, or indeed, any other four cards the numbers of which leave the remainders 1, 2, 3 and 4 on dividing by 6.

To save the labour of card shuffling use is often made of printed tables of random sampling numbers, in which, for example, all numbers of 4 figures are arranged in random order. The first such table was published by Tippett; another is available in *Statistical Tables*. Starting at any point in such a table and proceeding in any direction, such as up or down the columns, or along the rows, we may take each pair of digits to represent the number of a card in the pack of 100, disregarding any which may be superfluous for our purpose. Using such means the process of randomisation is extremely rapid, and a chart showing the arrangement of the experiment may be prepared as quickly as if the varieties had been set out in a systematic order.

23. Statistical Analysis of the Observations

The arithmetical discussion by which the experiment is to be interpreted is known as the analysis of variance. This is a simple arithmetical procedure, by means of which the results may be arranged and presented in a single compact table, which shows both the structure of the experiment and the relevant results, in such a way as to facilitate the necessary tests of their significance. The structure of the experiment is determined when it is planned, and before the content of its results, consisting of the actual yields from the different plots, is known. It depends on the number of varieties to be compared, on the number of replications of each obtainable, and on the system by which these are arranged; in our present example in randomised blocks. In its arithmetical aspect this structure is specified by the numbers of degrees of freedom, or of independent comparisons, which can be made between the plots, or relevant groups of plots. Between 40 plots 39 independent comparisons can be made, and so the total number of degrees of freedom will be 39. This number will be divided into 3 parts representing the numbers of independent comparisons (*a*) between varieties, (*b*) between blocks, and (*c*) representing the discrepancies between the relative performances of different varieties in different blocks, which discrepancies provide a basis for the estimation of error. We may specify the structure of our typical experiment by a partition of the total of 39 degrees of freedom into these three parts as under.

TABLE 6

Structure of an Experiment in Randomised Blocks

Varieties	4
Blocks	7
Error	28
Total	39

It is easy to see that the number of degrees of freedom for any group of simple comparisons, such as those between varieties or between blocks, must be 1 less than the number of items to be compared. In the present instance, in which the plots are assigned within the blocks wholly at random, the whole of the remaining 28 degrees of freedom are due simply to differences in fertility between different plots within the same block, and are therefore available for providing the estimate of error. As will be explained more fully later, many more complicated modes of subdivision of the total number of degrees of freedom may be employed, **and** will be appropriate to more complicated forms of experimental enquiry. The form we have set out is appropriate to the question whether the yields given by the different varieties show, as a whole, greater differences than would ordinarily be found, had only a single variety been sown on the same land. It is appropriate to test the null hypothesis that our 5 varieties give in fact equal yields.

The completion of the analysis of variance, when the yields are known, must be strictly in accordance with the structure imposed by the design of the experiment, and consists in the partition of a quantity known as the *sum of squares* (*i.e.* of deviations from the mean) into the same three parts as those into which we have already divided the degrees of freedom. Our data consists of 40 yields (y), 5 from each block and 8 from each variety which, for further calculation, can be conveniently arranged in a table of 5 columns and 8 lines.

TABLE 7

Scheme for calculation of totals and means

					Total	Mean
—	—	—	—	—	A	a
—	—	—	—	—	B	b
—	—	—	—	—	C	c
—	—	—	—	—	D	d
—	—	—	—	—	E	e
—	—	—	—	—	F	f
—	—	—	—	—	G	g
—	—	—	—	—	H	h
Total	P	Q	R	S	T	M
Mean	p	q	r	s	t	m

The totals of the five columns will then represent the totals of the yields obtained from the 5 varieties, and

ANALYSIS OF VARIANCE

may be designated by the capital letters P, Q, R, S and T. The mean yields from the five varieties, found by dividing these by 8, will be designated by small letters p, q, r, s and t. In like manner the totals of the rows will represent the total yields harvested from each of the blocks of land used; we shall denote these by A, B, C, D, E, F, G, H, and the corresponding means by a, b, c, d, e, f, g, h. Evidently, the totals of the rows and the totals of the columns are sub-totals of the same grand total, denoted by M, from which the general mean m is derived by dividing by 40. The arrangement is illustrated in Table 7; also Table 7A, p. 68, gives some numerical observations in this form.

The sum of squares of deviations from the mean, which, in the analysis of variance, is to be divided into portions corresponding to varieties, blocks and error, is found by adding together the squares of the 40 recorded yields, and deducting the product Mm. The difference, which corresponds to the total 39 degrees of freedom, is actually the sum of the squares of the 40 differences or deviations between the actual yields, y, and their general mean, m; it therefore measures the total amount of variation due to all causes, observed between our different plots. The method we have given, however, for obtaining this quantity is convenient for our purpose, for the product Mm is used also in our calculation of the other entries of the table, which are indeed rapidly obtainable once the total sum of squares is known. The portion, for example, corresponding to the 4 degrees of freedom between varieties is found simply by summing the products Pp+ Qq+ . . ., and deducting Mm. Similarly, the portion ascribable to the 7 degrees of freedom between blocks is obtained by summing the products Aa+Bb+Cc . . ., and deducting Mm. Knowing the contributions to the sum of squares of these 11 degrees of freedom, the amount corresponding to the remaining 28 degrees of freedom, due to error, may be found by subtraction from the total. In this way the total sum of squares, representing the total amount of variation due to all causes between the 40 yields of the experiment, is divided into the 3 portions relevant to its interpretation, measuring respectively the amount of variation between varieties, the amount of variation between blocks, and the amount of discrepancy between the performances of the different varieties in the different blocks. The greater part of the arithmetical labour is accomplished with the calculation of the total sum of squares.

Corresponding to the three sums of squares into which the total has been partitioned, we may now calculate the mean squares, by dividing each by the corresponding number of degrees of freedom. On the null hypothesis the mean squares for variety and error have the particularly simple interpretation that each may be regarded as an independent estimate of the same single quantity, the variance due to error of a single plot. If the varieties had in fact the same yield the mean square derived from the 4 degrees of freedom of varieties would have, on the average, the same value as that derived from the 28 degrees of freedom for error. In any one trial these values would indeed differ, but only by errors of random sampling. The relative precision of our estimates is determined solely by the number of degrees of freedom upon which each is based, so that, knowing these numbers, the ratio of any two estimates affords a test of significance. In other words, on the null hypothesis the random sampling distribution of this ratio is precisely known. Thus, in our example, it would happen just once in 20 trials that the estimate based on 4 degrees of freedom exceeded that based on

ANALYSIS OF VARIANCE

28 degrees of freedom in a ratio greater than 2·714. If, therefore, the observed ratio exceeds this level, we have a measurable basis for confidence that the differences observed between the yields of the different varieties are not due wholly to the differences in fertility of the plots on which they were grown. Again, in only 1 per cent. of trials will the ratio exceed 4·074, a value which thus marks the level of a more severe test of significance. Since tests of the same kind will be required for all possible pairs of numbers of degrees of freedom, it is convenient for purposes of tabulation to use a criterion which varies more regularly than the arithmetical ratio employed in the illustrations above. It is usual, therefore, to carry out the calculation by using the natural logarithms of the mean squares, and since the difference of the two logarithms specifies the ratios of the corresponding numbers, the tables used in this test of significance give the values of a quantity, z, defined as half the difference between the natural logarithms obtained.

The z test may be regarded as an extension of the t test, appropriate to cases where more than two variants are to be compared. Like it, it is derived from the theory of errors, and is exact when the normal law of errors is realised. It is even less affected than the t test by such deviations from normality as are met with in practice. As with the t test, its appropriateness to any particular body of data may be verified arithmetically. Such verification is not ordinarily necessary and is always laborious. Often the number of random arrangements available is far too great for them to be examined exhaustively, as was done with Darwin's experiment in Chapter III. Eden and Yates have, however, published a method of obtaining rapidly the results of a large random selection of these arrangements, and have demonstrated how closely the theoretical distribution was verified in material that was far from normal.

24. Precision of the Comparisons

If the yields of the different varieties in the experiment fail to satisfy the test of significance they will not often need to be considered further, for the results, as so far tested, are compatible with the view that all differences observed in the experiment are due to variations in the fertility of the experimental area, and this is the simplest interpretation to put upon them. If, however, a significant value of z has been obtained the null hypothesis has been falsified, and may therefore be set aside. We shall thereafter proceed to interpret the differences between the varietal yields as due, at least in part, to the inherent qualities of the varieties, as manifested on the conditions of the test, and shall be concerned to know with what precision these different yields have been evaluated. For this purpose the mean square, corresponding to the 28 degrees of freedom assigned to error, is available as an estimate of the variance of a single plot due to the uncontrolled causes which constitute the errors of our determinations. From this fundamental estimate we may derive a corresponding estimate of the variance of the sum of the yields from 8 plots by multiplying by 8, or, if we prefer, we may derive the variance of the mean yield of 8 plots by dividing by 8. In either case the square root of the variance gives the standard deviation, and provides therefore a means of judging which of the differences among our varietal yield values are sufficiently great to be regarded as well established, and which are to be regarded as probably fortuitous. If the experiment leaves any grounds for practical doubt, values may be compared by the t test mentioned in Chapter II., remembering that our estimate of error is based on 28 degrees of freedom.

It is an advantage of arrangements in randomised blocks that, corresponding to any particular comparison contrast, the components of error appertaining to this comparison may be isolated. This is done simply by finding the difference in yield, or performance, between the treatments, or groups of treatments, to be compared, in each replication of the experiment. The discrepancies between these differences obtained from different replications, taking account of their signs, constitute the components of error appropriate to this comparison, which may now be tested by a t test, independently of the other comparisons which the experiment affords. Although fewer degrees of freedom are available for the estimation of error from these components only, their isolation affords an additional safeguard when, as may sometimes occur, some comparisons are, in reality, less accurately evaluated than others.

When the z test does not demonstrate significant differentiation, much caution should be used before claiming significance for special comparisons. Comparisons, which the experiment was designed to make, may, of course, be made without hesitation. It is comparisons suggested subsequently, by a scrutiny of the results themselves, that are open to suspicion; for if the variants are numerous, a comparison of the highest with the lowest observed value, picked out from the results, will often appear to be significant, even from undifferentiated material. Properly, such unforeseen effects should be regarded only as suggestions for future experimentation, in which they can be deliberately tested. To form a preliminary opinion as to the strength of the evidence, it is sometimes useful to consider how many similar comparisons would have been from the start equally plausible. Thus, in comparing the best with the worst of ten tested varieties, we have chosen the pair with the largest apparent difference out of 45 pairs, which might equally have been chosen. We might, therefore, require the probability of the observed difference to be as small as 1 in 900, instead of 1 in 20, before attaching statistical significance to the contrast.

25. The Purposes of Replication

An examination of the structure of the standard type of agricultural experiment described above, and of the use made of its structure in the statistical process of interpretation, shows that the replication or repetition of the varieties tested on different plots of land serves two distinct purposes. It serves first to diminish the error, a purpose which has been widely recognised, though the manner in which it does so has not always been well understood. In our experiment the sampling variance of a mean yield was found by dividing the estimate of variance for a single plot by 8. Since the variance of a single plot would not be necessarily or systematically increased by increasing the number of blocks, the variance of our mean yields will generally fall off inversely to the number of replications included. In increasing the number of blocks, however, we should have increased the area of the experiment, and it is probable that this increase in area, even if we had used the same number of larger plots, would itself have served to diminish the experimental error. If the area of the experiment were kept constant and the replication increased by using smaller plots we should only gain in precision if, as abundant agricultural experiment shows to be generally the case, the greater proximity of the smaller areas led to a greater similarity in the fertility of the soil. The practical limit to plot subdivision is set, in agricultural experiments, by the necessity of discarding a strip at the edge of each plot. The width of the strip depends on the competition of neighbouring plants for moisture, soil nutrients and light, and is independent of the size of the plots. Consequently, as smaller plots are used, a larger proportion of the experimental area has to be discarded. The soil heterogeneity of most experimental land is, however, so pronounced that it is profitable to discard a considerable proportion of the area, in order to bring the experimental treatments or varieties to be contrasted more closely together than would otherwise be possible. With plants, such as potatoes and sugar-beets, where it is sufficient to discard a single row, one on each side of the plot, it has been repeatedly found that strips of 4 rows wide, of which only the central two are included in the yield, make a more economical use of a given experimental area than either wider or narrower plots would do.

Replication, therefore, in the sense of the comminution of the experimental area, down to plots of the most efficient size, has an important but limited part to play in increasing the precision of an experiment. It should not, in this connection, be overlooked that many other factors contribute to the same result, such as accuracy in harvesting and weighing the produce; in measuring the areas of land harvested; care in the choice of the experimental area; in insuring the similarity of the treatment of its different parts; in safeguarding the crop against damage, and its produce against loss. All these factors contribute to the precision of the experiment, and though, when the conditions are otherwise favourable, there can be no doubt that attention is rightly concentrated on diminishing the important causes of error due to variations in soil fertility, it is evident that, even in experiments in which these causes were almost wholly eliminated, neglect of common-sense precautions, which, none the less, require care and supervision, may lead to entirely unreliable results.

26. Validity of the Estimation of Error

Whereas replication of the experimental varieties or treatments on different plots, formed by the subdivision of the experimental area, is of value as one of the means of increasing the accuracy of the experimental comparisons, its main purpose, which there is no alternative method of achieving, is to supply an estimate of error by which the significance of these comparisons is to be judged. The need of such an estimate may be perceived by considering the doubts with which the interpretation of an experiment would be involved if it consisted only of a single plot for each treatment. The treatment giving the highest yield would of course appear to be the best, but no one could say whether the plot would not in fact have yielded as well under some or all of the other treatments. If, indeed, the difference in yield appeared large to the experimenters they might argue that so large a difference could not reasonably be ascribed to a difference in soil fertility, since it was contrary to their experience that neighbouring plots treated alike should differ so greatly. To enforce this argument they would in fact have to claim that their past experience had already furnished a basis for the estimation of error, which could be applied with confidence to the circumstances of the experiment under discussion. Even if this claim could be granted the experiment would carry with it the serious disadvantage that it would no longer be self-contained, but would depend for its interpretation from experience previously gathered. It could no longer be expected to carry conviction to others lacking this supplementary experience. How weak the evidence of such previous experience must always be will be seen by considering that, even if the identical area of the experiment, divided into the same plots, had been harvested under uniform treatment in previous years, it would need twenty years' experience to form even the roughest judgment as to how great a difference between the yields of any two plots would occur by chance as often as once in 20 trials. It is on the exactitude of our estimate of the magnitude of this difference that the precision of our test of significance must depend. Even such a tedious series of preliminary trials, moreover, could only supply a direct basis for the test of significance, if we could assume the absence of progressive changes, both in the weather and in the condition of the soil. This consideration effectively demonstrates that the accumulation of past experience, as a basis for testing significance, is as insecure in theory as it would be inconvenient in practice.

The impossibility of testing two or more treatments, in the same year, and on identically the same land, is not, however, an insuperable obstacle to exact experimentation. It is surmounted by testing the treatments not on identical land, but on random samples of the same experimental areas. From this aspect the appropriateness of a *random* assignment of the treatments to the different plots appears most inevitably. We shall need to judge of the magnitude of the differences introduced by testing our treatments upon different plots by the discrepancies between the performances of the same treatment in different blocks. Our estimate of error must be obtained by a comparison of plots treated alike, but it is to be applied to interpret the differences observed between sets of plots treated differently. The validity of our estimate of error for this purpose is guaranteed by the provision that any two plots, not in the same block, shall have the same probability of being treated alike, and the same probability of being treated differently in each of the ways in which this is possible. The purpose of randomisation in this, as in the previous experiments exemplified, is to guarantee the validity of the test of significance, this test being based on an estimate of error made possible by replication.

27. Bias of Systematic Arrangements

In any particular case it will probably be possible to assign sets of plots within an area to the several treatments so as to equalise their fertility more completely than is done by a random arrangement, and many systematic arrangements for doing this have from time to time been proposed. The effect of such a procedure on the test of significance may be seen by imagining it carried out on an area under uniform treatment, so that the actual yields are not at all affected by the reallocation of the plots. In the analysis of variance, therefore, the total sum of squares is unchanged, as is also the portion ascribable to blocks. If, therefore, the agronomist's ingenuity has been successful in diminishing the differences in fertility between treatments, the diminution of the sum of squares in that line of the table will have been exactly counterbalanced by an increase in the sum of squares upon which the estimate of error is based. The effect of the rearrangement will have been to diminish the real errors of the experiment, but at the expense of increasing the estimate of error; so that, although the comparisons have really been improved in precision they will appear to have been less accurate than before, and less reliance will be placed on the result. In the opposite case, likewise, if by bad luck or bad judgment the systematic arrangement adopted has increased rather than lessened the real errors of the experiment, then the estimate of error will be even diminished, and will be, for both reasons, an underestimate of the errors actually incurred. The results of using arrangements which differ from the random arrangement in either direction are thus in one way or the other undesirable. This is to be expected, since in both cases the estimate of error is vitiated, or rendered unreliable for the purpose for which it was made.

28. Partial Elimination of Error

It is to be noted that the restriction upon a purely random arrangement which has been imposed, by applying each treatment once only on each block of land, introduces no such disturbance of the validity of the estimate of error. For the differences in average fertility between the different blocks of land used, which have been, by this restriction, eliminated from our experimental comparisons, have been equally eliminated from our estimate of error in the analysis of variance. Prior to the introduction of this method it was, indeed, common for elements of error which had been carefully and thoroughly eliminated in the field to be reintroduced in the process of statistical estimation; so that successful experimental arrangements were made to appear to be unsuccessful and *vice versâ*. The essential fact governing our analysis is that the errors due to soil heterogeneity will be divided, by a good experiment, into two portions. The first, which is to be made as large as possible, will be completely eliminated, by the arrangement of the experiment, from the experimental comparisons, and will be as carefully eliminated in the statistical laboratory from the estimate of error. As to the remainder, which cannot be treated in this way, no attempt will be made to eliminate it in the field, but, on the contrary, it will be carefully randomised so as to provide a valid estimate of the errors to which the experiment is in fact liable.

29. Shape of Blocks and Plots

Having satisfied ourselves that replication, supplemented by randomisation, will afford a valid test of the significance of our comparisons, we may consider what modifications of our practical procedure will serve to increase the precision of these comparisons. If several areas of land are available for experiment some care may usually be given to choose one that appears to be uniform, as judged by the surface and texture of the soil, or by the appearance of a previous crop, though the value of such judgments by inspection, with which alone we are here concerned, appears to be very easily overrated. After choosing the area we usually have no guidance beyond the widely verified fact that patches in close proximity are commonly more alike, as judged by the yield of crops, than those which are further apart. Consequently, the division of the land into compact or approximately square blocks will usually result in the blocks being as much unlike as possible, while different areas within the same block will be more closely similar than if the blocks had been long and narrow. The effect of this upon the analysis of variance is to place as large a fraction as possible of the variance due to soil heterogeneity in the portion ascribable to variation between blocks, this portion being eliminated from our experimental error; and to leave as little as possible in the variation within blocks, which supplies both our experimental errors and our estimate of them. It is therefore a safe rule to make the blocks as compact as possible.

With respect to our subdivision of the blocks into plots our object is exactly the opposite. The experimental error arises solely from differences between the areas chosen as plots within the same block. These differences must be made as small as possible, or, in other words, each plot must, so far as may be, sample fairly the whole area of the block in which it is placed.

It is often desirable, therefore, when it does not conflict with agricultural convenience in other ways, to let the plots lie side by side as narrow strips, each running the whole length of its block. It is not, however, in every type of experiment an advantage to use such elongated plots. Some important causes of soil heterogeneity, dependent from agricultural operations, affect the land in stripes. Elongated plots will then only be advantageous if they can be laid transversely to these stripes. When this would entail inconvenience, or additional labour in cutting the correct area, as in the case of strip plots with cereals, running across the drill rows, the labour available for the experiment may often be better applied by using square plots, and improving the accuracy in other ways. Plots of compact shape are, indeed, commonly used in experiments in randomised blocks, not because the theoretical advantage of using elongated strips, in one direction or the other, is not appreciated, but for the purely practical reason that to realise it by laying strips in the required direction would be a more costly method of increasing precision than other methods at the experimenter's disposal.

30. Practical Example

The following data show a comparison of the yields of five varieties of barley in an experiment arranged in randomised blocks, carried out in the State of Minnesota in the years 1930 and 1931 and reported by F. R. Immer, H. K. Hayes and Le Roy Powers in the *Journal of the American Society of Agronomy*. The experiment really

TABLE 7A

Total Yields of Barley Varieties in Twelve Independent Trials

Place and Year.	Manchuria.	Svansota.	Velvet.	Trebi.	Peatland.	Total.	Mean.
1 {1931	81·0	105·4	119·7	109·7	98·3	514·1	102·82
1932	80·7	82·3	80·4	87·2	84·2	414·8	82·96
2 {1931	146·6	142·0	150·7	191·5	145·7	776·5	155·30
1932	100·4	115·5	112·2	147·7	108·1	583·9	116·78
3 {1931	82·3	77·3	78·4	131·3	89·6	458·9	91·78
1932	103·1	105·1	116·5	139·9	129·6	594·2	118·84
4 {1931	119·8	121·4	124·0	140·8	124·8	630·8	126·16
1932	98·9	61·9	96·2	125·5	75·7	458·2	91·64
5 {1931	98·9	89·0	69·1	89·3	104·1	450·4	90·08
1932	66·4	49·9	96·7	61·9	80·3	355·2	71·04
6 {1931	86·9	77·1	78·9	101·8	96·0	440·7	88·14
1932	67·7	66·7	67·4	91·8	94·1	387·7	77·54
Total	1132·7	1093·6	1190·2	1418·4	1230·5	6065·4	
Mean	94·3916	91·13	99·183	118·2	102·5416		101·09

dealt with ten varieties, of which five have been selected for this example. The blocks in the example are twelve separate experiments carried out at six locations in the State in the two years.

REFERENCES AND OTHER READING

T. EDEN and F. YATES (1933). On the validity of Fisher's z test when applied to an actual example of non-normal data. Journal of Agricultural Science, xxiii. 6-17.

R. A. FISHER (1924). On a distribution yielding the error functions of several well-known statistics. Proceedings of the International Mathematical Congress, Toronto, pp. 805-813.

R. A. FISHER (1925-1963). Statistical methods for research workers. Chap. VIII. and the Table of z, Table VI.

R. A. FISHER and F. YATES (1938-1963). Statistical Tables for biological, agricultural and medical research. Oliver and Boyd Ltd., Edinburgh.

F. R. IMMER, H. K. HAYES and LE ROY POWERS (1934). Statistical determination of barley varietal adaptation. Journal of the American Society of Agronomy, May, xxvi. 403-419.

L. H. C. TIPPETT (1927). Random sampling numbers. Tracts for computers, xv. Cambridge University Press.

V
THE LATIN SQUARE
31. Randomisation subject to Double Restriction

THE subdivision of the area of an agricultural experiment into compact blocks, in each of which all the experimental treatments to be compared are equally represented, has been found to add greatly to the precision of the experimental comparisons obtainable by the expenditure of a fixed amount of labour, and supervisory care, to a limited area of land. An equally great advantage is obtained, in other fields of research, by a similar subdivision of the material into relatively homogeneous series, to each of which the different experimental treatments are applied in equal proportion. The extent of this gain is limited only by the degree of homogeneity which can be obtained within each series. It is an essential condition of experimentation that the experimental material is known to be variable, but it is not known, in respect of any individual, in what direction his response to a given treatment will vary from the average. No direct allowance for this variability can, therefore, be made. The knowledge which guides us in increasing the precision of an experiment is not a knowledge of the individual peculiarities of particular experimental units, such as plots of land, experimental animals, coco-nut palms, or hospital patients, but a knowledge that there is less variation within certain aggregates of these than there is among different individuals belonging to different aggregates. The recognition of criteria by which the experimental material may be fruitfully subdivided thus plays an important part in all types of quantitative experimentation.

It was first shown in experimental agriculture, though the principle has since been applied in other fields, that the process of subdivision might profitably be duplicated. This experimental principle is best illustrated by the arrangement known as the Latin square, a method which is singularly reliable in giving precise comparisons when the number of treatments (or varieties, etc.) to be compared is from 4 to 8. Suppose we wish to compare 6 treatments. The experimental area (which need not be an exact square in form, but should be a relatively compact rectangle) is divided into 36 equal plots lying in 6 rows and 6 columns. It is then a combinatorial fact that we can assign plots to the 6 treatments such that for each treatment one plot lies in each row and one in each column of the square. It is possible generally to do this in a large number of ways, for if we start with one solution of the problem, and rearrange the rows in it as wholes, in any of the ways in which these may be arranged, a large number of new solutions will be found. With a 6×6 square the rows may be rearranged in 720 ways, including that from which we start, so we have at once a set of 720 solutions. Equally, or consecutively, we may arrange the columns in 720 ways; and, finally, we may do the same with the treatments, while still conserving the property which the Latin square was designed to possess. The process of transformation will generate a number of different solutions which varies according to the particular square from which we happen to start. The smallest transformation set comprises 1,728,000 solutions, while the 5 largest each comprise 93,312,000. If a solution is chosen at random from such a set each plot has an equal probability of receiving any of the possible treatments, and each pair of plots, not in the same row or column, has the same probability, namely one-fifth, of being treated alike. The process of randomisation, necessary to ensure the validity of the test of significance applied to the experiment, consists in choosing one at random out of the set of squares which can be generated from any chosen arrangement.

The object of arranging plots in a Latin square is to eliminate from the experimental comparisons possible differences in fertility which may exist between whole rows of plots, and between whole columns of plots, as they stand in the field. The need for such a double elimination was particularly apparent to agricultural experimenters owing to the fact that in many fields there is found to occur either a gradient of fertility across the whole area, or parallel strips of land having a higher or lower fertility than the average. But, for particular fields, it is not known whether such heterogeneity will be more pronounced in the one or the other direction in which the field is ordinarily cultivated. Such soil variations may be due in part to the past history of the field, such as the lands in which it has been laid up for drainage producing variations in the depth and present condition of the soil, or to portions of it having been manured or cropped otherwise than the remainder; but whatever the causes, the effects are sufficiently widespread to make apparent the importance of eliminating the major effects of soil heterogeneity, not only in one direction across the field, but at the same time in the direction at right angles to it. This double elimination is effected by the Latin square arrangement, which combines the combinatorial fact stated above with the possibility of basing estimates of error upon an effective randomisation.

32. The Estimation of Error

As has been already illustrated in the experiment in randomised blocks, the error will be properly estimated only if the same components of heterogeneity which have been successfully eliminated by the arrangement in the field are also eliminated in the interpretation of the results in the laboratory. This elimination is carried out by an analysis of variance closely similar to that used in the last chapter. The 35 independent comparisons possible among 36 yields give 35 degrees of freedom. Of these, 5 are ascribable to differences between rows, and 5 to differences between columns. Thus 10 degrees of freedom serve to represent the components of heterogeneity, which have been eliminated in the field, and must be excluded from our estimate of error. Of the remaining 25 degrees of freedom, 5 represent differences between the treatments tested, and 20 represent components of error which have not been eliminated, but which have been carefully randomised so as to ensure that they shall contribute no more and no less than their share to the errors of our experimental comparisons.

The table shows the subdivision of the degrees of freedom for a 6×6 square, and in general for an $s \times s$ square used to test s treatments.

TABLE 8

	6×6 square.	$s \times s$ square.
Rows	5	$s-1$
Columns	5	$s-1$
Treatments	5	$s-1$
Error	20	$(s-1)(s-2)$
Total	35	s^2-1

Corresponding to each part of the degrees of freedom the yield data of the experiment will provide a like portion of the total of what we have called sum of squares. The first 3 portions may be calculated in exactly the same way as are those for blocks and treatments in the randomised block arrangement. Thus, if the total yields in the rows are A, B, C, D, E, F, with a grand total M, while the corresponding mean yields are a, b, c, d, e, f, and m, the portion of the sum of squares ascribable to rows is

$$aA + bB + cC + dD + eE + fF - mM.$$

The portions ascribable to columns and to treatments are calculated similarly, while that ascribable to error is calculated by deducting the 3 other items from the total. This total sum of squares, as in other cases, is merely the sum of the squares of the deviations of the yields of all individual plots from their general mean, and may be calculated by subtracting mM from the sum of the squares of these yields.

33. Faulty Treatment of Square Designs

When the possibility of effecting a double elimination of errors due to soil heterogeneity was first realised, the mistake was sometimes made of judging the precision of the results merely from the observed discrepancies between plots treated alike. This would be correct only if the whole 36 plots had been assigned at random, and without restriction, to the 6 treatments. Its effect is that the 10 degrees of freedom corresponding to rows and columns are included in the estimate of error. Thus what experimental design had gained in the field arrangement was lost or thrown away in the statistical analysis. Indeed, by this method the apparent precision of the experiment, and the consequent reliance placed on its results, is less than if the treatments had been assigned wholly at random, disregarding rows and columns, for the large components due to these have been excluded from the 5 degrees of freedom ascribed to treatments, and therefore contribute more than proportionately to the 30 remaining degrees of freedom, which on this system is regarded as error.

A fault of purely statistical origin also appears in some of the earlier work, namely, the use of the total number of plots, in place of the number of degrees of freedom, as a divisor in obtaining the estimated variance of the mean square. This may lead to the error being seriously underestimated. Apart from its arithmetical simplicity, the great advantage of arranging the statistical work in an analysis of variance lies in the safeguard it affords against errors of these two kinds. Once the degrees of freedom are subdivided it is apparent that the residue after allowing for rows, columns, and treatments has only 20 degrees of freedom, and once the contributions to the sum of squares due to rows and columns are identified and set aside, no one would think of introducing these in attempting to arrive at an estimate of error. The mean square, obtained by dividing the residual sum of squares by the degrees of freedom available for the estimation of error, is a valid estimate of the variance of the yield of a single plot, due to the components of error which have been randomised. The sampling variance of the total of six plots having the same treatment is found by multiplying the mean square by 6, and that of the mean of such plots, by dividing it by 6. The variance so obtained is itself liable to sampling errors, dependent on the number of degrees of freedom on which it is based. Since this number is often small, we should use the exact z test for testing the significance of the group of treatments as a whole. If it is desired to compare any two particular treatments, or to make any other simple contrast among the treatments employed, whether based on the totals or on the means, we obtain the sampling variance appropriate to the expression, find the standard error by taking the square root, and the ratio of the differences to be tested to its appropriate standard error will be the value of t, appropriate to test its significance, with the same number of degrees of freedom as that on which the estimate of variance was originally based. Readers unfamiliar with the statistical procedure of these tests are referred to *Statistical Methods for Research Workers*.

In agricultural trials a single Latin square will frequently give precision high enough to reduce the standard error to less than 2 per cent. of the yield, and sometimes to less than 1 per cent. If experimentation were only concerned with the comparison of four to eight treatments or varieties, it would therefore be not merely the principal but almost the universal design employed. It is particularly fitted for the comparison at a number of different places of a small selected group of highly qualified varieties, the relation of which to varying conditions of soil and weather needs to be explored. Where it fails is to provide a means of testing simultaneously a large number of different treatments or varieties. The means used to obtain precision in such experiments will be developed in later chapters.

34. Systematic Squares

When the idea of effecting an elimination of errors due to soil heterogeneity in two directions at right angles was first appreciated, the necessity for randomisation in experimental trials was not realised. In consequence, certain systematic arrangements were adopted. One of these, which may be called a diagonal square, is shown below :—

```
A B C D E
E A B C D
D E A B C
C D E A B
B C D E A
```

It will be observed that the conditions that one plot of each kind lies in each row and one in each column are satisfied by this arrangement. The plots receiving treatment A are, however, all in a line along the diagonal of the square, and other lines parallel to this diagonal also receive the same treatment throughout their length. Consequently, there is ground to fear that if ridges or strips of fertility run obliquely acrc s the rows and columns they may give to some of the treatments a systematic advantage compared with the others. In other words, the components of soil fertility in which the areas assigned to different treatments may differ, may, not improbably, be larger than the remaining components, representing differences between plots treated alike, on which the estimate of error is based. Consequently, if a systematic arrangement of this kind is treated as though it were a Latin square two distinct effects may be anticipated. First, that the actual errors in the comparisons of the treatments will be greater than if a properly randomised Latin square had been used, and, second, that the estimate of error obtained from the experiment will be less, by reason of the exclusion of the more important components of soil variability, which have been confounded with the treatments.

The first of these dangers was readily recognised, though the second was ignored. Consequently, an improved systematic arrangement has been widely used.

It has been known in Denmark since about 1872, but is usually ascribed to the Norwegian, Knut Vik. The method is to move each row forward two places instead of one, giving the following pattern:—

```
A B C D E
D E A B C
B C D E A
E A B C D
C D E A B
```

In this arrangement the areas bearing each treatment are nicely distributed over the experimental area, so as to exclude all probability that the more important components of soil heterogeneity should influence the comparison between treatments. This was clearly the intention of the arrangement; but its fulfilment carries with it an unforeseen and unfortunate consequence. If, by the skill of the experimenter, the components of error, by which the comparisons between the treatments are affected are, on the average, less than those given by a random arrangement, it follows that those available for the estimation of error must be greater. This, as with randomised blocks, is easily seen by considering the subdivision of the sum of squares in the analysis of variance. The null hypothesis is that the treatments are without effect, and therefore that all the differences observed among the experimental results are due to experimental error. They are therefore unaffected by the arrangement adopted. The components ascribable to rows and columns, and eliminated from the experiment, are also the same however the plots may be arranged. Consequently, the total of the two components ascribed to treatments and to error, that is to say the true errors of our comparisons, and the estimate of these errors supplied by the experiment, have a total independent of the experimental arrangement. The sole effect of adopting one system of arrangement rather than another is on the manner in which the fixed total is divided between these two parts. The purpose of randomisation is to ensure that each degree of freedom shall have, on the null hypothesis, the same average content. Any method of arrangement, therefore, which diminishes the real errors must increase the apparent magnitude of these errors, by which the validity of the comparison is to be judged. The consequence is that not more but less reliance is placed, and must be placed, on the results, as a consequence of the experimenter's success in excluding the larger components of error from his comparisons.

It should be noted that this unfortunate consequence only ensues when a method of diminishing the real errors is adopted, unaccompanied by their elimination in the statistical analysis. Thus, when the treatments are equally distributed among the rows of an experiment, the real error is usually diminished, and the estimate of error may be diminished in the same measure by eliminating the differences between the different rows. The failure of systematic arrangements came from not recognising that the function of the experiment was not only to make an unbiased comparison, but to supply at the same time a valid test of its significance. This is vitiated equally whether the components affecting the comparisons are larger or smaller than those on which the estimate of error is based. The consequences of accepting an insignificant effect as significant, or of rejecting as insignificant one which, with sounder methods of experimentation, would have shown itself to be significant, are equally unfortunate. In fact, the calculation of standard errors is idle and misleading, if the method of arrangement adopted fails to guarantee their validity, and the same applies to all other means of testing significance.

The consequences surmised as to the effects of using the two systematic squares illustrated above have in fact been verified in detail by O. Tedin, by superimposing these arrangements, each 184 times, on the yields obtained in uniformity trials, in which the null hypothesis is known to be true; and by comparing the results with those obtained using random arrangements. The discrepancies found are just what might have been anticipated on theoretical grounds. The diagonal square gives larger real errors accompanied by greater apparent precision. The Knut Vik square gives lower real errors accompanied by less apparent precision. It is a curious fact that the bias of the Knut Vik square, which was unsuspected, appears to be actually larger than that of the diagonal square, which all experienced experimenters would confidently recognise.

35. Græco-Latin and Higher Squares

The number of arrangements in a Latin square is known for squares up to 7×7. Since the combinatorial properties which these illustrate are useful in experimental design apart from their application to the double elimination of error, it is well to know something of them. The 2×2 square

A B
B A

illustrates the fact that the three independent contrasts among 4 objects may be resolved into contrasts between pairs of them in the 3 ways in which such pairs can be chosen, such as the rows, columns, and letters of the square. Similarly, the 8 independent contrasts among 9 objects can be resolved into 4 independent sets of 2 degrees of freedom each, found by dividing the whole into 3 sets of 3 objects. This comes from the fact that not only is a 3×3 Latin square possible, but a 3×3 Græco-Latin square. A pair of letters, one Greek and one Latin, may be assigned to each cell of the square, so that each Latin letter appears once in each row and in each column, and each Greek letter appears once in each row, once in each column, and once with each Latin letter, as is shown below :—

$$\begin{array}{ccc} A\alpha & B\beta & C\gamma \\ B\gamma & C\alpha & A\beta \\ C\beta & A\gamma & B\alpha \end{array}$$

By rearranging the rows among themselves and also the columns, so that the letters in the first row and in the first column are in order, any Latin square can be reduced to a standard form. Since, after rearranging the columns so that A is on the left of the first row, only the remaining rows will be disturbed, each square of the standard form is capable of generating $s!(s-1)!$ different squares, where s is the number of letters in each row or column. When s is 2 or 3 there is only one solution in the standard position, so that the total numbers of Latin squares are 2 and 12 respectively. There is also essentially only one 3×3 Græco-Latin square, namely that shown above; but apart from the rearrangement of the Latin letters in the rows and columns in 12 ways, the Greek letters may be permuted among themselves in 6 ways, making 72 arrangements in all.

With 4×4 squares there are four arrangements of the Latin letters in the standard position, or 4×6×24 = 576 Latin squares. Only one of these 4, however, yields a Græco-Latin square, and that in two ways, so that there are $2 \times 6 \times 24^2$ (=6912) 4×4 Græco-Latin squares. The two arrangements of the Greek letters are, moreover, themselves orthogonal, so that the 15 degrees of freedom among 16 objects may be divided into 5 independent sets of 3 each, being the 3 independent comparisons among 4 sets of 4 objects each, into which the whole may be divided. An arrangement of this kind is shown below, in which numeral suffices are used in place of one set of Greek letters :—

$$\begin{array}{cccc} A_1\alpha & B_2\beta & C_3\gamma & D_4\delta \\ B_4\gamma & A_3\delta & D_2\alpha & C_1\beta \\ C_2\delta & D_1\gamma & A_4\beta & B_3\alpha \\ D_3\beta & C_4\alpha & B_1\delta & A_2\gamma \end{array}$$

There are in all $2 \times 6 \times 24^3$ such arrangements.

The 5×5 Latin squares in the standard position are 56 in number, and fall into two sets. One set of 50 yields no Græco-Latin square, but the set of 6, which are symmetrical about the diagonal, yield each 3 different squares which do not differ merely in a permutation of the Greek letters. There are therefore $3 \times 6 \times 24 \times 120^2$ 5×5 Græco-Latin squares. The three different arrangements are all mutually orthogonal, so that we may add a numeral suffix, as in the 4×4 square above, and obtain $6 \times 6 \times 24 \times 120^3$ solutions. And we may add a second suffix independent of the first, and of the letters, in $6 \times 6 \times 24 \times 120^4$ different ways. An example using two suffices is shown below :—

$$\begin{array}{ccccc} A_1\alpha_1 & B_2\beta_2 & C_3\gamma_3 & D_4\delta_4 & E_5\epsilon_5 \\ B_3\delta_5 & C_4\epsilon_1 & D_5\alpha_2 & E_1\beta_3 & A_2\gamma_4 \\ C_5\beta_4 & D_1\gamma_5 & E_2\delta_1 & A_3\epsilon_2 & B_4\alpha_3 \\ D_2\epsilon_3 & E_3\alpha_4 & A_4\beta_5 & B_5\gamma_1 & C_1\delta_2 \\ E_4\gamma_2 & A_5\delta_3 & B_1\epsilon_4 & C_2\alpha_5 & D_3\beta_1 \end{array}$$

Consequently, the 24 degrees of freedom among 25 objects can be subdivided into 6 independent sets of 4 corresponding to the rows, columns, Latin letters, first suffices, Greek letters and second suffices in the square above. Such a square may be said to be completely orthogonal.

Completely Orthogonal 8×8 *Square*

1 1 1 1 1 1 1	2 5 7 3 8 4 6	3 2 5 4 7 6 8	4 3 2 6 5 8 7	5 7 8 2 6 3 4	6 4 3 8 2 7 5	7 8 6 5 4 2 3	8 6 4 7 3 5 2
2 2 2 2 2 2 2	1 8 3 7 5 6 4	7 1 8 6 3 4 5	6 7 1 4 8 5 3	8 3 5 1 4 7 6	4 6 7 5 1 3 8	3 5 4 8 6 1 7	5 4 6 3 7 8 1
3 3 3 3 3 3 3	7 4 2 1 6 5 8	1 7 4 5 2 8 6	5 1 7 8 4 6 2	4 2 6 7 8 1 5	8 5 1 6 7 2 4	2 6 8 4 5 7 1	6 8 5 2 1 4 7
4 4 4 4 4 4 4	6 3 8 5 7 1 2	5 6 3 1 8 2 7	1 5 6 2 3 7 8	3 8 7 6 2 5 1	2 1 5 7 6 8 3	8 7 2 3 1 6 5	7 2 1 8 5 3 6
5 5 5 5 5 5 5	8 1 6 4 2 3 7	4 8 1 3 6 7 2	3 4 8 7 1 2 6	1 6 2 8 7 4 3	7 3 4 2 8 6 1	6 2 7 1 3 8 4	2 7 3 6 4 1 8
6 6 6 6 6 6 6	4 7 5 8 3 2 1	8 4 7 2 5 1 3	2 8 4 1 7 3 5	7 5 3 4 1 8 2	1 2 8 3 4 5 7	5 3 1 7 2 4 8	3 1 2 5 8 7 4
7 7 7 7 7 7 7	3 6 1 2 4 8 5	2 3 6 8 1 5 4	8 2 3 5 6 4 1	6 1 4 3 5 2 8	5 8 2 4 3 1 6	1 4 5 6 8 3 2	4 5 8 1 2 6 3
8 8 8 8 8 8 8	5 2 4 6 1 7 3	6 5 2 7 4 3 1	7 6 5 3 2 1 4	2 4 1 5 3 6 7	3 7 6 1 5 4 2	4 1 3 2 7 5 6	1 3 7 4 6 2 5

Although there are 9408 6×6 Latin squares in the standard position, belonging to 12 distinct types, yet none of these yields a Græco-Latin square, a conclusion arrived at by Euler after a considerable investigation, but only recently established for certain by the enumeration of the actual types which occur. Græco-Latin squares are easily formed with 7 units in a side, or any other odd number; the 7×7 squares were enumerated by Norton in 146 species; later A. Sade found the 147th species, adding 14,112 standard squares to the 16,927,968 found by Norton. It may be shown that with any prime number (p) the p^2-1 degrees of freedom among p^2 objects may be separated into $p+1$

independent sets of $p-1$ degrees of freedom each, each representing comparisons among p groups of p objects. Yates has shown that this is also true of 8^2 objects, and the same can be done with 9^2, as the examples illustrate. Stevens has demonstrated the possibility in general for powers of prime numbers.

After the rows and columns, the categories of subdivision are represented by the numbers in the cells of each square. The squares given may be randomised by permuting, (*a*) the rows, columns, and each set of numbers among themselves, (*b*) whole sets of numbers with each other and with the rows and columns.

Thus, in the 8×8 square shown above, the seven numbers represent seven different ways of dividing 64 objects into 8 groups of 8 each, making with the rows and columns 9 ways in all, so that no two objects are classified alike in any two of the nine ways.

Completely Orthogonal 9×9 *Square*

1111	2322	3233	9549	7457	8668	5975	6886	4794
1111	4546	7978	3834	6369	9492	2627	5753	8285
2246	3154	1365	7672	8583	9491	6717	4928	5839
8322	2454	5889	7715	1247	4673	9538	3961	6196
3378	1289	2197	8414	9625	7536	4843	5751	6962
6233	9665	3797	5926	8158	2581	4419	7842	1374
4435	5616	6524	3861	1742	2953	8399	9277	7188
9744	3279	6612	8567	2993	5135	7351	1486	4828
5567	6448	4659	1996	2874	3785	9132	7313	8221
4955	7187	1523	6448	9871	3316	5262	8694	2739
6693	4571	5482	2738	3919	1827	7264	8145	9356
2866	5398	8431	1659	4782	7224	3143	6575	9917
7729	8937	9818	6255	4166	5344	2681	3592	1473
5477	8813	2345	4291	7636	1768	6984	9129	3552
8852	9763	7941	4387	5298	6179	3426	1634	2515
3688	6721	9256	2172	5514	8949	1895	4337	7463
9984	7895	8776	5123	6331	4212	1558	2469	3647
7599	1932	4164	9383	3425	6857	8776	2218	5641

35·01 Configurations in Three or More Dimensions

Instead of considering a square, we may consider a configuration of n^3 elements, arranged in layers of n^2 elements each, there being three sets of such layers intersecting at right angles. Then any two layers of different sets will intersect in n common elements.

The first n letters of the Latin alphabet may be assigned to n^2 elements each, so that n of these fall in each layer of each of these sets. We thus form a Latin cube. If the process were repeated with n letters of the Greek alphabet, with the added restriction that each of the Latin letters coincides n times with each of the Greek letters, we should have a Græco-Latin cube. The greatest possible number of alphabets which can be accommodated in this way is $(n-1)(n+2)$, or n^2+n-2, but so many as this may not be possible for particular values of n. Together with the three modes of subdivision of the elements into n sets of n^2 each provided by the three sets of layers of the cube, the maximum number of modes of subdivision is n^2+n+1.

An elementary proof of this limitation, though somewhat an intricate one, is given below.

Let us suppose that there are q categories each dividing the n^3 elements into n sets of n^2 elements each. Then the number of pairs of elements alike in respect to any one category will be

$$\tfrac{1}{2}n^3(n^2-1),$$

and for all categories the number must be

$$\tfrac{1}{2}qn^3(n^2-1).$$

Now any pair of elements may be alike in respect to a certain number of categories, let us say p, which may differ from one pair to another, and may be zero for some pairs, but in any case we must have the relationship

$$\tfrac{1}{2}qn^3(n^2-1) = \Sigma(p);$$

where

$$\Sigma(p)$$

stands for the sum of the values of p for all possible pairs.

Further, each pair of categories divides the elements into n^2 sets of n elements each, such that members of the same set are alike in respect of both categories. Hence the total number of pairs alike in both categories is

$$\tfrac{1}{2}n^3(n-1),$$

and, as there are

$$\tfrac{1}{2}q(q-1)$$

pairs of categories in all, the total number of cases in which two elements are alike in two categories is

$$\tfrac{1}{4}q(q-1)n^3(n-1);$$

but a pair of elements alike in p categories will contribute

$$\tfrac{1}{2}p(p-1)$$

cases to this total. Hence

$$\tfrac{1}{2}q(q-1)n^3(n-1) = \Sigma(p^2-p).$$

We may now use the inequality

$$\tfrac{1}{2}n^3(n^3-1)\Sigma(p^2) \geqslant \Sigma^2(p),$$

since the number of pairs of elements, each contributing a value of p, is

$$\tfrac{1}{2}n^3(n^3-1);$$

the limiting equality being realised only when all values of p are equal.

Hence
$$\tfrac{1}{4}q^2n^6(n^2-1)^2 \leqslant \tfrac{1}{4}n^6(n^3-1)(n-1)q(q-1)+\tfrac{1}{4}n^6(n^3-1)(n^2-1)q$$
whence dividing by
$$\tfrac{1}{4}qn^6$$

$$q\{(n^2-1)^2-(n^3-1)(n-1)\} \leqslant (n^3-1)(n^2-n),$$
or
$$n(n-1)^2 q \leqslant n(n-1)^2(n^2+n+1)$$
$$q \leqslant n^2+n+1.$$

The largest possible number of categories is therefore
$$n^2+n+1,$$
and, if this number is realised, the number of categories in which any two elements are alike must be constant and equal to
$$\frac{n^2-1}{n^3-1}(n^2+n+1) = n+1.$$

If instead of a cube we have a configuration of n^r elements, the form of proof used above leads to the result that the greatest possible number of categories is

$$(n^r-1)/(n-1),$$

and that with this number every pair of elements will be alike in
$$(n^{r-1}-1)/(n-1)$$
different categories.

It is demonstrable that this maximum number can be realised whenever n is a power of a prime. The solution for a cube of two in seven categories solves a problem of confounding in 45·1. It is not known how many alphabets can be accommodated in a cube of six. An example of a cube with $n = 3$, in ten alphabets, represented by the ten positions in the triangles, is shown below. The first three letters of each alphabet are represented by 0, 1 and 2.

Cube of 3 in Ten Alphabets

0000	1111	2222	0012	1120	2201	0021	1102	2210
000	001	002	121	122	120	212	210	211
00	21	12	11	02	20	22	10	01
0	1	2	2	0	1	1	2	0
1200	2011	0122	1212	2020	0101	1221	2002	0110
111	112	110	202	200	201	020	021	022
12	00	21	20	11	02	01	22	10
1	2	0	0	1	2	2	0	1
2100	0211	1022	2112	0220	1001	2121	0202	1010
222	220	221	010	011	012	101	102	100
21	12	00	02	20	11	10	01	22
2	0	1	1	2	0	0	1	2

Such configurations in three or more dimensions greatly facilitate the solution of problems of confounding discussed in sections 35·1 to 47.

35·1. An Exceptional Design

The principal use of Græco-Latin and higher squares consists in clarifying complex combinatorial situations, and will be illustrated in later chapters. Occasionally, however, they may be used directly in experimental design. An example of this exceptional use has been given by L. H. C. Tippett.

In a cotton mill one of 5 spindles was found to be winding defective weft. The cause of the defect was unknown. Its origin could, however, be traced further by interchanging the component parts of the spindle. Four portions could thus be interchanged, designated by the symbols, w, x, y and z.

These parts were reassembled in twenty-five ways, so that each of the five components w was used once in combination with each of the five components x, and so with each pair of components. The reconstructed spindles were tested five at a time, over five periods, so that each component was used equally during each period.

The combinations used and the distribution of defects in the resulting weft are shown in Table 8·1.

TABLE 8·1

Spindle Composition and Defects Observed

Period.	Composition of Spindle.				
1	$w_1x_1y_1z_1$	$w_2x_2y_2z_2$ ++	$w_3x_3y_3z_3$ +	$w_4x_4y_4z_4$	$w_5x_5y_5z_5$
2	$w_1x_3y_4z_5$	$w_2x_1y_5z_4$ ++	$w_3x_2y_1z_5$	$w_4x_3y_2z_1$	$w_5x_4y_3z_2$
3	$w_1x_4y_2z_5$ +	$w_2x_5y_3z_1$ ++	$w_3x_1y_4z_2$	$w_4x_2y_5z_3$	$w_5x_3y_1z_4$
4	$w_1x_5y_5z_2$	$w_2x_4y_1z_3$ ++	$w_3x_5y_2z_4$	$w_4x_1y_3z_5$	$w_5x_2y_4z_1$
5	$w_1x_2y_3z_4$ +	$w_2x_3y_4z_5$ ++	$w_3x_4y_5z_1$	$w_4x_5y_1z_2$ +	$w_5x_1y_2z_3$

In the table moderate defect is marked +, and severe defect ++. It will be seen that severe defect is confined to the five spindles made up using the component w_2, and is invariably present when this component is used. The origin of the trouble was therefore traced to this part, and this knowledge led to the discovery of a cause of defect previously unsuspected.

Of the remaining 20 bobbins, 4 showed defect to a moderate degree. There is no sufficient reason to ascribe this partial failure to the parts used; the only parts which show defect more than once are w_1 and y_3.

In general, the danger of trying to test five different factors (the four components and the period) each susceptible of five variants, in only 25 trials, lies in the fact that we are examining only 25, out of 3125 possible combinations. In consequence any interactions between the effects of one component and those of another may appear in the results as due to a third component. The value of the test for its special purpose arises from the technologist's assurance that such consistent interaction is, in the special circumstances, very improbable. He has, therefore, confidence in interpreting Table 8·1 as condemning the component w_2. A theorist without such practical knowledge would, however, have no difficulty in accounting for the effect in terms of x, y and z, and without taking account of w. For example, if in the table we add the suffices of x and y and subtract that of z, severe defect is only found when the resulting number is 2 or 7, and is then invariably found. Such a result might be actually misleading if the suffices represented quantitative physical attributes of the components, and had not been assigned arbitrarily by the experimenter.

36. Practical Exercises

For experimental purposes it is, of course, the properties of the smaller squares that are most useful. The following data supply an example for readers who wish to familiarise themselves with the analysis of variance in its application to experiments designed in a Latin square. They represent the results of such an experiment carried out in potatoes at Ely in 1932. The six treatments designated by A B C D E F consist of different quantities of nitrogenous and phosphatic fertilisers. The results in lbs. of potatoes are quoted, with some simplification, from the 1932 Report of Rothamsted Experimental Station. The analysis may be checked by comparison with Tables 31 and 32.

TABLE 9
Arrangement and Yields of a Latin Square

E 633	B 527	F 652	A 390	C 504	D 416	3122
B 489	C 475	D 415	E 488	F 571	A 282	2720
A 384	E 481	C 483	B 422	D 334	F 646	2750
F 620	D 448	E 505	C 439	A 323	B 384	2719
D 452	A 432	B 411	F 617	E 594	C 466	2972
C 500	F 505	A 259	D 366	B 326	E 420	2376
3078	2868	2725	2722	2652	2614	16659

The following puzzle is of service in familiarising the mind with the combinatorial relationships underlying the use of the Latin square, and the like, in experimental design:—

Sixteen passengers on a liner discover that they are an exceptionally representative body. Four are Englishmen, four are Scots, four are Irish, and four are Welsh. There are also four each of four different ages, 35, 45, 55 and 65, and no two of the same age are of the same nationality. By profession also four are lawyers, four soldiers, four doctors and four clergymen, and no two of the same profession are of the same age or of the same nationality.

It appears, also, that four are bachelors, four married, four widowed and four divorced, and that no two of the same marital status are of the same profession, or the same age, or the same nationality. Finally, four are conservatives, four liberals, four socialists and four fascists, and no two of the same political sympathies are of the same marital status, or the same profession, or the same age, or the same nationality.

Three of the fascists are known to be an unmarried English lawyer of 65, a married Scots soldier of 55 and a widowed Irish doctor of 45. It is then easy to specify the remaining fascist.

It is further given that the Irish socialist is 35, the conservative of 45 is a Scotsman, and the Englishman of 55 is a clergyman. What do you know of the Welsh lawyer?

REFERENCES AND OTHER READING

Rothamsted Experimental Station Annual Report (1932).

R. A. FISHER (1925-1963). Statistical methods for research workers. Chap. VIII., § 49.

R. A. FISHER and F. YATES (1934). The 6×6 Latin squares. Proceedings of the Cambridge Philosophical Society, xxx. 492-507.

H. W. NORTON (1939). The 7×7 squares. Annals of Eugenics ix. 269-307.

A. SADE (1951). An omission in Norton's list of 7 × 7 squares. Annals of Mathematical Statistics, xxii. 306-7.

W. L. STEVENS (1939). The completely orthogonalised Latin square. Annals of Eugenics, ix. 82-93.

O. TEDIN (1931). The influence of systematic plot arrangement upon the estimate of error in field experiments. Journal of Agricultural Science, xxi. 191-208.

L. H. C. TIPPETT (1934). Applications of statistical methods to the control of quality in industrial production. Manchester Statistical Society.

VI

THE FACTORIAL DESIGN IN EXPERIMENTATION

37. The Single Factor

IN expositions of the scientific use of experimentation it is frequent to find an excessive stress laid on the importance of varying the essential conditions *only one at a time*. The experimenter interested in the causes which contribute to a certain effect is supposed, by a process of abstraction, to isolate these causes into a number of elementary ingredients, or factors, and it is often supposed, at least for purposes of exposition, that to establish controlled conditions in which all of these factors except one can be held constant, and then to study the effects of this single factor, is the essentially scientific approach to an experimental investigation. This ideal doctrine seems to be more nearly related to expositions of elementary physical theory than to laboratory practice in any branch of research. In experiments merely designed to illustrate or demonstrate simple laws, connecting cause and effect, the relationships of which with the laws relating to other causes are already known, it provides a means by which the student may apprehend the relationship, with which he is to familiarise himself, in as simple a manner as possible. By contrast, in the state of knowledge or ignorance in which genuine research, intended to advance knowledge, has to be carried on, this simple formula is not very helpful. We are usually ignorant which, out of innumerable possible factors, may prove ultimately to be the most important, though we may have strong presuppositions that some few of them are particularly worthy of study. We have usually no knowledge that any one factor will exert its effects independently of all others that can be varied, or that its effects are particularly simply related to variations in these other factors. On the contrary, when factors are chosen for investigation, it is not because we anticipate that the laws of nature can be expressed with any particular simplicity in terms of these variables, but because they are variables which can be controlled or measured with comparative ease. If the investigator, in these circumstances, confines his attention to any single factor, we may infer either that he is the unfortunate victim of a doctrinaire theory as to how experimentation should proceed, or that the time, material or equipment at his disposal is too limited to allow him to give attention to more than one narrow aspect of his problem.

The modifications possible to any complicated apparatus, machine or industrial process must always be considered as potentially interacting with one another, and must be judged by the probable effects of such interactions. If they have to be tested one at a time this is not because to do so is an ideal scientific procedure, but because to test them simultaneously would sometimes be too troublesome, or too costly. In many instances, as will be shown in this chapter, the belief that this is so has little foundation. Indeed, in a wide class of cases an experimental investigation, at the same time as it is made more comprehensive, may also be made more efficient if by more efficient we mean that more knowledge and a higher degree of precision are obtainable by the same number of observations.

38. A Simple Factorial Scheme

As an example, let us consider the case in which we require to study experimentally the effects of variations in composition of a mixture containing four active ingredients. It is indifferent for our illustration for what purpose the mixture may be required. It may be an industrial product, a medicinal prescription, a food ration or an artificial manure. It is sufficient that its efficacy in practical use cannot be calculated *a priori*, but can be measured by its effect in particular trials; and that the ideal quantitative composition is unknown in respect to all four ingredients. In principle, it matters little whether our doubts extend to wide variations in the quantities to be employed, or whether the variations in question are proportionately small; as they will be when wide experience has already determined the ideal proportions within narrow limits. Nor will it affect the question in principle, if, as in some cases, the cost of the items is an important consideration to be debited in assessing the net advantage of using more of them; or whether, as in other cases, the direct observations or measurements made in the test are alone to be considered.

So defined, the situation is clearly one of very wide occurrence. Let us consider an experiment in which 16 different mixtures are made up in the 16 combinations possible by combining either a larger or a smaller quantity of each of the 4 ingredients to be tested. The quantities may differ in some cases by a factor as large as 2, in which case each mixture will contain of each ingredient either a single or a double quantity. The factor need not be the same for all ingredients. For some one or more of them we may doubt whether its presence in any quantity is desirable; or whether it had not better be omitted altogether. We shall then test mixtures with or without this component. But, in any case, the general question as to whether, of each ingredient, more or less can be added with advantage, can be settled by making up all combinations, using either more or less of each component.

We will suppose now that 6 tests are made with each mixture, or 96 in all. The particular cases in which the different mixtures are to be tried will be assigned strictly at random; or, as in the agricultural experiment illustrated in Chapter IV. (randomised blocks), the tests will be divided into 6 series, each supposedly more homogeneous than the whole, and the 16 members of each will be assigned at random to different mixtures. Since no difference of principle arises here we will at first suppose that the tests are assigned entirely at random.

In respect of any one particular ingredient the first point to be noted is that we have for comparison 48 cases in which a larger, and 48 in which a smaller quantity has been employed. These 48 are not all alike in other respects. They are alike in sets of 6, but the eight sets differ from each other in the other ingredients used. Corresponding to each set of 6 in which a larger quantity of the first ingredient has been used, there will, however, be a set of 6 with a smaller quantity, and exactly similar to it in respect of the other ingredients. The difference, if any, in the effects observed in these two sets must be ascribed, apart from random fluctuations, to the particular ingredient in which they differ. Moreover, there are 8 such pairs of sets to supply confirmatory evidence of this effect if it exists. For each single factor, therefore, we have a direct comparison of the averages or totals of two sets of 48 trials; and this comparison will have the same precision as if the whole of the 96 trials had been devoted to testing the efficacy of one single component. The first fact contributing to the efficiency of experiments designed on the factorial system, is that every trial supplies information upon each of the main questions which the experiment is designed to examine.

The advantage of the factorial arrangement over a series of experiments, each designed to test a single factor is, however, much greater than this. For with separate experiments we should obtain no light whatever on the possible interactions of the different ingredients, even if we had gone to the labour of performing 96 experiments with each of them, and so, for each singly, had attained the same precision as that which the factorial experiment can give. If, for example, an increase in ingredient A were advantageous in the presence of B, but were ineffective or disadvantageous in its absence, we could only hope to learn this fact by carrying out our test of A both in the presence and in the absence of B ; and this, in fact, is what the factorial experiment is designed to do—but to do so thoroughly that the total system of possible interactions is explored in its entirety. To test if the effect of A in the presence of B is greater than in its absence, we may compare the total difference ascribable to A from one set of 24 pairs of otherwise comparable trials, in all of which B is present, with the corresponding effect derived from the remaining pair of sets of 24, in all of which B is absent. This is the same as comparing the results of 48 trials in which A and B are both employed in large, or both in small, quantity, with the other 48 trials in which the larger quantity of A and the smaller quantity of B, or *vice versâ*, have been combined. We have thus again a comparison between the results of two sets of 48 trials, comparable in all other respects save that in which we are interested, namely, whether the effect of an increase in A is, or is not, influenced by an increase in B. This difference, if it exists, might equally be expressed by saying that an increase in B is influenced in its effect by an increase in A. The difference, in fact, involves the two ingredients symmetrically, and is technically spoken of as the interaction of A with B. There are clearly 6 such interactions between pairs of the 4 ingredients, and each of these is evaluated by the experiment with the same precision.

The 4 contrasts for single ingredients, and the 6 interactions between pairs of them, still do not exhaust the possible interactions which may be present, and which the experiment is competent to reveal. It might be, for example, that the interaction between A and B is itself influenced by the quantity of the ingredient C. If we were to calculate the interaction of A and B for those mixtures which had a larger quantity of C, and subtract from this the corresponding interaction for mixtures having the smaller quantity of C, we should, in fact, be adding up all the results from mixtures having either all three or any one of the ingredients A, B and C present in the larger quantity, and subtracting from this total the effects of all the mixtures having either none or any two. The measure of discrepancy is, therefore, symmetrically related to the three ingredients A, B and C, and is known as the interaction of these three ingredients. If such an interaction exists the fact may be stated, as above, by saying that the interaction of A and B depends on the quantity of C present. Equally, we might make the equivalent statement that the interaction of B and C depends on the quantity of A present, or that the interaction of C and A depends on the quantity of B. The three statements are logically equivalent; and for any set of three ingredients there will be only one numerical measure of the interaction to be ascertained from the data. Since there are four sets of three which we might choose, there are four triple interactions which can be evaluated, each of which, like the interactions between two ingredients, is found by the comparison of a set of 48 chosen tests, with another set of 48 in other respects comparable with it.

Finally, if we ask whether the interaction of A, B and C is dependent on the quantity of D present, it appears that the answer to our question must be a comparison symmetrically related to all four ingredients. This is made by comparing the effects of those mixtures in which there is a larger quantity of 4, 2 or none of the ingredients with the effects of those mixtures in which there is a larger quantity of 3 or 1 of them. There is thus only one quadruple interaction in the system, and this will be evaluated with the same precision as all the others. We thus find that the 15 independent comparisons among the 16 different mixtures, which have been made up, may be logically resolved into 15 intelligible components, as shown in the following table :—

TABLE 10

Effects of single ingredients	4
Interactions of 2 ,,	6
,, ,, 3 ,,	4
,, ,, 4 ,,	1
Total	15

The numbers are the coefficients of the binomial expansion $(1+x)^4$, omitting the first. Each particular interaction may be clearly designated by the selection of letters, such as ABC, which represent the ingredients contributing to it. With respect to sign, it is a convenient convention to speak of an interaction as positive when the measured effects of the set of mixtures which contain all the ingredients involved in larger quantity exceed the measured effects of the remaining set.

The second advantage which we may note, therefore, in a factorially arranged experiment is that, in addition to measuring the effects of the four single ingredients with the same precision as though the whole of the experiment had been devoted to each of them, it measures also the 11 possible interactions between these ingredients with the same precision. These interactions may, or may not, be considerable in magnitude. It is none the less of importance in practical cases to know whether they are considerable or not.

The precision with which the 15 comparisons have been made is, of course, estimated from the variation between the results of the 6 trials of each mixture. Each mixture will therefore provide 5 degrees of freedom for making a pooled estimate of the precision of the experiment, thus giving 80 degrees of freedom in all. The analysis of variance of the 96 trials thus takes the simple form :—

TABLE 11

	Degrees of Freedom.
Treatments	15
Error	80
Total	95

The sum of squares corresponding to error is divided by 80 to obtain the estimated variance of a single experiment. This, multiplied by 96, gives the estimated variance of the difference between the sum of any chosen set of 48 results and the sum of the remainder. Thus we have only to add one-fifth of its value to the sum of squares for error, and take the square root, to find the standard deviation appropriate to each of the

15 comparisons that have been made, and in relation to which the significance of each may be judged. Since 80 is a relatively ample number of degrees of freedom for the estimation of error, those differences may conveniently be judged significant, which exceed twice their standard errors.

Had the experiment been arranged, for greater precision, in 6 blocks of trials, we should in the analysis eliminate the differences between these blocks from the estimate of error. We should then have :—

TABLE 12

	Degrees of Freedom.
Blocks	5
Treatments	15
Error	75
Total	95

leaving 75 degrees of freedom for the estimation of error; so that the sum of squares for error must be multiplied by 96/75 or 1·28 to obtain the variance of any comparison between 48 chosen tests and the remaining 48.

39. The Basis of Inductive Inference

We have seen that the factorial arrangement possesses two advantages over experiments involving only single factors : (i) Greater *efficiency*, in that these factors are evaluated with the same precision by means of only a quarter of the number of observations that would otherwise be necessary ; and (ii) Greater *comprehensiveness* in that, in addition to the 4 effects of single factors, their 11 possible interactions are evaluated. There is a third advantage which, while less obvious than the former two, has an important bearing upon the utility of the experimental results in their practical application.

This is that any conclusion, such as that it is advantageous to increase the quantity of a given ingredient, has a wider inductive basis when inferred from an experiment in which the quantities of other ingredients have been varied, than it would have from any amount of experimentation, in which these had been kept strictly constant. The exact standardisation of experimental conditions, which is often thoughtlessly advocated as a panacea, always carries with it the real disadvantage that a highly standardised experiment supplies direct information only in respect of the narrow range of conditions achieved by standardisation. Standardisation, therefore, weakens rather than strengthens our ground for inferring a like result, when, as is invariably the case in practice, these conditions are somewhat varied. As the analysis of variance clearly shows, such standardisation of conditions is only of value in increasing the precision of the experimental data when it is applied to the tests to be compared experimentally, differences between which affect the real errors, and the estimates of error with which they are randomised. Such standardisation is out of place when applied to parallel tests designed, by multiplying the observations, to increase the precision of all comparisons. In fact, as the factorial arrangement well illustrates, we may, by deliberately varying in each case some of the conditions of the experiment, achieve a wider inductive basis for our conclusions, without in any degree impairing their precision.

40. Inclusion of Subsidiary Factors

Factorial types of arrangement, by reconciling the desiderata of a relatively wide exploration with high precision, without sacrifice of either advantage, are eminently suitable when alternative procedures of any kind are apparently open to the experimenter. Thus it may be doubtful whether the cultivation of an orchard is most advantageously carried out in the spring or the autumn of the year. More than one method of cultivation, also, may be advocated with apparently equal force. Numerous similar examples will occur when the modification of any practical procedure is under consideration. Schools of opinion are formed on points which without systematic trial we cannot know to have, or to lack, real importance, as bee-keepers dispute whether the combs should lie parallel from the front of the hive to the back, or should lie transversely to this direction. These are all qualitative differences, but, provided that there is a quantitative method of measuring any real advantage which they may confer, they may be incorporated in experiments designed primarily to test other points ; with the real advantages that, if either general effects or interactions are detected, that will be so much knowledge gained at no expense to the other objects of the experiment ; and that, in any case, there will be no reason for rejecting the experimental results on the ground that the test was made in conditions differing in one or other of these respects from those in which it is proposed to apply the results.

There is another feature widespread in experimental work where the factorial arrangement relieves the investigator of a troublesome cause of anxiety ; namely, that the isolation of a single factor for separate experimentation can often be achieved only at the expense of a somewhat questionable arbitrariness of definition. To consider a very simple case, the experimenter may undertake to test the respective yields of two or more varieties of a cereal crop. At first sight it will readily be admitted that all other factors likely to affect the yield shall be made the same for the varieties to be compared. So the quantity of seed sown must be equalised. But if the seed-rates are measured in bushels per acre, the bushels, measured as equal volumes of seed of the different varieties, may differ in their weight; or, again, they may differ even more conspicuously in the numbers of seeds which they contain. If the test is to be carried out at the same single seed-rate for each variety, the equality of seed-rate must be defined, for the purpose of the test, with some degree of arbitrariness. Unless we have assurance that variations in seed-rate will not affect the yield obtained, this arbitrariness will infect the results of the experiment. Clearly, on the question at issue, as between the varieties, it would be desirable to make a comparison using for each variety that seed-rate which is most profitable for it. This may differ from equality, as measured by volume or by weight, or by number of seeds to the acre, by reason of the different capacities of the varieties to withstand causes of death, or, by forming additional shoots, to fill up vacant spaces. The experimenter who, in testing the varieties, at the same time makes a sufficient variation in seed-rates to embrace the optimal value, is clearly in a better position to meet the criticisms which may arise from these considerations, than is one who adopts any arbitrary conventions as to what the phrase " equality of conditions " is intended to convey.

In the conditions of cereal cultivation, moreover, variations in seed-rate inevitably raise with them further questions. In particular, what space should be left between the seed-rows or drills? At a heavier seed-rate it is reasonable to suppose that the drills could with advantage be placed nearer together than they could if less corn were sown. Consequently, the question as to what is the most advantageous seed-rate can never be answered experimentally without a simultaneous variation in the width of the drills. Equally, no investigation of the question of drill-width can be satisfactory unless the amount of seed sown be also varied. The simple question, therefore, of the comparison of the yields of different varieties of the same crop carries with it the simultaneous investigation of the effects of variations in other items of agricultural procedure. These do not, however, if a logical and comprehensive plan of experimentation be adopted, add to the cost or labour of an effective experimental programme. For extensive replication of the plots sown is in any case a necessity, if accurate results are to be attained; and this replication, at the same time as it increases the precision of comparisons, may be used simultaneously to supply the desired variations in all conditions likely to be bound up with that which is the primary object of the investigation.

For a given number of trials, the more experimental variants are tried the fewer will be the absolute replicates. Thus, with 96 trials, we may have 6 absolute replicates of 16 different experimental variants, and these, as we have seen, will still leave 80 degrees of freedom for the estimation of error. With the same material we might test 48 different mixtures, or treatments, and still have duplicate results for each. There would then be 48 degrees of freedom left for error. Although each test is only made in duplicate, yet all the primary questions, into which the differences among them may be resolved, are answered with the same precision as though the whole experiment had been devoted to each of these questions alone; the loss of absolute replication is made good by the hidden replication inherent in the factorial arrangement. The experimental error is no larger, although, being based on only 48 independent comparisons, it is known with a slightly lower precision than if 80 had been available. With factorial experiments designed to make a large number of comparisons, there will, in fact, usually be an ample number of degrees of freedom for the estimation of error.

41. Experiments without Replication

It may occasionally be desirable to dispense with absolute replication altogether. This occurs when it is required to test a large number of combinations simultaneously without enlarging the experiment so greatly as to make repeated use of each combination; and especially, when there is reason to believe that most of the interactions involving 3 or more factors will be unimportant experimentally, in the sense that their real effects, if any, will be too small to be statistically significant, in an experiment of the size contemplated. In such cases the whole of the independent comparisons which the experiment provides may be assigned to the factors tested and their interactions. There will in fact be none ascribable to pure error, but there will be numerous interactions the apparent effects of which are principally due to error, and these may be used to provide a measure of the precision of the more important comparisons.

For example, we may have 6 factors each providing two alternatives to be tested; and though we do not know *a priori* that these factors are unrelated, we may have reason to think that their action should be sufficiently nearly independent for all but the simple interactions between pairs of them to be experimentally negligible. If each of the combinations be tried once, the 63 independent comparisons which the experiment provides may be analysed as in the following table.

TABLE 13

	Degrees of Freedom
Single factors	6
Interaction between 2 factors	15
,, ,, 3 ,,	20 ⎫
,, ,, 4 ,,	15 ⎬ Error 42
,, ,, 5 ,,	6 ⎪
,, ,, 6 ,,	1 ⎭

The 6 primary effects of the individual factors will each be determined, as we have seen, by the whole weight of the evidence of 64 trials. To test the significance of the differences observed, we may use the 42 degrees of freedom for interactions involving more than 2 factors to supply an estimate of error; that is, an estimate of the variance, due to error, of a single trial. The same test may be applied to any of the 15 interactions between pairs of factors which may seem to be possibly significant. If none of these are very important compared with the average of the remainder we have an empirical confirmation of the supposition upon which the experiment was designed, that the 6 primary factors are not strongly related. If, however, contrary to expectation, it appeared that there was a large interaction between two of these factors, it would be advisable to examine separately the interactions between 3 factors which involve these two. We should thus pick out 4 particular suspects out of the 42 degrees of freedom provisionally ascribed to error, leaving 38 in comparison with which their significance could be tested.

Such a plan could, of course, fail; and would do so if a large number of the high order interactions were more important than the primary effects. It would, however, not likely be tried in these circumstances. More frequently we should find that the true error had been but slightly inflated, and the effective precision of the experiment slightly reduced, by the inclusion in the estimate of error of some small components really due to interactions of the factors tested.

REFERENCES AND OTHER READING

R. A. FISHER (1926). The arrangement of field experiments. Journal of Ministry of Agriculture, xxxiii. 503-513.

T. EDEN and R. A. FISHER (1927). The experimental determination of the value of top dressings with cereals. Journal of Agricultural Science, xvii. 548-562.

VII

CONFOUNDING

42. The Problem of Controlling Heterogeneity

It has been shown in the last chapter that great advantages may be obtained by testing experimentally an aggregate of variants, systematically arranged on the factorial scheme. The illustrations have shown that such aggregates may be very numerous. When, in Chapters III. and IV., the advantages of pairing, or of grouping, the material in relatively homogeneous blocks was discussed, it was seen that the precision attainable by a given amount of experimentation was liable to be reduced, when the number of comparisons to be made was large, by reason of the increased heterogeneity, which must in practice then be permitted, among the tests in the same group.

In agricultural experimentation this effect expresses itself very simply in the increased size of the blocks of land, each of which is to contain plots representative of all the different combinations to be tested. Thus if there are 48 different combinations, each block will have to be nearly an acre in extent, and it is common experience that within so large an area considerably greater soil heterogeneity will be found, than would be the case if the blocks could be reduced in size to a quarter of an acre or less. The same consideration applies to experimentation of all kinds. If large quantities of material are needed, or large numbers of laboratory animals, these will almost invariably be more heterogeneous than smaller lots could be made to be. In like manner, extensive compilations of statistical material often show evidence of such heterogeneity among the several parts which have been assembled, and are seriously injured in value if this heterogeneity is overlooked in making the compilation.

In many fields of experimentation quantitative knowledge is lacking as to the degree of heterogeneity to be anticipated in batches of material of different size, or drawn from more or less diverse sources. This is a drawback to precise planning, which increased care in experimental design will doubtless steadily remove. While, therefore, greater heterogeneity is always, on general principles, to be anticipated, when the scope of an experimental investigation is to be enlarged, this feature will often do but little to annul the advantages discussed in the last chapter. Nevertheless, the means by which such heterogeneity can be controlled are widely applicable, and will generally give a further increase in precision. In agricultural field trials, where the study of heterogeneity has been itself the object of a great deal of deliberate investigation, it is certain that the further advantages to be gained are very considerable.

In the last chapter, we have seen that a factorially arranged experiment supplies information on a large number of experimental comparisons. Some of these, such as the effects of single factors, will always be of interest. It is seldom, too, that we should be willing to forgo knowledge of any interactions which may exist between pairs of these factors. But, in the case of interactions involving 3 factors or more, the position is often somewhat different. Such interactions may with reason be deemed of little experimental value, either because the experimenter is confident that they are quantitatively unimportant, or because, if they were known to exist, there would be no immediate prospect of the fact being utilised. In such cases we may usefully adopt the artifice known as "confounding." This consists of increasing the number of blocks, or groups of relatively homogeneous material, beyond the number of replications in the experiment, so that each replication occupies two or more blocks; and, at the same time, of arranging that the experimental contrasts between the different blocks within each replication shall be contrasts between unimportant interactions, the study of which the experimenter is willing to sacrifice for the sake of increasing the precision of the remaining contrasts, in which he is specially interested. To do this it must be possible to evaluate these remaining contrasts solely by comparisons within the blocks. It is not necessary, however, that comparisons within any one block should provide the required contrast, but only that it should be possible to build this up by comparisons within all the blocks of a replication.

43. Example with 8 Treatments, Notation

A very simple example of confounding suggests itself when we have 3 factors, of only 2 variants each, and unite them in an experiment involving the 8 combinations, which they provide. It was in such an experiment that the principle of confounding was first used. As a convenient notation for these cases we may call the 3 factors A, B and C. Let us choose to regard one of these 8 variants as a standard, or "control," and denote it by the symbol (1). The variant which differs from this only in the factor A we will denote by (*a*), likewise (*b*) and (*c*) will stand for the two other treatments which differ from the control in respect of the factors B and C. The remaining treatments will differ from the control in either two, or in all three of the factors used, and may therefore be denoted unequivocally by the symbols (ab), (ac), (bc) and (abc). The same symbols may be used for the treatments, and for the quantitative measures of the results of these treatments, which the experiment is designed to ascertain. Thus (ab) will stand either for a particular treatment applied to an agricultural crop, or to the total yield of the plots which have received this treatment. Equally, it might be the average live weight, or the average longevity, of experimental animals which have been treated in this way, if the experiment is aimed at studying the conditions which influence weight or longevity.

In contradistinction to the treatments, or experimental variants, denoted by such symbols as (a) we shall use A, B, etc., to denote the experimental contrasts found for the factors, and for their interactions. Thus A will stand for the factor A and for its effect as measured by summing the treatments the symbols of which contain a, and deducting those which do not; *i.e.*

$$A = (abc)+(ab)+(ac)+(a)-(bc)-(b)-(c)-(1),$$

or, by analogy with the rules of multiplication of algebraic quantities,

Similarly,
$$A = (a-1)(b+1)(c+1).$$
$$B = (b-1)(a+1)(c+1),$$
and
$$C = (c-1)(a+1)(b+1).$$

The corresponding expressions for the interactions between pairs of factors are then easily seen to be

$$AB = (a-1)(b-1)(c+1),$$
$$BC = (b-1)(c-1)(a+1),$$
and
$$CA = (c-1)(a-1)(b+1).$$

Finally, the interaction of all 3 factors has the symbolic expression
$$ABC = (a-1)(b-1)(c-1).$$

The purpose of the symbolism is to make it easy to denote any particular contrast, or interaction, and to ascertain at once how the treatments should be compounded in evaluating it.

44. Design suited to Confounding the Triple Interaction

Such a set of eight treatments might be not inconveniently tested in groups of eight experimental trials, on relatively homogeneous material, or on blocks of land, each containing 8 plots. If, however, we decide in advance that the whole value of the experiment lies in the simple contrasts A, B, C, and in the interactions between pairs of factors, AB, BC, and CA, while the interaction between all 3 factors, ABC, is unimportant and may be neglected, then it is possible to divide the land into blocks of only four plots each, or, in general, to subdivide the experimental material in groups of four, choosing thus groups of greater homogeneity than could be obtained if 8 had to be included in each group. To do this we notice that the particular interaction ABC, which we are willing to sacrifice, is a simple contrast between one particular four of our eight treatments, namely,
$$(abc)+(a)+(b)+(c),$$
and the remaining four
$$(ab)+(bc)+(ca)+(1),$$
as in Fig. 1.

(c)	(bc)
(abc)	(ab)
(a)	(1)
(b)	(ac)

FIG. 1.—Diagram showing the arrangement of eight treatments in two complementary blocks belonging to the same replication.

If, therefore, these sets of four treatments are grouped together in blocks, and two such blocks be assigned to each replication, the contrast between the blocks within each replication is merely the treatment contrast which we are willing to sacrifice. This contrast has, therefore, been confounded indistinguishably with the differences between blocks, which are intended to be eliminated from the experimental errors of the comparisons we wish to make, and from the estimate of these errors. The remaining contrasts, representing single factors, or interactions between pairs of them, though none of them can be made by comparisons within a single block, are all built up by combining a contrast of one pair of treatments and another in the same block, with a similar contrast inside the other block of the replication. The reader should satisfy himself that this is so by examining each of these contrasts. It is then apparent that the errors to which these contrasts are subject arise solely from heterogeneity within the sets of four trials constituting the blocks, and that the differences between different blocks contribute nothing to the experimental error. By confounding the one unimportant contrast with the differences between blocks, it is therefore possible to evaluate the six more important contrasts with whatever added precision is attained by using more homogeneous material.

45. Effect on Analysis of Variance

As an example, let us suppose such a test were carried out with 5 replications. There would then be ten blocks, and 9 degrees of freedom belong to the contrasts between these blocks, which have been eliminated from the experimental error. Of these 9, one may be identified with the treatment contrast which has been confounded. The three independent comparisons, within each of the ten blocks of 4, must be assigned 30 degrees of freedom. Of these, 6 stand for the treatment contrasts evaluated, and the remaining 24 for the experimental error available for estimating the precision of the experiment, and for testing the significance of any particular result. The analysis of any such experiment is, therefore, of the form given below:—

TABLE 14

	Degrees of Freedom.
Blocks	9
Treatments	6
Error	24
Total	39

It is often instructive, and affords a useful check in more confusing examples, to see how the components ascribable to error may be obtained independently, rather than only by subtraction from the total. In this case there are two sets of five blocks, each containing identical treatments. The three differences among these treatments have, therefore, each been evaluated 5 times, and the 4 discrepancies between these 5 values will give 12 differences due wholly to error. Equally, the other set of five blocks contribute the other 12 degrees of freedom or error, making 24 in all, the total required.

Without confounding, the analysis of the experiment would read:—

TABLE 15

	Degrees of Freedom.
Blocks	4
Treatments	7
Error	28
Total	39

so that the effect of the subdivision of the replications has been to eliminate 5 additional degrees of freedom.

one from the treatments and four from the error. If greater homogeneity has in fact been obtained from the subdivision, the components by which the error has been diminished will have carried away a disproportionate share of the residual variation.

The subdivision of each replication into two or more blocks does not prevent, when this is desired, the isolation, among the degrees of freedom assigned to error, of the particular components of error which affect any chosen comparison within the blocks. Since, however, the comparisons in which we are interested, such as A, are built up of comparisons within blocks of different kinds, they are equally affected by the components of error within each kind of block. Thus, in our present example, instead of 4 degrees of freedom only being available for the estimation of error of the comparison A, 8 are available, and these 8 are the same as affect the precision of the interaction BC. Thus the 24 degrees of freedom for error are divisible into three sets of 8 each, appertaining to three pairs of comparisons among the degrees of freedom ascribed to treatments.

45·1. General Systems of Confounding in Powers of 2

If an experiment involves s factors at two levels each, there will be 2^s different treatments. As in Section 43, any particular treatment may be denoted by a combination of small letters in brackets, and any of the principal comparisons and their interactions by one or more capital letters. Now, just as any two primary comparisons A and B have an interaction AB, so any two comparisons such as ABC and ADE may be regarded as having an interaction. If we divide all the treatments into two halves for the contrast ABC, those in one half, including the control (1), will have an even number of letters in common with ABC, while those in the other half will have an odd number of letters in common with it. The same is true of the cross division of the same treatments into two halves for the comparison ADE. Consequently, if both subdivisions are made at once, the quarter containing the control will contain all treatments having an even number of letters in common both with ABC and with ADE. The opposite quarter will contain all treatments having an odd number of letters in common with both symbols. For this reason, if we combine the two symbols ABC and ADE by throwing them together and deleting any letters they have in common, a process which yields BCDE, this must be a comparison with which the treatments in both these quarters have an even number of letters in common, and is therefore the comparison of these two quarters taken together against the two which are either odd to ABC and even to ADE, or *vice versâ*. The comparison BCDE is thus the interaction of ABC with ADE.

The important consequence of recognising this relationship is that if, in an experiment with many factors, the comparisons ABC and ADE are both confounded, with comparisons between whole blocks, it necessarily follows that BCDE will also be confounded. In the language of the theory of groups, the comparisons confounded (together with an inert symbol I, the " identity ") constitute a " subgroup " selected from the " group " constituted by all the comparisons together with the identity. In fact, if we experiment with s factors in blocks of 2^{s-a}, so that there are 2^a blocks in each replication, the 2^s-1 comparisons to be made (with the identity) form a group of order 2^s, of which we may choose any subgroup of order 2^a to specify the 2^a-1 comparisons to be confounded.

For example, with six factors using four blocks of 16 in each replication, we could not choose as the three comparisons to be confounded the sixfold interaction ABCDEF, and two fivefold interactions, for these would not form a subgroup. The restriction under which the choice is to be made would, however, allow us to use ABC, DEF and their interaction ABCDEF; or, equally, to use ABCD, CDEF and their interaction ABEF. By this rule the available possibilities may easily be reviewed, and a choice made in accordance with what is known of the factors to be used.

It may be shown that, without confounding any interaction of less than three factors, we may use any number of factors which is less than the number of units in a block. So with blocks of eight we may use up to seven factors, and with blocks of sixteen s may be as large as fifteen.

When only one comparison is confounded, the contents of the two complementary blocks are easily written down. When, however, the subgroup of interactions to be confounded is of high order, to arrive at the contents of each block by successive subdivision is a tedious process. The block contents may, however, be written down at once, in the most complicated cases, by another method. It was observed in the first paragraph of this section that the treatment symbols of treatments in the same block with the "control" all have an even number of letters in common with every comparison confounded. (It is sufficient to apply this test to a members of the 2^a-1 in the subgroup confounded, provided these a are all independent, or, in other words, provided they do not belong to a subgroup of the subgroup confounded.) The set of treatment symbols which satisfy this condition themselves form a subgroup, since if any two possess the property, it is obvious that their interaction must do so. Consequently, we have only to find successively $s-a$ such treatment symbols in order to complete the particular block containing the control. This may be called the intrablock subgroup; it is of order 2^{s-a}. The contents of the block containing any other chosen treatment may then be found by writing down the interactions of this treatment with the treatments of the intrablock subgroup.

For example, to arrange the combinations of 7 factors in blocks of eight, we might choose for confounding first the interaction

$$ABC;$$

if, also, we choose ADE, we write down this new comparison and its interaction with what has gone before, thus:—

$$ADE, BCDE.$$

Taking next AFG, we have

$$AFG, BCFG, DEFG, ABCDEFG;$$

BDF has not yet been used, and yields the following eight

$$BDF, ACDF, ABEF, CEF, ABDG, CDG, BEG, ACEG,$$

completing the fifteen interactions of the subgroup confounded. For practical purposes we note that any elements from the four lines in which the interactions have been written down constitute a set of four independent generators, from which the whole can be, as indeed it has been, generated. We have now only to find in succession those treatments having an even number of letters in common with each of these four, in order to develop the intrablock subgroup.

It is easy to see that this condition is satisfied, not only by the control (1), but also by

$$(bcde),$$

equally by *(bcfg)* giving

$$(bcfg), (defg);$$

a third generator is supplied by *(acdf)* from which we have the last four,

$$(acdf), (abef), (abdg), (aceg)$$

These are, then, the seven treatments in the same block with (1). The contents of the block containing any treatment not in this block may at once be written down from its interactions with these seven.

It will be noticed that in this example the intrablock subgroup is also a subgroup of the subgroup confounded. These two subgroups, each of which can be uniquely derived from the other, may be identical, or one may be a subgroup of the other, or they may have a subgroup in common, or nothing in common but the identity.

In the example in the last section, of three factors in blocks of four, the intrablock subgroup was

$$(1), (ab), (bc), (ac),$$

while the subgroup confounded was

$$I, ABC.$$

In order to pass from any satisfactory solution to one involving one factor less, it is a convenient fact that we may delete all symbols of the subgroup confounded which involve any chosen letter, thus halving its order, and at the same time delete the same letter from the symbols in which it occurs in the intrablock subgroup. These changes introduce no new interaction to be confounded, and do not reduce the order of any, so that if in the initial solution none involves less than three factors, this will still be true of the new solution. Solutions for 6, 5 and 4 factors in blocks of eight may thus be obtained from the solution given above for seven factors.

45·2. Double Confounding

In the Latin square it has been shown possible to eliminate simultaneously the disturbances introduced by heterogeneity of two kinds arising from differences among rows and differences among columns. In the same way a set of 128 tests may be heterogeneous in one respect in blocks of eight, but in another respect in blocks of sixteen, orthogonal to the others. For example, a treatment might be applied at 8 sites on each of 16 cows, with the possibility that the reaction might be affected both by the individuality of the animals and by the differences between the sites chosen. In such cases one subgroup of 15 of the less important interactions could be chosen to be confounded with cows, while a second subgroup, which must have no element in common with the first, could be confounded with sites.

As an illustration, to be worked in detail, if the solution given above were chosen for the distribution of treatment combinations among the different cows, we might choose for confounding with sites the subgroup containing
ABCD,
ABEG, CDEG
BDFG, ACFG, ADEF, BCEF.

To make out a complete scheme showing what

treatment is to be applied at each site on each cow, we must form the intrablock subgroup for sites, namely

$$(efg)$$
$$(cdf), (cdeg)$$
$$(ace), (acfg), (adef), (adg)$$
$$(abf), (abeg), (abcd), (abcdefg), (bcef), (bcg), (bde), (bdfg).$$

These 15 treatments and the control are then assigned to any one site on the 16 cows, and the corresponding treatments at other sites are found from their interactions with the intrablock subgroup for cows. Thus the required treatment at any site may be written out at once in an 8×16 table.

A circumstance well worth taking into account in such cases is that the sites available might be so closely alike in pairs, *e.g.* on the Right and Left sides, that treatments applied at similar sites on the same animal might be regarded as perfectly comparable. On this view only 3 degrees of freedom instead of 7 have been confounded with difference likely to matter, and a set of four interactions, such as the four listed on the last line of the subgroup confounded with sites, will give reliable experimental comparisons. Consequently, we might have used important comparisons, involving only one or two factors, in their place, had this eased the difficulties of setting up a good experimental design.

46. Example with 27 Treatments

The principle of procedure illustrated in Section 44 may be extended and generalised in a large number of ways. Since only a few of these can be exemplified, the reader will find great advantage in investigating the possibilities of similar designs appropriate to the special problems in which he is interested. The variety of the subject is, in fact, unlimited, and probably many valuable possibilities remain to be discovered. We will consider next an experiment with 3 factors, in which each furnishes not two but three variants, so that there are in all 27 combinations to be investigated. Thus a large-scale investigation of the manurial requirements of young rubber plantations, in respect to the three primary manurial elements, nitrogen, potassium and phosphorus, combines, in addition to whatever basal treatment may be thought desirable, single or double applications of these three manures; making in all three levels for each ingredient, such as nitrogen only, nine combinations for any two ingredients, and 27 for all three together.

In order to reduce the size of the block below that needed to contain 27 plots, we have, to guide us in choosing a smaller size, the fact that blocks of nine will in any case be needed, if all the interactions between pairs of factors are to be conserved. The experiment will therefore have nine plots to a block, and three blocks in each complete replication. Everything now depends on the choice of the sets of nine treatments which are to be assigned to the same blocks, and, of course, within each block to strictly randomised positions. In order to conserve the main effects and the interactions between pairs, a set of nine treatments chosen to occupy the same block must fulfil the following requirements:—

(i) the three levels of each ingredient must be represented by three plots each,

(ii) the nine combinations of each pair of ingredients must be represented by one plot each.

If the set satisfies the second group of conditions it satisfies also the first. It is not, however, at first sight obvious that the second condition can be fulfilled at once for all three pairs of factors. The combinatorial relationship exhibited in the Latin square may here be applied most valuably.

Let us set out the nine combinations to be chosen in a diagrammatic square with three rows and three columns, and let us lay it down that treatments in the first row shall receive nitrogen at the first level, treatments in the second row at the second level, and treatments in the third row at the third level. Then the requirement that the three levels of nitrogen shall be equally represented in our selection is satisfied easily by having three plots in each row of our diagram. Similarly, we may lay it down that the columns of our square correspond to the levels of abundance of the second ingredient, potash, in the manurial mixture to be tested. It is then clear that we must have three plots in each column, and, if the interactions of nitrogen and potash are to be conserved, that there must be one plot at the intersection of each row with each column. With respect to the level at which phosphate is applied to any plot, we cannot now represent it by its position in our diagrammatic square, but shall simply use the numbers 1, 2, 3 inserted in any position in the diagram to represent the level of the phosphatic ingredient.

It now appears that a selection of nine treatments will satisfy the conditions laid down above, if it can be represented diagrammatically by a square containing nine numbers at the intersections of the three rows with the three columns, of which every row must contain 1, 2 and 3 once each, in order that the interaction of nitrogen and phosphorus should be conserved; and every column equally must contain a "1," a "2" and a "3," to make sure of conserving the interaction of potash and phosphorus. We have, in fact, merely to arrange the numbers 1, 2, 3 in a Latin square in order to obtain a single selection of the treatments, which might properly occupy a single block.

	k_1	k_2	k_3
n_1	1	2	3
n_2	2	3	1
n_3	3	1	2

There are only 12 solutions of the 3×3 Latin square. If we choose one of these to represent the contents of one block, we must next enquire whether any selection of treatments to occupy the other blocks in the same replications can be made to satisfy the conditions. We may convince ourselves on this point by considering the effect on our chosen selection of making a cyclic substitution of the levels of phosphate; that is by substituting 2 for 1, 3 for 2, and 1 for 3 throughout the diagram. Repeating such a substitution three times will clearly bring us back to the original selection; but the two new selections first produced will (i) both be represented by Latin squares, and (ii) will between them and the original from which they were derived contain all 27 treatments. This last essential fact becomes clear on perceiving that the number in any particular cell of the square must take the values 1, 2, 3 on successive applications of the substitution, whatever may be the initial value, while the aggregate of the 27 treatments used are represented simply by these three numbers, placed at all the nine points of the diagram.

The 26 independent comparisons among 27 treatments may be analysed according to the factors involved, as in the following table:—

TABLE 16

N	2
K	2
P	2
NK	4
PK	4
NP	4
NPK	8
Total	26

If, therefore, we make up the contents of the block in accordance with the solution provided by the Latin square diagram, we shall have sacrificed a particular 2 out of the 8 degrees of freedom for triple interaction.

It is, of course, always easy to recognise the particular components of treatment in which blocks in the same replication differ, and to obtain the aggregate sum of squares for the six triple interactions which have not been confounded, by subtraction. This residue of unconfounded interactions may then be tested for significance like other treatment effects, and will usually be of service in confirming experimentally the supposition upon which the experimental design was based, namely, that the triple interactions as a whole had not been quantitatively important.

There is, however, a certain advantage in being able to recognise which particular contrasts these unconfounded interactions represent. For then we can, if we wish, subdivide them to examine the significance of more particular effects. In the case of the design under discussion, based on a 3×3 Latin square, the combinatorial properties of such a square are such as to make this recognition easy. It has been mentioned above that there are only twelve 3×3 squares, and we have seen that each belongs to a set of 3 which can be generated from it by a cyclic substitution. There are,

ORTHOGONAL SETS

therefore, only 4 such sets, and the 4 squares below are representatives chosen one from each set.

I.	II.	III.	IV.
1 2 3	1 3 2	1 2 3	1 3 2
2 3 1	3 2 1	3 1 2	2 1 3
3 1 2	2 1 3	2 3 1	3 2 1

It may be observed that the second representative is formed from the first by interchanging the numbers 2 and 3; the third is formed from the first by interchanging the second and third rows, and the fourth is formed from the first by interchanging the second and third columns. If we consider, now, examples Nos. I. and II. it is to be observed that they agree in the three plots having phosphate at level No. 1, but differ in the other six. If, however, we applied the cyclic substitution to the first example, it would, after one operation, agree with the second example in the three plots at phosphate level No. 3, and differ in the other six; while after a second operation it would agree only in the three plots at phosphate level No. 2. Consequently, the nine treatments in any selection of set II. appear three each in the three selections of set I., these sets of treatments being those having the same quantity of phosphate. The treatment comparisons represented by the subdivision of the 27 treatments as in set II. are therefore wholly independent of the treatment comparisons of set I. Both represent 2 degrees of freedom out of the 8 available for triple interaction, and these 2 pairs of degrees of freedom have nothing in common. Consequently, when the pair of degrees of freedom of set I. are confounded, the pair represented by set II. are wholly conserved.

The same relationship subsists between set III. and set I. if we consider the treatments represented in the same row, *i.e.* those at an equal level of nitrogen, instead of those represented by the same number, at an equal level of phosphate. The examples shown above agree in the first row and differ in the two other rows. If the cyclic substitution be applied to the first example it will agree with the third successively in the second and third rows, while always differing in the remaining two. Consequently, the third set, like the second, represents a treatment contrast wholly independent of that represented by the first, and which is therefore entirely conserved if the latter is confounded. In like manner, the treatments in any selection of set IV. will be distributed by threes in the selections of set I., these threes lying in the same column, and having therefore equal quantities of potash. The 2 degrees of freedom of set IV. are also, therefore, wholly conserved. It should be noticed further that the three sets—II., III. and IV.—are not only independent of set I., but are also independent of each other. Thus in the examples shown, II. and III. agree in a single column, II. and IV. in a single row, and III. and IV. in a single number; and, in view of what has been said above, these facts suffice to show that the pairs of degrees of freedom represented by these sets are wholly independent and have nothing in common. Since all are included in the 6 degrees of freedom conserved, they must therefore constitute the whole of the 6 degrees of freedom, and constitute parts of it which may be separated in the analysis and examined separately in the test of significance.

Supposing, then, that the experiment were carried out with 12 replications, or, in all, 324 plots, we might choose one of the 4 pairs of degrees of freedom into which the 8 triple interactions have been divided, and decide to sacrifice these particular components of the triple interactions, in order to increase the precision of the comparisons to be made in the 6 components of single factor effects, the 12 components of interactions between pairs of factors, and the 6 components of triple interactions which have not been confounded. The experiment then consists of 36 blocks of 9 plots each, so that 35 degrees of freedom are eliminated as representing block differences. The sets of treatments in different blocks within the same replication are assigned by using one of the cyclic sets of 3×3 Latin squares; remembering, of course, that topographically these treatments are not arranged in a Latin square, but are assigned at random to the 9 plots in the block. Each set of 9 treatments replicated 12 times will provide 88 degrees of freedom for error, or 264 in all, so that the complete analysis of the experiment may be shortly represented as below :—

TABLE 17

	Degrees of Freedom
Blocks	35
Single factors	6
Interactions between pairs	12
Unconfounded triple interactions	6
Error	264
Total	323

47. Partial Confounding

In the example we have just considered with 27 treatments it has been shown that we can gain the great advantage of smaller blocks, or increased homogeneity of material, for all the primary comparisons, and their interactions by pairs, at the expense of some sacrifice of information about the triple interactions, which are presumed to be comparatively unimportant. The advantage of such a procedure would be great in many practical cases, even if all knowledge of the interactions of a higher order had to be forgone. Formally, however, the typical experiment discussed has shown that the sacrifice required is that of one only of four portions into which the triple interactions may be divided, and that we may sacrifice whichever one we please of these four portions. If, now, it is thought that knowledge of these interactions, though admittedly comparatively unimportant, is not wholly worthless, the fact that only one-quarter has to be sacrificed will appear to be a real advantage. This advantage is not, however, made fully accessible by the experiment proposed; for the 6 degrees of freedom conserved, while they afford satisfactory guidance as to the significance or insignificance of such triple interactions as may exist, represent manurial contrasts of a somewhat complex kind, and are not in fact the components we should choose for separate examination if the triple interactions had been conserved in their entirety.

When the quantity of an ingredient of a mixture has been tested at three different levels the two independent comparisons which these provide may often be usefully subdivided in a particular way. We may regard the difference between the highest level and the lowest as representing the principal effect of the ingredient, that is, as giving us the average effect brought about by a unit addition of this ingredient, averaged over the range of dosage studied. In conjunction with this principal degree of freedom we may introduce a second, orthogonal to, or statistically independent of, the first. This will be found by subtracting the sum of the effects of the first and third level from twice the effect of the second level of concentration. Thus if (n_1), (n_2), (n_3) stand for different levels of the amount of nitrogen in a manurial mixture, or for the measured effects of such treatments, the two components of the

CHOICE OF COMPONENTS

effect of nitrogen which may conveniently be separated are defined by

$$N_1 = (n_3) - (n_1)$$

and

$$N_2 = 2(n_2) - (n_1) - (n_3).$$

These forms at least will be convenient when the concentrations tested differ by equal steps, or by steps which, on any hypothesis under consideration, should produce equal effects. They may be modified to other orthogonal linear forms when the relationship between the quantities used experimentally is of a more complicated character. Here we are concerned only to illustrate the statement that when high order interactions are regarded as having any experimental importance, our interest will usually be centred on particular components into which such interactions may be analysed. The statistical independence of the two forms proposed above may be conveniently verified by multiplying together the coefficients of (n_1) in the two expressions, and adding the product so formed to the corresponding products for the coefficients of (n_2) and (n_3). If these products add up to zero the components designated are statistically independent and represent mutually exclusive degrees of freedom.

Considering the 4 degrees of freedom for the interaction of two ingredients, such as nitrogen and potash, it is now readily seen that these can be denoted by the four symbols, N_1K_1, N_2K_1, N_1K_2, and N_2K_2, any one of which may be interpreted in terms of the treatments concerned by algebraic expansion. Thus:—

$$\begin{aligned}N_1K_1 &= (n_3-n_1)(k_3-k_1)\\&= (n_3k_3)-(n_1k_3)-(n_3k_1)+(n_1k_1),\end{aligned}$$

and so with other expressions. It will be seen at once that, if our interest in the interaction between nitrogenous and potassic treatments arises principally from a suspicion that, with a larger supply of nitrogen, there may be a greater need for or opportunities for the utilisation of potash, then the particular component N_1K_1 will have an interest which the other components from which it has been separated do not share. Similarly, with triple interactions it might well be that the sole scientific interest of the eight independent comparisons which, in our experiment, these afford, lay in one particular component such as $N_1K_1P_1$.

The inconvenience of the confounding process used in the experiment consists, therefore, in the fact that if the triple interaction, or any component of it, is possibly of sufficient magnitude to be not wholly negligible, the components of triple interaction conserved by the experiment will not probably be themselves of any special interest, and, in the absence of the two components which have been confounded, will not afford the means of isolating the more interesting components for special study. It would seem, therefore, that it would have been preferable, if possible, to have spread such information, as the experiment is designed to give respecting the triple interactions, equally over the 8 degrees of freedom of which they are composed, unless the structure of the experiment is itself such as to isolate for conservation just those components which are of the greatest interest. The process of spreading the available information equally over the whole group of comparisons which are affected is known as partial confounding.

In our example there are 12 replications. If the number of replications, as in this case, is divisible by 4, then, instead of completely confounding a chosen pair of degrees of freedom out of the four pairs available, we might partially confound all four, by using each cyclic set three times instead of using the same set twelve times. The treatment comparison represented by any cyclic set will then have been conserved in 9 replications, while it has been sacrificed in 3. All these comparisons will therefore be capable of evaluation from the results of the experiment, though with only three-quarters of the precision with which interactions between pairs of factors and the effects of single factors have been evaluated. In such an arrangement the general advantage conferred by the principle of confounding may be most clearly seen, for the reduction in the size of the block from 27 plots to 9 will probably have increased the precision of the unconfounded comparisons in a higher ratio than that of 4 : 3 ; and, as the triple interactions are only confounded in 1 replication out of 4, they also will be evaluated with increased, but more moderately increased, precision, in spite of the quarter of the information respecting them which has been sacrificed in order that the blocks might be reduced to 9 plots each.

The information concerning the triple interactions which has been supplied by the experiment, if partial confounding has been practised, will be in the form of contrasts between the three sets of 9 treatments, each

$$\begin{pmatrix} 1 & 2 & 3 \\ 2 & 3 & 1 \\ 3 & 1 & 2 \end{pmatrix}, \begin{pmatrix} 2 & 3 & 1 \\ 3 & 1 & 2 \\ 1 & 2 & 3 \end{pmatrix}, \begin{pmatrix} 3 & 1 & 2 \\ 1 & 2 & 3 \\ 2 & 3 & 1 \end{pmatrix},$$

specified by one of the four cyclic sets of Latin squares. These contrasts must be gathered separately from those parts of the experiment from which they are available, without using those parts in which they are confounded. By this means the comparison is kept unaffected by the larger elements of soil heterogeneity, which distinguish the different blocks. Thus the contrast between the three sets of treatments may be obtained by adding up the aggregate response to these sets of treatments, in the 9 replications in which these contrasts of manurial treatment are not those which characterise whole blocks. Usually this will be done most easily by subtraction. A different set of 9 replications will provide the material for evaluating the contrasts between the sets of treatments of the second group, namely,

$$\begin{pmatrix} 1 & 3 & 2 \\ 3 & 2 & 1 \\ 2 & 1 & 3 \end{pmatrix}, \begin{pmatrix} 2 & 1 & 3 \\ 1 & 3 & 2 \\ 3 & 2 & 1 \end{pmatrix}, \begin{pmatrix} 3 & 2 & 1 \\ 2 & 1 & 3 \\ 1 & 3 & 2 \end{pmatrix},$$

while different sets of nine replications will give the contrasts between the cyclic derivatives obtained from the squares

$$\begin{matrix} 1 & 2 & 3 \\ 3 & 1 & 2 \\ 2 & 3 & 1 \end{matrix} \text{ and } \begin{matrix} 1 & 3 & 2 \\ 2 & 1 & 3 \\ 3 & 2 & 1 \end{matrix}$$

If, now, we are specially interested to evaluate a particular component of the triple interaction, such as that which has been denoted above by $N_1 K_1 P_1$, we must obtain this by a combination of the contrasts which the experiment provides directly. To do this may require some ingenuity. The solution in this case is found by using only those squares in which one diagonal, or the other, contains plots with the highest or lowest level of phosphate. Thus, if we compound with the proper positive and negative signs the yields given by the experiment for the 8 squares set out below, it appears that

$$\begin{pmatrix} 2 & 3 & 1 \\ 3 & 1 & 2 \\ 1 & 2 & 3 \end{pmatrix} - \begin{pmatrix} 1 & 2 & 3 \\ 2 & 3 & 1 \\ 3 & 1 & 2 \end{pmatrix} + \begin{pmatrix} 3 & 2 & 1 \\ 2 & 1 & 3 \\ 1 & 3 & 2 \end{pmatrix} - \begin{pmatrix} 2 & 1 & 3 \\ 1 & 3 & 2 \\ 3 & 2 & 1 \end{pmatrix}$$
$$+ \begin{pmatrix} 3 & 1 & 2 \\ 2 & 3 & 1 \\ 1 & 2 & 3 \end{pmatrix} - \begin{pmatrix} 1 & 2 & 3 \\ 3 & 1 & 2 \\ 2 & 3 & 1 \end{pmatrix} + \begin{pmatrix} 3 & 2 & 1 \\ 1 & 3 & 2 \\ 2 & 1 & 3 \end{pmatrix} - \begin{pmatrix} 1 & 3 & 2 \\ 2 & 1 & 3 \\ 3 & 2 & 1 \end{pmatrix}$$

is equal to $3 N_1 K_1 P_1$.

RECOVERY OF INFORMATION

With the aid of this example the reader will do well to consider how the data of the experiment should be combined to obtain other types of interaction, such as those denoted by $N_1K_1P_2$, $N_2K_2P_1$, and $N_2K_2P_2$, and to satisfy himself that these can each be derived by a similar choice of appropriate compounds from the data provided by the partially confounded experiment.

47·1. Practical Exercises

1. Show that, if each factor be tested at two levels, so many as fifteen factors can be tested in blocks of 16, without confounding any interaction of less than three factors.

2. In a linkage test with eight genetic factors, show that a selection of 8 out of the 128 possible types of multiple heterozygotes can be made so that each of the 28 pairs of factors is in "coupling" in 4, and in "repulsion" in 4.

3. Using a completely orthogonal 7×7 square, show that with eight replications a set of 49 varieties may be tested in blocks of 7, so that every possible pair of varieties occurs once only in the same block.

4. Show that the same may be done, in eight replications, with 57 varieties in blocks of 8, and that it cannot be done, in seven replications, with 43 varieties in blocks of 7.

5. Twenty-one experimental plants have each five leaves growing serially along the stem. Show how to allocate 21 treatments to the leaves, so that each pair of treatments occurs once on the same plant, and each treatment occurs once on a first leaf, and once on a leaf in the other ordinal positions. (Youden's Square.)

6. If 144 varieties be set out diagrammatically in a 12×12 square, and tested in blocks of 12, so that in one replication varieties in the same row fall in the

same block, and in a second replication the blocks contain varieties in the same column, show that the sampling variances of comparisons between pairs of varieties which (*a*) are, and (*b*) are not, tested in the same block, are in the ratio 13 : 14.

7. If 512 varieties be set out in an $8 \times 8 \times 8$ cube and tested in three replications in orthogonally chosen blocks of 8, compare the variances of the comparisons of any one variety with others which (*a*) occur in the same block as the first, (*b*) are connected with it at one remove through other varieties, and (*c*) are only connected at two removes (73 : 78 : 79).

REFERENCES AND OTHER READING

R. C. BOSE and K. KISHEN (1940). On the problem of confounding in the general symmetrical factorial design. Sankhya, v. 21-36.

R. C. BOSE and K. R. NAIR (1939). Partially balanced incomplete block designs. Sankhya, iv. 337-372.

R. A. FISHER and F. YATES (1938-1963). Statistical Tables for biological, agricultural and medical research. Oliver and Boyd Ltd., Edinburgh.

R. A. FISHER (1942). New cyclic solutions to problems in incomplete blocks. Annals of Eugenics, xi. 290-299.

R. A. FISHER (1942). The theory of confounding in factorial experiments in relation to the theory of groups. Annals of Eugenics, xi. 341-353.

R. A. FISHER (1945). A system of confounding for factors with more than two alternatives, giving completely orthogonal cubes and higher powers. Annals of Eugenics, xii. 283-290.

F. YATES (1933). The principles of orthogonality and confounding in replicated experiments. Journal of Agricultural Science xxiii. 108-145.

F. YATES (1935). Complex experiments. Supplement to Journal of the Royal Statistical Society, vol. ii. 181-223.

F. YATES (1936). A new method of arranging variety trials involving a large number of varieties. Journal of Agricultural Science, xxvi. 424-455.

F. YATES (1936). Incomplete randomised blocks. Annals of Eugenics, vii. 121-140.

VIII

SPECIAL CASES OF PARTIAL CONFOUNDING

48. TREATING of a subject such as experimental design in general, it is possible to give adequate space only to general principles leading to the more advantageous procedures which are available. These it is essential to grasp. Their applications to particular details that arise in practice are of endless variety and afford scope for a great deal of ingenuity. These require to be studied in detail by workers in different fields of experimentation in order to reap the full advantages which a clear grasp of general principles makes possible. It may be of use in this chapter if we consider some of the more special applications of the principle of partial confounding which were found to arise in their early application to field trials in agriculture.

49. Dummy Comparisons

It may happen that, in order that the different variants of each factor may occur with proportional frequency in combination with the variants of other factors, certain of the combinations used are actually indistinguishable. For example, in an experiment with four different nitrogenous manures we may also wish to vary the quantities used. We may wish to compare plots receiving no nitrogenous manure with others receiving a single or a double dressing. These single and double dressings will be applied to different plots, in each of the four nitrogenous materials to be tested, and the precision of the comparison between single and

138 CASES OF PARTIAL CONFOUNDING

double applications will be enhanced by the fact that each is represented on four kinds of plots. In order that the comparison with the plots receiving no nitrogenous manure may be of equal precision, it is necessary that these shall be as numerous as those receiving single or double dressings, and therefore four times as numerous as any one kind of these. To compare the efficacy of the four kinds or qualities of nitrogen simultaneously with the three quantities (0, 1, 2) with which they can be combined, we might make blocks of 12 plots each, in which the plots receiving single or double dressings will be manured differently, while the 4 plots receiving none will all be manured alike. The comparisons among these within each block will be ascribable solely to experimental error, including in that term, as is usual, variations in the fertility of different plots in the same block. Thus, if there were 5 replications, the analysis of the 59 independent comparisons among the 60 plots would not be

TABLE 18

	Degrees of Freedom.
Blocks	4
Treatments	11
Error	44
Total	59

but

TABLE 19

	Degrees of Freedom.
Blocks	4
Treatments	8
Error { between blocks	32
{ within blocks	15
Total	59

Here we have divided the 47 degrees of freedom available for the estimation of error into two parts, to show that 15 degrees of freedom come from a comparison of identical plots in the same block, 3 from each of the five blocks, while 32 come from the comparison of the differences among the 9 different treatments in the five blocks in which they are tried.

As between the two factors of quantity of nitrogen N, and quality Q, the 8 degrees of freedom between the 9 treatments will be allotted as follows. There will be 2 for the comparison of the three levels of the nitrogen, and 3 for the four qualitatively different mixtures in which it is applied, leaving 3 more for interactions between N and Q. In other words we have, as we would have if the four manures were applied only in single and double doses, 3 degrees for quality and 3 for interaction. The addition of the plots without the nitrogenous manure has left these two classes unaffected, but has added 1 to the degrees of freedom for quantity of nitrogen.

50. Interaction of Quantity and Quality

In this connection a modification is to be indicated in the manner in which the effects of quality and interaction are to be reckoned. If we were to consider N and Q as two independent factors, the 3 degrees of freedom for interaction would be obtained simply by comparison of the four quantities by which the double applications of each manure exceed the effects of single applications of the same materials. Equally, the simple qualitative effects would be represented by the contrast between the four totals of single and double dressings of these four materials. Such a subdivision is seen to be not wholly satisfactory, when we consider that the quantitative contrasts are differences caused by quantitative variations in the very substances which the qualitative comparisons are intended to compare. Thus, if a quantity of nitrogen applied as cyanamide differs in its effect on the crop from an equal quantity of nitrogen applied as urea, it is to be anticipated that with larger quantities of the manurial applications the difference would be enhanced. In fact, the hypothesis that the differences are proportional to the quantities of nitrogen applied is in many ways a simpler one, in the sense of being more natural and acceptable, than that the difference should be the same irrespective of the quantities of material added to the soil.

If we take this view, results in which the double dressings of two ingredients differ by twice as much as the single dressings, but in the same direction, would be regarded as exhibiting pure effects of quality Q, without interaction NQ. The interactions must, therefore, be identified with the three independent comparisons among the four quantities which would be obtained by subtracting the yield of the double dressing from twice the yield of the corresponding single dressing. For these four quantities would all be equal if interaction, in the sense in which we are now using this term, were completely absent. Equally, the primary effects of quality will now be reckoned by comparing the four sums found by adding twice the yield of the double application to the yield from the single application of the same manure; as in calculating the "regression" of the manurial response upon the manurial difference to which it is for the present purpose to be considered as proportional. The statistical principles and methods in the treatment of regression are developed in *Statistical Methods for Research Workers*.

That these two methods give different subdivisions of the same total follows from the algebraic identity

$$\tfrac{1}{2}(x+y)^2 + \tfrac{1}{2}(x-y)^2 = \tfrac{1}{5}(2x+y)^2 + \tfrac{1}{5}(x-2y)^2$$

If x, y stand for the yields of the double and single dressings of any manurial material, the two terms on the left represent the squares assigned to Q and NQ, using the convention that " interaction " means variation in the values $x-y$, while the two terms on the right represent the squares assigned to Q and NQ on the convention that " interaction " means variation among the values $x-2y$. The same method of subdivision with appropriate coefficients is evidently applicable whatever may be the ratio between the quantities used. Note that the divisor of each square is the sum of the squares of the coefficients, while the sum of the products of the coefficients in any two squares of the same set is zero.

51. Resolution of Three Comparisons among Four Materials

The 3 degrees of freedom in Q or in NQ are the three independent comparisons among four different materials, such as sulphate of ammonia (s), chloride of ammonia (m), cyanamide (c), and urea (u). These may be systematically subdivided, if it is thought convenient to do so, as the three possible comparisons between opposing pairs of materials. There are in fact just three ways of dividing four objects into two sets of two each; these are:—

$$s+m-c-u$$
$$s-m+c-u$$
$$s-m-c+u,$$

and these are all mutually independent, as may be verified by observing that the sum of the products of the coefficients ($+1$, or -1) of the symbols in any two of these three expressions is zero.

142   CASES OF PARTIAL CONFOUNDING

Regarded combinatorially, this is equivalent to the statement that a 2 × 2 Latin square is possible, namely,

A B
B A,

for in such a square the four objects are divided into pairs in three ways, as rows, as columns, and as letters, and the specification of a Latin square requires that these shall all be mutually independent.

52. An Early Example

An experiment with sulphate of ammonia, chloride of ammonia, cyanamide and urea, in quantities 0, 1, 2, with and without superphosphate, was carried out in barley at Rothamsted in 1927. Two replications, or 48 plots, were used. These were divided into four blocks of 12 plots each. In two blocks phosphate (p) was applied with chloride of ammonia and with urea, in both single and double dressings, while in the other two blocks it was applied with sulphate of ammonia and with cyanamide. Each block contained two plots without nitrogenous or phosphatic dressings, and two plots with phosphatic only. The plots were assigned to treatments at random within the blocks.

Among the 18 different treatments there will be 17 independent comparisons. One of these, however, namely,

$$(p-1)(s_1+s_2-m_1-m_2+c_1+c_2-u_1-u_2)$$

has been confounded with blocks. There are 16 degrees of freedom for treatments in the analysis, and 3 for blocks, leaving 28 for error. It would, however, be a mistake to assume from this that these 28 are all pure error, for it will appear that owing to the occurrence of dummy treatments, or more properly of plots treated alike in different blocks of the same replication, the degree of freedom destined to be sacrificed has in fact only been partially confounded.

It is instructive in such cases to consider exactly what comparisons will consist solely of error, unaffected by any treatment differences. Within each block there are two unmanured plots the difference between which is pure error, and two phosphatic plots of which the

	A				B		
(u_2p) 480	(m_2p) 542	(c_2) 373	(1) 186	(1) 268	(p) 297	(s_2p) 536	(s_1p) 471
(m_1p) 431	(c_1) 365	(s_2) 293	(s_1) 281	(u_1) 343	(c_2p) 443	$(u.)$ 498	(m_2) 522
(p) 284	(p) 313	(u_1p) 336	(1) 260	(m_1) 366	(c_1p) 412	(p) 250	(1) 239
(u_2) 475	(1) 275	(1) 242	(c_2p) 395	(p) 244	(c_2) 396	(s_2) 400	(p) 228
(p) 344	(p) 277	(s_1p) 359	(m_1) 368	(s_1) 413	(u_2p) 512	(1) 259	(m_1p) 453
(u_1) 401	(c_1p) 429	(s_2p) 464	(m_2) 542	(m_2p) 504	(c_1) 409	(u_1p) 389	(1) 267
	C				D		

FIG. 2.—Arrangement of treatments and yields of grain in experiment on quantity and quality of nitrogenous fertilisers in barley 1927.

same is true. Here, therefore, we have 8 degrees of freedom contributing only to our estimate of error. To make sure that these are not counted a second time, it is sufficient that all further comparisons to be made, if they involve these plots, shall involve only the pairs of plots treated alike taken together. Next, observe that there are two pairs of blocks with the same treatments, 10 in each. The 10 differences between the performances of these in the two blocks of a pair will be distributed about a mean representing the difference in fertility between the two blocks; but the 9 degrees of freedom of their variation about this mean will be pure error. There are thus 9 degrees of freedom from each pair of blocks, or 18 together, which, with the 8 from within blocks, make 26 in all. In subsequent comparisons we must, however, treat the two blocks of each pair together. Finally, the two pairs of blocks have two treatments in common, those unmanured and those having phosphate only. The differences between these two treatments in the two pairs of blocks will not be necessarily the same, and the discrepancy between them will be pure error; this last degree of freedom makes up the total to 27.

The yields of grain from each plot in units of 2 oz. are shown in Fig. 2; the contributions to the sum of squares ascribable to these 27 degrees of freedom of pure error are shown in Table 20.

Squares involving two plots only are divided by 2, others such as the first two entries in the second and third columns depend on 4 plots, and are divided by 4, and the squares of differences between pairs of blocks by 24. Finally, the discrepancy of 47 units between $(p)-(1)$ from the two pairs of blocks depends on 16 plots, and has its square divided by 16. The several ingredients are thus brought to a comparable basis.

It was mentioned above that the single manurial contrast

$$(p-1)(s_1+s_2-m_1-m_2+c_1+c_2-u_1-u_2)$$

in which the blocks differ, had not been totally confounded, meaning by this that it could be indirectly estimated by comparisons within blocks. The comparison within blocks, independent of all those used in the estimation of error, which depends only on this one manurial contrast, arises from the fact that though the

PRACTICAL EXAMPLE

TABLE 20

Analysis of Components of Error			
Squares of Differences between like Plots in the same Block.			Squares of Differences between like Treatments in like Blocks.
(p) . . . 841	(p) . . 7812·5	(1) . . 50	
(1) . . . 5476	(1) . . 3200	(p) . . 2738	
(p) . . . 2209	$(u_1 p)$ . . 2809	(u_1) . . 3364	
(1) . . . 841	(c_1) . . 1936	$(c_1 p)$ . . 289	
(p) . . . 4489·	(s_1) . . 17424	$(s_1 p)$ . . 12544	
(1) . . . 1089	$(m_1 p)$ . . 484	(m_1) . . 4	
(p) . . . 256	$(u_2 p)$ . . 1024	(u_2) . . 529	
(1) . . . 64	(c_2) . . 529	$(c_2 p)$ . . 2304	
	(s_2) . . 11449	$(s_2 p)$ . . 5184	
15265	$(m_2 p)$ . . 1444	(m_2) . . 400	
	Total . −9075	Total . −456·3	
	39036·5	26949·6	

Differences $(p)-(1)$ in Pairs of Blocks.		Summary of Contributions to Pure Error.	
		Degrees of Freedom.	Sum of Squares.
A	151	8	7632·5
D	−54	9	19518·25
		9	13474·83
A+D .	97	1 $47^2 \div 16$	138·0625
B	40		
C	104	27	40763·64583
B+C .	144		
A+D .	97		
B+C−A−D	47		

treatments concerned are not to be found in the same block, yet the different blocks in which they appear also contain some plots treated alike, with which each group can be compared. In each block, in fact, we may compare the plots having single and double dressings of nitrogen with twice the sum of the plots having none. Thus two blocks give

$$(ps_1)+(ps_2)+(m_1)+(m_2)+(pc_1)+(pc_2)$$
$$+(u_1)+(u_2)-2(p)-2(p)-2(1)-2(1) = 1483, 1157;$$
$$\text{B}\text{C}$$

while the other two give

$$(s_1)+(s_2)+(pm_1)+(pm_2)+(c_1)+(c_2)+(pu_1)$$
$$+(pu_2)-2(p)-2(p)-2(1)-2(1) = 1015, 1480;$$
$$\text{A}\text{D}$$

whence we obtain by subtraction

$$(p-1)(s_1+s_2-m_1-m_2+c_1+c_2-u_1-u_2) = 145.$$

The value for each block is a combination of 8 plots with coefficient $+1$, and 4 plots with coefficient -2, so the sum of the squares is 24, or for the four blocks, 96. The contribution to the sum of squares of this partially confounded manurial comparison is therefore $145^2 \div 96$, or 219·0104. To the divisor, 96 the plots having the treatments to be compared contribute only 32, so that the comparison is made with only one-third of the precision of the 16 unconfounded comparisons.

The other elements in the analysis may now be evaluated. The 3 degrees of freedom between blocks are easily found to account for a contribution of 12,215·75 to the sum of squares. The total effect of treatments could now be obtained by subtraction of the three items already evaluated from the total; the interest of the experiment, however, lies in evaluating the separate factors of the treatment differences. The total yields, contributed by 8 plots each, in the six classes of treatment formed by combining two levels of phosphate with three of nitrogen, are shown in Table 21.

TABLE 21

	No Nitrogen.	Single Nitrogen.	Double Nitrogen.	Total.
With phosphate	2237	3280	3876	9,393
Without phosphate	1996	2946	3499	8,441
Sum	4233	6226	7375	17,834
Difference	241	334	377	952

Plots receiving phosphate have exceeded those not receiving phosphate in all by 952 units, so that the 1 degree of freedom, P, contributes $952^2 \div 48$, or $18,881 \cdot \frac{1}{3}$.

We may next take the 2 degrees of freedom for quantity of nitrogen N, and 2 more for interaction with phosphate NP.

The 2 degrees of freedom for N are evidently found by comparing the sums 4233, 6226, 7375; clearly the principal effect, the contrast between double nitrogen and none, is the important part; the difference 3142 from 32 plots contributes the large item $3142^2 \div 32$, or $308,505 \cdot 125$ for N_1. The remaining degree of freedom, corresponding to diminishing return for the second dose of nitrogen, is found by subtracting the first and last totals from twice the total for single nitrogen, squaring, and dividing by 96. This gives $844^2 \div 96$, or $7420 \cdot 16$ for N_2, a much smaller, but still a significant, value. We may treat the differences in the same way. For N_1P we have $136^2 \div 32$, or 578, and for N_2P $50^2 \div 96$, or $26 \cdot 041\dot{6}$, both quite insignificant contributions, though in both cases of the expected sign. The items evaluated so far are :—

TABLE 22

	Degrees of Freedom.	Sum of Squares.
N_1	1	308505·125
N_2	1	7420·167
P	1	18881·333
NP	2	604·042

We may now consider the qualitative differences Q, and their interaction with quantity NQ. The totals from 4 plots each for the single and double application of the four nitrogenous nutrients are shown in Table 23.

TABLE 23

Quantity.	Material.				Differences between Pairs.		
	s.	m.	c.	u.	$s+m$ $-(c+u)$	$m+u$ $-(s+c)$	$m+c$ $-(s+u)$
(1)	1524	1618	1615	1469	...	...	...
(2)	1693	2110	1607	1965	...	...	...
2(2)+(1)	4910	5838	4829	5399	520	1498	358
2(1)−(2)	1355	1126	1623	973	−115	−879	421

The 3 degrees of freedom for Q and for NQ can now be found either by taking the sums of the squares of the deviations from their means, of the last two lines and dividing by 20, or by splitting the columns in the three possible ways into two opposed pairs, and dividing by 80. The latter process gives:—

TABLE 24

	Squares of Differences. Q	Squares of Differences. NQ
$s+m-c-u$	270400	13225
$m+u-s-c$	2244004	772641
$m+c-s-u$	128164	177241
	2642568	963107
Q . . 33032·1	NQ .	12038·8375

being each for 3 degrees of freedom. The separate evaluation of these three comparisons as above brings to light the somewhat suspicious circumstance that the largest contribution in each class is from the particular contrast between nitrogenous materials which has been used (in its interaction with phosphate) for confounding. If it is a coincidence that the two pairs of nutrients most contrasted in their effects on yield, and in their interaction with quantity of nitrogen have been chosen for the purpose, then the choice has been an unfortunate one. If not, then we may suspect that the conditions in the different blocks of land used have, in some obscure way, influenced the apparent reaction to these nutrients.

Had we adopted the subdivision between Q and NQ by means of the sums $(2)+(1)$ and differences $(2)-(1)$, we should have 21,739·094 for Q and 23,331·844 for NQ, making the same total, but giving a larger contribution for interaction than for the prime factors of quality. The subdivision employed above is therefore preferable, as based on a view of qualitative differences more in keeping with the facts.

The remaining interactions of phosphate with the quality of the nitrogenous application QP, and of phosphate with quality and quantity of nitrogen NQP, may be evaluated in a manner similar to Q and NQ, using the differences in place of the sums of the plots which have and have not received phosphate. In this group, however, we must remember that a particular component involving the contrast

$$(s-m+c-u)$$

has been confounded with blocks. The differences in yield between plots receiving phosphate and those receiving none are shown in the following table:—

TABLE 25

Quantity.	Material.				Differences between Pairs.		
	s.	m.	c.	u.	$m+s$ $-c-u$	$m+u$ $-s-c$	$s+u$ $-c-m$
(1)	136	150	67	−19	...	...	...
(2)	307	−18	69	19	...	...	...
2 (2)+(1)	750	114	205	19	640	(822)	450
2 (1)−(2)	−35	318	65	−57	275	(−231)	−475

The two unconfounded comparisons in the group QP make therefore a contribution evaluated by summing the squares of 640 and 450 and dividing by 80. This gives 7651·25 for these 2 degrees of freedom. The two corresponding components of NQP is the sum of the squares of 275 and 475 divided by 80, or 3765·625. The manurial comparison which has been confounded, namely,

$$(p-1)(s_1+s_2-m_1-m_2+c_1+c_2-u_1-u_2)$$

is not precisely a component either of QP or of NQP, as we have defined these groups. It would be a component of QP on the alternative definition discussed above, and the remaining unconfounded portion is the component of NQP of that definition, namely,

$$(p-1)(s_2-s_1-m_2+m_1+c_2-c_1-u_2+u_1).$$

This gives $303^2 \div 32$, or 2869·031.

The complete analysis of the variations observed among the yields of the 48 plots may therefore be set out as in the Table 26.

The total sum of squares for the 47 degrees of freedom, which have above been evaluated individually, must check with the sum of the squares of the deviations of the yields from the 48 plots from their mean without regard to the manurial treatments they have received,

PRACTICAL EXAMPLE 151

or to their topographical arrangement. This affords a check both on the arithmetic and on the logic of our procedure, at least so far as to show that it has consisted in a subdivision or partition of the different components of variation actually present.

TABLE 26

	Degrees of Freedom.	Sum of Squares.	Mean Square.	½ Log$_e$
Blocks	3	12215·750	...	...
N_1	1	308505·125	308505·125	2·8659
P	1	18881·333	18881·333	1·4691
Q	3	33032·100	11010·700	1·1995
N_2	1	7420·167	7420·167	1·0021
NQ	3	12038·838	4012·946	0·6948
QP	2	7651·250	3825·625	0·6709
NP	2	604·042	302·021	...
NQP	2	3765·625	1881·812	...
NQP } unconfounded	1	2869·031	2869·031	...
QP } confounded	1	219·010	219·010	...
Error	27	40763·646	1509·765	0·2060
	47	447965·917		

Next, it may be noticed that the confounding employed has involved a component of treatment recognisable as an interaction of P with one of the quality comparisons Q, but not identifiable with either of the particular aspects which we have thought it proper to recognise respectively as QP and NQP. It thus resembles the components confounded in the experiment with 27 treatments, discussed in Chapter VII., and, as in that case, would be a source of inconvenience if the unconfounded component observed were one of any importance. The table shows, however, that in the present case the component in question is of no practical interest.

53. Interpretation of Results

The treatment comparisons in the table have been arranged to show first those which have had a significant influence on the yield of grain, next those in which there may perhaps be an indication of real influence, but of a magnitude which could only be demonstrated by more precise experimentation, and finally those which in the present experiment appear to have exerted no appreciable effect. By far the largest contribution is made by what we have called the principal effect of nitrogen, this 1 degree of freedom containing indeed more than two-thirds of the total. The mean square is over 200 times the mean square for error, or, since $\sqrt{200}$ exceeds 14, we may see at once that the general effect of nitrogen in this experiment is over 14 times its standard error, and is therefore determined with comparatively high precision. The single degree of freedom for phosphate has a mean square over twelve times the average, showing that this effect also is certainly significant, though the quantitative value of this ingredient has been evaluated only roughly.

The statistical significance of each contribution to the total is most easily determined from the last column, which shows the half values of the natural logarithms of the mean squares. The table of z (*Statistical Methods*, Table VI.) shows that, with 27 degrees of freedom for error, the amount by which this entry may exceed that for error, at a 5 per cent. level of significance, is

·7187 for 1 degree of freedom
·6051 ,, 2 degrees ,,
and ·5427 ,, 3 ,, ,,

The corresponding values for significance at the 1 per cent. level are 1·0191, 0·8513 and 0·7631. The value for Q is therefore significant on the higher standard

(1 per cent.) and that for N_2 at the lower standard (5 per cent.). We may therefore take the values of Table 23 to indicate that chloride of ammonia was really more successful than sulphate of ammonia or cyanamide in stimulating grain production, with urea in an intermediate position, and that the second nitrogenous application was in general less fruitful than the first.

The mean squares for NQ and for QP, though considerably larger than that for error, do not reach the 5 per cent. level of significance. It therefore appears that the suggestion of the figures of Table 23 that chloride of ammonia and urea are not only more successful than sulphate of ammonia and cyanamide, but are disproportionately so in the double application, though supported by the data, is not demonstrated; and that the suggestion of Table 25 that when sulphate of ammonia is used superphosphate is more effective than with the other nitrogenous fertilisers, must also be regarded as doubtful. The remaining 6 degrees of freedom, ascribable to manurial treatment, are clearly insignificant to such an extent that it would have made no appreciable difference if their effects had been included with those of pure error. This circumstance shows that the principle used in the choice of a component for confounding was in fact justified by the result. Their separate evaluation serves to show how this can be done whenever necessary, and supplies the safeguard that our positive conclusions are based on an estimate of error uncontaminated by possible interactions among the treatments.

This example illustrates the fact that when quantitative and qualitative factors are combined in the same experiment, the special meaning of their interactions may well be taken into account in experimental design.

Especially, when the quantities involved include zero, some of the treatment comparisons vanish, leading, on the one hand, to an increase of the comparisons available for error, and sometimes also to the partial recovery of comparisons which would otherwise have been totally confounded. The reader should consider the effects of one simple modification of the design used, by supposing that the component

$$(p-1)(s_1+s_2-m_1-m_2+c_1+c_2-u_1-u_2)$$

were confounded with one pair of blocks, and the component

$$(p-1)(s_1-s_2-m_1+m_2+c_1-c_2-u_1+u_2)$$

with the other.

54. An Experiment with 81 Plots

In considering the experiment with 27 treatments in Chapter VII., it was shown that these could be arranged in blocks of 9 at the expense of confounding 2 of the degrees of freedom representing triple interactions. It was also shown that when replication can be carried out in multiples of 4, the confounding could be spread equally over the whole of these 8 degrees of freedom, so that all triple interactions could be recovered with some relative loss of precision, though possibly an absolute increase. When quantitative and qualitative factors are combined in the same experiment there is little point in restricting the effects of confounding to the triple interactions as defined for the purpose of that example. Moreover, it is often necessary to see what can be done in experiments of less than 108 plots. The following design, which was carried out in potatoes at Rothamsted in 1931, shows a method of utilising 81 plots so as to gain the principal advantages which the experiment was intended to secure.

The factors to be tested were three levels in the ratio 0, 1, 2 of nitrogenous manure in combination with three similar levels of potassic manures. The potassium was to be supplied in three qualitatively different materials, namely, potassium sulphate (s), potassium chloride (m), and a material known as potash manure salts (p), consisting of potassium chloride with a large admixture of common salt. The plots were divided in 9 blocks of 9 plots each, each block containing one plot with every possible combination of the three levels of nitrogen with the three levels of potash. The 3 plots without potash therefore received respectively 0, 1 and 2 doses of nitrogen. The same was true of the 3 plots receiving a single potassic dressing, and of the 3 plots receiving double potassic dressings; but in the case of these we have to choose in which form the potassium shall be supplied. In fact, at each level of potash one plot received sulphate, one chloride, and the third potash manure salts. The only ways in which the blocks can differ consist in the manner in which the three kinds of potash in each level are assigned to plots receiving 0, 1 or 2 quantities of nitrogen.

Considering only plots receiving single potassic dressings, we may designate those which receive sulphate of potash with 0, 1 and 2 quantities of nitrogen by s_0, s_1 and s_2. Then the set of plots at this level within any block will have some such formula as $s_0 m_1 p_2$, or, if we make the convention that the suffices are to be taken in their natural order, simply by $s\,m\,p$. If, now, corresponding to the block or blocks represented by the formula $s\,m\,p$, there are equal numbers of blocks represented by the formulæ $m\,p\,s$ and $p\,s\,m$, it is clear that the 9 kinds of plots which receive single potassic dressings will occur in the experiment in equal numbers;

and, in fact, that we may assign 3 of our blocks to each of these formulæ. We might equally have used the formulæ *s p m*, *p m s* and *m s p*, but our choice is limited to these two cyclic sets. The same is true of the specification of the blocks in respect of the plots within them which receive a double dressing of potash. The particular design we shall consider is that formed by choosing one of these cyclic sets for single potash, making a similar choice for the double potash, and finally deciding that each of the 3 blocks which have the same formula at one level of potash shall have three different formulæ at the other, so that the 9 blocks are all assigned to different sets of treatments. They are all alike, in sets of three, at one level of potash, and in different sets of three at the other level, like the rows and columns of a 3×3 square.

As in the previous example, let us now consider which comparisons are available for estimation of error, and which remain for estimations of the effects of treatments. Since the three treatments without potash are the same in every block, the comparisons among this group will at once yield 16 degrees of freedom. At the level of single potash the three treatments are the same in sets of three blocks each, so that each set yields 4 degrees of freedom for error, or 12 in all. A second group of 12 is provided by the level of double potash, bringing the total for comparisons made between plots at the same level of potash up to 40. We must now confine ourselves to comparisons in which plots with the same potassic dressing in the same block are treated together. Comparing the plots receiving single potash with those receiving none in the same block, we see that this comparison is the same in three sets of three blocks each, giving 6 degrees of freedom, while six more are obtained by comparing the double potash plots with those without potash in the same block.* There are thus 52 degrees of freedom ascribable solely to experimental error. Together with 8 for comparisons between whole blocks, and 20 for comparisons among the 21 different treatments, we have enumerated the whole of the 80 degrees of freedom in the experiment. There could be no more contributions to pure error, unless some one or more of the treatment comparisons had been totally confounded with block differences.

We may now consider the manurial comparisons. There are seven combinations of quantity and quality of potash, the six comparisons among which may be resolved into 2 for quantity K, 2 for quality Q, and the remaining 2 for interaction of quantity and quality KQ. The distinction between Q and KQ will be made by the same convention as in the last example. Variation of the quantity of nitrogen increases the number of manurial combinations to 21 and therefore introduces 14 new comparisons. Of these, 2 represent the effects of quantity of nitrogen only N, 4 the interactions of quantitative variations of nitrogen and potash NK, 4 the interactions of quantity of nitrogen with quality of potash NQ, and 4 more the triple interaction NKQ. All these groups of comparisons, except those denoted NQ and NKQ, are obviously free from confounding, for they can be made up directly by comparisons within blocks. It is only the last 8 degrees of freedom which require special consideration. As often happens, and as the previous examples have already illustrated, we shall best see what has happened

* The last two sets of six components each are not, however, independent, since the plots without potash are used in both. The sum of squares for all twelve is most simply obtained by deducting from the 26 degrees of freedom among the totals from each block for the 0, 1, and 2 levels of potash, the 2 degrees of freedom for K, the 8 for blocks and the 4 partially confounded effects of treatment which will be identified later.

to this group of comparisons by resolving them into components in a way specially appropriate to the structure of the experiment.

Since qualitative differences exist only in plots receiving either single or double potassic dressings, the eight comparisons, with which we are concerned, are equivalent to the four representing interactions of nitrogen with quality of potash on the plots receiving single potassic dressings, together with the similar four on the plots receiving double potassic dressings. Let us consider these two parts separately. Just as the three independent comparisons among the four nitrogenous materials of the last example were subdivided in the same manner as the contrasts between rows, columns and letters in a 2 × 2 Latin square, so we shall now use the analogous property which a 3 × 3 square possesses. Let the rows of such a square correspond to the quantities 0, 1 and 2 of nitrogen and the columns to the three sorts, s, m and p of potash. Then three of our blocks have, in respect to the single potash dressings, the formula $s\,p\,m$. These we may call blocks of type A and insert the letter A in the three corresponding cells of the square. There will also be blocks with the formula $p\,m\,s$, which we may call type B, and with the formula $m\,s\,p$ which we may call type C. If these letters be inserted we shall have a 3 × 3 Latin square as shown below:—

		Kind of Potash.		
Quantity of nitrogen	0	Aα	Cβ	Bγ
,, ,,	1	Cγ	Bα	Aβ
,, ,,	2	Bβ	Aγ	Cα

To the Latin letters in the square, Greek letters have been added, in such a way that each appears once in each column, once in each row and once with each Latin letter. The whole thus constitutes what is known as a Græco-Latin square. The fact that a Græco-Latin square is possible shows that the eight independent comparisons among nine objects can be resolved into four independent sets of 2 degrees of freedom each, each pair being the comparison between three sets of three chosen objects each. These are the two comparisons among rows, two among columns, two among Latin letters, and two among Greek letters. It will be observed that the comparisons among Latin letters have been chosen to correspond with the differences among the sets of blocks; consequently, the comparisons among the Greek letters are independent of these block differences, and, like the comparisons between rows N and columns Q, may be made by comparing yields within the same block. By adding up the yields of all plots having treatments of the combinations indicated by the letters a, β and γ, we may evaluate two treatment comparisons which have not been confounded. Two more are likewise obtained from the plots with double potassic dressings. In this way four of the eight comparisons represented by NQ and NKQ are isolated. The 4 remaining degrees have, however, been partially confounded.

The confounding of these four remaining comparisons with block differences is incomplete, owing to the fact that the blocks, which differ in respect of them, agree in containing other plots treated alike, with which they may be compared. Thus, in the three blocks of type A, the plots with single dressings of potash having the chosen constitutions s_0, p_1 and m_2 are situated in the same blocks with an aggregate of plots receiving no potash, and with an aggregate receiving double potash, both of which are the same as the aggregates which occur in the three blocks of type B, and in the three blocks of type C. We may therefore compare the total

CASES OF PARTIAL CONFOUNDING

yield from the treatments s_0, p_1 and m_2 with the totals from the other sets of treatments p_0, m_1 and s_2, and

TABLE 27

Arrangement and Yields of a Complex Experiment
$(s\ p\ m)_1\ (s\ m\ p)_2$

n_1m_2	n_0s_2	n_0s_1	n_1	n_2p_1	n_1s_2	n_2s_1	n_1p_2	n_2
751	733	686	851	890	874	1026	947	990
844	829	825	800	733	813	1050	871	1006
n_0	n_2	n_1p_1	n_0	n_2m_2	n_0p_2	n_2s_2	n_0m_2	n_0p_1
796	910	909	855	1026	865	1118	853	795
705	866	778	779	815	816	997	953	843
n_1	n_2m_1	n_2p_2	n_2	n_0m_1	n_1s_1	n_0	n_1	n_1m_1
895	1034	1052	913	756	892	1024	972	975
965	1046	830	752	830	930	979	1000	884
n_0	n_0m_1	n_0m_2	n_0s_2	n_2s_1	n_1m_1	n_1s_2	n_0	n_2m_2
1100	1014	975	956	1037	1035	1284	1012	1001
996	968	902	898	975	1027	1176	966	977
n_1s_1	n_1p_2	n_2s_2	n_0	n_0p_1	n_1m_2	n_2	n_0s_1	n_1
1029	1121	1252	1058	1038	1098	1316	1136	1087
1022	961	1127	1006	904	1206	1275	1049	1001
n_1	n_2p_1	n_2	n_2	n_2p_2	n_1	n_0p_2	n_2m_1	n_1p_1
959	1178	1317	1270	1234	1195	1307	1224	1069
930	1102	1145	1118	1134	1132	1348	1275	1128
n_2	n_2m_2	n_0p_1	n_1	n_1p_1	n_2s_2	n_2p_2	n_1m_2	n_0
1131	1140	1055	1151	1147	1156	1401	1214	1005
1034	1156	1026	1044	1056	1228	1391	1321	1011
n_1m_1	n_1s_2	n_1	n_2m_1	n_1p_2	n_2	n_0s_2	n_2	n_0m_1
980	1243	1224	1192	1225	1305	1310	1421	1190
1027	1065	1064	1199	1120	1276	1339	1417	1208
n_0	n_0p_2	n_2s_1	n_0m_2	n_0s_1	n_0	n_1s_1	n_1	n_2p_1
1020	653	935	629	947	1020	1361	1167	1222
605	999	1142	1056	1049	1102	1201	1215	1108

m_0, s_1 and p_2, by subtracting in each block the sum of the yields from plots without potash, and with double potash, from twice the yield of the plots with single potash, and adding together the results from blocks in which the plots with single potash have been treated alike. The manurial comparisons so made clearly represent two of those which have been confounded with blocks, but which can be made, in the manner explained above, by means of comparisons wholly within blocks, with a satisfactory precision. The sum of the squares of the coefficients of the expression

$$2k_1 - k_0 - k_2$$

is 6, and to this total the coefficient of k_1 contributes 4. Consequently, the comparison so made among the different sets of plots receiving single potash has two-thirds of the precision of the other comparisons of the experiment, and so, perhaps, a higher precision than they would have had, even if unconfounded, in an experiment with 27 plots to each block.

The student may familiarise himself with the process of analysis described above by applying it to the yields of tubers in quarter-lbs. shown in Table 27 of the arrangement of the treatments in different plots.

The upper and lower figures represent yields with and without phosphate, which was applied as an additional manure to half, chosen at random, of each plot. The sums of these yields may therefore be analysed as explained above, but their differences, representing the effects of the phosphatic manure, and its interactions, are already freed from all block effects, and will have their own standard error estimated directly from the discrepancies between these differences in plots treated alike.

REFERENCES AND OTHER READING

T. EDEN and R. A. FISHER (1929). Experiments on the response of the potato to potash and nitrogen. Journal of Agricultural Science, xix. 201-213.

R. A. FISHER and J. WISHART (1930). The arrangement of field experiments and the statistical reduction of the results. Imperial Bureau of Soil Science, Technical Communication, No. 10.

R. A. FISHER (1943). The theory of confounding in factorial experiments in relation to the theory of groups. Annals of Eugenics, xi. 4, 341-353.

R. A. FISHER (1945). A system of confounding for factors with more than two alternations, giving completely orthogonal cubes and higher powers. Annals of Eugenics, xii. 4, 283-290.

F. YATES (1937) The design and analysis of factorial experiments. Imperial Bureau of Soil Science Technical Communications No. 35.

IX

THE INCREASE OF PRECISION BY CONCOMITANT MEASUREMENTS. STATISTICAL CONTROL

55. Occasions suitable for Concomitant Measurements

IN the preceding chapters we have been principally concerned with the means whereby experimental precision may be increased through the knowledge that groups of material may be selected, the parts of which are more homogeneous than are the different groups. We have been using such facts as that animals more nearly related by blood are generally more alike than animals less nearly related, that men of the same race or district are likely to be more similar than men of different races, that plots of land resemble one another more nearly in fertility the closer together they lie, or that apparatus supplied by the same manufacturer will generally be more nearly comparable than the makes supplied by different firms. It has been shown that very great increases in precision are possible by utilising these and analogous facts, even when the amount of material which is closely homogeneous with any chosen unit is extremely limited, provided that within this limitation we may assign the treatments to be tested at will so as to build up a comprehensive experiment.

There is a second means by which precision may, in appropriate cases, be much increased by the elimination of causes of variation which cannot be controlled, which has the advantage of being applicable when we cannot exercise a free choice in the distribution of the treatments. For example, in a feeding experiment with animals, where we are concerned to measure their response to a number of different rations or diets, we may often be able to ensure that the animals entering on the treatments to be tested shall be of the same age, and often, also, that they shall be closely related, or of the same parentage. But such groups of closely related animals as are available will not, at the same age, have attained exactly to the same size, as measured by weight, or in any other appropriate manner. If we decide that they shall enter the experiment at the same age, it may well be that the differences in initial weight constitute an uncontrolled cause of variation among the responses to treatment, which will sensibly diminish the precision of the comparisons. If the animals are assigned at random to the different treatments, either absolutely or subject to restrictions of the kinds which have been discussed, the differences in initial weight will not, of course, vitiate the tests of significance, for, though they may contribute to the error of our comparisons, they will then also contribute in due measure to the estimates of error by which the significance of these comparisons are to be judged. They may, however, constitute an element of error which it is desirable, and possible, to eliminate. The possibility arises from the fact that, without being equalised, these differences of initial weight may none the less be measured. Their effects upon our final results may approximately be estimated, and the results adjusted in accordance with the estimated effects, so as to afford a final precision, in many cases, almost as great as though complete equalisation had been possible.

Similar situations frequently arise in other fields of work. In agricultural experiments involving the yield following different kinds of treatments, it may be apparent that the yields of the different plots have been much disturbed by variations in the number of plants which have established themselves. If we are satisfied that this variation in plant number is not itself an effect of the treatments being investigated, or if we are willing to confine our investigation to the effects on yield, excluding such as flow directly or indirectly from effects brought about by variations in plant number, then it will appear desirable to introduce into our comparisons a correction which makes allowance, at least approximately, for the variations in yield directly due to variation in plant number itself. In introducing such a correction it is important to make sure that our procedure shall not in any way invalidate the test of significance to be applied to the comparisons, and thought will often be required to assure ourselves that the effects eliminated shall really be only those which are irrelevant to the aim of the experiment.

Again, let us suppose that a number of remedial treatments are to be tested on an orchard or plantation, the trees of which show in varying measure the effects of disease. It might be possible to grade the individual plants prior to the application of the treatments, and to apply the treatments to equal numbers of plants showing each grade of injury. But this would not always be possible, especially if it is not to individual plants but to small plots, each containing several plants, that the remedial measures must be applied. Such a procedure would also, in any case, necessarily sacrifice the advantage of propinquity of the areas which it is desired to compare. To meet this difficulty it is open to us to apply the different treatments to plots randomised and adequately replicated, but chosen without regard to the initial grade of injury of the plants they contain. The grades of injury of these plants may however, be recorded both initially and finally, when the treatments may be supposed to have exerted such effects as they are capable of, and the comparison of the final condition of the plots which have received different treatments may be adjusted to take account of the degrees of injury initially shown by these same plots.

With perennial plantations the same principle may very advantageously be applied to studies of the effects of manuring, pruning, and other variable treatments, on the yield. Yield in such cases is evidently much influenced, not only by variations in soil fertility, but also by the individual capacities of different plants, which, whether hereditary or not, persist from year to year. Records of the yield of individual rubber trees, or of small areas of tea-plantation, thus show large and relatively permanent differences. In such cases records of yield for a preliminary period under uniform treatment provide a most valuable guide in interpreting the records after the treatments have been varied. It would in these cases be possible to choose areas for the different treatments such that their previous record was approximately equalised. But to do so is usually troublesome, inexact, and unnecessary. Moreover, as the plots so chosen cannot also be arranged in compact blocks, or in other advantageous arrangements, such as the Latin square, the loss of precision due to sacrificing this advantage is often considerable. It is now usual, therefore, to arrange the plots in some way which is topographically advantageous, irrespective of their previous records, and to utilise the information supplied by these as an adjustment or correction to the subsequent yields measured under varying treatments. It may be noted, however, that with annual agricultural crops, knowledge of the yields of the experimental area in a previous year under uniform treatment has not been found sufficiently to increase the precision to warrant the adoption of such uniformity trials as a preliminary to projected experiments. Such a procedure necessarily nearly doubles the experimental labour, and as it is not found to double the amount of information supplied by the experiment but to increase it, perhaps by 50 per cent., it is clearly unprofitable. For, by the application of twice the expenditure in time and money in the experimental year, the amount of information recovered may with confidence be expected to be approximately doubled. Consequently, on grounds of precision alone, such preliminary trials with annual crops are not to be recommended. The fact that they entail at least a year's delay in the experimental results adds to the force of this conclusion.

In many cases it may be possible to take two or more concomitant measurements, each of which severally may be expected, when proper allowance is made for it, to increase the precision of the comparisons to be made, and which, if used jointly, may increase them still further. Thus, if groups of school children be supplied, in addition to their home diet, with a ration of milk, either raw or pasteurised, children in the same school may be assigned properly at random to the groups receiving these different additions to their diet. With large numbers of subjects the age distributions of the two groups may be very nearly equalised, but with the smaller numbers attending a particular school such equalisation of age will necessarily be somewhat inexact, and, apart from age, it is certain that the two groups of children will differ somewhat in the initial height and weight. The variations in these initial values, moreover, may all be suspected of having, possibly, an appreciable influence on the apparent response to the nutrients as measured by increments in height or weight. The most thorough procedure for such a case would be to eliminate, or make allowance for, all these three variables jointly; and, though it might not in fact be necessary to take account of more than two, or of even one of them, we could only assure ourselves that such a simpler procedure was in reality effective by examining the effects of making allowance for all three jointly.

56. Arbitrary Corrections

In the examples outlined above, in which an observable but uncontrolled concomitant might reasonably be expected, if proper account can be taken of it, to add to the precision of the results, it is still a common practice to introduce corrections arrived at *a priori*, without reference to what the data themselves have to tell of the amount of the corrections to be applied. Thus in a feeding experiment with animals it might be thought proper to take account of the variation in their initial weights by calculating the responses of different individuals, not by their absolute increases in weight, but by their increase relative to their initial weight, or as percentage increases. Equally, in allowing for the effect of variation in plant number upon an agricultural yield it is possible, and has sometimes been thought appropriate, to calculate the yield per plant in place of the yield per unit area, as the measure of the efficacy of the treatments to be compared. In judging of the effects of treatments on the grade of visible damage caused by disease it might be thought sufficient to compare the differences between the average grades of the different plants receiving any treatment, before and after that treatment has been applied, in order to allow for the fact that the areas differently treated, though assigned properly at random, were not initially in exactly the same condition. When allowing for the differences between different plots observed in a preliminary trial of a plantation, either the proportional system, or that dependent on simple differences, might equally be advocated, and would, perhaps, not give greatly different results.

Such methods of discounting *a priori* the effects of concomitant variates, so utilising them to increase the precision of experimental comparisons, should not be rejected as invalid, even though we may know that the suppositions on which they are based are experimentally untrue. The experimenter, for example, has a perfect right to measure the efficiency of different feeding stuffs, either by the average percentage increase of different animals, or by the average absolute increase, as he pleases, and, with a properly designed experiment, he will ascertain whether the materials tested do or do not give significantly different results as measured in these alternative ways. He has this right, none the less, even if experiments with a uniform feeding mixture, and animals of varying initial weight, have shown that the increments in weight during the experiment period neither are independent of initial weight nor are proportional to it. What such experiments would make clear is that, for the purpose of detecting differences between the feeding stuffs tested, with the greatest possible precision in relation to the size of the trial, neither method of measuring weight increase is ideal, and that both are capable of some improvement. If, for example, in experiments in the course of which the average weight of the animals had doubled, it was found that an initial difference in weight of 1 lb. was followed at the end of the experiment by a difference on the average of $1\frac{1}{2}$ lbs., it is obvious that an allowance on this scale would be preferable, for the purpose of comparing different feeding stuffs, either to an allowance of a pound for a pound, as is the effect of taking simple weight increases without regard to the initial weight, or to an allowance of 2 lbs. for 1 lb., which would be approximately the effect of judging the experimental results by the proportional increases in weight.

Preliminary investigations of the correct allowance to make for concomitant variates are usually wanting, and are, fortunately, not a practical necessity, for the results of a replicated experiment may themselves be used to supply what is wanted. Let us suppose that five feeding stuffs are to be tested, each on ten pigs, the animals being assigned to the different rations entirely at random. The average initial weights of the groups assigned to the different feeding stuffs will therefore vary somewhat by chance, though this variation will not be so great as the variation between the initial weights of the different animals receiving the same feeding stuff. The assignment being at random in fact gives an assurance that the average differences between the different lots of ten shall be smaller than the individual differences in the ratio $1 : \sqrt{10}$ or, in fact, should be rather less than one-third as great. A direct comparison, within the groups receiving the same mixture, of the extent to which greater initial weight is followed by greater final weight, will, therefore, generally supply an estimate of the true allowance to be made of amply sufficient accuracy for the small adjustments which are to be based upon it. Moreover, such an allowance based on the very same data to which it is to be applied, is generally preferable to one based on other experiments, even if these are much more extensive, since it is certain that the conditions in which different experiments are made vary greatly, and in many unknown and uncontrolled ways. We have no assurance that the allowance appropriate to one set of conditions or to one type of material shall still be even approximately appropriate when the conditions and the material are varied. Consequently, even if the appropriate allowance for each concomitant variable had been previously ascertained by sufficiently extensive experimentation, it would still be advantageous to rely, in each particular case, on the internal evidence of the experiment in question. It may also be noted that by doing so the experiment conserves its property of being self-contained, and, therefore, adequate to supply genuinely independent testimony on any point in dispute, and that such complete independence is attenuated, if not lost, if extraneous data are introduced in the process of its interpretation.

57. Calculation of the Adjustment

The process of calculating the average apparent effect on the experimental value of the increase of one unit of the concomitant measurement is, in principle, extremely simple. Statistically, such values fall into the class of what are known as "regression coefficients," and the variety of methods, appropriate to calculating such regressions, forms an extensive subject, which is treated more fully in the author's book, *Statistical Methods for Research Workers*. To illustrate the principle used, the detailed working for a simple case will be given here, from which the reader who is unfamiliar with regressions will be able to see exactly what the calculation amounts to, though a fuller study would be needed to recognise how the operations should best be carried out in particular cases. We will suppose that five feeding mixtures are being tested in respect to the live weight increase produced by them, between fixed limits of age, on groups of ten pigs, assigned at random to each of the five mixtures. If no account whatever were to be taken of the initial weight (x), we might deal with the final weights (y) as follows :—The ten final weights for each treatment are added to give totals corresponding to each treatment (A, B, C, D, E), and divided by 10 to give the corresponding mean values (a, b, c, d, e). To judge of the significance of the differences between these totals, or between these means, we must make an estimate of the magnitude of the variations due to uncontrolled causes, including initial weight, and this we may do by examining the variation in final weight among pigs fed with the same mixture. Each set of ten pigs treated alike will supply 9 degrees of freedom for this purpose, or 45 degrees of freedom in all, for the estimation of error. The sum of squares, corresponding to each 9 degrees of freedom, is found by squaring the ten final weights, adding the squares and deducting Aa, the product of the total and mean weight for the treatment concerned. The sum of squares corresponding to the 4 degrees of freedom for variance among treatments is likewise obtained by adding together the products of means and totals for the several treatments, and deducting the product of the grand total and the general mean, *i.e.* by

$$A a + B b + C c + D d + E e - M m \, ;$$

where M stands for the grand total, and m for the general mean. The analysis of variance is thus of the simple form :—

TABLE 28

	Degrees of Freedom.
Treatments	4
Error	45
Total	49

Obviously, an exactly similar analysis can be made of the initial weights (x), though this is of no direct experimental interest. To do so is, however, the first step to take in utilising the values x in order to adjust the final weights (y) in making a closer comparison of the responses to different feeding mixtures. The next step is to make a third table of the same kind, utilising now, at each stage, instead of squares, the products of the numbers x and y. Thus, for each set of 9 degrees of freedom recognised as error, we take the product of the initial and final weight of each individual pig, add the products for the ten pigs treated alike, and deduct the product of the initial total and the final mean, or, what comes to the same thing, of the initial mean and the final total. We thus have a sum of products for the 45 degrees of freedom ascribed to error, comparable in every respect with the sum of squares belonging to the same degrees of freedom for the initial weights, or for the final weights. Equally, for the 4 degrees of freedom ascribed to treatments we may find the appropriate sum of products by multiplying the initial total weight for any treatment by the final mean weight, adding the five products so obtained and deducting the initial total weight of all the pigs, multiplied by their final mean weight.

The three corresponding tables derived from the squares of the final weights, the squares of the initial weights and the products of the two series, contain all that is needed for the adjustment of the final weight, and for the further study of the adjusted values. In particular, the appropriate adjustment to be subtracted from each final weight to allow for each additional pound in initial weight, as judged from the internal evidence of the experiment, by a comparison among pigs treated alike is found simply by dividing the error term of the sum of products by the corresponding term in the sum of squares of initial weights.

This procedure is of quite general application. If, for example, the experiment had been of a more intricate design we might have chosen sets of five pigs each, from ten different litters, and assigned one pig of each litter to each treatment, so that the treatments should be tried on animals of more nearly equal genetic constitution than if a lot of fifty had been distributed wholly at random. The analysis would then have taken the form

TABLE 29

	Degrees of Freedom.
Litters	9
Treatments	4
Error	36
Total	49

The differences between litters being thus eliminated from the experiment, both in the effects of treatment and in the estimation of error, we should in consequence derive the adjustment by dividing the error component of the sum of products by the corresponding component in the sum of squares in initial weights, because it is now only the relation between initial and final weight among pigs of the same litter that is wanted in adjusting the results. We may, therefore, in all cases obtain the empirical adjustment, indicated by the particular results of the experiment, by dividing the error component of the sum of products by that of the sum of squares of the concomitant observation.

In cases in which it is desired to make allowance simultaneously for two or more concomitant measurements, separate analyses in the same form should be made for each of these, and for the sum of products of each with the dependent variate to be adjusted, and with each other. The error terms of these tables will then provide a system of two or more linear equations in accordance with the general procedure of partial regression, the solutions of which represent the average effects of unit changes in the several independent variates. The principle of the adjustment is thus exactly the same whether we have to do with one concomitant variate or with many, and the use of two or more such concomitants involves no unmanageable increase in the labour of computation. The limiting factor in the utility of concomitant observations lies rather in the labour of additional measurements, which may not, even when the best possible use is made of them, lead to so great an increase in precision as could be obtained by increasing the size of the experiment on a simpler plan, or, in other ways, by the expenditure of an equivalent amount of time and attention. In cases, however, as with the initial weights of experimental animals, where the measurement to be used as a concomitant is one which would not in any case be omitted, the precision which can be gained by a direct evaluation of their actual effects is entirely profitable to the experiment.

58. The Test of Significance

We have now to evaluate this gain in precision so that the significance of the responses to different treatments may be tested after adjustment. Since the adjustment has been obtained from the error term we may regard 1 of the degrees of freedom ascribed to error as having been utilised in evaluating it. Supposing, that is, that only one concomitant variate has been used, and therefore only one coefficient has been evaluated. In general, the number of degrees of freedom utilised is equal to the number of concomitant variates. After allowing, therefore, for the initial weights of the animals in our experiment there will remain only 44 degrees of freedom for the estimation of error if the animals have been assigned wholly at random, and only 35 degrees of freedom if they have been assigned at random within the litters. The deduction to be made from the sum of squares ascribed to error in the analysis of the final weights due to the removal of this 1 degree of freedom, is easily calculated. It consists of the square of the error component in the analysis of covariance, divided by the error component of the analysis of variance of the initial weights. After deducting this portion the sum of squares ascribed to error may be divided by the degrees of freedom, to obtain the mean square appropriate to testing the significance of differences among the adjusted final weights. This use is entirely appropriate only if, as should be the case in a properly randomised experiment, the differences among the mean initial weights of the different groups of animals are small compared with the differences amongst animals of the same group from which the adjustment has been evaluated. If this were not so, the adjusted values would in some measure be also affected by the errors of estimation of the value of the adjustment applied. It is therefore a useful resource to apply a test of significance to the adjusted values, or any component of them of special interest, which shall take full account of the inexactitude of our estimate. We may illustrate the procedure for the case in which sets of five pigs from ten different litters have been assigned at random to the five feeding mixtures.

In this case 9 degrees of freedom representing differences between litters have been eliminated from the experimental comparisons, and from the estimate of error. With these we are no longer concerned. The sum of squares corresponding to the 35 degrees of freedom for the estimation of error, after adjustment, has been evaluated by means of deducting the square of a term from the analyses of covariance, divided by the corresponding term in the analysis of variance of the initial yields. The same process is now applied using the sum of the components for treatment and error from the same tables. This gives us the sum of squares corresponding to 39 degrees of freedom, for 1 has been deducted from the 40 originally available. Subtracting now the portion obtained for error, the difference represents the 4 degrees of freedom ascribable to treatments, after exact allowance has been made for the sampling error of the coefficient used in their adjustment. The sum of squares for these 4 degrees of freedom may therefore be compared with that for the 35 degrees of freedom due to error, as in an ordinary analysis of variance, in which no concomitant variate has been eliminated. The sum of squares ascribed to treatment by this method will be found to be somewhat less than the corresponding value derived from the adjusted means and totals, although these adjusted values are the best available from the experiment, only because a calculable portion of the variance among them is ascribable to the sampling error of the estimated rate of allowance, which portion it is proper to remove in making an exact test of the significance of the variation observed. In cases where the concomitant variation has not been properly randomised, the omission of this precaution may lead to serious errors, but in such cases the possibility of testing significance accurately is always questionable.

58.1. Missing Values

It sometimes happens, in an experiment in which some cause of disturbance has been carefully equalised, as are the rows and columns in a Latin square, that, by some unforeseen accident, one of the experimental values is missing. This may happen through the death of an individual, injury to a portion of a growing crop, a gross error in recording, or to any such cause. Without the missing value equalisation is no longer complete, and it is sometimes thought that the whole experiment has been wasted. Indeed, the possibility of such mishaps has been held to be a reason for avoiding all experiments having intricate or complex structures. The technique applicable to concomitant observations does not, however, rest upon any assumption of equalisation, and may be used to recover the information available in the values that remain. The experiment so repaired will not, of course, be so good as if it had not been injured; but there is no reason to suppose that the loss of information suffered will be disproportionate to the value of the experiment as a whole.

Instead of estimating an adjustment based on a regression coefficient, it is convenient to estimate the missing value itself. The principle employed is simply to insert an algebraic symbol (x) for the missing value. The ordinary process of analysis of variance will then yield not wholly numerical expressions for the sums of squares, but algebraic expressions quadratic in x. If, for example, the sum of squares for error is
$$A - 2Bx + Cx^2,$$
this will be minimised for the value,
$$x = \frac{B}{C}$$
which is the required estimate of the missing observation. For this value of x, the sum of squares ascribable to error is
$$A - \frac{B^2}{C}$$
which corresponds to a number of degrees of freedom

one less than would have been available had no value been missing.

It will be noticed that the coefficients of x^2 in the different lines of the analysis are all positive, and correspond with the analysis of variance of the concomitant observation, while the coefficients of $-2x$ correspond with the analysis of covariance. As in the case of concomitant measurements, where allowance must be made for the sampling error of the regression, the results of inserting the estimated value are not so accurate as if that value had actually been observed; nevertheless, an unbiased test of significance may be made by minimising the sum of squares for the total of errors and treatments, and from this minimised value subtracting the minimised value for error only.

The reader is advised to practise this procedure, using such an example as that shown in Table 9, omitting any one of the 36 values there given. If more than one value is omitted, there will be two or more unknowns, and the process of minimising the sum of squares will yield as many equations, analogous to the simultaneous equations in partial regression.

It is instructive also to follow through algebraically the process given above, *e.g.* to show that when a single value is missing from a 6×6 Latin square,

(a) the reconstructed value is

$$\frac{1}{20}(6R' + 6C' + 6T' - 2M)$$

when R', C' and T' are the incomplete totals for row, column and treatment from which the observation was missing, and M is the incomplete total of all observations.

(b) The sum of squares ascribable to treatments in testing significance is

$$\frac{1}{6}S(T-\bar{T})^2 + \frac{1}{100}(5T' + R' + C' - M)^2$$

180 INCREASE OF PRECISION

where T stands for the total of any of the completely observed treatments, and $\bar{T}$ for the mean of these totals. Evidently, the value lost has not affected the precision of comparisons among these treatments, representing 4 degrees of freedom. The second term gives the comparison of the incompletely recorded treatment with the mean of the other five.

59. Practical Examples

In Table 29·1 the pairs of numbers are the initial and final weights (after 12 weeks) of four groups of Northumbrian sheep, treated respectively with A nothing, B phenothiazine, C minerals and D both phenothiazine

TABLE 29·1

Initial and Final Weights in lbs. of Experimental Sheep
(*W. Lyle Stewart's data*)

A.		B.		C.		D.	
57,	94	46,	80	45,	70	38,	79
54,	84	40,	68	55,	90	59,	100
43,	81	36,	62	49,	80	45,	72
58,	87	40,	79	37,	58	44,	77
44,	70	49,	79	49,	82	47,	75
42,	67	50,	79	52,	91	52,	95
38,	78	44,	86	43,	79	37,	66
54,	71	44,	70	63,	84	49,	90
54,	80	44,	75	56,	96	38,	83
47,	63	42,	79	28,	59	33,	59
51,	90	45,	68	44,	69	65,	106
52,	86	43,	80	49,	90	35,	61
47,	79	42,	77	47,	79	41,	79
56,	79	44,	74	45,	74	43,	73
44,	74	54,	84	40,	70	41,	74
50,	82	43,	77	40,	62	41,	85
34,	56	46,	70	39,	71	47,	85
30,	49	31,	67	35,	62	51,	91
31,	28	41,	70	48,	84	48,	84
32,	58					31,	66
918,	1456	824,	1424	864,	1450	885,	1600

and minerals. Analyse the variance and covariance of initial and final weights, within and between groups, and compare the average final weights after allowance for variations in initial weight.

Table 30 shows the arrangement of an experiment, carried out in sugar-beet, at Good Easter, Chelmsford, in 1932, by the National Institute of Agricultural Botany. Three varieties, a, b and c, are tested in combination with eight manurial mixtures, respectively containing and lacking sulphate of ammonia, at the rate of 0·6 cwt. of nitrogen per acre, superphosphate at the rate of 0·5 cwt. P_2O_5 per acre, and chloride of potash at the rate of 0·75 cwt. K_2O per acre. The twenty-four combinations of the three varieties with the eight manurings are arranged in four randomised blocks in the order shown. In each plot the first number is the number of plants lifted, while the second is the weight in pounds of washed roots.

In carrying out the analysis it should be observed that the varieties show significant differences in plant number, so that the yields in root weight adjusted for plant number will not necessarily represent varietal differences in yield under any uniform system of field treatment, but should represent yield differences for equal plant establishment.

When variation in plant number is not large a proportional allowance, based on a simple regression coefficient, is often entirely adequate. Theoretically, however, we should not expect the relationship between yield and plant number to be represented by a straight line over a wide range, but rather by a curve, having a maximum within or outside the range of the observations. To deal with curved regression, when it seems to be advantageous, it is only necessary to introduce not only the plant number but its square also as a second

TABLE 30
Arrangement, Plant Number and Yield, of Combined Manurial and Varietal Experiment with Sugar-beet
(Rothamsted Experimental Station Report, 1932)

I.		II.		III.		IV.	
cnp	103,112	bnp	142,150	anpk	107,139	ap	109,162
cn	121,118	anpk	147,155	a	114,127	bpk	143,139
ank	134,112	c	138,132	c	119,123	ck	129,151
bk	156,117	cn	141,152	cp	127,120	anpk	148,192
ak	131,152	cpk	126,115	bnp	133,118	bn	159,174
cpk	129,140	ap	134,175	cn	127,149	bp	120,143
cp	123,118	bnk	142,144	an	127,168	ak	145,188
bnpk	146,144	bn	138,159	anp	119,157	bnp	138,157
b	145,133	bpk	145,132	bn	140,166	a	127,158
an	136,184	cnp	144,175	bnpk	138,155	cpk	142,152
bnp	140,168	apk	133,158	b	139,138	cnp	143,173
cnk	126,148	ank	156,193	cpk	129,130	anp	132,193
bp	152,140	b	147,130	cnpk	129,173	bk	147,147
bnk	136,143	cp	139,142	cnk	107,147	cp	124,138
bn	124,163	bp	138,101	bnk	133,142	cn	127,165
a	124,162	cnpk	125,160	ak	130,141	ank	138,191
c	113,122	ak	161,164	ck	118,142	bnpk	140,153
cnpk	120,162	an	134,178	ap	134,142	b	127,128
anpk	120,175	a	133,162	bp	125,124	an	139,199
apk	126,140	bnpk	135,160	bpk	125,132	cnpk	137,185
anp	132,190	cnk	128,152	cnp	102,152	apk	132,160
ap	115,173	bk	152,137	bk	107,121	c	127,146
ck	91,107	ck	149,171	apk	106,148	bnk	148,159
bpk	137,127	anp	104,166	ank	101,171	cnk	110,136

concomitant observation, and to treat these exactly as though they were two independent variables. The fact that one of these may be calculated from the other does not in any respect interfere with their use in this way, and, of course, in special cases, the same principle may be used to introduce more complicated curves. Sound judgment as to the probable value of such elaborations, in comparison with the work required, can only be gained by trying them on bodies of actual observations, such as those shown in the table.

REFERENCES AND OTHER READING

T. EDEN (1931). The experimental errors of field experiments with tea. Journal of Agricultural Science, xxi. 547-573.

R. A. FISHER (1925-1963). Statistical methods for research workers. Chap. V., §§ 25-29; Chap. VIII., § 49·1.

Rothamsted Experimental Station Annual Report (1932).

H. G. SANDERS (1930). A note on the value of uniformity trials for subsequent experiments. Journal of Agricultural Science, xx. 63-73.

X

THE GENERALISATION OF NULL HYPOTHESES. FIDUCIAL PROBABILITY

60. Precision regarded as Amount of Information

The foregoing Chapters, III. to IX., have been devoted to cases to which the theory of errors is appropriate; that is to say, to cases in which the experimental result sought is found by testing the significance of the deviations shown by the observations, from a null hypothesis of a particular kind. In this kind of hypothesis all discrepancies classified as error, and not eliminated from our comparisons by equalisation or regression, are due to variation, in the material examined, following the normal law of errors with a definite and constant, but unknown, variance.

Granting the appropriateness of null hypotheses of this kind, our purpose has been to diminish the magnitude of the error components in the comparisons, and a number of devices have been illustrated by which this can be done, while at the same time the requirement can be satisfied that the experiment shall supply a valid estimate of the magnitude of the residual errors, by which the comparisons are still affected. In general, it has been seen that, with repeated experimentation on like material, the variance ascribable to error falls off inversely to the number of replications, so that in measuring the effectiveness of methods of reducing the error, an appropriate scale is provided by the inverse of the variance, or the *invariance*, as it is sometimes called, of the averages determined by the experiment.

If, therefore, any such average is determined with a sampling variance V, we may define a quantity I such that $I = 1/V$, and I will measure the quantity of information supplied by the experiment in respect of the particular value to which the variance refers. Information, of course, like other quantities, may be measured in units of different sizes, according to the subject under discussion. Thus, with agricultural yields it is convenient to consider an experiment giving a standard error of 10 per cent. as supplying one unit of information. One giving a standard error of only 5 per cent. will, therefore, supply four units. An experiment with a standard error of 2 per cent. will yield twenty-five, and one with a standard error of 1 per cent. will yield a hundred of such units. The amount of information is thus measurable on a scale inverse to the variance, or inverse to the square of the standard error.

One immediate consequence of this method of evaluation is that when an experimental programme is enlarged by simple repetition on like material, the amount of information gained is proportional to the labour and expense incurred. Consequently, we may ascertain the cost, per unit of information gained, of any type of experimentation of which we have adequate experience; or, if we wish, we may so use the data from any single large experiment. The cost of attaining any desired level of precision, or of gaining any desired amount of information by the same method is thus easily calculable. What is also important, the relative costliness of different methods of experimentation may be directly compared, and the saving effected by improved methods of design, or by the use of concomitant observations, may be given an entirely objective and tangible value.

In such calculations it is appropriate that the items of labour and skilled supervision chargeable to a particular method of experimentation shall be fairly and carefully recorded and calculated. For any time and labour devoted to experimental work must be regarded as having been diverted from other work of scientific value, to which they might otherwise have been given. Even rough costings of this kind will usually show that the efficiency with which limited resources can be applied is capable of relatively enormous increases by careful planning of the experimental programme, and there is nothing in the nature of scientific work which requires that the allocation of the resources to the ends aimed at should be in any degree rougher, or less scrupulous, than in the case of a commercial business. The waste of scientific resources in futile experimentation has, in the past, been immense in many fields. One important cause at least of this waste has been a failure to utilise past experience in evaluating the precision attainable by an experiment of given magnitude, and in planning to work on a scale sufficient to give a practically useful result.

A serious consequence of the neglect to make systematically estimates of the efficiency of different methods of experimentation is the danger that satisfactory methods, or methods which with further improvement are capable of becoming satisfactory, may be overlooked, or discarded, in favour of others enjoying a temporary popularity. Fashions in scientific research are subject to rapid changes. Any brilliant achievement, on which attention is temporarily focused, may give a prestige to the method employed, or to some part of it, even in applications to which it has no special appropriateness. The teaching given in universities to future research workers is often particularly unbalanced in this respect, possibly because the university teacher cannot give his whole time to the study of the practical aspects of research problems, possibly because he unwittingly emphasises the importance of the particular procedures with which he is best acquainted.

61. Multiplicity of Tests of the same Hypothesis

The concept of quantity of information is applicable to types of experimentation and of observational programmes other than those for which the theory of errors supplies the appropriate null hypotheses. Before considering these, as will be done in the following chapter, it is advisable to consider a somewhat more elaborate logical situation than that introduced in Chapter II. It was there pointed out that, in order to be used as a null hypothesis, a hypothesis must specify the frequencies with which the different results of our experiment shall occur, and that the interpretation of the experiment consisted in dividing these results into two classes, one of which is to be judged as opposed to, and the other as conformable with the null hypothesis. If these classes of results are chosen, such that the first will occur when the null hypothesis is true with a known degree of rarity in, for example, 5 per cent. or 1 per cent. of trials, then we have a test by which to judge, at a known level of significance, whether or not the data contradict the hypothesis to be tested.

We may now observe that the same data may contradict the hypothesis in any one of a number of different ways. For example, in the psycho-physical experiment (Chapter II.) it is not only possible for the subject to designate the cups correctly more often than would be expected by chance, but it is also possible that she may do so less often. Instead of using a test of significance which separates from the remainder a group of possible occurrences, known to have a certain small probability when the null hypothesis is true, and characterised by showing an excess of correct classifications, we might have chosen a test separating an equally infrequent group of occurrences of the opposite kind. The reason for not using this latter test is obvious, since the object of the experiment was to demonstrate, if it existed, the sensory discrimination of a subject claiming to be able to distinguish correctly two classes of objects. For this purpose the new test proposed would be entirely inappropriate, and no experimenter would be tempted to employ it. Mathematically, however, it is as valid as any other, in that with proper randomisation it is demonstrable that it would give a significant result with known probability, if the null hypothesis were true.

Again, in Darwin's experiment on growth rate discussed in Chapter III., it has been shown that the test of significance using " Student's " t is appropriate to the question with a view to which the experiment was carried out. Many other tests, however, less appropriate in this regard, or quite inappropriate, might have been applied to the data. Such tests may be made mathematically valid by ensuring that they each separate, for purposes of interpretation, a group of possible results of the experiment having a known and small probability, when the null hypothesis is true. For this purpose any quantity might have been calculated from the data, provided that its sampling distribution is completely determined by the null hypothesis, and any portion of the range of distribution of this quantity could be chosen as significant, provided that the frequency with which it falls in this portion of its range is ·05 or ·01, or whatever may be the level of significance chosen for the test.

Some such tests would be of no interest in any circumstances with which experimenters are familiar. Others, though not appropriate to the object Darwin had in view, might be appropriate to an experimenter studying a different subject. Thus, if the aim of the experiment had been, not to ascertain whether the average height of the cross-fertilised plants was, or was not, greater than that of the self-fertilised, but whether the difference in height between the cross- and self-fertilised plants of any pair was distributed normally, or in an unsymmetrical distribution, a valid test appropriate to this point could be devised. In addition to calculating, as in Chapter III., the sum of the squares of the deviations of these differences from their mean, we might calculate the sum of the cubes of these differences, having regard to their signs, and the ratio of the latter sum to the former raised to the power of 3/2 may be shown, on the null hypothesis, to have a determinate distribution for a given number of pairs of plants. The exact form of this distribution is at present unknown, since the distributional problem here considered is not one of those that have been solved. Nothing, however, but lack of mathematical knowledge prevents us from stating exactly outside what limits the ratio must lie to have a given level of significance. This test would pick out as statistically significant quite different sets of experimental results from those selected by the t test. It is in no sense a substitute for that test, or suited to perform the same functions. It is designed to answer a different question, although in both cases the question is answered by selecting a group of possible experimental results deemed to contradict the same null hypothesis. They may properly be thought of as testing different features of this hypothesis. The hypothesis tested in both cases states that the distribution of differences in height is centred at zero and is normal in form. The one test is appropriate when we are interested especially in the possibility that it is not centred at zero. In this case the question of normality is, as has been shown, of quite trivial importance. The other is appropriate when we are interested in the possibility that the distribution is skew, or unsymmetrical about its mean, and in this case the value of the mean is entirely irrelevant.

The notion that different tests of significance are appropriate to test different features of the same null hypothesis presents no difficulty to workers engaged in practical experimentation, but has been the occasion of much theoretical discussion among mathematical statisticians. The reason for this diversity of view-point is perhaps that the experimenter is thinking in terms of observational values, and is aware of what observational discrepancy it is which interests him, and which he thinks may be statistically significant, before he enquires what test of significance, if any, is available appropriate to his needs. He is, therefore, not usually concerned with the question : To what observational feature should a test of significance be applied ? This question, when the answer to it is not already known, can be fruitfully discussed only when the experimenter has in view, not a single null hypothesis, but a class of such hypotheses, in the significance of deviations from each of which he is equally interested. We shall, later, discuss in more detail the logical situation created when this is the case. It should not, however, be thought that such an elaborate theoretical background is a normal condition of experimentation, or that it is needed for the competent and effective use of tests of significance.

62. Extension of the t Test

In hypotheses, based on the theory of errors, there is, however, one extension which is normally held in view, and which, for the great simplicity of its consequences, is well fitted to introduce the more complex situations in which methods of statistical estimation require to be discussed. In Chapter III. we illustrated

"Student's" t test of significance with Darwin's data on the growth of young maize plants. The hypothesis to be tested was that the difference in height, between the cross-fertilised and the self-fertilised plant of the same pair, was distributed in some normal distribution about zero as its mean. We might, however, have considered a similar hypothesis, giving to the mean difference any other number, positive, negative, or fractional, of inches. If, instead of testing whether or not the mean could have been zero, we had chosen to test whether or not it had any unspecified value, μ, measured in eighths of an inch, then the deviation of our observed mean, 20·93, from the hypothetical value, μ, is

$$20\cdot93 - \mu$$

and this quantity, on the hypothesis to be tested, will be distributed normally about zero with a standard deviation, of which we have an estimate based on 14 degrees of freedom, the value of which is 9·746.

Consequently, if

$$t = \frac{20\cdot93 - \mu}{9\cdot746}$$

then t will be distributed in the distribution given by "Student" for 14 degrees of freedom, a distribution which is known with exactitude independently of the observations. We have the important logical situation, in which a quantity, t, having a sampling distribution known with precision, is expressible in terms of an unknown and hypothetical quantity, μ, together with other quantities known exactly by observation. We say known exactly, because the mathematical relations stated are true of the actual values derived from the observations, and not of the hypothetical values of which they might be regarded as estimates. Such actual values derived from the observation are distinguished by the term *statistics*, from the parameters, or hypothetical quantities introduced to specify the population sampled.

An important application, due to Maskell, is to choose the values of t appropriate to any chosen level of significance, and insert them in the equation. Thus t has a 5 per cent. chance of lying outside the limits $\pm 2 \cdot 145$. Multiplying this value by the estimated standard deviation, 9·746, we have 20·90 and may write

$$\mu = 20 \cdot 93 \pm 20 \cdot 90$$
$$= 0 \cdot 03, \text{ or } 41 \cdot 83$$

as the corresponding limits for the value of μ.

One familiar way of viewing this result is, that the experiment has provided an estimate, 20·93 eighths of an inch, of the average difference, μ, between the heights of two sorts of plants; that this estimate has an estimated standard error 9·746, and that "Student's" distribution shows that, for 14 degrees of freedom, the probability is only 5 per cent. of t lying outside the limits $\pm 2 \cdot 145$. An alternative view of the matter is to consider that variation of the unknown parameter, μ, generates a continuum of hypotheses each of which might be regarded as a null hypothesis, which the experiment is capable of testing. In this case the data of the experiment, and the test of significance based upon them, have divided this continuum into two portions. One, a region in which μ lies between the limits 0·03 and 41·83, is accepted by the test of significance, in the sense that values of μ within this region are not contradicted by the data, at the level of significance chosen. The remainder of the continuum, including all values of μ outside these limits, is rejected by the test of significance. This region has been chosen so that the probability of μ actually lying in the outer zone is only 5 per cent.; any other probability could equally have been chosen.

It can now be seen that the t test is not only valid for the original null hypothesis that the mean difference is zero, but is particularly appropriate to an experimenter who has in view the whole set of hypotheses obtained by giving μ different values. The reason is that the two quantities, the sum and the sum of squares, calculated from the data together contain all the information supplied by the data concerning the mean and variance of the hypothetical normal curve. Statistics possessing this remarkable property are said to be *sufficient*, because no others can, in these cases, add anything to our information. The peculiarities presented by t, which give it its unique value for this type of problem, are :—

   (i) Its distribution is known with exactitude, without any supplementary assumptions or approximations.
   (ii) It is expressible in terms of the single unknown parameter, μ, together with known statistics only.
   (iii) The statistics involved in this expression are sufficient.

In fact, it leads to an exact specification of μ as a Random Variable; probability statements about the unknown are correct, in the light of the observation, at all levels.

Without sufficient, or, more generally, exhaustive estimation, the mere fact that the true value will lie between calculated limits with a known frequency, realisable by repeated sampling, is not equivalent to a probability statement about the unknown, for information included in the data may have been lost in calculating these limits. True probability statements derived from the observations require all the information available, and this implies in particular either (a) that no probability statements can be made prior to these observations or

194 GENERALISATION OF NULL HYPOTHESES

(b) that the estimation of the limits shall have been exhaustive and have included all that the data supply. When there really is exact knowledge *a priori*, Bayes' method is available.

62·1. Fiducial Limits of a Ratio

The flexibility and directness of the fiducial argument is well illustrated by its application to find the fiducial limits of a ratio between quantities having normally distributed estimates. Galton, whom we quote on page 31, was interested in the ratio of the measurements of self-fertilised to cross-fertilised plants. He gives the ratio for maize as 84 per cent., though the totals of the fifteen pairs of plants given in his table, 2109 and 2423, in eighths of an inch, have a ratio 87 per cent. It is often convenient to the experimenter to state such ratios, and it is useful also to be able to state limits within which the true ratio probably lies.

In some cases, where the measurements are necessarily positive, it is satisfactory, though somewhat laborious, to make the test of the preceding section using the logarithms of all measurements in place of the measurements themselves. The difference between the logarithms of measurements of the same pair would be treated as normally distributed about some unknown mean, representing the true mean ratio. More generally, however, negative ratios are possible, and it is of more interest to consider the ratio of the true means rather than the mean of the logarithms. Instead of finding the fiducial limit of μ, by applying the t test to a set of quantities

$$x - y - \mu,$$

and finding the values of μ for which t has a given level of significance, we may instead apply the t test to the set of quantities

$$y - ax,$$

so that the fiducial limits of a, at, for example, the 5 per cent. level, are such as to make $t = 2\cdot 145$, for 14 degrees of freedom.

The total of the fifteen values of $y-ax$ will then be $2109-2423\,a$, and the term for the discrepancy between the means in the analysis of variance is

$$(2109-2423a)^2/15.$$

The sum of the squares of the deviations from their mean of the individual values of $y-ax$ is

$$3771\cdot 6 + 2225\cdot 8(2a) + 11721\cdot 73(a^2);$$

a quadratic expression in which the coefficients are the sums of squares and products (with reversed sign) of the deviations from their means of the measurements of the self-fertilised and the cross-fertilised plants. If the first expression bears to the second the ratio $t^2/14$, then a must satisfy the quadratic equation

$$387543a^2 - 341405(2a) + 295286 = 0,$$

the roots of which,

$$a = \cdot 99980 \text{ and } a = \cdot 76209,$$

show that the data contradict, at the 5 per cent. level of significance, any statement of the form " The true average for self-fertilisation is the fraction a of the true average for cross-fertilisation," whenever a lies outside the limits 76·209 per cent. and 99·980 per cent. The probability that a lies between these limits is 95 per cent.

63. The χ^2 Test

What is meant by choosing a test of significance appropriate to a special purpose, may now be illustrated by considering what should be done if the experimenter were interested, not in whether the mean of the distribution could exceed a given value, or could lie in a given range, but in the value of the variance of the same distribution. What is now needed is a test of significance provided by a quantity (i) having a precisely known distribution, and (ii) expressible in terms of the unknown variance, ϕ, of the distribution sampled, together with sufficient statistics only.

Now, if χ^2/n is the ratio of the variance, as estimated from the sample for n degrees of freedom, to the true variance, ϕ, it is known that χ^2 is distributed, independently of the mean and variance of the population sampled, in a distribution which is known when n is known. If we wish to set a probable upper limit to the value of ϕ, we note that for $n = 14$, χ^2 is less than 6·571 * in only 5 per cent. of trials. Putting this value for χ^2 in the equation,

$$\chi^2 = \frac{19945}{\phi},$$

we have

$$\phi = 3035.$$

In other words, variances exceeding 3035, or standard deviations exceeding 55·09, are rejected at the 5 per cent. level of significance.

Equally, had we wished to set a probable lower limit to the value of ϕ we should have noted that χ^2 exceeds the value 23·685 in only 5 per cent. of trials. Consequently, the rejection of the 5 per cent. of values of χ^2 which are highest will exclude values of less than 19945/23·685, or 844. We may thus reject values of the variance below 844, or of the standard deviation below 29·06, at the 5 per cent. level of significance. If, however, we rejected values for the standard deviation both below 29·06 and above 55·09, we should be rejecting both of two sets of contingencies each having

* For Table of χ^2 see *Statistical methods for research workers*, Table III.

a probability of 5 per cent., and so should be working, not at the 5 per cent., but at the 10 per cent. level of significance. If he wishes to work at the 5 per cent. level, the experimenter has the choice, according to the purpose of his researches,

- (i) of ascertaining an upper limit for the unknown variance without rejecting any lower values,
- (ii) of ascertaining a lower limit without rejecting any higher value, or
- (iii) of ascertaining a pair of limits beyond which values are rejected, representing two frequencies totalling 5 per cent. together.

Using the *probability statements* obtained by the fiducial argument as suggested below such various forms are seen as aspects of a single inference, a probability distribution for the unknown, covering all levels of significance.

The tests appropriate for discriminating among a group of hypothetical populations having different variances are thus quite distinct from those appropriate to discriminating among distributions having different means. Within the limits of the theory of errors, the mean and the variance are the only two quantities needed to specify the hypothetical population. It is the circumstance that statistics sufficient for the estimation of these two quantities are obtained merely from the sum and the sum of squares of the observations, that gives a peculiar simplicity to problems for which the theory of errors is appropriate. This simplicity appears in an alternative form of statement, which is legitimate in these cases, namely, statements of the probability that the unknown parameters, such as μ and ϕ, should lie within specified limits. Such statements are termed statements of *fiducial* probability, to distinguish them from the statements of *inverse* probability, by which mathematicians formerly attempted to express the results of inductive inference. Statements of inverse probability have a different logical basis from statements of fiducial probability, in spite of their similarity of form, for they require for their truth the postulation of knowledge beyond that obtained by direct observation.

Such knowledge is indeed sometimes available, notably in genetics, where observations on earlier generations may provide probability statements valid prior to any test-matings in which an animal is used. More usually it is absent and its introduction by axiom is in the nature of mathematical sleight-of-hand. When knowledge *a priori* in the form of mathematically exact probability statements is available, the fiducial argument is not used, but that of Bayes. Usually exact knowledge is absent, and, when the experiment can be so designed that estimation can be exhaustive, similar probability statements *a posteriori* may be inferred by the fiducial argument.

In the discussion above of the results of Darwin's experiment, instead of saying that at the 5 per cent. level of significance we should reject hypothetical variances exceeding 3035, it would be equivalent to say that the fiducial probability is 5 per cent. that the variance should exceed 3035. Equally, the fiducial probability is 5 per cent. that it should be less than 844; consequently, it is 10 per cent. that it should lie outside the range between these two numbers. With respect to the mean, it may in the same way be said that it has a fiducial probability of $2\frac{1}{2}$ per cent. of being less than 0·03, or of being greater than 41·83, and, in the same sense, a probability of 95 per cent. of lying within these fiducial limits. The word fiducial in these statements may now usually be omitted as understood, for there is no other method ordinarily available for making correct statements of probability about the real world.

64. Wider Tests based on the Analysis of Variance

In the more general type of problem, to which the z test is applicable, and in which we may have an analysis of the variance into a considerable number of subdivisions, we have a wide choice of tests of significance, each appropriate to answering a different question. Logically, these questions refer to the acceptance or rejection of different hypotheses or sets of hypotheses, and it will be useful to discuss them explicitly from this point of view. The practically useful variations are those which concern the hypothetical means of the different classes of observations which have been made.

As an example, let us consider the 6 × 6 Latin square, of which numerical observations were given in Chapter V. The arithmetical analysis obtained by the method there described is set out below.

TABLE 31

	Degrees of Freedom.	Sum of Squares.	Mean Square.	½ Log$_e$.
Rows	5	54,199	...	...
Columns	5	24,467	...	...
Treatments	5	248,180	49,636	1·9524
Error	20	30,541	1,527	0·2117
Total	35	357,387	...	1·7407

It will be seen that the value obtained for z was 1·7407. The 1 per cent. level for 5 degrees of freedom against 20 is ·7058. Consequently, the data very significantly contradict the hypothesis that all treatments were giving the same yield. We might, *if it seemed appropriate*, go further and say that if ζ stands for the true value of which z is an estimate, then all hypotheses which make ζ less than 1·0349 are contradicted by the data at the 1 per cent. level of significance. The hypothesis that the treatments do not affect the yield makes $\zeta = 0$. The wider hypothesis, that the yields produced by the different treatments are a random sample from a normal distribution, will provide an indeterminate positive value for ζ. If ζ were 1·0349 the mean square ascribed to treatments would be 7·923 times that ascribed to error. Since the mean square ascribed to treatments includes also the variability due to sampling error, the portion due to the effects of treatments themselves cannot be less than 6·923 times as great as the variance due to error in our estimates of the mean yields from the several treatments.

The mean square due to error has been found to be 1527, and this is the variance ascribable to error of a single plot. Dividing by 6, we find that the variance of the mean of six plots is 254·5. Multiplying this by 6·923 we have 1761·9 as the least admissible value for the variance due to treatments. The standard deviation corresponding to this variance is 41·97, or just over 9 per cent. of the mean yield, 462·75, observed in the experimental plots. It may be noticed that, apart from the inappropriateness, in the present instance, of the hypothesis that the *treatment* effects constitute a sample from a normal distribution, the calculation above departs from strict rigour in accepting the estimate of error based on 20 degrees of freedom, without making special allowance for the fact that this estimate is itself liable to sampling errors.

The exact solution of the distribution of the true treatment variance, supposing the true treatment means can properly be regarded as a random sample of a normal distribution, was first given in 1935, but has only recently been computed by M. J. R. Healy and tabulated in *Statistical Tables* (Table VI).

We have treated the experiment above as though nothing were known of the treatments applied, or as though these were regarded merely as causes disturbing the yields with an unknown variance. Actually, it is known that the treatments D, E, F differ from A, B, C in including an additional nitrogenous dressing, while A, B, C and, in like manner, D, E, F differ among themselves in receiving respectively 0, 1, 2 units of a phosphatic dressing. The 5 degrees of freedom ascribed to treatments are, therefore, not plausibly to be considered as homogeneous among themselves, but may properly be subdivided, as we have seen in previous chapters, into unitary elements of very different agricultural importance.

The total yields of the six treatments are set out below in relation to the manurial treatment received. The eighteen plots receiving the nitrogenous dressing

TABLE 32

	No Nitrogen.	Nitrogen.	Total.	Difference.
No phosphate	2070	2431	4,501	361
Single phosphate	2559	3121	5,680	562
Double phosphate	2867	3611	6,478	744
Total	7496	9163	16,659	1667

exceed the remaining eighteen plots in yield by 1667, so that this degree of freedom, N, contributes $1667^2/36$ or 77,191, to the total of 248,180 ascribed to treatments. The second degree of freedom of primary importance, P_1, is found by subtracting the yield of the twelve plots without phosphate from that of the twelve plots with double phosphate. The difference is 1977, so that the contribution of this degree of freedom is $1977^2/24$, or 162,855. This is the primary effect of phosphate. The second degree of freedom of this ingredient, P_2, may be found by observing that the increment in the yield of twelve plots, due to a single application, is 1179, while the additional increment due to the second application is 798. The difference is 381, and represents the excess of twice the yields for a single application over those for 0, or 2 units. The contribution of this degree of freedom is therefore $381^2/72$, or 2016. Similarly, using the differences between the yields with and without nitrogen, instead of the sums, we find for NP_1, $383^2/24 = 6112$, and for NP_2, $19^2/72 = 5$.

We are now at liberty to discuss the significance of each degree of freedom severally, and, because the experiment is a well-designed one, we shall find that to each corresponds a system of appropriate hypotheses relevant to the aims of the experiment. Thus, the 1 degree of freedom due to nitrogen has a mean square 50·56 times as great as that due to error. The value of "Student's" t is therefore about 7·110. The 5 per cent. value of t for 20 degrees of freedom is 2·086, so that at this level of significance we may exclude all hypotheses ascribing to the nitrogenous dressing less than 5·024/7·110, or more than 9·196/7·110 of the apparent benefit observed. This conclusion refers directly to the manurial comparison on which it is based, and is entirely independent of the other conclusions to be drawn from the experiment as to the effects of phosphate or of its interactions with nitrogen. Naturally, also, if there were no such interactions, the conclusion would be applicable at levels of phosphatic manuring other than those used in the experiment. The data here indicate that the return from nitrogen would be definitely higher with higher phosphatic dressings than those used. The inference as to the return with the actual phosphatic dressings used is, however, direct and independent of the interaction.

Each of the other elements into which the effects of treatments have been analysed may be treated independently, as we have treated N. A glance at the items of the expanded analysis of variance will show on which of these decisive evidence has been obtained.

TABLE 33

Component of Treatment.	Degrees of Freedom.	Mean Square.
N	1	77,191
P_1	1	162,855
P_2	1	2,016
NP_1	1	6,112
NP_2	1	5
Total	5	248,179
Error	20	1,527

Thus the primary effect of phosphate, like that of nitrogen, is demonstrated with unquestioned significance, and the magnitude of the return evaluated with fair accuracy. On the other hand, the contribution of the component NP_2 is much less than might have appeared as the result of random errors. The results are entirely compatible with the theoretical possibility that the response to nitrogen at different levels of phosphatic application changes in strict proportion to the amount of phosphate applied. The two remaining items, for P_2 and NP_1, are intermediate in magnitude, indicating that the evidence of the experiment on two corresponding modes of varying the null hypothesis is of an intermediate character. For P_2 the contribution, 2016, is statistically insignificant. The experiment does not prove that the additional response to the second dose of phosphatic manure is certainly less than that to the first. It is in fact less, in accordance with common agricultural experience, but the experiment could not suffice by itself to demonstrate the reality of this decrease. If we consider a series of hypotheses with different values for the diminishing return, and determine which of these values are compatible, at any given level of significance, with the observed yields, some of the values which would appear to be acceptable would be negative, *i.e.* would represent increasing returns, though in the greater part of the acceptable range positive values would prevail. Even if the test of significance were chosen, so as to determine not both limits, but the lower level only, this lower limit would be found to be negative. For t about $1·15$, the fiducial probability of " increasing return " is about 13 per cent.

In the case of NP_1, which measures the extent to which the response to nitrogen is increased by an increased phosphatic dressing, the state of the evidence is somewhat different. The value of " Student's " t is $2·0004$, while the value which is exceeded either positively or negatively in 5 per cent. of trials is $2·068$. If, therefore, the experimenter had no more reason to expect an increasing than a decreasing response the observed value would have fallen just short of 5 per cent. significance. Since, however, normal agricultural experience would lead us to anticipate an increase, while a decrease would be somewhat anomalous, this test tells us not only that the observed magnitude of the effect is nearly significant, but also that it is in the right direction. These two independent pieces of evidence are combined by choosing a test, which determines a lower fiducial limit only, *i.e.* by taking for comparison the value of t, $1·725$, tabulated as corresponding to the probability 10 per cent. This test shows that, at the 5 per cent. level of significance, any hypothesis which gives to the increase, in response to nitrogen, a negative value is contradicted by the experimental results, or, in other words, that a positive effect is demonstrated, at this level of significance, by the experiment. The probability, on these data only, of the interaction being really negative is only about 2·98 per cent. or, less than 3 per cent. As in other cases, where an effect is little more than barely significant, the precision with which its value is estimated is, of course, extremely low.

It would have been legitimate to choose other comparisons among the treatments employed, and to make with them other tests of significance. We might, for example, have compared the plots receiving double directly with those receiving single phosphate, and have discussed the significance of this difference, in isolation from the other experimental results. The only inconvenience of such a course is that if, as is usually the case, the result is to be used in the examination of scientific theory, in the framing of practical advice, or in the designing of future experiments, in conjunction with facts of the same kind as the remainder of the experiment provides, it is clearly preferable that the whole of these should be recognised by means of a series of independent tests, each having some agricultural relevance. Although a series of tests can always be chosen, independent of the one with which we may start, the supplementary information provided by them will often be of too complicated a kind for its bearing on our effective conclusions to be readily appreciated. Consequently, it will usually be preferable, as in the example chosen, to design the experiment so as to lead uniquely to a single series of tests chosen in advance.

Where a number of independent tests of significance have been made, on data from the same experiment, each test allowing of the rejection of the true hypotheses in 5 per cent. of trials, it follows that a hypothesis specifying all the differences in yield between the treatments tested will, although true, be rejected with a higher frequency. If, therefore, it were desired to examine the possible variations of any hypothesis which specified all these differences simultaneously while maintaining a given level of significance, a different procedure should be adopted. Actually, in biology or in agriculture, it is seldom that the hypothetical background is so fully elaborated that this is necessary. It is therefore usually preferable to consider the experiment, as we have done above, as throwing light upon a number of logically independent questions. There is, however, no difficulty, when required, in making a comprehensive test on all questions simultaneously, using the z test first employed, and extending this test so as to specify the aggregate of compound hypotheses which are contradicted by the experiment at any assigned level of significance.

In analysing the 5 degrees of freedom ascribable to treatments in the 6×6 Latin square, certain differences were obtained from the experimental yields, such as the 1667 units of yield by which the plots receiving nitrogen exceeded the remainder. Any hypothesis respecting the differences in yield of the six treatments used may be specified by the hypothetical values which it gives corresponding to these observed differences. Thus if a_1 were the hypothetical value corresponding to 1667, a_2 corresponding to 1977, and so on, the sum of squares for the 5 degrees of freedom representing the deviations of the observed responses to treatments from those predicted by hypothesis would be

$$\frac{(1667-a_1)^2}{36} + \frac{(1977-a_2)^2}{24} + \frac{(381-a_3)^2}{72} + \frac{(383-a_4)^2}{24} + \frac{(19-a_5)^2}{72}.$$

We may now find how large this expression must

be in order that z should be equal to its 1 per cent. value. This value as given by the table is 0·7058. Adding to this ½ log, for error, ·2117, we have ·9175 corresponding to a mean square 6265, or to a sum of squares 31,325. The 1 per cent. test of significance for a hypothesis specifying all the values $a_1, a_2, \ldots, a_5$ will therefore reject any hypothesis for which the quadratic expression set out above exceeds 31,325, and will accept all hypotheses for which it has a lower value.

This exact test for a hypothesis specifying the true differences between every two of the treatments used, is equally applicable to other cases, such as variety trials. It appears to be both simpler and more satisfactory than to make so-called " multiple decision " tests. It supplies the simultaneous probability distribution of the whole set of n unknown differences.

It does not imply, or contradict, the view that the several treatment means constitute a sample from some normal population.

65. Comparisons with Interactions

The last class of variation to be considered in the tests of significance derivable from the analysis of variance consists of cases in which we compare primary effects with interactions, or interactions with interactions of a higher order. If, for example, a test were carried out of five varieties of an agricultural plant, using a Latin square laid down at each of ten representative farms, in a region to which the five varieties tested have all some claim to be thought appropriate, the experiment at each farm will provide an analysis of the form :

TABLE 34

	Degrees of Freedom.
Rows	4
Columns	4
Varieties	4
Error	12
Total	24

If we have corresponding data for each of ten places the whole series will yield together 40 degrees of freedom for rows and 40 degrees for columns, all of which represent components of heterogeneity which have been eliminated. There will also be 120 degrees of freedom in all for error. But the remaining 40, composed of the components ascribed to variety at each of the ten places, is divisible into 4 degrees of freedom for variety V, and 36 for interaction between variety and place VP. There would, of course, also be 9 degrees of freedom, representing the contrast between places, but with these we are not concerned. The complete analysis of such a record of 250 yields would therefore be as follows :—

TABLE 35

	Degrees of Freedom.
Rows	40
Columns	40
Places	9
Varieties	4
V × P	36
Error	120
Total	249

It would be proper, of course, to examine the record from each farm for significant differences between the varieties, for even if these were not concordant they might indicate a greater aptitude of some varieties compared with others to the soil conditions of a particular site. Even in the absence of significant differences on individual farms, the results of the different experiments might be sufficiently concordant to give a significant comparison in the analysis of the entire experiment between varieties and error. This might not, however, be the most appropriate comparison to make, for since the varieties might react differently to different types of soil, it is not improbable that the mean square corresponding to the 36 degrees of freedom VP is greater than the mean square due to error. If the precision of the individual experiments were high, the difference between the aggregate yields of two varieties might be significant compared with error, although one was the better at only six places, while the other was better at the remaining four. In fact, if our concern is to ascertain not merely the best variety on the aggregate of the ten fields actually used, but to ascertain which is the best over the whole area deemed suitable for this type of crop, within the region from which the sites of the experiment have been selected, the comparison between varieties V and interaction of varieties and places VP will be the more appropriate. For, if the ten sites have been chosen at random from this area, a significant difference in this comparison would indicate, at the level of significance used, varietal differences applicable to the whole area. The precision of this comparison may not be greatly increased by higher precision in the individual experiments, especially if the mean square corresponding to VP is considerably greater than that ascribable to experimental error. To increase its precision we may rather require an increase in the number of sites used, or in other words, if the area sampled is considerably heterogeneous with respect to varietal response, it may be necessary to sample it more thoroughly. The hypothetical population with which we are principally concerned will then be the population of possible sites available for growing the crop under consideration, rather than the population of possible yields of plots within a given site. The test employed is, in fact, equivalent to considering, from each farm, only the aggregate yield of each variety, and the estimation of error within each individual Latin square is of value, apart from the local information supplied, only in providing assurance that the experimentation has been carried out with an exactitude sufficient to guarantee the adequacy of the comparisons between different places.

Cases in which it is one of the higher order interactions, rather than error proper, that should appropriately be used as a basis for tests of significance, are relatively numerous. The data of Table 7A (p. 68) are of this kind. Agricultural experiments, whether with manures, implements of cultivation or varieties of crop plants, are much affected by the weather. If a treatment effect is significant, compared with error in any one year, the experiment will have indicated what treatments have in that year proved most advantageous. But, if independent experiments over a series of years show a significant difference between treatments on the one hand and the interaction between treatments and years on the other, the experiment has shown what treatments are the most successful in an aggregate of seasons, of which those experienced may be taken as a random sample. There seems, in fact, in no part of the world to be any such similarity between successive seasons as would make the experience of a sequence of trials unreliable for future application in the absence of genuine secular changes of the climate.

The same principle is of wide application in economic and sociological enquiries, where, in comparisons of rates of death, morbidity, births, prices and so on, the effective unit is far more often a district, or a town, than an individual. The supposition that rates, based on the registration of individuals, possess the precision which would be appropriate if all the individuals concerned could be regarded as independent in their sociological reactions, is clearly inappropriate when we are interested in the effects on these reactions of economic or legislative causes, or other agencies derived from social organisation, liable to affect large numbers of individuals in a similar manner. The effective samples available for administrative decisions, even though based ultimately on millions of individual persons, are often much smaller than those available in biological experimentation, and for this reason require, even more than the latter, the accurate methods of analysis by which small samples may be interpreted. Statisticians should be able to supply probability statements, such as are used in tests of significance, relevant to such decisions. They are " selling the moon " if they claim to have a mathematical theory for the decisions themselves.

It should not, however, be supposed that whenever an experimental effect shows significant interaction with some other category into which the data are divisible, it is always this interaction which provides the relevant estimate of error. Reflection should be given to individual cases. Distinction must be made between different kinds of categories.

The categories in the examples chosen above, " farms " and " seasons," may be described as **indefinite**. By this is meant that although the experiment may have shown that the conditions on the different farms are not equivalent, yet we have no means of reproducing such different conditions at will. If the results of the experiment are used to predict the responses to be expected at other farms, chosen at random from the same population, these will probably differ to much the same extent from those of the original experiment, and the interaction with farms supplies the appropriate estimate of error.

In an industrial process interactions may be found with categories such as are provided by varying the temperature, or a concentration, at some stage of the test. Such categories are **definite**, in that the particular temperature or concentration used in a future operation can be controlled, or, at least, measured. The accuracy of prediction of performance expected at a given temperature depends on the agreement observed between tests run at the same temperature, and is not affected by the change in response observed as the temperature is changed.

The interaction of a definite with an indefinite category is necessarily indefinite. Thus if p procedures are being tested, at t temperatures, using for each combination specimens from m batches of material, the significance of the observed effects of procedure at a known temperature would be tested by $p-1$ degrees of freedom, against $(p-1)(m-1)$, or by appropriate components of these; while the significance of the interaction of procedure and temperature would be tested using $(p-1)(t-1)$ degrees of freedom, against $(p-1)(m-1)(t-1)$.

When two indefinite categories appear in the analysis, they must, in one sense, be regarded as one. For example, the results of an experiment carried out on 10 farms for five years, are intended to be applied to other farms in other years. The precision of a predicted response is affected by its variation over the totality of all farms in all years. Had the data consisted of fifty experiments, each conducted on a different farm in a different year, the variance of this totality would have been directly estimated, and would supply the proper basis of a test of significance. Our actual information consists of three mean squares based on 4, 9 and 36 degrees of freedom respectively, which readily yield a test of significance, and fiducial limits, for the response of the same farms in future years, but is awkward material for doing the same for future tests on other farms.

When the significance neither of the differences between farms, nor of those between years, is in doubt, a satisfactory estimate of the variance of the general mean is found by adding the mean squares for farms and for years, subtracting that for interaction, and dividing by 50. The error is not exactly of "Student" 's type, so that no number of degrees of freedom should be assigned to it. In the circumstances stipulated, the departure from normality need not be great.

REFERENCES AND OTHER READING

R. A. FISHER (1933). The concepts of inverse and fiducial probability referring to unknown parameters. Proceedings of the Royal Society, A, cxxxix., 343-348.

R. A. FISHER (1933). Two new properties of mathematical likelihood. Proceedings of the Royal Society, A, cxliv., 285-307.

R. A. FISHER (1933). The contributions of Rothamsted to the development of statistics. Rothamsted Experimental Station Report.

R. A. FISHER (1934). Probability, likelihood and quantity of information in the logic of uncertain inference. Proceedings of the Royal Society, A, cxlvi., 1-8.

R. A. FISHER (1935). The fiducial argument in statistical inference. Annals of Eugenics, vi, 391-8.

R. A. FISHER (1956, 1959). Statistical methods and scientific inference. Oliver and Boyd Ltd., Edinburgh.

M. J. R. HEALY (1963). Fiducial limits for a variance component. Journal of the Royal Statistical Society, Series B, XXV, 128-130.

E. J. MASKELL (1929). Experimental error: a survey of recent advances in statistical method. Tropical Agriculture, vi, 5-11.

XI
THE MEASUREMENT OF AMOUNT OF INFORMATION IN GENERAL

66. Estimation in General

THE situations we shall now examine are of a more general character than those considered in the classical theory of errors, which have been dealt with in previous chapters. It has been seen in the last chapter that we may be interested to interpret the data as arising, subject to errors of unknown magnitude, but distributed normally, from one or more unknown quantities, *parameters*, of which we are interested to form *estimates*, of known precision, and to make this precision as great as possible. In the most general situation of this kind, all the different kinds of individual events which it is possible to observe are regarded as occurring with frequencies functionally dependent in any way on one or more of such unknown parameters. This is the general situation considered in the Theory of Estimation. From the purely statistical standpoint, they present the problem of how best the observations can be combined, in order to afford the most precise estimates possible of the unknowns. The mathematical principles of this process of combination are now satisfactorily understood, and have been illustrated in detail in the ninth chapter of the author's book, *Statistical Methods for Research Workers*. From the point of view of the practical design of experiments, or of observational programmes, we shall here be concerned only indirectly with the technique of the calculation of efficient estimates, and can turn attention at once to the problem of assessing, in any particular case which arises, the quantity of information which the data supply, and which we may assume will be efficiently utilised.

The reason for this standpoint, so contrary to that traditional among statisticians, deserves some explanation. During the period in which highly inefficient methods of estimation were commonly employed, and, indeed, strongly advocated by the most influential authorities, it was natural that a great deal of ingenuity should be devoted, in each type of problem as it arose, to the invention of methods of estimation, with the idea always latent, though seldom clearly expressed, of making these as accurate as possible. The attainment of a result of high accuracy was, in fact, evidence not only of the intrinsic value of the data examined but also to some extent of the skill with which it had been treated. The extent to which this was so was the greater the more inefficient were the methods ordinarily recommended; but, clearly, in any subject in which the statistical methods ordinarily employed leave little to be desired, the precision of the result obtained will depend almost entirely on the value of the data on which it is based, and it is useless to commend the statistician, if this is great, or to reproach him if it is small. At the present time any novice in the theory of estimation should be able to set out the calculations necessary for making estimates, almost, if not quite, as good as they can possibly be. Any improvement which can be made by further refinements of computational technique are, in ordinary cases, and setting gross incompetence aside, exceedingly small compared to the improvements which may be effected in the observational data.

The amount of information to be expected in respect

of any unknown parameters, from a given number of observations of independent objects or events, the frequencies of which depend on that parameter, may be obtained by a simple application of the differential calculus. It may be worth while to consider a few easy examples in detail, in order to obtain a clear grasp of the process generally involved.

67. Frequencies of Two Alternatives

Let us suppose that only two kinds of objects or events are to be distinguished, and that we are concerned to estimate the frequency, p, with which one of them occurs as a fraction of all occurrences; or, what comes to the same thing, the complementary frequency, q ($= 1-p$), with which the alternative event occurs. We might, for example, be estimating the proportion of males in the aggregate of live births, or the proportion of sterile samples drawn from a bulk in which an unknown number of organisms are distributed, or the proportion of experimental animals which die under well-defined experimental conditions. The experimental or observational record will then give us the numbers of the two kinds of observations made, a of one kind and b of another, out of a total number of n cases examined. We wish to know how much information the examination of n cases may be expected to provide, concerning the values of p and q, which are to be estimated from the data.

A general procedure, which may be easily applied to many cases, is to set down the frequencies to be expected in each of the distinguishable classes in terms of the unknown parameter. For each class we then find the differential coefficient, with respect to p, of this expectation. The squares of these, divided by the corresponding expectations, and added together, supply the amount of information to be anticipated from the observational record. That such a calculation will give a quantity of the kind we want, may be perceived at once by considering that the differential coefficients of the expectations, with respect to p, measure the rates at which these expectations will commence to be altered if p is gradually varied; and the greater these rates are, whether the expectations are increased or diminished as p is increased, or in other words, whether the differential coefficients are positive or negative, the more sensitively will the expectations respond to variations of p. Consequently, it might have been anticipated that the value of the observational record for our purpose would be simply related to the squares of these differential coefficients.

We may now set out the process of calculation for the simple case of the estimation of the frequency of one of two classes.

TABLE 36

Observed Frequency. (x)	Expected Frequency. (m)	Differential Coefficient. dm/dp	$\frac{1}{m}\left(\frac{dm}{dp}\right)^2$
a	pn	n	n/p
b	qn	$-n$	n/q
n	n	0	n/pq

The frequencies expected are found by multiplying the number of observations, n, by the theoretical frequency, p, which is the object of estimation, and by its complementary frequency, q. The differential coefficients of these expectations with respect to p are simply n and $-n$. The sum of these is zero, as must be the case whenever, as is usual, the number of observations made is independent of the parameter to be estimated. It is obviously, therefore, not the total of the differential coefficients which measures the value of the data, but effectively the extent to which these differ in the different distinguishable classes, as measured by their squares appropriately weighted, as shown in the last column.

The total amount of information is found to be

$$I = \frac{n}{pq},$$

and we may now note the well-known fact that, if our sample of observations were indefinitely increased, the estimate of p, obtained from the data, tends in the limit to be distributed normally about the true value with variance $\frac{pq}{n}$. The general method here given of measuring quantity of information thus agrees with the concept, which has been formed of this quantity in previous chapters, where we were concerned only with normally distributed errors.

68. Functional Relationships among Parameters

It is often true that the frequency of a particular event among different events of a like kind is itself an object of enquiry, as is the case, for example, with the sex-ratio of births. More often the frequency is itself only of value because it is believed to be functionally related to some other quantity of more direct importance. The frequency, p, of sterile samples from a vessel containing an unknown density of organisms is related to the average number, m, of organisms in the sampling unit, by the relation,

$$p = e^{-m},$$
$$m = -\log p,$$

where the logarithm is taken from a table prepared on the natural or Napierian system.

If now the object of making a count discriminating only the two types of sample, viz., sterile samples which contain no organism, and fertile samples which contain at least one, is to make an estimate of the density in the material sampled, or in the material from which the dilution sampled was prepared, we shall be interested, not directly in the amount of information about p, but rather in the amount of information about m, which the sample provides. Since m and p are functionally related, this can be obtained by using the relationship directly, from the amount of information about p.

If in the table set out above, showing the calculation of the amount of information respecting p supplied by n observations, we had differentiated with respect to m instead of in respect to p, the process would have led to the amount of information with respect to m. The component terms of this calculation would each have differed from those we obtained only in containing, as an additional factor, the square of the differential coefficient of p with respect to m. In general, if I_m stands for the amount of information with respect to m, and I_p for the amount of information with respect to p, we have the transformation formula

$$I_m = \left(\frac{dp}{dm}\right)^2 I_p.$$

In the present case
$$p = e^{-m}$$
whence
$$\frac{dp}{dm} = -e^{-m} = -p.$$

And since
$$I_p = \frac{n}{pq}$$
it follows that
$$I_m = \frac{np}{q} = \frac{n}{e^m - 1}.$$

As in the case of p, if the number of samples examined is increased, our estimates of m derived from a given number of samples tend to be normally distributed about the true value, and the variance of this limiting distribution is given by the reciprocal of the amount of information,

$$V(m) = \frac{q}{pn} = \frac{e^m - 1}{n}.$$

The errors of estimation are least when m is near to zero, and increase rapidly if m is made large. This, however, does not mean that the determination will be most accurately carried out with very high dilutions, or with very small sampling units, by which means m may be made as small as we please; for it must be remembered that if we reduce the sampling errors of m by making m smaller, we will not necessarily diminish the relative magnitude of these errors when compared with m. To minimise the relative magnitude of the sampling errors, we need to consider the variance of m divided by m^2 or, in fact the variance of $\log_e m$; thus

$$V(\log m) = \frac{1}{m^2} V(m) = \frac{1}{n} \frac{e^m - 1}{m^2}.$$

This quantity tends to infinite values when m is made either very small or very large. This expression for the limiting value of the relative variance in large samples, corresponds with the amount of information supplied by an experiment, however small, relative to $\log m$; for by the general transformation we find

$$I(\log m) = m^2 I_m;$$

whence it appears without any large-sample approximation that the amount of information supplied by the experiment relative to $\log m$ is given by the expression

$$I = n \frac{m^2}{e^m - 1}.$$

This quantity vanishes as m tends to zero or infinity, but is finite at all intermediate values; and the relative precision of our estimate of the number of organisms will be greatest if the dilution or the sampling unit were adjusted, so as to maximise this quantity.

Fig. 3 shows the quantity of information, for all values of m, for which this quantity is not very small.

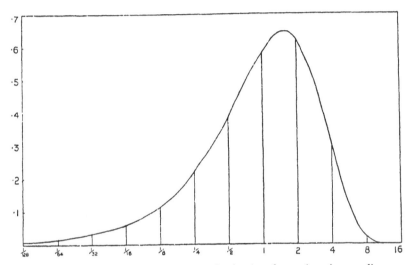

FIG. 3.—Quantity of information as to the density of organisms in a medium according to the average number m of organisms per sample.

The horizontal scale is logarithmic, so that values of m indicated at equal intervals are in geometric progression. It is evident from the figure that the most useful values are between about 1·1 and 2·2. The absolute maximum of information is given when m is about 1·6, or, more precisely, a number $\hat{m}$ with rather remarkable properties, such that

$$1 - e^{-\hat{m}} = \tfrac{1}{2}\hat{m} \;;$$

since

$$\hat{p} = e^{-\hat{m}}$$

it follows that the ratio of p to q is the same as the ratio of $(2-\hat{m})$ to $\hat{m}$. Numerically, it appears that

$$\hat{m} = 1 \cdot 593{,}624{,}26$$
$$1-e^{-\hat{m}} = \cdot 796{,}812{,}13 = \tfrac{1}{2}\hat{m} = \hat{q}$$
$$2-\hat{m} = \cdot 406{,}375{,}74$$
$$e^{-\hat{m}} = \cdot 203{,}187{,}87 = 1-\tfrac{1}{2}\hat{m}=\hat{p}.$$

The ideal proportion of sterile samples for estimating by this method the density of the organisms is, therefore, just over 20 per cent. Any proportion between 10 per cent. and 33 per cent. sterile will, however, supply nearly as much information, and the aim in adjusting the sampling process should be to obtain a percentage of sterile samples between these limits. The maximal amount of information per sample is

$$\hat{i} = \cdot 647{,}610{,}24 = \hat{m}(2-\hat{m}) = 4\hat{p}\hat{q}.$$

To find the minimal number of samples needed to estimate m with any given precision, we may now equate $n\hat{i}$ to the invariance of log m required. Thus if we required to reduce the standard error of m to about 10 per cent. of its value, we might put the standard error of log m equal to 0·1; the variance of log m would then be 0·01 and its invariance would be 100. We should then have

$$n\hat{i} = 100,$$

or

$$n = 154 \cdot 4.$$

Even in the most favourable circumstances, therefore, it would need 155 samples to reduce the standard error below 10 per cent. of the estimated density. Since, owing to our ignorance of the true density, the dilution cannot be adjusted exactly so as to give the ideal proportion of sterile samples, it would usually be wise to divide the amount of information required by a smaller divisor than the maximal value ·6477, *e.g.* by ·6, which would raise our estimated requirement to 167 samples. The whole calculation shows that the method of estimating the density of organisms by discriminating only their presence, or absence, in samples is of low precision, compared with methods in which individuals or colonies may be counted.

In many types of research a series of dilutions is employed, giving densities falling off in geometric progression, with a constant factor most commonly of 2 or 10. The amounts of information supplied by each of these is represented in the diagram (Fig. 3) by the heights of a series of equally spaced ordinates. If the series is extended so as to cover all densities which supply an appreciable amount of information, the sum of the ordinates for two-fold dilution is nearly constant in value and has an average value

$$\frac{\pi^2}{6 \log_e 2} = 2 \cdot 373,138.$$

This is, therefore, the amount of information supplied by a single sample at each dilution, and this may be used to calculate the precision to be expected, using any number of samples at each dilution, or to calculate the number of samples required to attain any stipulated level of precision. For four-fold dilutions the average amount of information supplied is, of course, a half, and for eight-fold dilutions one-third of the number found above. For ten-fold dilutions it is about three-tenths, but for the higher dilution ratios the sum of the ordinates shows a rapidly increasing variation, with the consequence that the amount of information actually obtained becomes less reliable the larger the dilution-ratio employed.

69. The Frequency Ratio in Biological Assay

Use is often made of a frequency ratio between two distinguishable classes, to supply a measure of an underlying variate, as when the toxic content of a drug is inferred from the mortality of experimental animals receiving a known dosage, or from the dosage required to cause a given mortality. The underlying theory is illustrated in Fig. 4. The curve represents a normal distribution with unit standard deviation, divided by

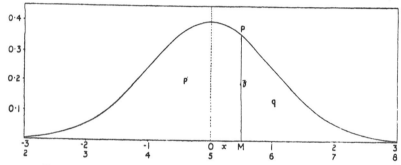

FIG. 4.—Normal distribution curve to illustrate the interpretation of percentage mortality as a probit value.

the ordinate P M into two portions. The area to the left of the ordinate represents the proportion, p, which die under the treatment, the area to the right the proportion, q, which survive. The height of the ordinate is represented by z, and its distance from the central axis of the curve by x, taken positive to the right of the axis. As x increases from $-\infty$ to ∞, the proportion dying increases from 0 to 1. Knowing any of the three quantities x, p or z, the other two can be obtained from available tables. An experimental determination of the fraction, p, will therefore supply a corresponding determination of the deviation, x, and this is found, in a large number of cases, to increase or decrease proportionally with the logarithm of the toxic content of the dose. If, therefore, the relationship between dosage and mortality has been established for a standard preparation, the toxicity of any material under test may be gauged by observing the mortality which supervenes on a known dosage of the material to be tested. Moreover, so long as a linear relation holds between the toxic content measured logarithmically and the deviation, x, the precision of the assay will be proportional to the precision with which x is estimated.

As is seen from the figure, if x is increased by a small quantity dx, the initial increase of p is zdx. Hence

$$\frac{dp}{dx} = z.$$

But we know that the amount of information with respect to x is given by the equation

$$I_x = I_p \left(\frac{dp}{dx}\right)^2$$
$$= \frac{nz^2}{pq},$$

where n is the number of animals employed. Although $\frac{1}{pq}$ is least when p equals q, or at 50 per cent. mortality, the quantity of information, $\frac{nz^2}{pq}$, is greatest at this point. Hence for a single test, the highest precision is obtained for a given number of animals by adjusting the dosage approximately to the 50 per cent. death point. The quantity

$$I = \frac{nz^2}{pq}$$

is used further as the weight to be assigned to the estimated value of x when a number of tests at different dosages are to be combined. The quantity $5+x$, known as the "probit value," is used as a practical measure of mortality, and Dr C. I. Bliss has given tables of the weighting factor, and other relationships needed for the more complex problems which arise in toxicological research. Fig. 5 shows the amount of information respecting the probit value supplied by each animal observed, for different percentage mortalities. It will

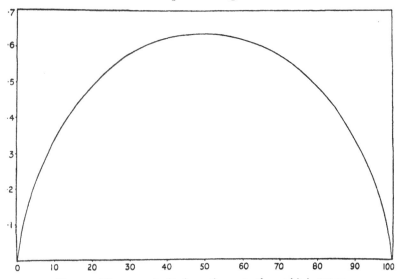

FIG. 5.—The amount of information as to the probit in terms of the percentage mortality

be seen that when the mortality is between one-third and two-thirds, the information gained falls little short of the highest possible.

Tables and illustrations of their use are given in *Statistical Tables*.

70. Linkage Values inferred from Frequency Ratios

When an organism receives from its two parents corresponding genes of different kinds, it generally hands on one kind to half its offspring and the other kind to the remainder. The numbers of the two kinds of offspring observed may, however, differ, either by chance or owing to the unequal viability of these two kinds. The parent is said to be heterozygous for the Mendelian factor in question. If the parent is heterozygous for two different factors, which are not linked in inheritance, he may make contributions of four different kinds to the germinal constitution of the offspring, and these will occur in equal numbers. If, however, the factors are linked, or carried in the germ-plasm by the same chromosome, the two gene-combinations received by the heterozygote from his parents will be handed on to the offspring more frequently than the remaining two combinations formed by interchanging the pairs of genes. The intensity of the linkage is measured, in an inverse sense, by the frequency among all the offspring of those receiving recombinations. When the recombination frequency is small, the linkage is close ; when it is large, approaching 50 per cent., the linkage is loose. The type of mating which best tests the intensity of linkage is one with an organism distinguishable from the heterozygote in respect of both factors, *i.e.* when there is dominance, with a double recessive.

If there is no difference in mortality among the four distinguishable types of offspring, up to the time at which they can be recorded, any such mating will determine the linkage value, with precision limited only by the number of offspring. Thus, if a fraction, p, of recombinations is estimated from a count of n offspring, the amount of information available as to the value of p is

$$I = \frac{n}{pq}.$$

If, however, the two types counted as recombinations have an average viability different from that of the two types of parental combinations, the linkage value so estimated will be distorted by the differential mortality. It is possible to overcome this difficulty by making up heterozygotes of the two kinds possible, so that the recombinations from one set of matings are genetically similar to the parental combinations from the other. Thus, if in one set the apparent recombination value has been raised by differential viability, it will have been lowered in the other set. If, therefore, we have a record of two such sets of matings as shown in the following table

	Recombination.	Parental Combination.	Total.
1st set	a_1	b_1	n_1
2nd set	a_2	b_2	n_2

we may argue that the ratio a_1/b_1 has been raised (or lowered) in the first set in the same proportion as the ratio a_2/b_2 has been lowered (or raised) in the second. Hence, if we take the geometric mean of these two ratios, and use, as our equation of estimation,

$$\frac{p}{q} = \sqrt{\frac{a_1 a_2}{b_1 b_2}},$$

we shall obtain an estimate unbiased by differential mortality, in so far as it is caused by the factors studied.

To determine the precision of such an estimate, we may consider first the precision with which the quantity

$$\log \frac{p}{q} = \log p - \log q$$

is derived from a simple frequency ratio $a : b$; since

$$\frac{d}{dp} \log \frac{p}{q} = \frac{1}{p} + \frac{1}{q} = \frac{1}{pq},$$

it follows that

$$I_{\log p/q} = p^2 q^2 I_p = p^2 q^2 \frac{n}{pq} = npq;$$

IDEAL PROPORTIONS

and since, as our estimate, we shall put

$$p = \frac{a}{n}, \; q = \frac{b}{n},$$

the amount of information may be written,

$$I = \frac{ab}{n};$$

or, in large samples, the sampling variance of $\log \frac{p}{q}$ is

$$\frac{a+b}{ab} = \frac{1}{a} + \frac{1}{b}.$$

Now, in estimating $\log \frac{p}{q}$ from the geometric mean of two observed ratios, we are taking half the sum of two estimates of $\log \frac{p}{q}$, and the sampling variance will therefore be one-quarter of the sum of the four reciprocals.

Hence,
$$V = \frac{1}{4}\left(\frac{1}{a_1} + \frac{1}{b_1} + \frac{1}{a_2} + \frac{1}{b_2}\right).$$

$$I = \frac{4}{\frac{1}{a_1} + \frac{1}{b_1} + \frac{1}{a_2} + \frac{1}{b_2}} = h,$$

where h is the harmonic mean of the four frequencies observed.

The information respecting the recombination fraction, p, estimated in this way, may be calculated, as before, from that respecting $\log \frac{p}{q}$, and is evidently

$$I_p = h/p^2 q^2.$$

We may now ask in what proportions the two types of mating should be used, in order to secure the greatest precision for a given number of organisms bred and examined.

If p_1, q_1 stand for the proportions observed from

matings of the first kind, and p_2, q_2 from matings of the second kind, the amount of information has been shown to be inversely proportional to

or to
$$\frac{1}{n_1 p_1} + \frac{1}{n_1 q_1} + \frac{1}{n_2 p_2} + \frac{1}{n_2 q_2},$$
$$\frac{1}{n_1 p_1 q_1} + \frac{1}{n_2 p_2 q_2}.$$

We wish to make this quantity as small as possible, consistently with a fixed total of organisms observed, $n_1 + n_2$. If the numbers are such as to make this quantity a minimum, it will be unaltered by a small decrement in the number n_1, accompanied by a corresponding small increment in the number n_2. But if n_1 is diminished and n_2 increased by a small change dn, the quantity above is increased by

$$\frac{dn}{n_1^2 p_1 q_1} - \frac{dn}{n_2^2 p_2 q_2}.$$

Hence for the most satisfactory proportion,

$$n_1^2 p_1 q_1 = n_2^2 p_2 q_2.$$

We should then endeavour to adjust our observational numbers so that

$$a_1 b_1 = a_2 b_2 ;$$

in other words, the product of the observed frequencies from one set of matings should be approximately equal to the product from the other set.

By reversing the manner in which the frequencies are combined, the data may be used to determine differential viability, in the same way as they are used to estimate the recombination frequency. The consequence is that the same proportionate numbers from the two types of mating, which are ideal for the estimation of linkage, are also ideal for the estimation of differential viability.

71. Linkage Values inferred from the Progeny of Self-fertilised or Intercrossed Heterozygotes

With many plants it is easier to ensure self-fertilisation than to execute controlled crossings, consequently much of the information available as to linkage in plants is derived from the families of self-fertilised heterozygotes. With animals also such data are obtained in the course of combining two recessives not yet available in combination. Methods of estimating the intensity of linkage have been examined in Chapter IX. of the author's *Statistical Methods*, and analogous but more complex cases have been discussed by J. B. Hutchinson and F. R. Immer. We are here only concerned with the evaluation, in such problems, of the quantity of information as to the linkage value postulated, which the data make available. If there is reason to suspect differential viability, there is no satisfactory substitute for back-crossing, so we shall discuss only the case in which this complication is absent.

The frequencies of the four distinguishable types to be expected may best be inferred from that of the double recessives, for this is only produced when both of the uniting gametes lack both dominant genes. When the two dominant genes have been received by the parents from different grandparents (repulsion), the proportion of such doubly recessive gametes will be $\frac{1}{2} p$, where p is the recombination fraction. The probability that both the uniting gametes are of this kind is therefore $\frac{1}{4} p^2$, or $\frac{1}{4} p p'$ if the recombination fractions should be different in male and female gametogenesis, and are represented by p and p'. We may, therefore, represent this fraction by $\frac{1}{4} \theta$, noting that for repulsion $\sqrt{\theta}$ will be the recombination fraction, or at least the geometric mean of the two recombination fractions, if there are two different values. The frequencies, in any case, are expressible in terms of θ. Consequently, it is only of this quantity that the data provide information. The data provide no means of detecting any difference that may exist between p and p', and we shall from this point use the symbol χ merely as an equivalent to $\sqrt{\theta}$. In the case of coupling, on the other hand, the doubly recessive gametes will be of the parental combination, and the recombination fraction will be $1-\chi$.

From the expected proportion of double recessives the proportions of the other classes may be easily inferred from the fact that each recessive separately must appear in one-quarter of the offspring, irrespective of linkage. The two singly recessive genotypes have, therefore, each a proportional expectation of $\frac{1}{4}(1-\theta)$, leaving $\frac{1}{4}(2+\theta)$ for the last, or doubly dominant type.

Having evaluated the expectations, we may now, as before, calculate directly the amount of information which a record of n offspring will supply as to the value of θ. The table below shows this calculation.

TABLE 37

Offspring expected. (m).	$dm/d\theta$.	$\frac{1}{m}\left(\frac{dm}{d\theta}\right)^2$.	Total.
$\frac{n}{4}\theta$	$\frac{n}{4}$	$n/4\theta$	
$\frac{n}{4}(1-\theta)$	$-\frac{n}{4}$	$n/4(1-\theta)$	$\dfrac{2n(1+2\theta)}{4\theta(1-\theta)(2+\theta)}$
$\frac{n}{4}(1-\theta)$	$-\frac{n}{4}$	$n/4(1-\theta)$	
$\frac{n}{4}(2+\theta)$	$\frac{n}{4}$	$n/4(2+\theta)$	

from which it appears that

$$I_\theta = \frac{n(1+2\theta)}{2\theta(1-\theta)(2+\theta)}$$

for all values of θ. It will be noted that the second and third classes of offspring, the expectations of which are the same functions of θ, might have been treated together without altering the result. In fact, we are only concerned with the total number in these two classes, and not with the parts of which this total is composed, in estimating the value of θ. The fact that these two classes are usually distinguishable adds nothing to our information. The same applies wherever distinguishable classes have proportional frequencies.

Knowing the information available respecting θ, we can now obtain the quantity of information respecting χ. For, since
$$\theta = \chi^2,$$
$$d\theta/d\chi = 2\chi,$$
and
$$\left(\frac{d\theta}{d\chi}\right)^2 = 4\chi^2 = 4\theta.$$
Hence
$$i_\chi = 4\theta\, i_\theta = \frac{2(1+2\theta)}{(1-\theta)(2+\theta)}.$$

This quantity rises steadily from the value unity when $\theta = 0$, the closest possible linkage in repulsion, through 16/9 when $\theta = \frac{1}{4}$, linkage being absent, to an infinite value when $\theta = 1$, the limit of close linkage in coupling. When linkage is at all close, therefore, interbreeding of heterozygotes in coupling is immensely more informative, for the same number of offspring, than the interbreeding of heterozygotes in repulsion. Roughly speaking, with 10 per cent. recombination, coupling matings are worth about ten times as much as repulsion matings; and if the recombination fraction is as small as 5 per cent., they are worth about twenty times as much.

Lack of recognition of this great contrast between the amounts of information supplied by these two types of progenies has led, on several occasions in the genetical literature, to curious misinterpretations of the genetical results. Indeed, it greatly delayed the discovery of the phenomenon of linkage itself, for English geneticists, discussing undoubted cases of linkage in plants, while observing the occurrence of recombination among the coupling progenies, failed to recognise its occurrence in the progenies from heterozygotes in repulsion, and were led to believe that these two different aspects of the same problem followed different laws. The discovery of linkage was thus delayed until animal geneticists, working with a biparental organism, *Drosophila*, in which back-crossing is as convenient as the interbreeding of heterozygotes, demonstrated that the recombination fraction was the same, irrespective of whether the two dominant genes entered the cross from the same or from different parents. Had the plant geneticists been aware that a progeny of 200 offspring in repulsion might be equivalent, in evidential value, to some 25 offspring in coupling, they would, perhaps, have grown sufficiently numerous repulsion progenies to have demonstrated the identity of the two phenomena, which had attracted their attention.

A number of further inferences of practical interest follow from the evaluation of the amount of information to be derived from progenies by self-fertilisation. which the reader may usefully verify for himself.

(1) With close linkage, progenies obtained by self-fertilising heterozygotes in coupling are of nearly equivalent value with back-cross progenies. Thus the advantage of back-crossing when it is possible, lies, in cases of close linkage, principally in the opportunity it affords of eliminating, and of evaluating, differential viability, and of detecting any difference there may be in the recombination fraction in male and female gametogenesis.

(2) When no double recessives are available, the only double heterozygotes that can be formed are in repulsion. Self-fertilising or interbreeding these supplies very little information when the linkage is close. When, however, on growing such a progeny, this situation is found to have occurred, the plant geneticist has usually the choice of two alternative methods of adding to his information in the next generation. (*a*) He may repeat his previous procedure on a large scale, and (*b*) he may grow selfed progenies from the last generation, and so ascertain which are homozygous and which heterozygous, and among the double heterozygotes, which are in coupling and which in repulsion. Supposing the land and labour required to grow each such family to be equivalent to that of growing 25 self-fertilised plants of the kind first obtained, procedure (*b*) will be the more profitable when linkage is very close, and less profitable when it is looser. It is an instructive problem to ascertain at what linkage value the two methods are equally advantageous. In considering this problem it should be noted that in procedure (*b*) the geneticist may choose to form families from the singly recessive plants, or from the double dominants, or from both, but has clearly nothing to learn from the doubly recessive plants. The value of the second season's work will lie, not in the total information gained by a complete classification, but only in information additional to what has been gained by the first season's work. In the second season, however, there will be some further information, from the progenies of 25 plants each from those self-fertilised plants which happen to be double heterozygotes, and of these a certain proportion must be expected to be in coupling.

72. Information as to Linkage derived from Human Families

The greatest obstacle to the study of linkage in man is that it is seldom possible to test or examine for known factors so many as three generations of a family showing any hereditary peculiarity. Consequently, when double heterozygotes are found among parents, it is not known, supposing there is linkage, whether they are in coupling or repulsion. Apart from recent race mixture, however, and other causes of disturbance, these two phases may be expected to occur in equal numbers and, indeed, this fact, when true, can be verified from the family records of only two generations. The possibility of obtaining from such records indications of linkage was first proposed by Bernstein by the use of methods, however, which do not in general utilise the whole of the information in the record. The problem has since been more fully discussed by Haldane and others. We shall here only illustrate the general method of assessing the amount of information obtainable by a classification of the different kinds of families in the record.

Many rare anomalies are transmitted from generation to generation by persons heterozygous for the mutants responsible. If these and their spouses and children are examined for some known factor, such as the capacity for tasting phenylthiocarbamide, a certain number will be heterozygous tasters. Since homozygous tasters cannot be discriminated from heterozygotes, this will only be known if the affected parent is a taster, and if at least one of the children is a non-taster. Only such families can, therefore, be included in the record. Apart from the classification of the children, such families are of two kinds: (*a*) in which the normal parent is a non-taster, for which Bernstein's method is satisfactory; and (*b*) in which the normal parent is a heterozygous taster, for which it is less successful, and which we may take as an example.

The families of two in such a record are of seven possible kinds, which are shown in Table 38 below, where distinguishable individuals are denoted as follows: the affected A, normals a, tasters T, non-tasters t. The first six kinds of family are arranged in the table in

TABLE 38

Types of Family. AT At aT at	Frequency expected. Coupling.	Repulsion.	m.	$dm/d\xi$.	m. $\xi=\frac{1}{4}$	$\frac{1}{m}\left(\frac{dm}{d\xi}\right)^2$. $\xi=\frac{1}{4}$
0 0 0 2 0 2 0 0	x^2 $(1-x)^2$	$(1-x)^2$ x^2	$1-2\xi$	-2	$\frac{1}{2}$	8
1 0 0 1 0 1 1 0	$2x(1+x)$ $2(1-x)(2-x)$	$2(1-x)(2-x)$ $2x(1+x)$	$4(1-\xi)$	-4	3	16/3
0 0 1 1 1 1 0 0	$2x(2-x)$ $2(1-x^2)$	$2(1-x^2)$ $2x(2-x)$	$2(1+2\xi)$	$+4$	3	16/3
0 1 0 1	$2x(1-x)$	$2x(1-x)$	2ξ	$+2$	$\frac{1}{2}$	8
			7	0	7	80/3

pairs, each member of which has the same frequency for heterozygotes in coupling as the other has for repulsion. The combined frequency of these two kinds of family is thus independent of the relative frequency of these two kinds of heterozygotes, while the equality of frequency of members of the same pair will serve to confirm the view that the two types of heterozygote are equally frequent, or, if this were not so, to estimate their relative frequency. We shall here be concerned only with the combined frequency of these pairs. This combined frequency, being a symmetrical function of the recombination fraction, χ, and its complement $1-\chi$, may be simply expressed in terms of the product,

$$\xi = \chi(1-\chi).$$

The expected frequencies are shown in the table for a total of seven suitable families observed.

In order to assess the efficacy of the classification in detecting linkage, we need to know the amount of information which it provides in the limit for loose linkage, when $\chi = \frac{1}{2}$, and $\xi = \frac{1}{4}$. After calculating, therefore, the values of $dm/d\xi$ for the four types of family to be distinguished, the frequencies are rewritten for the particular value $\xi = \frac{1}{4}$, and the amount of information calculated for this value in the last column. It is easily seen that the total amount of information is 80/3 for seven families, or the information per family is

$$i = 80/21.$$

The loss of information in Bernstein's method arises from the fact that he draws no distinction between the types of family in the first and second pairs, or between the third pair and the last type of family. If we were to throw these together, so distinguishing only two groups of families, and relying on the relative frequencies of these two groups only for the detection of linkage, we should have the table set out below.

TABLE 39

	m.	$dm/d\xi$.	m. $\xi = \frac{1}{4}$	$\frac{1}{m}\left(\frac{dm}{d\xi}\right)^2$. $\xi = \frac{1}{4}$
First group	$5-6\xi$	-6	$3\frac{1}{2}$	72/7
Second group . . .	$2+6\xi$	$+6$	$3\frac{1}{2}$	72/7
Totals . .	7	0	7	144/7

The amount of information available from seven families, using Bernstein's classification, is therefore only 144/7, in place of 80/3 available when the families are fully classified. The fraction of the information utilised by Bernstein's method is the ratio of these two quantities, or 27/35. This ratio is termed the efficiency of the method. For larger families its value is found to be somewhat, but not much, lower, the limiting value for large families being 9/16. There is, however, no difficulty in utilising the whole of the information available in the record for families of any size, once the loss of information, and its cause, are recognised. The reader may find it instructive to examine in like manner the classification of families of three children.

73. The Information elicited by Different Methods of Estimation

The foregoing example illustrates the fact, of very general importance, that methods of estimation which proceed without reference to the possibility of evaluating the quantity of information actually contained in the data, are liable to be defective in the quantity that they utilise. When, as is usual, many methods of estimation are available, it becomes important to be able to distinguish which use less, which more, and which, if any, use all. Since the method of measuring information, which has been illustrated, is applicable to data of all kinds, it is only necessary, in order to ascertain how much information is utilised by any proposed method, to determine the sampling distribution of the estimates obtained by that method from quantities of data of the same value as those observed. It is often possible, though sometimes a matter of great mathematical difficulty, to obtain the exact sampling distribution of the estimate arrived at by any particular method, and in such cases the amount of information elicited by the estimate is that of a single observation drawn from this distribution, calculated exactly as in the cases illustrated above.

In many cases in which the exact distribution of an estimate derived from a finite body of data is unknown, it is easy to show that as the sample is increased in magnitude, the sampling distribution tends to the normal form with a calculable variance, V, inversely proportional to the size of the sample, so that

$$V = \frac{v}{n},$$

where v is calculable for any chosen method of estimation.

We shall now show, by a direct application of the general method of calculating the information to such a distribution of a proposed statistic, that the amount of information elicited by the statistic is $\frac{n}{v}$.

Since, in the limiting case considered, the distribution of the statistic becomes continuous and all observable values of it are distinguishable, instead of a summation over a number of classes, we shall be concerned with an integration over all the elementary ranges, dT, in which the statistic T may be found to lie. T, then, is known to be distributed about the true value θ of the parameter, whatever it may be, of which T is an estimate, in a normal distribution with known variance, V. The probability that it will be found to lie in the infinitesimal range, dT, is therefore,

$$df = \frac{1}{\sqrt{2\pi V}} e^{-\frac{(T-\theta)^2}{2V}} dT$$

Differentiating this with respect to θ, in order to ascertain how much information about θ the value of T, regarded now as a single observation, provides, we have

$$\frac{T-\theta}{V} df.$$

The square of this divided by df is now seen to be

$$\frac{(T-\theta)^2}{V^2} df,$$

and the integration of this over all values of T gives simply

$$\frac{V}{V^2} = \frac{1}{V},$$

since, as is well known, the average value of $(T-\theta)^2$ is equal to V, V being, in fact, the mean square deviation, or variance, of the normal distribution.

Consequently, we have found that the amount of information provided by an estimate, normally distributed with variance V, is equal to $1/V$, the *invariance* of that normal distribution. It is thus easy to test whether in the limit for large samples any proposed method of estimation tends to elicit the whole of the information supplied by the data, or a lesser amount. We have only to compare the quantity $1/V$ with I, the amount known to be available; or, dividing both of these quantities by n, to compare $1/v$ with the amount of information, i, provided by each individual observation. The ratio of the amount elicited to the amount available is called the "efficiency" of the method of estimation under discussion, and it has been demonstrated, as the common sense of the method requires, that the efficiency can in no circumstances exceed unity.

74. The Information lost in the Estimation of Error

In the limit for large samples it is always possible to obtain estimates of 100 per cent. efficiency, but with small samples, when treated exactly, this is not found to be generally possible. In some simple cases, however, estimates may be made, which in themselves contain the whole of the information available for finite samples. These especially valuable and comprehensive estimates are called *sufficient* statistics, and the great simplicity of the problems, which fall under the head of the theory of errors, is due to the fact that with the normal distribution both of the quantities requiring estimation, the mean, and the variance, possess sufficient estimates. It is for this reason that in so much experimental work we need only be concerned with the precision of the total, or mean, of the values observed, and with the estimation of this precision from the sum of the squares of the residual deviations. Moreover, as experimenters it is important to plan experiments so as to obtain genuinely independent values of equal precision, to which normal theory can properly be applied.

There is, however, one further point in connection with experiments involving measurements, to which the theory of errors is applicable, which may be cleared up by the methods of this chapter.

When, as the result of an experiment, a value x has been assigned a sampling variance, s^2, validly and correctly estimated from n degrees of freedom, the position is not the same as if the variance were known with exactitude. Our estimate of the variance is itself subject to sampling error, and exact allowance for such error is made by using the true distribution of t, instead of the normal distribution, when testing the significance of the deviation of our observed value from any proposed hypothetical value. In view of this procedure, it must be considered to be inexact to state the amount of information supplied by the experiment respecting the true value of which x is an estimate, merely as $1/s^2$, as though our estimate were known to be normally distributed with this variance. We need, in fact, in considering the absolute precision of an experimental result, to take into account, not only the estimate s^2 derived from the data, but also the number of degrees of freedom upon which our estimate, s^2, was based.

Now the probability that the quantity t, defined by the relationship

$$x - \mu = st,$$

where x is the observed value and μ the hypothetical value of which it is an estimate, shall lie in any assigned range, dt, is given by the formula

$$df = \frac{\frac{n-1}{2}!}{\frac{n-2}{2}! \sqrt{\pi n}} \cdot \frac{dt}{\left(1 + \frac{t^2}{n}\right)^{\frac{1}{2}(n+1)}};$$

or in terms of x and μ, by

$$df = \frac{\frac{n-1}{2}!}{\frac{n-2}{2}! \, s\sqrt{\pi n}} \cdot \frac{dx}{\left(1 + \frac{(x-\mu)^2}{ns^2}\right)^{\frac{1}{2}(n+1)}}.$$

From this we can evaluate the amount of information supplied by an observed value, x, relative to the unknown parameter, μ, as we have done with the normal curve above, by differentiating with respect to μ. This gives

$$\frac{n+1}{n\,s^2} \cdot \frac{(x-\mu)df}{1 + \frac{(x-\mu)^2}{n\,s^2}};$$

squaring this, and dividing by df, we find

$$\frac{(n+1)^2}{n^2 s^4} \cdot \frac{(x-\mu)^2 df}{\left\{1+\frac{(x-\mu)^2}{ns^2}\right\}^2}.$$

When integrated over all possible values of the observable quantity, x, this amounts to

$$\frac{n+1}{(n+3)s^2}.$$

It appears that the true precision of our estimate is somewhat lower than it would have been, had the variance been known with exactitude to be s^2. In the extreme case, when $n = 1$, and the estimate is based on only 1 degree of freedom, the precision is halved. And in general, the true precision is less than it might be thought, if the uncertainty of our estimate of the variance were ignored, by the fraction $2/(n+3)$. It may thus be worth while to sacrifice, to some small extent, the aim of diminishing the value of s^2, if this diminution carries with it any undue reduction in the number of degrees of freedom, available for the estimation of error.

Writers averse to the use of fiducial probability (Neyman, Walsh) have attempted solution of this problem by other paths. The results are immensely complex, and these writers seem to overlook the fact that for *given* levels of significance, " Student " had already solved the problem in 1908. What is needed in experimental design is knowledge of the allowance to be made for the limitation of the number of degrees of freedom, before the level or levels of interest are known. The allowance found in this Section has been in use now for 25 years without any serious alternative. It is not an approximation, or a provisional artifice, and has been unintelligible only to those who over a long period resisted the cogency of the fiducial argument.

REFERENCES AND OTHER READING

C. I. BLISS (1935). The calculation of the dosage mortality curve. Annals of Applied Biology, xxii. 134-167.

C. I. BLISS (1935). The comparison of dosage mortality data. Annals of Applied Biology, xxii. 307-333.

R. A. FISHER (1922). On the mathematical foundations of theoretical statistics. The Philosophical Transactions of the Royal Society, A, ccxxii. 309-368.

R. A. FISHER (1925-1963). Statistical methods for research workers. Chapter IX.

R. A. FISHER and BHAI BALMUKAND (1928). The estimation of linkage from the offspring of selfed heterozygotes. Journal of Genetics, xx. 79-92.

R. A. FISHER (1934). The amount of information supplied by records of families as a function of the linkage in the population sampled. Annals of Eugenics, vi. 66-70.

R. A. FISHER and F. YATES (1938-1963). Statistical tables for biological, agricultural and medical research. Oliver and Boyd Ltd., Edinburgh.

J. B. HUTCHINSON (1929). The application of the "Method of Maximum Likelihood" to the estimation of linkage. Genetics, xiv. 514-537.

F. R. IMMER (1934). Calculating linkage intensities from F_3 data. Genetics, xix. 119-136.

J. NEYMAN et al. (1935). Statististical problems in agricultural experimentation. Supplement to the Journal of the Royal Statistical Society, ii. 114-136.

J. E. WALSH (1949). On the information lost by using a t-test when the population variance is known. Journal of the American Statistical Association, xliv. 122-125.

INDEX

Acceptance procedures, 25
Amount of information, 184-187, 214-245
Analysis of covariance, 171-177
Analysis of variance, 52-58, 73-75, 100, 101, 114-116, 129, 138, 151, 172, 174, 176, 178, 179, 198-213
Annual crops, 166
Arbitrary corrections, 168

Balmukand, 245
Barley, 67-69, 142
Basis of inference, 101-106
Bayes, 5-7, 10, 194, 198
Bernstein, 236, 238, 239
Bias, 64, 65, 74
Biological assay, 224-226
Bliss, 226, 245
Boole, 4
Bose, 136
Boyle, xv

Cabinet Cyclopædia, 4
χ^2 test, 195-198
Chrystal, 4
Complete orthogonalisation, 135
Components of error, 145
Concomitant measurements, 163-183
Confounding, 109-136
Costing, 186
Cotton spinning, 88
Cyclic sets, 156
Cyclic substitution, 125

Darwin, 27, 30, 32-38, 40, 44, 49, 57, 188, 191, 198
De Morgan, 4, 6, 10

Digitalis, 28
Dilution method, 219-223
Double confounding, 121, 122
Drosophila, 234
Dummy comparisons, 137-139

Eden, 57, 69, 108, 162, 183
Efficiency, 241
Ely, 90
Estimation, 214
Euler, 83

Factorial design, 93-108
Fiducial probability, 184, 197-198

Galton, 28-32, 39-41, 44, 194
General hypothesis, 44-48
Good Easter, 181
Græco-Latin Square, 80-85, 159
Groups, 117

Haldane, 236
Hayes, 67, 69
Healy, 201, 213
Hutchinson, 231, 245

Immer, 67, 69, 231, 245
Induction, 3, 101
Inductive inference, 25, 198
Interaction, 97-101, 107, 110, 135, 207-213
Interpretation, 1, 12, 152-154, 188
Invariance, 184, 241
Inverse probability, 6, 7, 198
Ipomœa, 28

Laplace, 4

INDEX

Latin Square, 70-92, 121, 124, 125, 126, 129, 142, 158, 166, 178, 179, 199, 206-209
Lavoisier, xv
Limnanthes, 28, 31
Linkage, 226-239

Maskell, 192, 213
Minnesota, 67
Missing values, 177-180

Nair, 136
Neyman, 244, 245
Norton, 83, 92
Null hypothesis, 15-17, 20-23, 35-38, 48, 54, 56, 184-213

Orthogonal sets, 126-129
Orthogonal squares, 80-83

Pairing, 32-36
Parameters, 214-218
Partial confounding, 129-162
Perennial crops, 166
Petunia, 28
Pigs, 171
Potatoes, 90
Powers, 67, 69
Precision, 21-25, 58-60, 184, 220-245
Price, 6
Probit, 224-236
Problem of distribution, 16
Psycho-physical experiment, 11

Quantity and quality, 139-141

Randomisation, 17-21, 41-44, 51-52, 62-66, 70-72, 188
Randomised blocks, 50-52
Ratio, fiducial limits of, 194-195
References, 10, 25, 49, 69, 92, 108, 136, 162, 183, 213, 245
Regression, 140, 171

Replication, 60-62, 90, 108
Reseda lutea, 28
Rothamsted, 90, 92, 142, 154, 182, 183

Sade, 92
Sanders, 183
Sex ratio, 214-216
Shape of blocks and plots, 66-67
Significance, 13, 24, 56-59, 63, 75, 107, 115, 179, 187-190, 196, 198, 199
Skewness, 190
Spindles, 85
Statistical control, 163-183
Stevens, 84, 92
Stewart, 180
" Student," 34, 38, 39, 44, 45, 49, 188, 191, 192, 202, 204, 213, 244
Subsidiary factors, 102
Sugar-beet, 181, 182
Systematic designs, 64, 76-80

t test, 34-40, 44-48, 57-59, 76, 188-194, 204, 242
Tea, 11, 166
Tedin, 80, 92
Theory of errors, 27, 34, 44
Theory of estimation, 214
Tippett, 52, 69, 88, 92

Venn, 4
Vik, 78, 80
Viola, 28

Walsh, 244, 245
Wishart, 162

Yates, 47, 57, 69, 84, 92, 136, 162, 245
Youden, 135

z test, 57-59, 75, 152, 199, 200, 206, 207
Zea, 28-31

Statistical Methods and
Scientific Inference

Statistical Methods and Scientific Inference

by

Sir Ronald A. Fisher, Sc.D., F.R.S.

D.Sc (Adelaide, Ames, Chicago, Harvard, Indian Statistical Institute, Leeds, London)
LL.D. (Calcutta, Glasgow)

Honorary Research Fellow, Division of Mathematical Statistics, C.S.I.R.O., University of Adelaide; Foreign Associate, United States National Academy of Sciences; Foreign Honorary Member, American Academy of Arts and Sciences; Foreign Member, American Philosophical Society; Honorary Member, American Statistical Association; Honorary President International Statistical Institute; Foreign Member, Royal Swedish Academy of Sciences; Member, Royal Danish Academy of Sciences; Member, Pontifical Academy; Member, Imperial German Academy of Natural Science; formerly Fellow of Gonville and Caius College, Cambridge; formerly Galton Professor, University of London; and formerly Balfour Professor of Genetics, University of Cambridge.

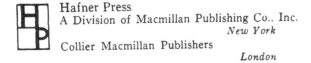

Hafner Press
A Division of Macmillan Publishing Co., Inc.
New York
Collier Macmillan Publishers
London

First Published 1956
Second Edition, revised . . . 1959
Third Edition, revised and enlarged . 1973

Third Edition Copyright © 1973 by the
University of Adelaide

All rights reserved. No part of this book may be reproduced or transmitted in any form or by any means, electronic or mechanical, including photocopying, recording, or by any information storage and retrieval system, without permission in writing from the publisher.

Hafner Press
A Division of Macmillan Publishing Co., Inc.
866 Third Avenue, New York, N.Y. 10022

Collier-Macmillan Canada Ltd., Toronto, Ontario

Library of Congress Catalog Card Number: 72-97990
ISBN: 0-02-844740-9

NOTE TO THIS REPRINT EDITION

THE THIRD EDITION, 1973, which is reprinted here, introduced much new material which the author had entered in his interleaved copy of the book for this purpose, some time before his death on 29 July, 1962. This included a Section on Isaac Todhunter in Chapter II, an extended discussion of the problem involving observations of two kinds in Chapter V, and numerous smaller additions and clarifications throughout the text.—J.H.B.

We use Reason for improving the Sciences;
whereas we ought to use the Sciences for
improving our Reason.
ANTOINE ARNAULD, 1662
(*The Port-Royal Logic*)

Another use to be made of this Doctrine of
Chances is, that it may serve in Conjunction
with the other parts of the Mathematicks, as a
fit Introduction to the Art of Reasoning.
DE MOIVRE, 1718

If logic investigates the general principles of
valid thought, the study of arguments, to which
it is rational to attach *some* weight, is as much
part of it as the study of those which are
demonstrative.
J. M. KEYNES, 1921

CONTENTS

CHAPTER		PAGE
I	FOREWORD	1
II	THE EARLY ATTEMPTS AND THEIR DIFFICULTIES	8
	1. Thomas Bayes	8
	2. George Boole	17
	3. John Venn and the Rule of Succession	24
	3.1 Isaac Todhunter	31
	4. The meaning of probability	33
III	FORMS OF QUANTITATIVE INFERENCE	40
	1. The simple test of significance	40
	2. More general hypotheses	49
	3. The fiducial argument	54
	4. Accurate statements of precision	60
	5. Discontinuous observations	63
	6. Mathematical Likelihood	71
IV	SOME MISAPPREHENSIONS ABOUT TESTS OF SIGNIFICANCE	79
	1. Tests of significance and acceptance decisions	79
	2. "Student" 's Test	82
	3. The case of linear regression	86
	4. The two-by-two table	89
	5. Excluding a composite hypothesis	92
	6. Behrens' Test	97
	7. The "randomization test"	101
	8. Qualitative differences	103
V	SOME SIMPLE EXAMPLES OF INFERENCES INVOLVING PROBABILITY AND LIKELIHOOD	110
	1. The logical consequences of uncertainty	110
	2. Bayesian prediction	115
	3. Fiducial prediction	117
	4. Predictions from a Normal Sample	119
	5. The fiducial distribution of functions of the parameters	125

viii   CONTENTS

CHAPTER		PAGE
	6. Observations of two kinds	127
	7. Inferences from likelihoods	132
	8. Variety of logical types	138
VI	THE PRINCIPLES OF ESTIMATION	145
	1. Relations to other work	145
	2. Criteria of estimation	148
	3. The concept of efficiency	151
	4. Likelihood and information	154
	5. Grouping of samples	156
	6. Simultaneous estimation	158
	7. Ancillary information	162
	8. The location and scale of a frequency distribution of known form	165
	9. An example of the Nile problem	169
	10. The sampling distribution of the estimate	171
	11. The use of an ancillary statistic to recover the information lost	172
	12. Simultaneous distribution of the parameters of a bivariate Normal distribution	175
INDEX		181

CHAPTER I

FOREWORD

The very large output, characteristic of the present time, of works on various aspects of Statistics, many of them of a much higher standard than were available in the past, not all in the scientific field, but covering the requirements of technological, commercial, educational and administrative purposes, is a recent efflorescence following as a natural and perhaps inevitable consequence on the efforts towards abstract understanding largely set on foot long ago by that versatile and somewhat eccentric man of genius, Francis Galton. Although many of the purposes, to which statistical methods and ideas are in our time successfully applied, are not primarily scientific in aim, that is, are not directed specifically towards an improved understanding of the natural world, yet the fruitfulness and success of the train of studies initiated by Galton were, I submit, due to his own outlook of untrammelled scientific curiosity, and to his confidence that it was in regard to scientific problems that a more penetrating statistical methodology was required. Though all branches of statistical science have profited and been revivified by its influence, it is the course of progress achieved on the scientific front which requires recapitulation, if the nature of the whole movement is to be grasped in spite of its growing complexity and diversity.

Galton's great gift lay in his awareness, which grew during his life, of the vagueness of many of the phrases in which men tried to express themselves in describing natural phenomena. He was before his time in his recognition that such vagueness could be removed, and a certain precision of thought attempted by finding quantitative definitions of concepts fit to take the place of such phrases as "the average man", "variability", "the strength of inheritance", and so forth, through the assembly of objective data, and its systematic examination. That the methods he himself used were often extremely crude, and sometimes seriously faulty, is, indeed, the strongest evidence of the eventual value to the progress of science of his unswerving faith that objectivity and rationality were accessible, even in such elusive fields as psychology, if only a factual basis for these qualities were diligently sought. The systematic improvement of statistical methods and the development of their utility in the study of biological variation and inheritance were the aims to which he deliberately devoted his personal fortune, through the support and endowment of a research laboratory under Professor K. Pearson.

The peculiar mixture of qualities exhibited by Pearson made this choice in some respects regrettable, though in others highly successful. Pearson's energy was unbounded. In the course of his long life he gained the devoted service of a number of able assistants, some of whom he did not treat particularly well. He was prolific in magnificent, or grandiose, schemes capable of realization perhaps by an army of industrious robots responsive to a magic wand. In a sense he undoubtedly appreciated Galton's conception of the greatness of the potential contribution of Statistics in the service of Science, and as a means of rendering strictly scientific a range of studies not traditionally included in the Natural Sciences, but, as perceived through his eyes, this greatness was not easily to be distinguished from the greatness of Pearson himself.

The terrible weakness of his mathematical and scientific work flowed from his incapacity in self-criticism, and his unwillingness to admit the possibility that he had anything to learn from others, even in biology, of which he knew very little. His mathematics, consequently, though always vigorous, were usually clumsy, and often misleading. In controversy, to which he was much addicted, he constantly showed himself to be without a sense of justice. In his dispute with Bateson on the validity of Mendelian inheritance he was the bull to a skilful matador. His immense personal output of writings, his great enterprise in publication, and the excellence of production characteristic of the Royal Society and the Cambridge Press, left an impressive literature. The biological world, for the most part, ignored it, for it was indeed both pretentious and erratic. Yet the intrinsic magnitude of some of the problems brought into discussion, the high prestige that mathematical writing always carries, and a certain imaginative boldness, did suffice to save this material from complete neglect. Little as Pearson cared for the past—for example, for the Gaussian tradition—and much as he would have disliked the future of statistical science, his activities have a real place in the history of a greater movement.

Though Pearson did not appreciate it, quantitative biology, especially in its agricultural applications, was beginning to need accurate tests of significance.

4 STATISTICAL METHODS AND SCIENTIFIC INFERENCE

So early as Darwin's experiments on growth rate the need was felt for some sort of a test of whether an apparent effect "might reasonably be due to chance". At the same time it was recognized that the available test based on the conventional "probable error" was not always to be relied on. I have discussed this particular case in *The Design of Experiments* (Chapter III).[1] It was characteristic of the early period, and of Pearson, that such difficulties were habitually blamed on "paucity of data", and not ascribed specifically to the fact that mathematicians had so far offered no solution which the practitioner could use, and indeed had not been sufficiently aware of the difficulty to have discussed the problem. As is well known, it was a research chemist, W. S. Gosset, writing under the designation of "Student",[3] who supplied the test which, important as it was in itself, was of far greater importance in inaugurating the first stage of the process by which statistical methods attained sufficient refinement to be of real assistance in the interpretation of data. As a result of his work, problems of distribution were one after another given exact solutions; by about 1930 all statistical problems which were thought to deserve careful treatment were being discussed in terms of mathematically exact distributions, and the tests of significance based upon them. The logical basis of these scientific applications was the elementary one of excluding, at an assigned level of significance, hypotheses, or views of the causal background, which could only by a more or less implausible coincidence have led to what had been observed. The "Theory of Testing Hypotheses" was a later attempt, by authors who had taken no part in the development of these tests, or in their scientific application, to reinterpret them in terms of an imagined process of acceptance sampling, such as was beginning to be used in commerce; although such processes have a logical basis very different from those of a scientist engaged in gaining from his observations an improved understanding of reality.

The exact solutions of the series of problems of distribution left unsolved by the Pearsonian school had not only as its immediate fruit the refinement, by accurate tests of significance, of the experimenter's facility in examining his data critically; at a deeper logical level they allowed of the development of objective principles of estimation, and so revealed the misleading character of many of the methods of estimation commonly advocated. The variety of concepts relevant to the logical basis of a process of estimation of hypothetical quantities by the aid of observational material is considerable, and some account is given of them in Chapter VI. It was early necessary to distinguish Mathematical Likelihood from Mathematical Probability, and the concept of quantity of information (different from the meaning later given to the same phrase in Communication Theory) itself intimately related to the Likelihood, was found to measure effectively the competence of any proposed method of estimation, even in its application to small samples.

From a larger viewpoint than that of merely refining and perfecting the statistical processes used in the examination of a fixed body of data, the concepts of the theory of estimation lent themselves to the effectual comparison of different bodies of data, and therefore of the experimental procedures, or observational programmes capable of giving rise to such observational foundations. This is the leading consideration in the branch of statistical science known as Experimental Design, on which during the last few years very comprehensive and substantial works have appeared. The practical and theoretical study of Experimental Design, with which should be included that of sampling for the purpose of factual ascertainment as developed by Mahalanobis,[2] and by Yates,[4] may be regarded as the second great movement in the development of Statistics for the clarification of scientific thought.

In the author's view, however, there has already appeared a need for the exposition and consolidation of the specifically logical concepts which have emerged as it were as by-products of both (a) the purely mathematical elucidation of statistical problems in the first phase, and (b) in the second phase the development of experimental designs, logically coherent with the processes used in their discussion, and with the scientific inferences of which they are to supply the basis, so as to form with them a complete illustration of the mode in which new scientific knowledge is generated. Once recognized and applied, there is little danger of such an advance in procedure being lost. Practitioners, however, are not the natural repositories of logical niceties, and teachers especially, engaged in introducing students to these new fields, may value an attempt to consolidate the specifically logical gains of the past half-century, and will perhaps tolerate a certain amount of necessary hair-splitting.

In the introduction to the *Design of Experiments* (p. 9)[1] I have stressed my conviction that the art of framing cogent experiments, like that of their statistical analysis, can each only establish its full significance as parts of a single process of the improvement of natural knowledge; and that the logical coherence of this whole is the only full justification for free individual thought in the formation of opinions applicable to the real world. I would wish again now to reiterate this point of view. To one brought up in the free intellectual atmosphere of an earlier time there is something rather horrifying in the ideological movement represented by the doctrine that reasoning, properly speaking, cannot be applied to empirical data to lead to inferences valid in the real world. It is undeniable that the intellectual freedom that we in the West have taken for granted is now successfully denied over a great part of the earth's surface. The validity of the logical steps by which we can still dare to draw our own conclusions cannot therefore, in these days, be too clearly expounded, or too strongly affirmed.

REFERENCES

1. R. A. Fisher (1935-1966). *The Design of Experiments.* Oliver and Boyd, Edinburgh.
2. P. C. Mahalanobis (1946). On large-scale sample surveys. *Phil. Trans. Roy. Soc.*, B, vol. 231, pp. 329-451.
3. "Student" (1908). The probable error of a mean. *Biometrika*, vol. 6, pp. 1-25.
4. F. Yates (1949). *Sampling Methods for Censuses and Surveys.* Charles Griffin, London.

CHAPTER II

THE EARLY ATTEMPTS AND THEIR DIFFICULTIES

1. Thomas Bayes

For the first serious attempt known to us to give a rational account of the process of scientific inference as a means of understanding the real world, in the sense in which this term is understood by experimental investigators, we must look back over two hundred years to an English clergyman, the Reverend Thomas Bayes, whose life spanned the first half of the eighteenth century. It is indeed only in the present century, with the rapid expansion of those studies which are collectively known as Statistics, that the importance of Bayes' contribution has come to be appreciated. *The Dictionary of National Biography*,[13] representing opinion current in the last quarter of the nineteenth century, does not include his name. The omission is the more striking since this work of reference does include a notice of his father, Joshua Bayes (1671-1746). While the father was no doubt a learned and eloquent preacher, still, in his own time, his son Thomas was for twenty years a Fellow of the Royal Society, and therefore known also as a not inconsiderable mathematician. Indeed, his mathematical contributions to the *Philosophical Transactions* show him to have been in the first rank of independent thinkers, very well qualified to attempt the really revolutionary task opened out by his posthumous paper "An Essay towards solving a problem in the doctrine of chances", which appeared in the *Philosophical Transactions*[1] in 1763, not long after his death in 1761.

It is entirely appropriate that this first attempt should have been made at this time. For more than a century the learned world had been coming to regard deliberate experimentation as the fundamental means to "The Improvement of Natural Knowledge", in the words chosen by the Royal Society. With Isaac Newton, moreover, and such men as Robert Boyle, the possibility of formulating natural law in quantitative terms had been brilliantly exhibited. The nature of the reasoning process by which appropriate inferences, or conclusions, could be drawn from quantitative observational data was ripe for consideration. The prime difficulty lay in the uncertainty of such inferences, and it was a fortunate coincidence that the recognition of the concept of probability, and its associated mathematical laws, in its application to games of chance, should at the same time have provided a possible means by which such uncertainty could be specified and made explicit. In England such a publication as Abraham de Moivre's *Doctrine of Chances*[9] must have been a very immediate stimulus to Bayes' reflexions on this subject.

Bayes' Essay was communicated to the Royal Society some time after his death by his friend Richard Price. Price added various demonstrations and illustrations of the method, and seems to have replaced Bayes' introduction by a prefatory explanation of his own. It is to be regretted that we have not Bayes' own introduction, for it seems clear that Bayes had recognized that the postulate proposed in his argument (though not used in his central theorem) would be thought disputable by a critical reader, and there can be little doubt that this was the reason why his treatise was not offered for publication in his own lifetime. Price evidently laid less weight on these doubts than did Bayes himself; on the other hand he very fully appreciated the importance of what Bayes had done, or attempted, for the advancement of experimental philosophy; although the central theorem of the essay is framed in somewhat academic and abstract terms, without expatiating on the large consequences for human reasoning which would flow from his axiom.

It contains also an element of artificiality, which has obscured its understanding, and which, I believe, is capable of being removed (see Chapter V, Section 6, and reference 6 to this Chapter).

The most important passage of Price's introductory letter is as follows ([1] p. 370):

> In an introduction which he has writ to this Essay, he says, that his design at first in thinking on the subject of it was, to find out a method by which we might judge concerning the probability that an event has to happen, in given circumstances, upon supposition that we know nothing concerning it but that, under the same circumstances, it has happened a certain number of times, and failed a certain other number of times. He adds, that he soon perceived that it would not be very difficult to do this, provided some rule could be found according to which we ought to estimate the chance that the probability for the happening of an event perfectly unknown, should lie between any two named degrees of probability, antecedently to any experiments made about it; and that it appeared to him that the rule must be to suppose the chance the same that it should lie between any two equidistant degrees; which, if it were allowed, all the rest might be easily calculated in the common

EARLY ATTEMPTS AND THEIR DIFFICULTIES 11

method of proceeding in the doctrine of chances. Accordingly, I find among his papers a very ingenious solution of this problem in this way. But he afterwards considered, that the *postulate* on which he had argued might not perhaps be looked upon by all as reasonable; and therefore he chose to lay down in another form the proposition in which he thought the solution of the problem is contained, and in a *scholium* to subjoin the reasons why he thought so, rather than to take into his mathematical reasoning anything that might admit dispute.

The actual mathematics of Bayes' theorem may be expressed very briefly in modern notation.

If in $a+b$ independent trials it has been observed that there have been a successes and b failures, then if p were the hypothetical probability of success in each of these trials, the probability of the happening of what has been observed would be

$$\frac{(a+b)!}{a!\,b!} p^a(1-p)^b\,; \qquad (1)$$

but if in addition we know, or can properly postulate, that p itself has been chosen by an antecedent random process, such that the probability of p lying in any infinitesimal range dp between the limiting values 0 and 1 is equal simply to

$$dp, \qquad (2)$$

then the probability of the compound event of p lying in the assigned range, and of the observed numbers of successes and failures occurring will be the product of these two expressions, namely

$$\frac{(a+b)!}{a!\,b!} p^a(1-p)^b\,dp\,. \qquad (3)$$

But, from the data, such a compound event has happened for some element or other of those into which the total range from 0 to 1 may be divided, so that the probability that any particular one should have happened in fact is the ratio,

$$\frac{\frac{(a+b)!}{a!\,b!}\, p^a(1-p)^b\, dp}{\frac{(a+b)!}{a!\,b!}\int_0^1 p^a(1-p)^b\, dp}, \qquad (4)$$

of which the denominator, involving a complete Eulerian integral, is equal $1/(a+b+1)$.

The finite probability that p should lie between any assigned limits u and v may therefore be expressed as the incomplete integral

$$\frac{(a+b+1)!}{a!\,b!}\int_u^v p^a(1-p)^b\, dp. \qquad (5)$$

The postulate which Bayes regarded as questionable is represented above by the expression (2). The greater part of Bayes' analysis is concerned with approximate forms for the discussion of these integrals, and is of historical rather than mathematical interest for the modern reader. How explicit Bayes is in introducing the critical datum is shown by his own introductory remarks.

The crucial theorem, proposition 8, comes in the second section of the essay, and is preceded by a special explanatory foreword[1] (p. 385):

SECTION II

Postulate 1. I suppose the square table or plane $ABCD$ to be so made and levelled, that if either of the balls O or W be thrown upon it, there shall be the same probability that it rests upon any one equal part of the plane as another, and that it must necessarily rest somewhere upon it.

EARLY ATTEMPTS AND THEIR DIFFICULTIES 13

2. I suppose that the ball W shall be first thrown, and through the point where it rests a line os shall be drawn parallel to AD, and meeting CD and AB in s and o; and that afterwards the ball O shall be thrown $p+q$ or n times, and that its resting between AD and os after a single throw be called the happening of the event M in a single trial. These things supposed,

Lemma 1. The probability that the point o will fall between any two points in the line AB is the ratio of the distance between the two points to the whole line AB. (The proof occupies two pages, with an examination of incommensurability after the manner of the fifth book of Euclid's elements.)

Lemma 2. The ball W having been thrown, and the line os drawn, the probability of the event M in a single trial is the ratio Ao to AB.

After a short proof there follow the enunciation of proposition 8 and its demonstration. A single figure is used to represent the square table and the construction upon it, and also outside the square, a graph representing the function,

$$\frac{(a+b)!}{a!\,b!}\,p^a q^b, \qquad p+q=1,$$

for all values of p from 0 to 1. The latter is used to give geometrical significance to the analytic integrals used above.

In broaching the boundaries of an entirely new field of thought by means of a single illustrative theorem, pregnant as it was, Bayes left untouched many distinctions of importance to its discussion in the future. In respect to the nature of the concept of probability very diverse opinions have been expressed. In particular, although perhaps all would agree that the word denotes a measure of the strength of an opinion or state of judgement, some have insisted that it should properly be used only for the expression of a state of rational judgement based on sufficient objective evidence, while others have thought that equality of probability may be asserted merely from the indifference of, or the absence of *differentiae* in, the objective evidence, if any, and therefore from the total absence of objective evidence, if there were none.

Bayes evidently held the first of these opinions and frames a definition suited—in my view—to show (1) that he was not thinking merely of games of chance, and, (2) at the same time that his concept of probability was that of the mathematicians, such as Montmort[10] and de Moivre, who had treated largely of gambling problems, in which the equality of probability assigned to numerous possible events and combinations of events is a consequence of the assumed perfection of the apparatus and operations employed.

Bayes' definition is[1] (p. 376): "5. The *probability of any event* is the ratio between the value at which an expectation depending on the happening of the event ought to be computed, and the value of the thing expected upon its happening. 6. By *chance* I mean the same as probability." There is no room for doubt that Bayes would have regarded the "expectation" to which he referred as capable of verification to any required approximation, by repeated trials with sufficiently perfect apparatus. Subject to the latent stipulation of fair use, or of homogeneity in the series of tests, his definition is therefore equivalent to the limiting value of the relative frequency of success.

On the contrary Laplace, who needed a definition

wide enough to be used in the vastly diverse applications of the *Théorie analytique*, manifestly inclined to the second view[7] (1820).

> La théorie des hasards consiste à réduire tous les évènemens du même genre, à un certain nombre de cas également possibles, c'est-à-dire, tels que nous soyons également indécis sur leur existence; et à déterminer le nombre de cas favorables à l'évènement dont on cherche la probabilité. Le rapport de ce nombre à celui de tous les cas possibles, est la mesure de cette probabilité qui n'est ainsi qu'une fraction dont le numérateur est le nombre des cas favorables, et dont le dénominateur est le nombre de tous les cas possibles.

This differs a little from the form used in 1812:

> La théorie des probabilités consiste à réduire tous les évènemens qui peuvent avoir lieu dans une circonstance donnée, à un certain nombre de cas également possibles, c'est-à-dire tels que nous soyons également indécis sur leur existence, et à déterminer parmi ces cas, le nombre de ceux qui sont favorables à l'évènement dont on cherche la probabilité. Le rapport de ce nombre à celui de tous les cas possibles, est la mesure de cette probabilité qui n'est donc qu'une fraction dont le numérateur est le nombre des cas favorables, et dont le dénominateur est celui de tous les cas possibles.

It is seen that Laplace effectively avoids any objective definition, first by using the term *possible* in a context in which *probable* could be used, without explaining what difference, if any, he intends between the two words, and secondly by his indication that equal possibility could be judged without cogent evidence of equality.

In consequence of this difference of concept, Bayes' attempt is exposed to different types of criticism in his own hands and in those of Laplace. While I have, for myself, no doubt that Bayes' definition is the more satisfactory, being not only in accordance with the ideas upon which the *Doctrine of Chances* of his own time was built, but in connecting the comparatively modern notion of *probability*, which seems to have been unknown to the Islamic and to the Greek mathematicians, with the much more ancient notion of an *expectation*, capable of being bought, sold and evaluated, nevertheless it would merely confuse the discussion to give further reasons for this opinion; the difficulties to which Bayes' approach was eventually found to lead can be easily expressed in terms of the notions which he himself favoured.

Whereas Laplace defined probability by means of the enumeration of discrete units, Bayes defined a continuous probability distribution, by a formula for the probability between any pair of assigned limits. He did not, however, consider the metric of his continuum. For in stating his prime postulate in the form that the chance *a priori* of the unknown probability lying between p_1 and p_2 should be equal to p_2-p_1 he might, so far as cogent evidence is concerned, equally have taken any monotonic function of p, such as

$$\phi = \tfrac{1}{2}\cos^{-1}(1-2p), \quad p = \sin^2\phi,$$

and postulated that the chance that ϕ should lie between ϕ_1 and ϕ_2 should be

$$\frac{2}{\pi}(\phi_2-\phi_1),$$

so that, instead of inserting the probability *a priori* as
$$dp,$$
it would have appeared in the analysis as

$$\frac{2}{\pi}d\phi = \frac{1}{\pi\sqrt{pq}} \cdot dp, \quad q = 1-p, \tag{6}$$

a postulate or assumption rather more favourable to extreme values of the unknown p, near to 0 or 1, at the expense of more central values.

Bayes' introduction of an expression representing probability *a priori* thus contained an arbitrary element, and it was doubtless some consciousness of this that led to his hesitation in putting his work forward. He removes the ambiguity formally by means of the auxiliary experiment with the ball W, and in Chapter V it is shown that such an auxiliary experiment can sometimes be realized in practice.

A more important question, however, is whether in scientific research, and especially in the interpretation of experiments, there is cogent reason for inserting a corresponding expression representing probabilities *a priori*. This practical question cannot be answered peremptorily, or in general, for certainly cases can be found, or constructed, in which valid probabilities *a priori* exist, and can be deduced from the data. More frequently, however, and especially when the probabilities of contrasted scientific theories are in question, a candid examination of the data at the disposal of the scientist shows that nothing of the kind can be claimed.

2. George Boole

The superb pre-eminence of Laplace as a mathematical analyst undoubtedly inclined mathematicians for nearly fifty years to the view that the logical approach adopted by him had removed all doubts as to the applicability in practice of Bayes' theorem. That this was indeed Laplace's view may be judged from his reference to the position of Bayes in the history of the subject[7] (p. cxxxvii).

Bayes dans les *Transactions philosophiques* de l'année 1763, a cherché directement la probabilité que les possibilités indiquées par des expériences déjà faites, sont comprises dans des limites données; et il y a parvenu d'une manière fine et très ingénieuse, quoiqu'un peu embarrassée.

I imagine that the hint of criticism in the last phrase is directed against Bayes' hesitation to regard the postulate he required as axiomatic. It will be noticed in the sequel that discussion turned on just this question of its axiomatic nature, and not on the question, more natural to an experimental investigator, of whether, in the particular circumstances of the investigation, the knowledge implied by the postulate was or was not in fact available. It is the submission of the author that actual familiarity with the processes of scientific research helps greatly in the understanding of scientific data, and has in the present century clarified the issue by bringing into prominence the factual question, rather than the abstract question of axiomatic validity. It was not, however, until the weight of opinion among philosophical mathematicians had turned against the supposed axiom that the controversy could come to be examined in this more realistic manner.

Simple examples are provided by genetic situations. In Mendelian theory there are black mice of two genetic kinds. Some, known as homozygotes (BB), when mated with brown yield exclusively black offspring; others, known as heterozygotes (Bb), while themselves also black, are expected to yield half black and half brown. The expectation from a mating between two heterozygotes is 1 homozygous black, to 2 heterozygotes, to 1 brown. A black mouse from such a mating has thus, prior to any test-mating in which it may be used, a known probability of 1/3 of being homozygous, and of 2/3 of being heterozygous. If, therefore, on testing with a brown mate it yields seven offspring, all being black, we have a situation perfectly analogous to that set out by Bayes in his proposition, and can develop the counterpart of his argument, as follows:

The prior chance of the mouse being homozygous is 1/3; if it is homozygous the probability that the young shall be all black is unity; hence the probability of the compound event of a homozygote producing the test litter is the product of the two numbers, or 1/3.

Similarly, the prior chance of it being heterozygous is 2/3; if heterozygous the probability that the young shall be all black is $1/2^7$, or 1/128; hence the probability of the compound event is the product, 1/192.

But, one of these compound events has occurred; hence the probability after testing that the mouse tested is homozygous is

$$1/3 \div \left(\frac{1}{3} + \frac{1}{192}\right) = 64/65,$$

and the probability that it is heterozygous is

$$1/192 \div \left(\frac{1}{3} + \frac{1}{192}\right) = 1/65.$$

If, therefore, the experimenter knows that the animal under test is the offspring of two heterozygotes, as would be the case if both parents were known to be black, and a parent of each were known to be brown, or, more strictly, if both had either a brown parent or a brown offspring, cogent knowledge *a priori* would have been available, and the method of Bayes could properly be applied. But, if knowledge of the origin of the mouse tested were lacking, no experimenter would feel he had warrant for arguing as if he knew that of which in fact he was ignorant, and for lack of adequate data Bayes' method of reasoning would be inapplicable to his problem.

It is evidently easier for the practitioner of natural science to recognize the difference between knowing and not knowing than this seems to be for the more abstract mathematician. The traditional line of thought running from Laplace to, for example, Sir Harold Jeffreys in our own time would be to argue that, in the absence of relevant genealogical evidence, there being only two possibilities, mutually exclusive, and with no prior information favouring one rather than the other, it is axiomatic that their probabilities *a priori* are equal, and that Bayes' argument should be applied on this basis. This is to treat the problem, in which we have no genealogical evidence, exactly *as if* the mouse to be tested were known to have been derived from a mating producing half homozygotes and half heterozygotes.

In spite of the high prestige of all that flowed from Laplace's pen, and the great ability and industry of his expositors, it is yet surprising that the doubts which such a process of reasoning from ignorance must engender should begin to find explicit expression only in the second half of the nineteenth century, and then with caution. That extraordinary work,

EARLY ATTEMPTS AND THEIR DIFFICULTIES

The Laws of Thought, by George Boole appeared in 1854.[2] Its twentieth chapter is given to problems of causes, and of this the second half to problems in which (p. 320) we "may be required to determine the probability of a particular cause, or of some particular connection among a system of causes, from observed effects,". A hint of Boole's point of view appears in the opening words of Section 20:

It is remarkable that the solutions of the previous problems are void of any arbitrary element. We should scarcely, from the appearance of the data, have anticipated such a circumstance. It is, however, to be observed, that in all those problems the probabilities of the *causes* involved are supposed to be known *a priori*. In the absence of this assumed element of knowledge, it seems probable that arbitrary constants would *necessarily* appear in the final solution.

The cases chosen by Boole to illustrate this view are two: (*a*) The Reverend J. Michell[3] had calculated that if stars of each magnitude were dispersed at random over the celestial sphere, there would very rarely occur so many apparent double stars or clusters as those actually observed by astronomers. (*b*) The planes of revolution of the planets of the solar system are more nearly coincident than could often occur if these planes had been assigned at random. Instead of treating these calculations, as would now generally be done, as *tests of significance* overthrowing the theory of random dispersal, and therefore all cosmological theories implying random dispersal (disposing of this hypothesis without reference to, or consideration of, any alternative hypothesis which might be actually or conceivably brought forward); instead of this, it had been thought proper to discuss each question as one in inverse probability, and

22 STATISTICAL METHODS AND SCIENTIFIC INFERENCE

Boole has no difficulty in showing that as such it requires two elements really unknown, namely the probability *a priori* of random dispersal, and, secondly, the probability, in the aggregate of alternative hypotheses, of the observed frequency of conjunctions being realized. As he says on p. 367, "Any solutions which profess to accomplish this object, either are erroneous in principle, or involve a tacit assumption respecting the above arbitrary elements."

Again in Section 22:

Are we, however, justified in assigning to [these two unknowns] particular values? I am strongly disposed to think that we are not. The question is of less importance in the special instance than in its ulterior bearings. In the received applications of the theory of probabilities, arbitrary constants do not explicitly appear; but in the above, and in many other instances sanctioned by the highest authorities, some virtual determination of them has been attempted. And this circumstance has given to the results of the theory, especially in reference to questions of causation, a character of definite precision, which, while on the one hand it has seemed to exalt the dominion and extend the province of numbers, even beyond the measure of their ancient claim to rule the world; on the other hand has called forth vigorous protests against their intrusion into realms in which conjecture is the only basis of inference. The very fact of the appearance of arbitrary constants in the solutions of problems like the above, treated by the method of this work, seems to imply, that definite solution is impossible, and to mark the point where inquiry ought to stop.

On page 370:

It has been said, that the principle involved in the above and in similar applications is that of the equal distribution of our knowledge, or rather of our ignorance—the assigning to

EARLY ATTEMPTS AND THEIR DIFFICULTIES 23

different states of things of which we know nothing, and upon the very ground that we know nothing, equal degrees of probability. I apprehend, however, that this is an arbitrary method of procedure.

And finally, on page 375:

These results only illustrate the fact, that when the defect of data is supplied by hypothesis, the solutions will, in general, vary with the nature of the hypotheses assumed; so that the question still remains, only more definite in form, whether the principles of the theory of probabilities serve to guide us in the election of such hypotheses. I have already expressed my conviction that they do not—a conviction strengthened by other reasons than those above stated . . . Still it is with diffidence that I express my dissent on these points from mathematicians generally, and more especially from one who, of English writers, has most fully entered into the spirit and the methods of Laplace; and I venture to hope, that a question, second to none other in the Theory of Probabilities in importance, will receive the careful attention which it deserves.

These quotations, which I have picked out from the rather lengthy mathematical examples which Boole developed, are sufficient to exhibit unmistakably his logical point of view. He does not, indeed, go so far as to say that no statements in terms of mathematical probability can properly be based on data of the kind considered, but he is entirely clear in rejecting the application to these cases of the method of arriving at such statements which, in the absence of appropriate data, introduced values of the probabilities *a priori* supported only by a questionable axiom.

His phrase, moreover, on supplying by hypothesis what is lacking in the data, points to an abuse very congenial to certain twentieth-century writers.

3. John Venn and the Rule of Succession

An immediate inference from Bayes' theorem assigning the frequency distribution

$$\frac{(a+b+1)!}{a!\,b!}\, p^a\,(1-p)^b\,dp \tag{7}$$

to the probability of success of an event, supposed constant, after a successes have been observed in $a+b$ independent trials, is to calculate the probability of success of a new trial, of the same kind as the others and, like them, independent. For this we have only to multiply the frequency element above by p and integrate between the limits 0 and 1. The result takes the simple form

$$(a+1)/(a+b+2), \tag{8}$$

and this inference came to be known as the Rule of Succession; it is often quoted in the form taken when $b=0$, as leading to the probability $(a+1)/(a+2)$.

It should be emphasized, as it has sometimes passed unnoticed, that such a rule can be based on Bayes' theorem only on certain conditions. It requires that (i) the record of a successes out of $a+b$ trials constitutes the whole of the information available; (ii) the successive trials are independent in the sense that the success or failure of one trial has no effect in favouring the success or failure of subsequent tests, which have in each case the same probabilities.

During the long period over which its correctness was unquestioned, the Rule of Succession had been eagerly seized upon by logicians as providing a solid mathematical basis for inductive reasoning. In his *Logic of Chance*,[14] Venn, who was developing the concept of probability as an objective fact, verifiable by observations of frequency, devotes a chapter to demolishing the Rule of Succession, and this, from a writer of his weight and dignity, had an undoubted effect in shaking the confidence of mathematicians in its mathematical foundation.

Venn, however, does not discuss its foundation, and perhaps was not aware that it had a mathematical basis demonstrated by Laplace; that like other mathematical theorems it contained stipulations specific for its validity; and that in particular it rested upon the supposed, though disputable, axiom used for the demonstration of Bayes' proposition. As in other cases in which a work of demolition is undertaken with great confidence, there is no doubt that Venn in this chapter uses arguments of a quality which he would scarcely have employed had he regarded the matter as one open to rational debate.

After giving instances of not very unreasonable inferences drawn by Laplace and De Morgan with the aid of the Rule, Venn writes[15] (p. 180):

> Let us add an example or two more of our own. I have observed it rain three days successively,—I have found on three separate occasions that to give my fowls strychnine has caused their death,—I have given a false alarm of fire on three different occasions and found the people come to help me each time.

These examples seem to be little more than rhetorical sallies intended to overwhelm an opponent with ridicule. They scarcely attempt to conform with the conditions of Bayes' theorem, or of the rule of succession based upon it. In the last case the reader is to presume, the same neighbours having been deceived on three occasions, that on the fourth they will be for this reason less ready to exert themselves; that is to say, the successive trials are not even conceived to be independent. Objection could be made on the same ground to the first example, which is perhaps particularly unrealistic in that three rainy days are postulated to comprise the whole of the subject's experience of days wet or fine. Perhaps the example could be repaired by making him arrive by air in a region of unknown climate; if so, Bayes' postulate implies that the region has been chosen at random from an aggregate of regions in each of which the probability of rain is constant and independent from day to day, while this probability varies from region to region in an equal distribution from 0 to 1. If applied to cases in which this information is lacking the inference is not indeed ridiculous, though I should agree with Venn that it would often be found to be mistaken, if put to the test of repeated trials, and it can scarcely be doubted that Bayes would have taken the same view. A climate without the unnatural feature of independence of weather from day to day, and therefore without conforming to the conditions of Bayes' theorem, might yet justify the Rule of Succession, in the limited form here used, if the proportion of all rainy days falling in spells of n successive rainy days were

$$\frac{4n}{(n+1)(n+2)(n+3)}; \qquad (9)$$

and it is clearly a question of ascertainable fact, and

not of personal predilection, whether the climate of any part of the world conforms to such a rule.*

The standardization of drugs by an experimental assay of their potency often involves the determination of the 50% lethal dose, or, that which, with the population of animals sampled for testing, will have a probability of 50% of killing in each case. The rhetorical force of Venn's example lies in the presumption that much more than the 50% lethal dose of strychnine was employed, but a valid criticism of Bayes' theorem through the failure of the Rule of Succession requires a less cavalier treatment of the example. If, for example, 50 strengths of dose were made up at concentrations capable of killing 1%, 3%, ..., 99% of the animals to be tested, and if each experiment consisted in choosing one of these doses, with equal probability, and applying this dose to each of four hens chosen at random; of ascertaining

* While this simple distribution suffices to justify the Rule of Succession when applied to experience of only wet or of only fine days, the general form of the rule requires that the spells of wet and fine weather must be arranged to fulfil further conditions. Thus the frequency with which two successive spells are of u and v days respectively must be

$$\frac{6(u+1)!\,(v+1)!}{(u+v+3)!}, \qquad (10)$$

consistent with the marginal frequency for u and v

$$\frac{12(u!)}{(u+3)!}. \qquad (11)$$

Three successive spells of lengths u, v and w, in that order, have frequency

$$\frac{6(v+2)!\,(u+w)!}{(u+v+w+3)!}, \qquad (12)$$

while four spells of lengths t, u, v, w will have a frequency

$$\frac{6(t+v+1)!\,(u+w+1)!}{(t+u+v+w+3)!}. \qquad (13)$$

first if each of three of these hens had died, and if so of predicting that the fourth would die also, the proportion of successes would agree closely with the fraction 4/5 given by the rule of succession. If, on the contrary, the doses were spaced in toxic content between the 1% and the 99% dosages, either in arithmetical or in geometrical progression, the forecast, though by no means a ridiculous estimate, would doubtless be somewhat in error. *Knowledge* of the experimental conditions might thus justify the rule, though it cannot rationally be based on ignorance of them.

It seems that in this chapter Venn was to such an extent carried away by his confidence that the rule of induction he was criticizing was indefensible in many of its seeming applications, and by his eagerness to dispose of it finally, that he became uncritical of the quality of the arguments he used. A most serious lapse of a general character appears on page 181 at the beginning of Section 9:

> It is surely a mere evasion of the difficulty to assert, as is sometimes done, that the rule is to be employed in those cases only in which we do not know anything beforehand about the mode and frequency of occurrence of the events. The truth or falsity of the rule cannot be in any way dependent upon the ignorance of the man who uses it. His ignorance affects himself only, and corresponds to no distinction in the things.

Taken in its sweeping generality such an argument seems to imply that the extent of the observational data available can have no bearing on the nature or precision of our inferences from them; that a jury ignorant of certain facts ought to give the same verdict as one to whom they have been presented! The precise specification of our knowledge is, however, the same as the precise specification of our ignorance. Certainly, the observer's knowledge or ignorance may have no effect on external objects, but the extent of the observations to which his reasoning is applied does make a selection of those material systems to which he imagines his conclusions to be applicable; and the objective frequencies observed in such selected systems may depend, and indeed must depend if inductive reasoning have any validity, on the observational basis by which the selection is effected, and on which the reasoning is based.

It is certain that Venn understood this in respect of the inductive process generally, and that nothing but inadvertence can have led him to develop, in criticizing "the rule", a mode of argument fatal equally to all inferences based on experience.

Perhaps the most important result of Venn's criticism was the departure made by Professor G. Chrystal in eliminating from his celebrated textbook of *Algebra*[3] the whole of the traditional material usually presented under the headings of *Inverse Probability* and of the *Theory of Evidence*. Chrystal does not discuss the objections to this material, but expresses the opinion that "many of the criticisms of Mr. Venn on this part of the doctrine of chances are unanswerable. The mildest judgement we could pronounce would be the following words of De Morgan himself, who seems, after all, to have 'doubted': 'My own impression derived from this and many other circumstances connected with the analysis of probabilities, is, that the mathematical results have outrun their interpretation.'" (Chapter xxxvi, p. 604.)

It should be noted that De Morgan's remark has

been quoted clean out of its context; that he was not writing on inverse probability, nor even on the theory of evidence, but about the curiously ubiquitous success of methods based on the Normal law of errors, even when applied to cases in which such a law is not accurately plausible. In fact the passage (from the Fourth Appendix, the title of which is "On the average result of a number of observations") goes on:

and that some simple explanation of the force and meaning of the celebrated integral, whose values are tabulated at the end of this work, will one day be found to connect the higher and lower parts of the subject with a degree of simplicity which will at once render useless (except to the historian) all the works hitherto written.

In reality, the introduction of the inverse method was to De Morgan[11] (p. vi) one of the most important advances to be recorded in the history of the theory of probability. I have already quoted his opinion to this effect, in the introduction to my book on *The Design of Experiments*[6]:

There was also another circumstance which stood in the way of the first investigators, namely, the not having considered, or, at least, not having discovered the method of reasoning from the happening of an event to the probability of one or another cause. The questions treated in the third chapter of this work could not therefore be attempted by them. Given an hypothesis presenting the necessity of one or another out of a certain, and not very large, number of consequences, they could determine the chance that any given one or other of those consequences should arrive; but given an event as having happened, and which might have been the consequence of either of several different causes, or explicable by either of several different hypotheses, they could not infer the probability with which the happening of the event should cause the different hypotheses to be viewed.

EARLY ATTEMPTS AND THEIR DIFFICULTIES 31

But, just as in natural philosophy the selection of an hypothesis by means of observed facts is always preliminary to any attempt at deductive discovery; so in the application of the notion of probability to the actual affairs of life, the process of reasoning from observed events to their most probable antecedents must go before the direct use of any such antecedent, cause, hypothesis, or whatever it may be correctly termed. These two obstacles, therefore, the mathematical difficulty, and the want of an inverse method, prevented the science from extending its views beyond problems of that simple nature which games of chance present.

If he ever checked the reference to his quotation, therefore, Chrystal was scarcely playing fair. His case as well as Venn's illustrates the truth that the best causes tend to attract to their support the worst arguments, which seems to be equally true in the intellectual and in the moral sense.

3.1 Isaac Todhunter

In his *History of the Mathematical Theory of Probability* (1865) Todhunter devotes the fourteenth chapter to Bayes. The brief opening paragraph states, no doubt correctly, what was regarded as common ground at this period (p. 294).

"539. The name of Bayes is associated with one of the most important parts of our subject, namely, the method of estimating the probabilities of the causes by which an observed event may have been produced. As we shall see, Bayes commenced the investigation, and Laplace developed it and enunciated the general principle in the form which it has since retained".

Little fault need be found with the first sentence, save that it is unnecessarily vague. Bayes had proposed a method of *calculating*, with mathematical accuracy, and without uncertainty, the probability that the unknown probability of an observable event should lie between definite limits.

"541. Bayes proposes to establish the following theorem:". Todhunter states the conclusion unconditionally much as in expression (5) p. 12. However, he adds the reservation "Moreover we shall see that there is an important condition implied which we have omitted in the above enunciation, for the sake of brevity: we shall return to this point in Art. 552".

"552. It must be observed with respect to the result of Art. 549, that in Bayes' own problem we *know* that *a priori* any position of EF between AB and CD is equally likely; or at least we know what amount of assumption is involved in this supposition. In the applications which have been made of Bayes' theorem, and of such results as that which we have taken from Laplace in Art. 551, there has however often been no adequate ground for such knowledge or assumption".

Todhunter thus dissociates himself from the Laplacian generalization, and stresses the factual rather than the axiomatic basis proposed for Bayes' knowledge of probabilities *a priori*. It is indeed regrettable that, when he was on the point of having grasped this essential distinction, he should have wavered so far as to add "or at least we know what amount of assumption is involved in this supposition". One is inclined to ask:—Just how does this come to be known, and, if it does, what purpose is served by such knowledge? Bayes' knowledge is based on his imagined experiment with the billiard balls. It is essential for the demonstration of his Theorem. In chapter V Section 6 it will be shown that somewhat different experimental knowledge leads to a somewhat different conclusion.

It will be noticed that Todhunter, like Venn, whose book was published in the following year (1866), picks on the Rule of Succession as particularly open to criticism. However, the Rule of Succession follows as a direct deductive consequence from Bayes' expression,

$$\frac{(a+b+1)!}{a!b!} p^a q^b dp,$$

for the probability that the true value p should lie in the infinitesimal range dp, and is therefore true whenever Bayes' result is justified. It is the lack of experimental knowledge behind many seeming applications of Bayes' theorem, not the derivation from it of the Rule of Succession, that invalidates such arguments.

Near as he came to clarifying the situation, Todhunter's name cannot properly be added to those who finally succeeded in extricating the mathematical thought of the mid-nineteenth century from its bewildering difficulties. Though he knew of and expressed admiration for Boole's book of 11 years earlier, by slurring over the essential distinction which Boole had striven to make clear, his influence in effect was to confirm the textbook expositions current in his own time.

4. The meaning of probability

Whatever view may be preferred on the controversial issues which the quotations set out above have been selected to illustrate, it is evident beyond question that highly competent, or even illustrious, mathematicians had formed upon them quite irreconcilable opinions; and this appearance of inability to find a common ground is not lessened by a perusal of what has been written in our own century.

Since there is no reason to doubt the purely mathematical ability of these writers, it is natural to suspect a semantic difficulty due to an imperfect analysis of words regarded as being too simple to be elucidated by further examination, such as the word "probability" itself. Of course, each writer has "defined" this word to his own satisfaction. Mathematical definition is, however, often no more than a succinct statement of the axioms to be applied when the word occurs in deductive mathematical reasoning, and may pay less attention than is needed to the conditions of the correct applicability of the term in the real world. It is these conditions of applicability which are properly the concern of those responsible for Applied Mathematics.

Indeed, I believe that a rather simple semantic confusion may be indicated as relevant to the issues discussed, as soon as consideration is given to the meaning that the word probability must have to anyone so much practically interested as is a gambler, who, for example, stands to gain or lose money, in the event of an ace being thrown with a single die. To such a man the information supplied by a familiar mathematical statement such as: "If a aces are thrown in n trials, the probability that the difference in absolute value between a/n and $1/6$ shall exceed any positive value ϵ, however small, shall tend to zero as the number n is increased indefinitely", will seem not merely remote, but also incomplete and lacking in definiteness in its application to the particular throw in which he is interested. Indeed, by itself it says nothing about that throw. It is obvious, moreover, that many subsets of future throws, which may include his own, can be shown to give probabilities, in this sense, either greater or less than 1/6. Before the limiting ratio of the whole set can be accepted as applicable to a particular throw, a second condition must be satisfied, namely that before the die is cast no such subset can be *recognized*. This is a necessary and sufficient condition for the applicability of the limiting ratio of the entire aggregate of possible future throws as the probability of any one particular throw. On this condition we may think of a particular throw, or of a succession of throws, as a *random* sample from the aggregate, which is in this sense subjectively homogeneous and without recognizable stratification.

Makers of the standard apparatus of games of chance, dice, cards, roulettes, etc., take great care to satisfy both the requirements of a sufficiently specific statement of what is meant by probability. If either the long-run frequencies were faulty, or, in particular, if there were any means of foreseeing, even to a limited extent, the outcome of their use in a particular case, the apparatus, or, perhaps, the method of using them, would be judged defective for the purpose for which they were made.

This fundamental requirement for the applicability to individual cases of the concept of classical probability shows clearly the role both of well specified ignorance and of specific knowledge in a typical probability statement. It has been often recognized that any probability statement, being a rigorous statement involving uncertainty, has less factual content than an assertion of certain fact would have, and at the same time has more factual content than a statement of complete ignorance. The *knowledge* required for such a statement refers to a well-defined aggregate, or population of possibilities within which the limiting frequency ratio must be exactly known. The necessary *ignorance* is specified by our inability to discriminate any of the different sub-aggregates having different frequency ratios, such as must always exist. Laplace's definition of probability, in which he actually speaks of "évènemens", is so worded that this necessary stipulation of ignorance in respect of particular events can be transferred to *hypotheses*, so as to imply in Boole's words "the assigning to different states of things of which we know nothing, and upon the very ground that we know nothing, equal degrees of probability". Has Laplace not in fact passed unawares from proposition (*a*) below to proposition (*b*)?

(*a*) A possible outcome must be assigned equal probabilities in different future throws, because we can draw no relevant distinction between these in advance.

(*b*) Hypotheses must be judged equally probable *a priori* if no relevant distinction can be drawn between them.

How extremely conservative was the tradition of mathematical teaching is shown by the slowness with which the opinions of Boole, Venn and Chrystal were appreciated. The reluctance naturally felt to abandoning a false start was certainly enhanced by the fact that, so far as the problem of scientific induction was concerned, nothing had been put forward to replace that which had been taken away. The gap seems to have been felt only subconsciously. In many cases it must have been clear that it was possible for data of great value to the formation of

EARLY ATTEMPTS AND THEIR DIFFICULTIES 37

our scientific ideas to be presented, and yet for there to be no defensible basis, in the light of the criticisms which had been made, for the application of Bayes' theorem. Many mathematicians must have felt that with a proper restatement, the theorem, or one fulfilling the same purpose in inductive reasoning, could be set on its feet again. Indeed, the two leading statisticians in England at the beginning of the twentieth century, K. Pearson (1920)[12] and F. Y. Edgeworth (1908)[4] (p. 387) both put forward attempts, discordant indeed and both abortive, to justify the mode of reasoning in which no doubt each had been brought up, but which had since been discredited.

The reader of this preliminary chapter will have seized my meaning if he perceives that the different situations in which uncertain inferences may be attempted admit of logical distinctions which should guide our procedure. That it may be that the data are such as to allow us to apply Bayes' theorem, leading to statements of probability; or secondly, that we may be able validly to apply a test of significance to discredit a hypothesis the expectations from which are widely at variance with ascertained fact. If we use the term rejection for our attitude to such a hypothesis, it should be clearly understood that no irreversible decision has been taken; that, as rational beings, we are prepared to be convinced by future evidence that appearances were deceptive, and that in fact a very remarkable and exceptional coincidence had taken place. Such a test of significance does not authorize us to make any statement about the hypothesis in question in terms of mathematical probability, while, none the less, it does afford direct guidance as to what elements we may reasonably incorporate in any theories we may be attempting to form in explanation of objectively observable phenomena. Thirdly, the logical situation we are confronted with may admit of the consideration of a series, or, more usually, of a continuum of hypotheses, one of which *must* be true, and among which a selection may be made, and that selection justified, so far as may be, by statistical reasoning. The stasis or deadlock which had set in by the end of the century has been, I shall hope to show, in fact, released by the consideration of these diverse possibilities.

REFERENCES

1. T. Bayes (1763). An essay towards solving a problem in the doctrine of chances.
 Phil. Trans. Roy. Soc., vol. 53, p. 370.
2. G. Boole (1854). *The Laws of Thought.*
 Macmillan and Co., London. Reprinted by Dover Publications, Inc., New York, 1951.
3. G. Chrystal (1886). *Algebra.*
 Adam and Charles Black, London.
4. F. Y. Edgeworth (1908). On the probable errors of frequency-constants.
 J. Roy. Stat. Soc., vol. 71, pp. 381-397.
5. R. A. Fisher (1935-1966). *The Design of Experiments.*
 Oliver and Boyd, Edinburgh.
6. R. A. Fisher (1958). Mathematical probability in the natural sciences.
 18th Int. Congr. of Pharm. Sciences. Brussels, Sept. 1958.

 R. A. Fisher (1962) Some examples of Bayes' method of the experimental determination of probabilities *a priori*. *J. Royal Stat. Soc., B,* vol. 24, pp. 118–124.
7. P.-S. marquis de Laplace (1812, 1820). *Théorie analytique des probabilités.*
 Paris. 1st ed., p. 178. 3rd ed. (1820), preface, p. iv.

8. J. Michell (1767). An inquiry into the probable parallax, and magnitude of the fixed stars, from the quantity of light which they afford us, and the particular circumstances of their situation.
Phil. Trans., vol. 57, pp. 234-264.

9. A. de Moivre (1718, 1738, 1756). *Doctrine of Chances.*

10. P. De Montmort (1708, 1714). Essai d'analyse sur les jeux de hazards.

11. A. De Morgan (1838). *An Essay on Probabilities and on their Application to Life Contingencies and Insurance Offices.*
Longman and Co., London.

12. K. Pearson (1920). The fundamental problem of practical statistics.
Biometrika, vol. 13, pp. 1-16.

13. *The Dictionary of National Biography.* Oxford University Press.

14. I. Todhunter (1865). *A History of the Mathematical Theory of Probability.*
Macmillan and Co., London.

15. J. Venn (1866, 1876, 1888). *The Logic of Chance.*
Macmillan and Co., London.
Reprinted by Chelsea Publishing Co., New York, 1962.

CHAPTER III

FORMS OF QUANTITATIVE INFERENCE

1. The simple test of significance

While, as Bayes perceived, the concept of Mathematical Probability affords a means, in some cases, of expressing inferences from observational data, involving a degree of uncertainty, and of expressing them rigorously, in that the nature and extent of the uncertainty is specified with exactitude, yet it is by no means axiomatic that the appropriate inferences, though in all cases involving uncertainty, should always be rigorously expressible in terms of this same concept. Although this belief seems to have been unquestioned over the period of 150 years covered by the discussion of Chapter II, familiarity with the actual use made of statistical methods in the experimental sciences shows that in the vast majority of cases the work is completed without any statement of mathematical probability being made about the hypothesis or hypotheses under consideration. The simple rejection of a hypothesis, at an assigned level of significance, is of this kind, and is often all that is needed, and all that is proper, for the consideration of a hypothesis in relation to the body of experimental data available. It is therefore desirable to examine the logical nature of this sort of uncertain inferences.

The example chosen by Boole of Michell's calculation with respect to the Pleiades will serve as an illustration. He demonstrates that Bayes' method can be applied to this case only by assuming arbitrary

FORMS OF QUANTITATIVE INFERENCE 41

values not provided by the data, and therefore that no probability *a posteriori* can be assigned to the hypothesis that the stars, down to the sixth magnitude, are distributed at random over the celestial sphere. He does not emphasize that nevertheless Michell had by his calculations presented a strong reason for rejecting this hypothesis, or attempt to exhibit just how such a rational inference should be correctly stated.

Michell supposed that there were in all 1500 stars of the required magnitude and sought to calculate the probability, on the hypothesis that they are individually distributed at random, that any one of them should have five neighbours within a distance of a minutes of arc from it. I find the details of Michell's calculation obscure, and suggest the following argument.

The fraction of the celestial sphere within a circle of radius a minutes is, to a satisfactory approximation,

$$p = \left(\frac{a}{6875 \cdot 5}\right)^2 \qquad (14)$$

in which the denominator of the fraction within brackets is the number of minutes in two radians. So, if a is 49, the number of minutes from Maia to its fifth nearest neighbour, Atlas, we have

$$p = \frac{1}{(140 \cdot 316)^2} = \frac{1}{19689}. \qquad (15)$$

Out of 1499 stars other than Maia of the requisite magnitude the expected number within this distance is therefore

$$m = \frac{1499}{19689} = \frac{1}{13 \cdot 1345} = \cdot 07613. \qquad (16)$$

The frequency with which 5 stars should fall

within the prescribed area is then given approximately by the term of the Poisson series

$$e^{-m}\frac{m^5}{5!},\qquad(17)$$

or, about 1 in 50,000,000, the probabilities of having 6 or more close neighbours adding very little to this frequency. Since 1500 stars have each this probability of being the centre of such a close cluster of 6, although these probabilities are not strictly independent, the probability that among them any one fulfils the condition cannot be far from and certainly cannot exceed 30 in a million, or 1 in 33,000. Michell arrived at a chance of only 1 in 500,000 but the higher probability obtained by the calculations indicated above is amply low enough to exclude at a high level of significance any theory involving a random distribution.

The force with which such a conclusion is supported is logically that of the simple disjunction: *Either* an exceptionally rare chance has occurred, *or* the theory of random distribution is not true.

In view of the efforts which have been made to force a frequency interpretation on to such a disjunction, it is to be noted that the mental reluctance to accept an event intrinsically improbable would still be felt if, for example, a *datum* were added to Michell's problem to the effect that it was a million to one *a priori* that the stars should be scattered at random. We need not consider what such a statement of probability *a priori* could possibly mean in the astronomical problem; all that is needed is that if this datum were introduced into the calculation, then, in view of the observations, a probability statement could be inferred *a posteriori*, to the effect that the odds were 30 to 1 that the stars really had been scattered at random. The inherent improbability of what has been observed being observable on this view still remains in our minds, and no explanation has been given of it. It has been overweighted, not neutralized, by the even greater supposed improbability of the universe chosen for examination being of the supposedly exceptional kind in which the stars are *not* distributed at random. The observer is thus not left at all in the same state of mind as if the stars had actually displayed no evidence against a random arrangement, although he would have been forced logically to admit that (so far as statements in terms of probability went) such a theory was probably true, and that the remarkable features that had attracted his attention were, incredible as it might seem, wholly fortuitous.

The example shows that the resistance felt by the normal mind to accepting a story intrinsically too improbable is not capable of finding expression in any calculation of probability *a posteriori*. The variety of ways in which this resistance does express itself very well exhibits its reality. Common reactions are:

(*a*) The whole thing is a fabrication.

(*b*) There is no sufficient reason to think that the facts were observed and put on record accurately.

(*c*) There has been exaggeration, and the omission of circumstances that would help to explain what is claimed.

(*d*) Some occult cause, beyond our present understanding, must be invoked.

In the studies known as parapsychology enormous

odds are often claimed, evidently with a view to raising the resistance felt to accepting what is intrinsically improbable to such a pitch that conclusion (*d*), although itself repugnant, shall be accepted in preference. The incredulous, however, tend to prefer explanations of types (*a*), (*b*) or (*c*) either to accepting such a claim as, let us say, "precognition", or, what seems almost always to be the last choice, to the acceptance as genuine of a very rare contingency.

The fact, important for the understanding of logical situations of this kind, that reluctance to accept a hypothesis strongly contradicted by a test of significance is not removed, though it may be outweighed, by information *a priori*, is exhibited also by the consideration that if the proposed datum, "The odds are a million to one *a priori* that the stars should really be distributed singly and at random"— if this datum were considered as a *hypothesis*, it would be rejected at once by the observations at a level of significance almost as great as the hypothesis, "The stars are really distributed at random", was rejected in the first instance. Were such a conflict of evidence, as has here been imagined under discussion, not in a mathematical department, but in a scientific laboratory, it would, I suggest, be some prior assumption, corresponding to an axiom or a datum in a mathematical argument, that would certainly be impugned.

The attempts that have been made to explain the cogency of tests of significance in scientific research, by reference to supposed frequencies of possible statements, based on them, being right or wrong, thus seem to miss the essential nature of such tests. A man who "rejects" a hypothesis provisionally, as a matter of habitual practice, when the significance is at the 1% level or higher, will certainly be mistaken in not more than 1% of such decisions. For when the hypothesis is correct he will be mistaken in just 1% of these cases, and when it is incorrect he will never be mistaken in rejection. This inequality statement can therefore be made. However, the calculation is absurdly academic, for in fact no scientific worker has a fixed level of significance at which from year to year, and in all circumstances, he rejects hypotheses; he rather gives his mind to each particular case in the light of his evidence and his ideas. It should not be forgotten that the cases chosen for applying a test are manifestly a highly selected set, and that the conditions of selection cannot be specified even for a single worker; nor that in the argument used it would clearly be illegitimate for one to choose the actual level of significance indicated by a particular trial as though it were his lifelong habit to use just this level. Further, the calculation is based solely on a hypothesis, which, in the light of the evidence, is often not believed to be true at all, so that the actual probability of erroneous decision, supposing such a phrase to have any meaning, may be, for this reason only, much less than the frequency specifying the level of significance. A test of significance contains no criterion for "accepting" a hypothesis. According to circumstances it may or may not influence its acceptability.

On the whole the ideas (*a*) that a test of significance must be regarded as one of a series of similar tests applied to a succession of similar bodies of data, and (*b*) that the purpose of the test is to discriminate or "decide" between two or more hypotheses, have greatly obscured their understanding, when taken not as contingent possibilities but as elements essential to their logic. The appreciation of such more complex cases will be much aided by a clear view of the nature of a test of significance applied to a single hypothesis by a unique body of observations.

Though recognizable as a psychological condition of reluctance, or resistance to the acceptance of a proposition, the feeling induced by a test of significance has an objective basis in that the probability statement on which it is based is a fact communicable to, and verifiable by, other rational minds. The level of significance in such cases fulfils the conditions of a measure of the rational grounds for the disbelief it engenders. It is more primitive, or elemental than, and does not justify, any exact probability statement about the proposition.

When a prediction is made, having a known low degree of probability, such as that a particular throw with four dice shall show four sixes, an event known to have a mathematical probability, in the strict sense, of 1 in 1296, the same reluctance will be felt towards accepting this assertion, and for just the same reason, indeed, that a similar reluctance is shown to accepting a hypothesis rejected at this level of significance. There is the logical disjunction: Either an intrinsically improbable event will occur, or, the prediction will not be verified. The psychological resistance has been, I think wrongly, ascribed to the fact that the event in question *has*, in the proper sense of the Theory of Probability, the low probability assigned to it, rather than to the fact, very near in this case, that the correctness of the assertion would *entail* an event of this low probability. The probability statement is a sufficient, but not a necessary, condition for disbelief in this degree. Disbelief is equally justified when the probability is hypothetical. The difficulty of traditional forms of expression, in this as in other cases, flows from the assumption, too widely disseminated, that Mathematical Probability, being the first well-defined concept, and for a long while the only one available, for the expression of statements of uncertainty, must necessarily be by itself competent for the adequate specification of uncertainty, that is of the grounds for belief or disbelief, in all logical situations. The logical consequences of a statement of Mathematical Probability are clear and well-known. They allow of the calculation of long-run frequency-ratios, and therefore to the habitual gambler, of long-run policies in laying bets, in the hypothetical and restricted field of games of chance, played fairly with perfect apparatus; but these logical consequences may, as in the examples under discussion, be in themselves of little importance to the bearing of observable facts on the acceptibility of possible hypotheses.

In general, tests of significance are based on *hypothetical* probabilities calculated from their null hypotheses. They do not generally lead to any probability statements about the real world, but to a rational and well-defined measure of reluctance to the acceptance of the hypotheses they test.

Closely related to the assumption that all expressions of uncertain knowledge must have the same logical form, namely that of a statement of probability, is the assumption that all kinds of evidence used as data for such inferences have the same kind of logical consequences. Even a writer so shrewd as J. M. Keynes (1921) has exposed himself to this criticism. In an illuminating passage he objected to the traditional way of speaking, which he ascribed to "Laplace and his followers", of a probability as "unknown". "Do we mean unknown through lack of skill in arguing from given evidence, or unknown through lack of evidence? The first is alone admissible, for new evidence would give us a new probability, not a fuller knowledge of the old one"[8] (p. 31).

In this he must not, as might at first be thought, be taken to assert that in all cases a probability statement can be made, for he says later on the same page "We ought to say that in the case of some arguments a relation of probability does not exist, and not that it is unknown". This admission greatly reduces the force of his objection, for if, at one stage, a probability does not exist, but with further observations it becomes possible to assert a definite probability, it is not a misuse of words to say that it was at first unknown, and has later been ascertained. However, for the cases of which he was thinking, I have the same preference as Keynes, and welcome his statement that in some cases no probability exists. In other cases which go back historically so far as Bayes, we undoubtedly have, or can logically conceive of having, partial knowledge of a probability, so that probability statements can be made about its value, and it would be stretching language unprofitably to say that in such cases the probability only partially exists.

Keynes' difficulty arises, I think, partly from a desire to use the word probability primarily to refer to the truth of propositions, and only to the occurrence of events in the sense that the probability of an event can be identified with the probability of the truth of the proposition that the event shall occur. In view of this identification there is no occasion to be fastidious, and certainly events, such as the disintegration within a determinate time of an atom of a radio-active element may be reasonably thought of as having an objective probability, independent of the state of our evidence, and this may be unknown, or known with a limited and determinate precision. The kind of evidence of which Keynes is thinking is the kind which, as it is increased indefinitely, would lead the probability inferred to tend to 0 or 1, that is to a statement without uncertainty. He does not notice that other kinds of evidence may lead to the estimation of an objective probability with increasingly high precision, so as to tend in the limit to an exact knowledge of its value; or, that those other kinds of evidence which can lead to no statement of probability whatever, may be of direct inferential value.

2. More general hypotheses

Scientific hypotheses usually differ from the simple hypothesis (the random distribution of the stars) considered by Michell, in that they allow of one or more parameters, or adjustable "constants", any value of which, or, any value within assigned limits, is consistent with the hypothesis. With respect to such hypotheses tests of significance may be applied in two ways. In the first place, a test of significance may be developed capable of rejecting the hypothesis as a whole, if any relevant feature of the observational record can be shown to depend on a contingency which is sufficiently rare whatever may be the value of the parameter. Secondly, if no such feature obtrudes itself, or, if any such as can be found is far-fetched and artificial, so that the general hypothesis is judged provisionally acceptable, the question arises of the Estimation of the parameter's value. In some such cases, but not all, as will be seen, inferences can be drawn assigning a calculable mathematical probability to any assertion to the effect that the parameter lies between assigned limits.

In choosing the grounds upon which a general hypothesis should be rejected, personal judgement may and should properly be exercised. The experimenter will rightly consider all points on which, in the light of current knowledge, the hypothesis may be imperfectly accurate, and will select tests, so far as possible, sensitive to these possible faults, rather than to others. Sometimes also he may make a comprehensive test, such as Pearson's test of Goodness of Fit, applied to observed frequencies, which though strictly speaking approximate, and inapplicable to small frequencies, is, albeit not specifically sensitive to particular faults, yet, when the frequencies are not too small, capable of detecting sufficiently pronounced discrepancies of any kind.

In experimental breeding the data available for theoretical discussion are the frequencies observable when a progeny, or group of progenies, is classified into a number of classes (phenotypes) the members of which are distinguishable by inspection, or by other tests. Thus in an intercross between organisms each heterozygous in respect of two genetic factors showing complete dominance, the expectations out of n observed will be

$$\frac{n}{16} (9, 3, 3, 1), \tag{18}$$

FORMS OF QUANTITATIVE INFERENCE 51

if the factors are unlinked, but in general

$$\frac{n}{4}(2+\theta,\ 1-\theta,\ 1-\theta,\ \theta)\,, \tag{19}$$

where θ is a parameter, depending on the closeness of linkage, any value of which between 0 and 1 would be intelligible.

In the general χ^2 method, if a_1, a_2, a_3, a_4 are frequencies observed corresponding with frequencies m_1 to m_4 expected, the measure of discrepancy is

$$\chi^2 = \underset{1}{S^4}\left\{\frac{(a-m)^2}{m}\right\} = \underset{1}{S^4}\left(\frac{a^2}{m}\right) - n\,, \tag{20}$$

and the probability with which the value of χ^2 observed will be exceeded by chance for three degrees of freedom can be found from the well-known tables. When the expectation is calculated for factors inherited independently, the χ^2 value is divisible into three positive and independent parts.

$$\left.\begin{array}{l}\dfrac{1}{3n}(a_1+a_2-3a_3-3a_4)^2\,,\\[6pt]\dfrac{1}{3n}(a_1-3a_2+a_3-3a_4)^2\,,\\[6pt]\text{and}\quad\dfrac{1}{9n}(a_1-3a_2-3a_3+9a_4)^2\,,\end{array}\right\} \tag{21}$$

each of which is distributed, on the hypothesis tested, as χ^2 for one degree of freedom, that is, as the square of a normal deviate having unit variance. The last of these three parts is specifically sensitive to a disturbance of the expectations of the kind due to linkage; the first two are sensitive to disturbances of the single-factor ratios, and if the only disturbance really present is a small effect of linkage, the test using three degrees of freedom including them also, will be less sensitive to the linkage effect than that based on only one.

To test the general hypothesis the usual procedure for large samples is to estimate the value of θ, by an efficient estimate, to find the corresponding expectations, and to ascertain the level of significance of the χ^2 value obtained, using the Table at two degrees of freedom. A low probability indicates that the general hypothesis is to be rejected at the level of significance found; in other words, there is no value of θ for which the hypothesis is acceptable. We do not simply use the sum of the components of χ^2 for the two single-factor ratios in this test, for if linkage were present these two components would no longer be independent, and to combine the evidence of the two requires an estimate of the parameter. This problem has been discussed in detail in *Statistical Methods*, section 55.[3]

Although in testing the acceptability of a hypothesis in general, no reference to the theory of estimation is required, at least in this important class of cases, beyond the stipulation that expectations can be fitted efficiently to the data, yet when the general hypothesis is found to be acceptable, and accepting it as true, we proceed to the next step of discussing the bearing of the observational record upon the problem of discriminating among the various possible values of the parameter, we are discussing the theory of estimation itself. In this theory a case of peculiar simplicity arises when an estimate exists which, perhaps in conjunction with ancillary statistics, subsumes the whole of the information, relevant to the parameter, supplied by the observational record. Such estimates are termed exhaustive, and their

FORMS OF QUANTITATIVE INFERENCE 53

special property may be expressed in various ways: (i) that, given the exhaustive statistic, every other estimate possible has a sampling distribution completely independent of the parameter to be estimated, and (ii) that the Likelihood Function of the parameter inferred from the sampling distribution of the exhaustive estimate and its ancillaries is exactly the same as that inferred from the original observations. In fact for all purposes of inference an exhaustive statistic, in association perhaps with certain ancillary values, which themselves have distributions independent of the parametric value, can replace the entire observational record from which it was calculated.

Exhaustive estimates do not always exist; it is therefore important to note that their existence is the first requirement of the mode of inductive reasoning to be developed below as the Fiducial Argument. A second condition to be noted is that the observations should not be discontinuous, as are frequencies, but should be measurements sufficiently accurate to be regarded without significant error as observed values of continuous variates, so that the statistics calculable from them shall have continuous distributions. In making these distinctions, I do not wish to deny either that measurements, however accurate, are in the strictly mathematical sense discontinuous, or that counts may be of numbers so large that they could without sensible error be treated as measurements of continuous variables. It is only necessary to point out that cases commonly occur to which this distinction is relevant. In the same way, it may be said that it can always be imagined that statistical samples are made so large that the distinction between exhaustive and other

54 STATISTICAL METHODS AND SCIENTIFIC INFERENCE

efficient estimates shall become unimportant. All that is needed is to recognize that samples are not always so large as this, and that in such cases the logical consequences which flow from this distinction are not irrelevant.

3. The fiducial argument *

The term fiducial has been introduced to distinguish a particular form of inductive reasoning from that of Bayes, which for contrast may be termed the Bayesian argument. The distinction was needed because, like the method of Bayes, it leads to probability statements applicable in the light of the observations to an unknown parameter. Whereas, however, the argument of Bayes requires a distribution *a priori* involving probability statements of the same logical form as those finally obtained *a posteriori*, the application of the fiducial argument can only be made in the absence of such information *a priori*. In the Bayesian argument the observations are used to convert a random variable having a well-defined distribution *a priori* to a random variable having an equally well-defined distribution *a posteriori*; and it is well known that, if the observations are increased in number their importance grows relatively to that of the information supplied *a priori*, so that the latter becomes less and less influential upon the conclusions. By contrast, the fiducial argument uses the observations only to change the logical status of the parameter from one in which nothing is known

* Probability statements derived by arguments of the fiducial type have often been called statements of "fiducial probability". This usage is a convenient one, so long as it is recognized that the concept of probability involved is entirely identical with the classical probability of the early writers, such as Bayes. It is only the mode of derivation which was unknown to them.

of it, and no probability statement about it can be made, to the status of a random variable having a well-defined distribution.

If direct and exact observations could be made on the parameter itself, a similar change of logical status would be effected by the observation of its value, from one in which it was wholly unknown, or had perhaps a known frequency distribution, to one in which it could be assigned a definite value. It is, therefore, perhaps not surprising that similar exact observations, though not on the parameter itself yet on variates having distributions known in terms of the parameter, should be able in favourable cases to effect, at a lower level, a similar change of status.

As an example of the mode of reasoning, consider a radio-active source emitting particles with unknown frequency at instants completely independent of each other. The interval of time between two successive emissions will then be distributed at random in the exponential distribution

$$df = \theta e^{-\theta x}\, dx\,, \qquad (22)$$

in which θ is the unknown average number of emissions in each time unit. We may conceive such time intervals to be accurately measurable, and that a record of n of them has shown intervals

$$x_1,\ x_2,\ x_3,\ \ldots,\ x_n\,.$$

We suppose that these conform sufficiently well to expectations based on the estimate

$$\left. \begin{array}{c} \theta = T\,, \\ \text{where} \qquad T = \dfrac{n}{X}\,, \end{array} \right\} \quad (23)$$

and X stands for the sum of the times observed, for it to be agreed that the general hypothesis is acceptable, and that what remains is only to make such statements about the value of θ as the data allow.

From the original data, the n observed time intervals being independent, the Mathematical Likelihood of any value θ which the parameter may take is seen to be proportional to

$$\theta^n e^{-\theta X},$$

which is maximized at the value

$$\hat{\theta} = \frac{n}{X};$$

so that the estimate T chosen above is an estimate of maximum likelihood. It is also a Sufficient Estimate, that is to say an exhaustive estimate without ancillary statistics, for the sampling distribution of X is seen easily to be

$$df = \theta^n e^{-\theta X} \cdot \frac{X^{n-1}}{(n-1)!} dX, \qquad (24)$$

giving exactly the same likelihood function for θ as was given by the original data. The distribution of X also is continuous over all positive values, uniformly for all values of θ.

This distribution of X for any given θ is, in fact, equivalent to the distribution of the quantity χ^2 for $2n$ degrees of freedom, if χ^2 is equated to

$$\chi^2 = 2\theta X = 2n\frac{\theta}{T}. \qquad (25)$$

In this case the χ^2 distribution is exact, and not only approximate as is Pearson's measure of discrepancy for frequencies; consequently if we choose any probability P, and write

$$\chi^2_{2n}(P)$$

for that value which is exceeded for $2n$ degrees of freedom with frequency P, a value which is calculable with exactitude for all n and P, it appears that the statement

$$\theta > \frac{T}{2n} \chi^2_{2n}(P) \qquad (26)$$

is verified with the frequency P, for all values of P chosen, and therefore that we have derived formally a frequency distribution of the unknown parameter θ appropriate to the observations available.

The applicability of the probability distribution to the particular unknown value of θ sought by an experimenter, without knowledge *a priori*, on the basis of the particular value of T given by his experiment, has been disputed, and certainly deserves to be examined, especially as in the first case in which I exhibited this form of argument, namely to the correlation coefficient (1930),[4] though the example was appropriate, my explanation left a good deal to be desired.

The reasoning developed so far has been entirely deductive; the example was chosen, however, to bring out some necessary characteristics of inductive reasoning. The probability statement first developed above (24) had as reference set all the values of X, and therefore of T, which might have occurred in unselected samples for a particular value of θ. It has, however, been proved for all values of θ, and so is applicable to the enlarged reference set of all pairs of values (T, θ) obtained from all values of θ. The particular pair of values of θ and T appropriate to a particular experimenter certainly belongs to this enlarged set, and within this set the proportion of cases satisfying the inequality (26)

$$\theta > \frac{T}{2n} \chi^2_{2n}(P)$$

is certainly equal to the chosen probability P. It might, however, have been true, as in the case of a gambler throwing a single die, discussed on page 32, that in some recognizable sub-set, to which his case belongs, the proportion of cases in which the inequality was satisfied should have some value other than P. It is the stipulated absence of knowledge *a priori* of the distribution of θ, together with the exhaustive character of the statistic T, that makes the recognition of any such subset impossible, and so guarantees that in his particular case, as in the case of a single particular throw contemplated by the gambler, the general probability is applicable.

Had knowledge *a priori* been available, the argument developed above would have been precluded by the consideration that some of the relevant data had been omitted. For, although in the deduction of statements of certainty it is legitimate to draw inferences from some of the axioms available while ignoring others, or, in other words to base a valid argument on a chosen subset only of the available axioms, no such liberty can be taken with statements of uncertainty, where it is essential to take the whole of the data into account, though some part of it may be shown on examination to be irrelevant, and not to affect the result.

Again, had there been knowledge *a priori*, the argument of Bayes could have been developed, which would have utilized all the data, and which would in general have led to a distribution *a posteriori* different from that to which the fiducial argument leads. Bayes' method in fact calculates the distribution of θ in a particular subset of pairs of values (T, θ), defined by T, and to which therefore the observation belongs. Consequently, it is essential to introduce the absence of knowledge *a priori* as a distinctive datum in order to demonstrate completely the applicability of the fiducial method of reasoning to the particular real and experimental cases for which it was developed. This last point I failed to perceive when, in 1930, I first put forward the fiducial argument for calculating probabilities. For a time this led me to think that there was a difference in logical content between probability statements derived by different methods of reasoning. There are in reality no grounds for any such distinction.

Various writers, including Sir Harold Jeffreys[7] and A. Kolmogorov,[9] recognizing the rational cogency of the fiducial form of argument, and the difficulty of rendering it coherent with the customary forms of statement used in mathematical probability, have proposed the introduction of new axioms to bridge what was felt to be a gap. The treatment in this book involves no new axiom; it does, however, rely on a property inherent in the semantics of the word "probability", though not required explicitly so long as the applicability in the real world of the logical relationship denoted is not in question. Purely abstract studies of the formal mathematics of probability can, in fact, be developed without reference to this aspect of the word's meaning. It is not, of course, unusual that mathematical definition should have, as it often does, axiomatic implications. The distinction should be made, in this case, that the completion of the word's definition specifies the nature and extent, not of the knowledge but explicitly of the ignorance, required in the logical situation envisaged, and that so long as it is assumed, as in

60 STATISTICAL METHODS AND SCIENTIFIC INFERENCE

purely deductive reasoning is proper, that valid deductions can be drawn from every subset of the axiomatic material available, it can be argued, as by Venn, that "his ignorance affects himself only, and corresponds to no distinction in the things". Mathematical probability, however, as conceived by the early writers, was applicable to the real world, and to make it available not only in deductive, but also in inductive reasoning a more complete definition is required. The subject of a statement of probability must not only belong to a measurable set, of which a known fraction fulfils a certain condition, but every subset to which it belongs, and which is characterized by a different fraction, must be unrecognizable.

4. Accurate statements of precision

The possibility of making exact statements of probability about unknown constants of Nature supplies a need long felt of making a complete specification of the precision with which such constants are estimated. For example, if, in the case considered in the foregoing section, there had been 500 accurately measured time intervals, calculations based on the distribution of χ^2 for 1000 degrees of freedom would show that the probability was 25% of the true value lying below ·96957 of the estimate, and 25% of lying above 1·02988, times the same quantity. These values then bracket a central region of about 6% within which the true value will lie with a probability of just one half. They therefore fulfil the same function as the traditional "probable error" often given in respect of astronomical observations. The concept of the probable error indicates the desire felt for such probability statements, although the great complexity of the observational material to be reduced in many astronomical calculations has stood in the way of the refinement of the concept, and has indeed often introduced great difficulties in the way of obtaining a reliable figure. When, as in the example chosen, the data are simple and the meaning of the calculations completely clear, other relevant probability statements may be made with equal confidence and exactitude. For example, the probability is 5% each way of the true value lying outside the limiting ratios ·92731 and 1·07439, and it is only 1% of it lying below ·89819 and another 1% of lying above 1·10622, so that the odds are 49 to 1 that it should lie within these last limits. The fiducial distribution in this way comprises a complete set of probability statements appropriate to any chosen level of probability, or to any chosen limits. In such cases the precision of the estimate has been completely specified.

Had there been a smaller number of time-measurements, such as 15, the precision of the estimate would have been lower, but the fiducial limits and corresponding statements of probability would still have been exact. Compendiously the corresponding results may be stated as follows:

Excluding at each end	Lower limit	Upper limit
1%	·4984	1·6964
5%	·6164	1·4591
25%	·8159	1·1600

There is here a probability of over 1% that the true value should be less than half the estimate.

The objection has been raised that since any statement of probability to be objective must be verifiable as a prediction of frequency, the calculations set out above cannot lead to a true probability statement referring to a particular value of T, for the data do not provide the means of calculating this. This seems to assume that no valid probability statement can be made except by the use of Bayes' theorem. However, the aggregate of cases of which the particular experimental case is one, for which the relative frequency of satisfying the inequality statement is known to be P, and to which all values of T are admissible, could certainly be sampled indefinitely to demonstrate the correct frequency. In the absence of a prior distribution of population values there is no meaning to be attached to the demand for calculating the results of random sampling among *populations*, and it is just this absence which completes the demonstration that samples giving a particular value T, arising from a particular but unknown value of θ, do not constitute a distinguishable sub-aggregate to which a different probability should be assigned. Probabilities obtained by a fiducial argument are objectively verifiable in exactly the same sense and in exactly the same way as are the probabilities assigned in games of chance.

It has, as was shown in the previous chapter, been a hope or ambition among many mathematicians of the last two hundred years that the concept of Mathematical Probability should be found to be applicable, not only to idealized games of chance, but to practical affairs, and in particular to inferences in the Natural Sciences. The fiducial argument demonstrates at least one meaningful application beyond that for which it was originally defined, and without needing the knowledge *a priori* required for Bayes' method of reasoning.

It may be noticed in the example chosen above, and in other cases in which a probability distribution can be calculated by the fiducial argument, that the region containing, for example, the lowest 1% of the frequency distribution is exactly that comprising values of the parameter which would have been rejected as too low by a valid test of significance at the 1% level. The two concepts should none the less be distinguished, for valid tests of significance at all levels may exist without the possibility of deducing by an accurate argument, a probability distribution for the unknown parameter. The direct step from the test of significance to a probability distribution cannot be sustained, and this circumstance has been responsible for some misunderstanding, and confusion of the terminology.

5. Discontinuous observations

The data discussed by Bayes in which a successes have been observed out of $(a+b)$ trials are discontinuous in character, unlike all examples suitable for exhibiting the fiducial argument, in the simple form so far exhibited. As has been mentioned, as the frequencies counted are increased, the mathematical niceties necessary for the accurate treatment of small samples, and the corresponding logical distinctions, become unimportant in their effects, and may therefore be ignored in practice. Nevertheless, no apology is needed for examining, as will be done later in this section, what inferences can properly be drawn when such data are treated quite exactly.

If the frequencies observed are very large, for

example, if both a and b are counts of many millions or hundreds of millions of independent instances, the probability of success might, if only ordinary levels of precision were in view, be recognized to have effectively the status of a directly observable quantity namely,

$$p = a/(a+b), \qquad (27)$$

in which we may ignore the logical fact that p is equated to an estimate affected by errors of random sampling. At least we may ignore this fact after having satisfied ourselves, by reference to a more exact treatment, that the precision is such that for the purposes for which we need it, the observed value is sufficiently accurate, with sufficiently high probability.

If the frequencies are of intermediate size, of the order, let us say, of 1000 or 10,000, we probably should not be willing to ignore the sampling error, but might be content, as in the case of the χ^2 test of Goodness of Fit, to ignore the discontinuity of the observations. Comparing the expectations for any assigned value of p, with the observations, we should have, with $a + b = N$,

	Expected	Observed
	pN	a
	qN	b

leading to

$$\left. \begin{aligned} \frac{(qa-pb)^2}{pq(a+b)} &= \chi^2, \\ \text{or,} \quad \frac{qa-pb}{\sqrt{pq(a+b)}} &= \chi, \end{aligned} \right\} \qquad (28)$$

as the test of significance for one degree of freedom. χ is then normally distributed with unit variance.

FORMS OF QUANTITATIVE INFERENCE 65

Choosing any appropriate significance level, such as 2%, the value of χ^2 is known, in this case to be 5·412, and substituting this value, we have a quadratic equation, giving the two values of p for which the deviation from the observations is exactly at the 2% level of significance. The probability that p should be less than the lower root of the equation is 1%, as is that of it exceeding the higher root. Similar values, excluding 5% at each end, could be obtained by taking 2·706 for χ^2. When the discontinuity of the observations can be ignored, the fiducial argument justifies statements of probability, concerning the unknown value p, which is thus estimated as a random variable with a precision defined by a consistent aggregate of such probability statements.

Since the fiducial argument in full strictness cannot be applied, owing to the actual discontinuity of the data, it would be improper to regard this distribution, though precise in form, as more than an asymptotic solution of Bayes' problem, when knowledge *a priori* is absent. As such, it is, however, relevant that the mean value of p for given observational frequencies a, b may be expressed as

$$\bar{p} = \frac{a}{N} + \frac{b-a}{2N}\left\{1 - e^{\frac{1}{2}N}\sqrt{N}\int_{\sqrt{N}}^{\infty} e^{-\frac{1}{2}u^2} du\right\}, \quad (28.1)$$

or

$$\frac{a}{N} + \frac{b-a}{2N^2} - \frac{3(b-a)}{2N^3} + \cdots . \quad (28.2)$$

Now the mean of Bayes' distribution *a posteriori* admits of the expansion

$$\frac{a+1}{N+2} = \frac{a}{N} + \frac{b-a}{N^2} - \frac{2(b-a)}{N^3} + \cdots \quad (28.3)$$

whereas had Bayes' calculation been carried out with the probability element *a priori* given by (6)

$$\frac{1}{\pi\sqrt{pq}}\,dp,$$

the mean would have been

$$\bar{p} = \frac{a+\tfrac{1}{2}}{N+1} = \frac{a}{N} + \frac{b-a}{2N^2} - \frac{b-a}{2N^3} + \frac{b-a}{2N^4} - \cdots$$

(28.4)

For asymptotic agreement with the fiducial distribution, so far as the term in N^{-2}, Bayes' postulated distribution *a priori* should have been that given by expression (6), derived from the angular transformation. To this extent a particular given distribution *a priori* may be nearly equivalent to complete ignorance *a priori*.

Equation (28.1) is not, indeed, exact, since in the third term of the expansion (28.2) the effects of the non-normality of the binomial distribution become appreciable. If allowance be made for these by the method of Cornish and Fisher[1] (1937), an expansion may be obtained of the parameter p in terms of a normal deviate x, of which the earlier terms, carried now a few steps further than in the first edition, and arranged, after the first three in tabular form, are

$$a/N - \sqrt{ab}\,x/N^{3/2} + (b-a)(2x^2+1)/6N^2$$

$$+ \begin{Bmatrix} x^3 & x & \\ -2 & -7 & N^2 \\ 26 & 34 & ab \end{Bmatrix} \div 72N^{5/2}\sqrt{ab}$$

$$+ (b-a) \begin{Bmatrix} x^4 & x^2 & x^0 & \\ -12 & 17 & 19 & N^2 \\ -276 & -644 & -148 & ab \end{Bmatrix} \div 3240 N^3 ab$$

$$+\left\{\begin{array}{cccc} x^5 & x^3 & x & \\ -36 & 470 & 265 & N^4 \\ 936 & 5564 & 5188 & N^2ab \\ -11268 & -39712 & -23804 & a^2b^2 \end{array}\right\} \div \frac{48 \times 3240}{N^{7/2}(ab)^{3/2}}$$

$$+(b-a)\left\{\begin{array}{cccc} x^6 & x^4 & x^2 & x^0 \\ 24 & 207 & -418 & -563 \quad N^4 \\ 240 & -261 & -2059 & -758 \quad N^2ab \\ 5784 & 28026 & 29882 & 3400 \quad a^2b^2 \end{array}\right\}$$
$$\div 126 \times 3240 N^4 a^2 b^2$$

$$+\left\{\begin{array}{cc} x^7 & x^5 \\ 30024 & -17016 \\ -51192 & -570492 \\ -1{,}014120 & -9{,}896200 \\ 12{,}139416 & 76{,}126776 \\ x^3 & x \\ -396616 & 792709 \quad N^6 \\ -435102 & -3{,}185262 \quad N^4 ab \\ -22{,}500120 & -10{,}404420 \quad N^2 a^2 b^2 \\ 122{,}206216 & 37{,}191656 \quad a^3 b^3 \end{array}\right\} \div \frac{112 \times 3240^2}{N^{9/2}(ab)^{5/2}}$$

From this expansion, substituting its average value for each power of x, it can be ascertained that the mean of the fiducial distribution is

$$\bar{p} = \frac{a}{N} + \frac{b-a}{2N^2} - \frac{b-a}{2N^3} + \frac{b-a}{2N^4} \ldots$$

agreeing so far as the fourth term with (28·4), the mean of the Bayesian distribution *a posteriori*, using the element *a priori*

$$\frac{1}{\pi\sqrt{pq}}\,dp\,.$$

I do not know for how many terms this agreement continues. It should, however, be noted that the variances of the two distributions are not the same. That for the fiducial distribution is the larger by $1/12N^2$, or more accurately by $1/12(N+1)(N+2)$.

An odd consequence of the analysis developed above is that the Rule of Succession derivable from the particular distribution of probability *a priori*

$$\frac{dp}{\pi\sqrt{pq}},$$

namely that the probability of success in the next trial is

$$\frac{a+1/2}{a+b+1}$$

is justifiable, at least to a remarkably high approximation, in the absence of any knowledge *a priori*; and this although the corresponding complete distribution *a posteriori* is not so justifiable.

For smaller numbers of independent observations than those for which the effects of discontinuity are negligible, a lower logical status is recognizable in which neither effectively exact definitive statements, nor statements in terms of Mathematical Probability, are possible, yet in which some information is available, and we are not in a state of complete ignorance. I have, indeed, recently[6] discussed a quasi-probabilistic specification of the inferences in this case, in terms which are, however, still too unfamiliar for inclusion here. Evidently, also, in such cases tests of significance are available. Thus, to take an example employed in illustrating the Table of z, in the Introduction to *Statistical Tables*,[5] if 3 successes have been observed out of 14 trials, then probabilities

FORMS OF QUANTITATIVE INFERENCE 69

of success exceeding ·557 may be excluded by the consideration that for such values the total probability of observing 0, 1, 2 or 3 successes is only 1%, while values below ·0331 may be excluded on the ground that the total probability of values from 3 to 14 would similarly fall below 1%. In this way tests of significance are available capable of finding at any level of significance limits outside which all values of the parameter are to be rejected. These have been called "Confidence Limits", and though they fall short in logical content of the limits found by the fiducial argument, and with which they have often been confused, they do fulfil some of the desiderata of statistical inferences.

(i) With the aid of the Table of the z-distribution, or of other tables serving the same purpose of determining the partial sum of the terms of the binomial expansion,

$$(q+p)^n,$$

the Confidence Limits can be sufficiently easily calculated.

(ii) They serve to divide the range of possible values of the unknown into a series of zones of more or less acceptable values.

(iii) The results of the calculations are readily communicable, and the method is sufficiently known to be widely understood.

Nevertheless, from the point of view of making the most of limited data, and of drawing from them conclusions as strong as they can properly be made, the system of Confidence Limits seems to provide less than even this comparatively uninformative type of data would support.

It has been frequently stated, as though it were a

characteristic property of Confidence limits, that the interval between them will in repeated samples cover the true value with the exact frequency corresponding with the level of significance chosen. E.g. that in 98% of trials the true value would be found to lie between the two 1% points. This, if true, when exhaustive estimation has been used, would give them the force of a statement of probability. However, actually, the true value will lie between the assigned limits generally in more than 98% of such trials, and no exact statement of probability can be inferred. Exactly verifiable probability statements are not a characteristic of Confidence limits, as they are of the limits that can be assigned when the fiducial argument is available.

As the probability defining the level of significance is raised the width of the central zone still deemed acceptable is narrowed, but it is not closed until the probability level is raised to considerably over 50%. At the value of p suggested by the data, namely 3/14 (21·42857%), the probability of observing 3 or less is 64·832% and that of observing 3 or more is 60·402%. At $p=·22$, the probability of 3 or less is 62·807% and that of 3 or more is 62·394. This system of zones therefore closes at some value slightly greater than $p=·22$, and at this value the level of significance is over 62%. Neither value seems to have any sufficient inferential content to be worthy of record and report. The method of zoning by Confidence Intervals does not pick out as of any importance the value of p for which the Mathematical Likelihood is greatest, or, in other words, that which would have the highest probability of leading to the result observed.

6. Mathematical Likelihood

Objection has sometimes been made that the method of calculating Confidence Limits by setting an assigned value such as 1% on the frequency of observing 3 or less (or at the other end of observing 3 or more) is unrealistic in treating the values less than 3, which have not been observed, in exactly the same manner as the value 3, which is the one that has been observed. This feature is indeed not very defensible save as an approximation. It should be pointed out that when the probability of 3 or less is small, most of this small probability will be due to the case "exactly 3", and that the contribution of the other three cases is not very important, although it does increase or decrease with varying p at a relative rate different from the contribution of "exactly 3" itself. Similarly, at values of p at which "3 or more" has a low probability, a large part of this probability will be due to the particular case that has been observed, and the calculation will have been perhaps little influenced by the frequencies of the cases, included in the calculation, which have not in fact been observed.

It would, however, have been better to have compared the different possible values of p, in relation to the frequencies with which the actual values observed would have been produced by them, as is done by the Mathematical Likelihood, a function of the unknown parameter proportional to these frequencies, or in this case to

$$p^a(1-p)^b,$$

having its maximum value at

$$p = a/(a+b),$$

and therefore expressible in terms of its maximum, as

$$\left(\frac{p}{a}\right)^a \left(\frac{1-p}{b}\right)^b (a+b)^{a+b},$$

$$\text{or} \quad \frac{(a+b)^{a+b}}{a^a b^b} \cdot p^a (1-p)^b. \tag{29}$$

The Mathematical Likelihood assignable to every value of the unknown parameter p supplies a zoning of the admissible range of values more directly appropriate to the observations than that provided by the system of Confidence belts. Mathematical Likelihood is not, of course, to be confused with Mathematical Probability. It is, like Mathematical Probability, a well-defined quantitative feature of the logical situations in which it occurs, and like Mathematical Probability can serve in a well-defined sense as a "measure of rational belief"; but it is a quantity of a different kind from probability, and does not obey the laws of probability. Whereas such a phrase as "the probability of A or B" has a simple meaning, where A and B are mutually exclusive possibilities, the phrase "the likelihood of A or B" is more parallel with "the income of Peter or Paul"—you cannot know what it is until you know which is meant.

In relation to the logical situations so far discussed Mathematical Likelihood has already appeared as the factor, appropriate to each possible parametric value, by which each element of probability *a priori* is converted in Bayes' method to the corresponding element of probability *a posteriori*. It represents that part of Bayes' calculation provided by the data themselves. With regard to simple tests of significance, if such tests were performed on the same data against several mutually exclusive hypotheses, since the likelihood of any hypothesis is proportional to the probability, accepting that hypothesis as true, of such observations occurring as have been made, the greatest reluctance will be felt to accepting the least likely hypothesis, and the least reluctance to the most likely. Such comparisons can be made even though only relative values of the likelihood function are meaningful. The likelihood supplies a natural order of preference among the possibilities under consideration. It is not surprising, therefore, though independently demonstrable, that in the Theory of Estimation, all rational criteria of what is to be desired in an estimate converge on the particular value for which the likelihood is maximized. The Method of Maximum Likelihood is indeed much used and widely appreciated in the statistical literature, without, I fancy, so much appreciation of the significance of the system of likelihood values at other possible values of the parameter. In the theory of estimation[2] it has appeared that the whole of the information supplied by a sample, within the framework of a given sampling method, is comprised in the likelihood, as a function known for all possible values of the parameter.

The relation between probability and likelihood in the case in which probabilities are accessible by the fiducial argument, is intimate. If θ stand for an unknown parameter and T for a Sufficient, or Exhaustive, estimate of it, and if for all values of θ and T, the function,

$$P = F(T, \theta),$$

stand for the probability that a sample drawn from a population with parameter θ, shall yield a statistic less than T, both θ and T being continuous functions over the same range, and F monotonic for both variables, then the distribution of T for given θ has the frequency element,

$$\frac{\partial F}{\partial T} dT,$$

so that the likelihood of any value θ is determined by the relation

$$e^{L(\theta)} \propto \frac{\partial F}{\partial T},$$

for the particular value of T observed, and for all values of θ. Concurrently, we have seen that the frequency distribution of θ to be inferred from the data has the frequency element

$$-\frac{\partial F}{\partial \theta} d\theta.$$

The likelihood function and the probability distribution thus supply complementary specifications of the same situation.

"Confidence Limits" and "Confidence Belts" were I think developed and advocated under the impression that in a wider class of cases they could provide information similar to that of the probability statements derived by the fiducial argument. It is clear, however, that no exact probability statements can be based upon them, and this seems now to be understood. They may be taken to supply statements of inequality. The tests of significance on which they are based are, of course, valid, but if these are used for zoning the possible values of the parameter, the zones they give do not assign exactly the same order of preference as that supplied by the likelihood function. This is due to the use of heterogeneous

FORMS OF QUANTITATIVE INFERENCE 75

groups of possibilities of which some have been observed, and others have not, in making up the blocks for which the probabilities are calculated.

For all purposes, and more particularly for the *communication* of the relevant evidence supplied by a body of data, the values of the Mathematical Likelihood are better fitted to analyse, summarize, and communicate statistical evidence of types too weak to supply true probability statements; it is important that the likelihood always exists, and is directly calculable. It is usually convenient to tabulate its logarithm, since for independent bodies of data such as might be obtained by different investigators, the "combination of observations" requires only that the log likelihoods from different sources should be added.

In the case under discussion a simple graph of the values of the Mathematical Likelihood expressed as a percentage of its maximum, against the possible values of the parameter p, shows clearly enough what values of the parameter have likelihoods comparable with the maximum, and outside what limits the likelihood falls to levels at which the corresponding values of the parameter become implausible. In Fig. 1 the likelihood is plotted against p. If instead of p, a transformed value, such as the normal deviate commonly used in Biological Assay is employed, the curve is transformed using invariant ordinates, and not, as would be the case with a frequency curve, with invariant areas. This is shown in Fig. 2. The areas under these curves are irrelevant. In each diagram zones are indicated showing the limits within which the likelihood exceeds $1/2$, $1/5$, and $1/15$ of the maximum. Values of the parameter outside the last

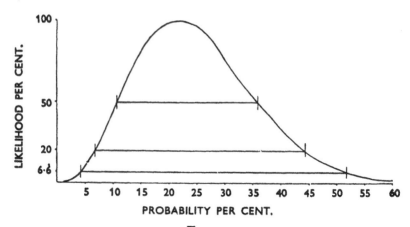

Fig. 1

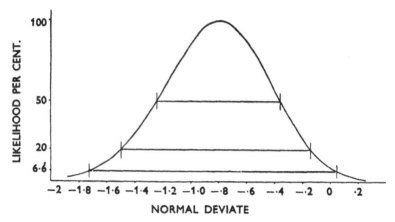

Fig. 2

limits are obviously open to grave suspicion. The actual limits found are:

Likelihood ratio	Lower limit %	Upper limit %
50%	10·5889	35·9225
20%	6·6652	44·3301
6·6%	4·2001	51·6876

The simplicity of the data chosen for this example is an unessential accident. Very extensive observations may best be summarized in terms of the likelihood function calculable from them. A schedule of log likelihoods such as that shown below, in a form suitable for the information of a later worker, may be much more compact than the data from which it has been derived, and yet convey all that is needed from them:

$p\%$	$-3 \log p$ $-11 \log (1-p)$	$p\%$	$-3 \log p$ $-11 \log (1-p)$
3	4·714147	20	3·162920
4	4·388837	25	3·180506
5	4·148130	30	3·272558
6	3·961139	35	3·425749
8	3·689064	40	3·634156
10	3·503332	45	3·896373
12	3·373147	50	4·214420
14	3·282132	55	4·593574
17	3·198794	60	5·042886

Apart from the simple test of significance, therefore, there are to be recognized and distinguished, between the levels of certain knowledge and of total nescience, two well-defined levels of logical status for parameters lying on a continuum of possible values, namely that in which the probability is known for the parameter to lie between any assigned values, and that in which no probability statements being possible, or only statements of inequality, the Mathematical Likelihood of all possible values can be determined from the body of observations available.

By means of appropriate observations a quantity may conceivably pass discontinuously from one status to another implying fuller knowledge. Alternatively, the mere accumulation of data of the same kind on a sufficient scale may induce a kind of asymptotic approach to a higher status. The implications of such a classificatory framework may be made clear by a greater variety of examples such as will be discussed in later chapters.

REFERENCES

1. E. A. Cornish and R. A. Fisher (1937). Moments and cumulants in the specification of distributions. *Rev. Inst. int. Statist.*, vol. 5, pp. 307–320.
2. R. A. Fisher (1925). Theory of statistical estimation. *Proc. Camb. Phil. Soc.*, vol. 22, pp. 700-725.
3. R. A. Fisher (1925-1958). *Statistical Method for Research Workers.* Oliver and Boyd, Edinburgh.
4. R. A. Fisher (1930). Inverse probability. *Proc. Camb. Phil. Soc.*, vol. 26, pp. 528-535.
5. R. A. Fisher and F. Yates (1938-1963). *Statistical Tables.* Oliver and Boyd, Edinburgh.
6. R. A. Fisher (1957) The underworld of probability. *Sankhya*, vol. 18, pp. 201-210.
7. H. Jeffreys (1940). Note on the Behrens-Fisher formula. *Ann. Eugen.*, vol. 10, pp. 48-51.
8. J. M. Keynes (1921). *A Treatise on Probability.* Macmillan and Co., London.
9. A. N. Kolmogorov (1942). The estimation of the mean and precision of a finite sample of observations. Section 5—Fisher's fiducial limits and fiducial probability. *Bull. Acad. Sci. U.S.S.R., Math. Ser.*, vol. 6, pp. 3-32.

CHAPTER IV

SOME MISAPPREHENSIONS ABOUT TESTS OF SIGNIFICANCE

1. Tests of significance and acceptance decisions

The common tests of significance, familiarly known as Pearson's χ^2 test of goodness of fit (1900),[22] "Student" 's t-test (1908),[24] the z (or F) test of the analysis of variance (1924),[4] and many others designed on the same principles, have come in the first two quarters of the twentieth century to play a rather central part in statistical analysis. In the day-to-day work of experimental research in the natural sciences, they are constantly in use to distinguish real effects of importance to a research programme from such apparent effects as might have appeared in consequence of errors of random sampling, or of uncontrolled variability, of any sort, in the physical or biological material under examination. They are used to recognize, among innumerable examples that could be given, the genuineness of a genetic linkage, the reality of the response to manurial treatment of a cultivated crop, the deterioration of a food product in storage, or the difference between machines in the frequency of defective parts produced by them. The conclusions drawn from such tests constitute the steps by which the research worker gains a better understanding of his experimental material, and of the problems which it presents.

It is noteworthy, too, that the men who felt the need for these tests, who first conceived them, or later made them mathematically precise, were all actively concerned with researches in the natural sciences. More recently, indeed, a considerable body of doctrine has attempted to explain, or rather to reinterpret, these tests on the basis of quite a different model, namely as means to making decisions in an acceptance procedure. The differences between these two situations seem to the author many and wide, and I do not think it would have been possible to overlook them had the authors of this reinterpretation had any real familiarity with work in the natural sciences, or consciousness of those features of an observational record which permit of an improved scientific understanding, such as are particularly in view in the design of experiments. The misapprehensions, indeed, appear to go deeper than would be expected from a mere transference of techniques from one field of study to another.

In various ways what are known as acceptance procedures are of great importance in the modern world. When a large concern such as the Royal Navy receives material from its makers, it is, I suppose, subjected to sufficiently careful inspection and testing to reduce the frequency of the acceptance of faulty or defective consignments. The instructions to the officers carrying out the tests must also, I conceive, be such as to keep low both the cost of testing and the frequency of the rejection of satisfactory lots. Much ingenuity and skill must be exercised in making an acceptance procedure a really effectual and economical one. It is not therefore at all in disdain of an artifice of proved value, in commerce and technology, that I shall emphasize some of the various ways in which this operation differs from that by which improved theoretical knowledge is sought in experimental research. This emphasis is primarily necessary because the needs and purposes of workers in the experimental sciences have been so badly misunderstood and misrepresented. It is, of course, also to be suspected that those authors, such as Neyman and Wald, who have treated these tests with little regard to their purpose in the natural sciences, may not have been more successful in the application of their ideas to the needs of acceptance procedures. It is, however, to the evident advantage of both kinds of application that the theories developed and taught to mathematicians should not confuse their several requirements.

In attempting to identify a test of significance as used in the natural sciences with a test for acceptance, one of the deepest dissimilarities lies in the population, or reference set, available for making statements of probability. Confusion under this head has on several occasions led to erroneous numerical values; for, where acceptance procedures are appropriate, the population of lots of one or more items, which could be chosen for examination, is unequivocally defined. The source of supply has an objective empirical reality. Whereas, the only populations that can be referred to in a test of significance have no objective reality, being exclusively the product of the statistician's imagination through the hypotheses which he has decided to test, or usually indeed of some specific aspect of these hypotheses. The demand was first made, I believe, in connection with Behrens' test of the significance of the difference between the means of two populations of unknown variances, that the level of significance should be determined by "repeated sampling from the same population", evidently with no clear realization that the population in question is hypothetical, that it could be defined in many ways, and that the first to come to mind may be quite misleading; or, that an understanding, of what the information is which the test is to supply, is needed before an appropriate population, if indeed we must express ourselves in this way, can be specified. This particular case will be examined more fully in a later section, after illustrating the more general effects of the confusion between the level of significance appropriately assigned to a specific test, with the frequency of occurrence of a specified type of decision.

2. "Student"'s Test

In the test of significance due to "Student" (W. S. Gosset), and generally known as the t-test, the data are taken to consist of N values of a single observable variate x, which are to be interpreted on the hypothesis that they are independent values of a variate normally distributed, with both mean and variance unknown, and of which no probability statements *a priori* are available. The two statistics

$$\bar{x} = \frac{1}{N} S(x)$$

$$s^2 = \frac{1}{N-1} S(x-\bar{x})^2 \tag{30}$$

are known to be jointly Sufficient for the estimation of the true mean, μ, and the true variance, σ^2. Their sampling distributions in random samples of N may be expressed exactly in terms of the population parameters. Thus, as was shown by Gauss, the mean $\bar{x}$ has a Normal distribution about the true

TESTS OF SIGNIFICANCE

mean, μ, with the population variance divided by N. I.e. the frequency element is

$$\sqrt{\frac{N}{2\pi\sigma^2}}\, e^{-N(\bar{x}-\mu)^2/2\sigma^2}\, d\bar{x}\,, \tag{31}$$

while the distribution of s^2, due first I believe to Helmert[17] (1875) is independent of μ, and of $\bar{x}$, though involving σ^2. If u stand for

$$u = \tfrac{1}{2}\frac{(N-1)s^2}{\sigma^2}\,, \tag{32}$$

it takes the form

$$\frac{1}{\frac{N-3}{2}!}\, u^{\frac{N-3}{2}}\, e^{-u}\, du\,, \tag{33}$$

an Eulerian distribution, identifiable with that of χ^2 for $N-1$ degrees of freedom, by the equivalence

$$u = \tfrac{1}{2}\chi^2\,. \tag{34}$$

On the basis of these distributions "Student" set himself to ascertain the exact sampling distribution of the ratio between the sampling error of the mean, and its standard error *as estimated*, namely

$$t = \frac{\sqrt{N}(\bar{x}-\mu)}{s} \tag{35}$$

when $\bar{x}$ and s are calculated from a finite sample, and he was successful in establishing the frequency element

$$\frac{1}{\sqrt{\pi(N-1)}}\, \frac{\frac{N-2}{2}!}{\frac{N-3}{2}!}\cdot \frac{dt}{\left(1+\frac{t^2}{N-1}\right)^{N/2}}\,, \tag{36}$$

a distribution depending on N only, independent of both parameters of the Normal distribution sampled, and which can, therefore, be tabulated so as to give the value of t exceeded in absolute value, with any given probability, and for any number, $n = (N-1)$, of the degrees of freedom.

It will be recognized that "Student"'s distribution allows of induction of the fiducial type, for the inequality

$$\mu < \bar{x} - \frac{1}{\sqrt{N}} ts , \qquad (37)$$

will be satisfied with just half the probability for which t is tabulated, if t is positive, and with the complement of this value if t is negative. The reference set for which this probability statement holds is that of the values of μ, $\bar{x}$ and s corresponding to the same sample, for all samples of a given size of all normal populations. Since $\bar{x}$ and s are jointly Sufficient for estimation, and knowledge of μ and σ *a priori* is absent, there is no possibility of recognizing any sub-set of cases, within the general set, for which any different value of the probability should hold. The unknown parameter μ has therefore a frequency distribution *a posteriori* defined by "Student"'s distribution.

Although his was the first exact test of significance, characteristic of the modern period, "Student" did not go so far as to claim that he was introducing a new mode of reasoning, and perhaps would have been unwilling to believe it had he been told so; for he was only applying his own good sense to a logical situation with which he was quite familiar. He was usually content to leave the inference in the form of a test of significance, namely that, on the hypothesis, for example, that the true mean of the population was zero, the value of t observed, or a greater value, would have occurred in less than 1% of trials, and was therefore significant at, for example, the 1% level. It does appear, however, at one point, that he certainly was thinking of the unknown mean as having, in the light of the observations, a definable frequency distribution, for in the extreme case of only two observations, x_1 and x_2, when the distribution of t reduces to Cauchy's distribution,

$$\frac{1}{\pi} \cdot \frac{dt}{1+t^2}, \tag{38}$$

and $$s^2 = \tfrac{1}{2}|x_1-x_2|^2, \tag{39}$$

so that

$$t = \frac{2(\bar{x}-\mu)}{|x_1-x_2|}, \tag{40}$$

or exactly $+1$ or -1 if μ takes one or other of the two observed values, he does point out that these are the quartiles of Cauchy's distribution, or the points cutting off a quarter of the area on each side.

Thus by integration of his general formula "Student" had shown, in this case, that the mean of the population sampled had exactly the probability of one half of lying between the two values observed; this conclusion is notable as illustrating a mode of inference entirely independent of the form of the distribution, save that it be continuous, namely that as each observation has a half chance of being above and a half chance of being below the median, and as these chances are independent, it could have been demonstrated that the median should lie between the two observations, these being our sole source of information about its position, with probability exactly one half. This is a typically fiducial argument, which would have been vitiated by the existence of information *a priori*. "Student", indeed, guards himself against this possibility by stipulating that his sample "itself affords the only indication of the variability". I take this to make clear also that it is not one of an objective series of similar samples from the same population existing in reality, though it can be regarded by an act of imagination as one of a hypothetical reference set.

This case is more complex than that dealing with radioactive emission discussed in the last Chapter (Section 3), for here the simultaneous estimation of two parameters is required. It is more simple than will generally be the case in statistical work, for in this case no characteristic of the sample (i.e. of the whole body of observations available) can be found to define a subset to which our sample belongs, and which might exhibit a different, and more relevant, frequency distribution. It is this simplicity that has deceived those writers who have considered this one alone of the practically useful tests of significance into ignoring such subsets, or thinking that when such subsets are available their existence can be ignored; and that a mere consideration of repeated sampling, in one of the many forms this may take, is sufficient to specify the level of significance appropriate.

3. The case of linear regression

A case which illustrates well how misleading the advice is to base the calculations on repeated sampling from the same population, if such advice were taken literally, is that of data suitable for the estimation of a coefficient of linear regression. Suppose we have N pairs of values (x, y) constituting the numerical data, and suppose also that it is given that, for each value of x, the values of y are normally distributed; that the mean y for any value of x is a linear function of that variable,

$$E_x(y) = Y_x = a + \beta x, \tag{41}$$

and that the variance of each distribution, though unknown numerically, is known to be the same for all values of x; i.e.

$$E_x(y - Y_x)^2 = \sigma^2. \tag{42}$$

The distribution of x may also be given, with or without parameters, but this information is, as will be seen, irrelevant.

At least since the time of Gauss it has been known that if A, B, C stand for the sums of squares and products

$$S(x-\bar{x})^2, \; S(x-\bar{x})(y-\bar{y}), \; S(y-\bar{y})^2, \tag{43}$$

then the best estimate of β, the slope of the regression line, is

$$b = \frac{B}{A}. \tag{44}$$

The best estimate of σ^2 is

$$s^2 = \frac{1}{N-2}\left(C - \frac{B^2}{A}\right), \tag{45}$$

and, for samples having the same fixed value for A, the estimates b will be normally distributed about the true value, β, with sampling variance

$$V(b) = \frac{\sigma^2}{A}. \tag{46}$$

We may therefore identify "Student" 's t with

$$t = \sqrt{A} \cdot \frac{b-\beta_0}{s}, \qquad (47)$$

where β_0 is any theoretical value of the coefficient, such as zero, proposed for comparison. Equally, the unknown β may be assigned a known frequency distribution in the light of the observations, based on

$$\beta = b + \frac{st}{\sqrt{A}}, \qquad (48)$$

where t is distributed in "Student" 's distribution for $(N-2)$ degrees of freedom.

This simple and well-known form of analysis is not, I believe, disputed, save in the interpretation of the fiducial inferences. It should be noted none the less that it does violate the criterion of judging by repeated sampling, for in repeated sampling from the bivariate distribution of x and y, the value of A would vary from sample to sample. The distribution of $(b-\beta)$ would no longer be normal, and, before we knew what it was, the distribution of A, which in turn depends on that of x, would have to be investigated. Indeed, at an early stage Karl Pearson did attempt the problem of the precision of a regression coefficient in this way, assuming x to be normally distributed. The right way had, however, been demonstrated many years before by Gauss, and his method only lacked for completeness the refinement of the use of "Student" 's distribution, appropriate for samples of rather small numbers of observations.

To judge of the precision of a given value of b, by reference to a mixture of samples having different values of A, and therefore different precisions for the values of b they supply, is erroneous because these other samples throw no light on the precision of that value which we have observed. If we must think in terms of random sampling, it is only that *selection* of random samples which agree exactly with our own in respect to the value of

$$A = S(x-\bar{x})^2, \tag{49}$$

that is relevant to assessing its real precision. Such a selection might be quite inaccessible in sampling for acceptance. The function of selection is to transform the crude aggregate obtained by "repeated sampling" into a sample genuinely representative of the reference set appropriate to the test of significance. It is not obvious that this can always be done by selection from the material obtainable by mechanical repetition; when the test of significance is understood, such a roundabout course is unnecessary.

4. The two-by-two table

Although in the case of simple regression there has not, so far as I know, been any tendency to calculate erroneous values through mistaking the logical nature of the test, at least since the time of Karl Pearson, the very similar and equally fundamental test of proportionality in a two-by-two table has on more than one occasion become a matter of dispute. The data in this case consist of a number of cases doubly dichotomized, as people may be classified as male or female, or, again as tasters or non-tasters of phenyl-thiocarbamide, which a proportion of people cannot taste at concentrations which to others taste distinctly bitter. The statistical test is intended to find out whether the four frequencies observed are in proportion, or, in other words, whether the two classifications are independent. It will be noticed that, as in most tests, what is here to be rejected by a significant result is a whole class of hypotheses. These will have various values for the expected marginal frequencies, each, however, having proportional frequencies in the contents.

In this problem, as was pointed out concurrently by Yates[29] and Fisher[7] in 1934, a great simplification is available in that the subset of possible samples having the same marginal frequencies as that observed will have, whenever the two classifications are independent, the same frequency distribution. So, if in the records of a controlled experiment we find that three treated animals all die, and three control animals all survive, the two-by-two table

	Died	Survived	Total
Treated . .	3	0	3
Control . .	0	3	3
Total . .	3	3	6

is recognized to be one of a subset of possibilities, being double dichotomies having the same marginal totals, represented briefly by

0	3		1	2		2	1		3	0
3	0		2	1		1	2		0	3

Frequency   1   9   9   1 ÷ 20

These possibilities have relative probabilities independent of the unknown frequencies of death or survival, and calculable merely by the algebraic method of permutations; so that within this subset the observed result is seen to be the most successful out of twenty possibilities, all equally probable on any view that the treatment is without effect.

In other cases of compound hypotheses it does not always occur that when all the possibilities to be excluded are excluded by a single test, the level of significance is the same as the frequency of erroneous exclusion, for all the possible hypotheses to be tested. It may then be that when some particular case of the compound hypothesis is true, the proportion of samples capable of dismissing the whole range of hypotheses under test will not be so great as the level of significance would suggest. It would indeed be unreasonable in general to expect it to be otherwise; it is therefore worth calling special attention to the exceptional feature of the subset available for testing significance in the two-by-two table, that the frequency of rejection is the same for all the simple hypotheses included.

On two occasions (Wilson, 1941,[26, 27, 28] Barnard, 1945[1], 1949[2]) distinguished mathematical statisticians have tried to improve the test by including in the enumeration cases in which the marginal totals differ from those in the sample observed. On both occasions, after discussion and elucidation of the logical basis of the test, these attempts have been abandoned. The argument was well illustrated by Barnard using the case discussed above, wherein, if we had to consider repetition of the experiment on the assumption that the treated and control animals have the same expectation p of dying, the frequency of what has been observed is exactly

$$p^3 q^3. \qquad (50)$$

For any real values of p and q, adding to unity, this is a small fraction. Its *maximum* value, attained when $p=\frac{1}{2}$, is only 1/64. So, if a repetition of the experiment were the right criterion, as has been very dogmatically asserted, significance at least at this level could be claimed. It was on this basis asserted that the method of Neyman and Pearson, relying on the formula of repeated sampling, had led to "a much more powerful test than Fisher's". The 64 cases enumerated in this argument include, however, not only the 20 having the same marginal totals, but 44 others belonging to six other subsets, of which it should be noted (i) that what has been observed does not belong to any of these six subsets, and (ii) within each of these questionable subsets there is no configuration which could be judged significant. Their inclusion in the enumeration can therefore do no other than enhance the apparent significance by inflation of the denominator.

Professor Barnard has since then frankly avowed that further reflection has led him to the same conclusion as Yates and Fisher, as indeed Wilson, with equal generosity, had done earlier. It is, therefore, obvious that he had at first been misled by the form of argument developed by Neyman and Pearson. E. S. Pearson himself has, so far as I know, made no such disavowal; but the numerical values given in Table 38 of his *Biometrika Tables* (1954)[21] do seem *in fact* to have been calculated by the method of Fisher and Yates, though no credit is given to these authors. It is to be feared, therefore, that the principles of Neyman and Pearson's "Theory of Testing Hypotheses" are liable to mislead those who follow them into much wasted effort and disappointment, and that its authors are not inclined to warn students of these dangers.

5. Excluding a composite hypothesis

It was remarked above that very commonly a test

of significance is used to exclude any one of a class of hypotheses, or, as it is sometimes called, a composite hypothesis. Some of the examples given have the exceptional feature that a single test can be found appropriate for the purpose, in which, were any of this class of hypothesis true, the criterion of rejection would be satisfied with the same frequency as the appropriate level of significance. The errors due to equating these concepts which have been indicated so far have been only of the kind which flow from ignoring the appropriate subset of cases to which the observed sample belongs, and seeking to ascertain the frequency of occurrences in a more inclusive set containing elements of a different kind, the variations of which are irrelevant to the observed case under consideration.

Composite hypotheses in general, however, contain another reason for ignoring the assumption that the frequency of rejection should be equated to the level of significance; which reason flows from the very fact that they are composite, i.e. that two or more distinct possibilities are to be rejected, each on sufficiently strong evidence. It may be that samples of the kinds available do not so easily dismiss the whole range of hypotheses to be tested even at a moderate level of significance.

A simple, though artificial, example is the following. A number of cards is made up into a pack, in which the proportions of the four suits are unknown. The composite hypothesis to be tested is that the proportions of the two red suits do not both exceed 25%. The data on which the test is to be based are a sample of 100 random draws each followed by replacement and reshuffling.

The possibility that there were no more than 25%

Hearts could be excluded at a reasonable level of significance, if 34 Hearts appeared in 100 chosen. If the true proportion were 25%, the probability of observing 34 or more is found to be 2·759%. However, the composite hypothesis is disproved only if it is demonstrable that the proportion of Diamonds also is more than 25%, and this requires that the sample should contain at least 34 Diamonds. It is not difficult to anticipate that both these conditions together will be fulfilled very rarely, even in the case in which both Hearts and Diamonds contribute a full quarter to the material sampled. In fact, an apparent disproof of the composite hypothesis at the moderate level of significance chosen would in this case be obtained in less than 34 trials in a million.

Even if the test had been based on 1000 randomly chosen cards, so that 278 were required of each red suit, a pack with exactly 25% of each would provide an apparent disproof with a frequency of only about 38 trials in a million.

It is, of course, no inconvenience that the frequency of rejecting the hypothesis in some cases when it is true should be low, but the calculation indicates also that even if both suits did really occupy somewhat more than 25% of the material sampled, it would not be easy to demonstrate this fact, even at a moderate level of significance, if neither of them were greatly in excess. Sufficiently large samples could indeed make such a demonstration probable; but the frequency of attaining a significant sample in the limiting case where both red suits have exactly 25%, is always less than the square of the fraction measuring the level of significance.

In scientific work it is necessary to be able to assess the strength of the evidence that a particular hypothesis, single or composite, appears to be untenable. The example has been chosen to show that strong evidence may sometimes be hard to obtain. Warnings that the strength of the evidence is not to be measured by the frequency observed in "repeated sampling from the same population" have not, on the whole, been well received by the authors of this formula. E. S. Pearson (*Biometrika*, 1947)[20] quotes me as writing (*Sankhya*, 1945)[14]:

In recent times one often repeated exposition of the tests of significance, by J. Neyman, a writer not closely associated with the development of these tests, seems liable to lead mathematical readers astray, through laying down axiomatically, what is not agreed or generally true, that the level of significance must be equal to the frequency with which the hypothesis is rejected in repeated sampling of any fixed population allowed by hypothesis. This intrusive axiom, which is foreign to the reasoning on which the tests of significance were in fact based seems to be a real bar to progress. . . .

On this E. S. Pearson remarks, "But the subject of criticism seems to me less an intrusive mathematical axiom than a mathematical formulation of a practical requirement which statisticians of many schools of thought have deliberately advanced". It would, however, be difficult to find "schools of thought" other than that of Neyman and Pearson themselves, which have deliberately advanced anything of the kind. The rather wooden attitude adopted by this school seems to stem only from their having committed themselves to an unrealistic formalism.

Obvious as it might seem, it is evidently necessary to point out that it is no remedy to construct a test of significance with a firm intention that the hypothesis *shall* be rejected when true in a fixed proportion of trials. For this may well be mathematically impossible, for the whole range of cases; and consequently, what is of much greater importance, a test which is made to conform in one case may go widely astray in others. For example, if the test were chosen in the example discussed above, that the sum of the numbers of the two red suits observed in a sample of 100 should exceed 60 (which, if both red suits contributed just 25%, would reject the hypothesis to be tested with a "satisfactory" frequency, 2·623%), it would reject it very much more frequently in other cases, in which the null hypothesis was true, as for example if Hearts were 50% and Diamonds only 25%. Such a "test" while purporting to examine the truth of the null hypothesis would in fact reject it readily without the evidence of the observations being appreciably adverse. At least one of the Tables published by Professor E. Pearson and H. O. Hartley is indeed misleading in just this way (No. 11).[21] The editors' justification is: "It is the result of an attempt to produce a test which *does* satisfy the condition that the probability of the rejection of the hypothesis tested will be equal to a specified figure". The logical basis of a test of significance as a means of learning from experimental data is here completely overlooked; for the potential user has no warning of the concealed laxity of the test.[15]

In fact, as a matter of principle, the infrequency with which, in particular circumstances, decisive evidence is obtained, should not be confused with the force, or cogency, of such evidence. It is the highest frequency, within the range of the class of hypotheses tested, that is logically relevant.

6. Behrens' Test

The immediate effect of "Student" 's work, as its use came to be appreciated, was to supply a test of significance for the deviation, from some theoretically expected value, of the mean of a sample of observations, or of a regression, even though based on a small sample. It was quickly shown that the same Table could be used for the comparison of means, or regressions, based on small samples, provided the two sets of observations to be compared had the same precision, or to take a rarer but more general case, precisions in a known ratio.

These early results covered immediate practical requirements rather fully, for with large samples capable of supplying, on internal evidence, accurate estimates of precision, the large-sample procedure of estimating the two sources of error independently could be relied on; and, in a great deal of practical experimental work with small samples, although different lots of material might in reality have somewhat unequal variances, there were good reasons for supposing the real differences to be small compared with the errors of estimation from the small samples individually; so that better comparisons would be obtainable by pooling the variances of the different lots. The mathematical problem of the comparison of means of samples, not only small in size, but for which there is no reason *a priori* to dismiss the largest imaginable differences in precision, was of mathematical interest, and potential experimental importance, though it is not common to find realistic data which present this problem, partly because they are rarely sought.

For samples from a single population, the effect of eliminating the unknown variance, σ^2, by "Student"'s method, on the distribution of the error of the mean, is to replace, in the specification of this error,

$$\frac{\sigma}{\sqrt{N}} x, \qquad (51)$$

where x is normally distributed with unit variance, but σ is unknown, by

$$\frac{s}{\sqrt{N}} t, \qquad (52)$$

where t is distributed in "Student"'s distribution, for the appropriate number of degrees of freedom, $n(=N-1)$, and s is the estimate of σ available from n degrees of freedom.

For two samples from populations having a common mean, the deviations will be independent, and the data will supply values s_1, based on n_1 degrees of freedom, and s_2 based on n_2. The difference between the observed means is then the sum (or difference) of the two deviations from the true mean, so that on the null hypothesis considered, namely that the two population means are equal, we have

$$\bar{x}_1 - \bar{x}_2 = \frac{s_1}{\sqrt{N_1}} t_1 - \frac{s_2}{\sqrt{N_2}} t_2, \qquad (53)$$

where t_1 and t_2 are distributed independently in the two "Student" distributions.

If the frequency is small, such as 1%, that the expression on the right, which has a known distribution, for the observed values s_1 and s_2, shall exceed the observed difference in the sample means, this difference may be judged significant.

Effectively this solution, in a somewhat different

notation, was given by W.-U. Behrens[3] in 1929; Behrens also gave a short numerical table. A paper of mine in 1935 confirmed and somewhat extended Behren's theory. A much more substantial table was supplied by P. V. Sukhatmé[25] in 1938 covering values of n_1 and n_2 of 6, 8, 12, 24 and ∞, and values of the ratio $s_1/s_2 (=\tan \theta)$ for values of θ of 0°, 15°, 30°, 45°, 60°, 75° and 90°. Apart from very small samples this was sufficient to make the test easily available. In 1941 the author gave[13] asymptotic expansions for calculating the probabilities with accuracy in any particular case, and a further range of tables for the case when either n_1 or n_2 is large. More recently Fisher and Healy[16] have given, for a number of levels of significance, the exact values when n_1 and n_2 are small and odd. The numerical values seem to make allowance nicely for the fact that a composite hypothesis, in which all ratios of σ_1/σ_2 are possible, is being tested, for it is required to set a limit which will rarely be passed by random samples of populations having the same mean, whatever may be the true variance ratio (*Statistical Tables* fifth edition VI, VI$_1$ and VI$_2$).[12]

In the extreme case in which both samples are of only two readings such as x_1 and x_2 for the first sample, and y_1 and y_2 for the second, Behrens' test takes the simple form of calculating

$$\frac{x_1+x_2-y_1-y_2}{|x_1-x_2|+|y_1-y_2|} = T, \qquad (54)$$

where the level of significance is determined by giving T "Student"'s distribution for one degree of freedom, so that the level of significance is

$$\frac{2}{\pi} \tan^{-1}(1/T). \qquad (55)$$

This extreme case, academic as it is, is particularly suitable for exhibiting the logic of the test. In 1937 I gave[11] the frequency distribution of T in repeated samples from populations having a fixed variance ratio,

$$\phi = \sigma_2^2/\sigma_1^2 \qquad (56)$$

in the exact form,

$$dp = \frac{dT}{\pi(1+T^2)} \left\{ \sqrt{\frac{\phi}{(1+\phi+T^2)}} + \sqrt{\frac{1}{1+\phi+\phi T^2}} \right\}, \quad (57)$$

with the probability integral,

$$\frac{1}{2} + \frac{1}{\pi}\left\{\sin^{-1}\frac{T\sqrt{\phi}}{\sqrt{(1+\phi)(1+T^2)}} + \sin^{-1}\frac{T}{\sqrt{(1+\phi)(1+T^2)}}\right\} \qquad (58)$$

from which it is evident that, if we take the 5% value of $|T|$ as

$$\tan 85° 30' = 12{\cdot}7062$$

this value will be exceeded in 5% of such repeated trials only at the limits $\phi=0$, or ∞, while midway between these, if ϕ were equal to unity, the criterion would be exceeded in less than 1% of random trials.

This circumstance, indeed, caused me no surprise, for the reference set in (58) has not been limited to the subset having the ratio s_1/s_2 observed, but was eagerly seized upon by M. S. Bartlett, as though it were a defect in the test of significance of a composite or comprehensive hypothesis, that in special cases the criterion of rejection is less frequently attained by chance than in others. On reflexion I do not think one should expect anything else, and it was perhaps only because at this time Bartlett was confidently putting forward an alternative attempt to solve the same problem that he made so much of a circumstance which is, indeed, generally to be expected.

7. The "randomization test"

How important it seemed to Bartlett that, whatever the true nature of the population sampled, the null hypothesis should be rejected when true with exactly the frequency suggested by the level of significance, is shown by the fact that he did for a time put forward as an alternative, presumably thought to be better than Behrens', a test involving a deliberately introduced element of hazard.

Observing the formal resemblance of the problem discussed by Behrens, in this extreme case when both samples are of two values only, to a paired comparison of "Student" 's type with only two pairs, it is easy to see that if in addition to Behrens' value T, we calculate also

$$\frac{x_1 + x_2 - y_1 - y_2}{\big||x_1 - x_2| - |y_1 - y_2|\big|} = T', \qquad (59)$$

then, if one of the values T, T' is chosen by an equal chance, the one chosen will exceed an appropriate criterion such as 12·7, in absolute value with a frequency equal to the level of significance adopted; and this for all possible variance ratios of the population sampled.

This proposal, which has perhaps now been abandoned (though at the time an equally faulty proposal was quickly put forward by Neyman) has two conspicuous objections, one of which is of general importance in that it applies to all "randomization tests" in the Natural Sciences. Namely, that if T and T' lie on opposite sides of the criterion, T' being always the larger, and a coin is thrown to decide which shall be chosen, it is then obvious at the time that the judgement of significance has been decided not by the evidence of the sample, but by the throw of the coin. It is not obvious how the research worker is to be made to forget this circumstance; and it is certain that he ought not to forget it, if he is concerned to assess the weight only of objective observational facts against the hypothesis in question. A real experimenter, in fact, so far from being willing to introduce an element of chance into the formation of his scientific conclusions, has been steadily exerting himself, in the planning of his experiments, and in their execution, to decrease or to eliminate by randomization all the causes of fortuitous variation which might obscure the evidence.

Consequently, whereas in the "Theory of Games" a deliberately randomized decision[8] (1934) may often be useful to give an unpredictable element to the strategy of play; and whereas planned randomization[10] (1935-1966) is widely recognized as essential in the selection and allocation of experimental material, it has no useful part to play in the formation of opinion, and consequently in the tests of significance designed to aid the formation of opinion in the Natural Sciences.

The second and specific objection to Bartlett's T', as a test of significance, is that it does not increase or decrease monotonically for changes in the weight of the evidence. For example, if y_1-y_2 is less than x_1-x_2, then an equal change in y_1 and y_2, taking them farther apart, will diminish the denominator of T', and actually increase its value; so indicating a higher level of significance, due to a *greater* discrepancy between two parallel observations.

In fact a practical worker who had calculated T, and T', could only regard them as providing evidence of significance, if both exceeded the minimum level, and since T' is never less than T, this implies simply the use of Behrens' test.

It is an indication of the remoteness from scientific application with which this problem must have been discussed that soon after Bartlett put forward his randomization test an equivalent test using pairing of the n_1 values of one sample, at random with an equal number from the other sample, as can be done in no less than

$$\frac{n_2!}{(n_2-n_1)!} \qquad (60)$$

different ways, was put forward by J. Neyman as a general solution of the problem. That "solution" also has never, I believe, been applied in practice.

8. Qualitative differences

The examples elaborated in the foregoing sections of numerical discrepancies arising from the rigid formulation of a rule, which at first acquaintance it seemed natural to apply to all tests of significance, constitute only one aspect of the deep-seated difference in point of view which arises when Tests of Significance are reinterpreted on the analogy of Acceptance Decisions. It is indeed not only numerically erroneous conclusions, serious as these are, that are to be feared from an uncritical acceptance of this analogy.

An important difference is that Decisions are final, while the state of opinion derived from a test of significance is provisional, and capable, not only of confirmation, but of revision. An acceptance procedure is devised for a whole class of cases. No particular thought is given to each case as it arises, nor is the tester's capacity for learning exercised. A test of significance on the other hand is intended to aid the process of learning by observational experience. In what it has to teach each case is unique, though we may judge that our information needs supplementing by further observations of the same, or of a different kind. To regard the test as one of a series is artificial; the examples given have shown how far this unrealistic attitude is capable of deflecting attention from the vital matter of the weight of the evidence actually supplied by the observations on some theoretical question, to, what is really irrelevant, the frequency of events in an endless series of repeated trials which will never take place. The rejection of Behrens' test, and the series of futile attempts to find an alternative, based on the Neyman and Pearson theory of testing hypotheses, well exhibit the sterility of such formalism.

The concept that the scientific worker can regard himself as an inert item in a vast co-operative concern working according to accepted rules, is encouraged by directing attention away from his duty to form correct scientific conclusions, to summarize them and to communicate them to his scientific colleagues, and by stressing his supposed duty mechanically to make a succession of automatic "decisions", deriving spurious authority from the very incomplete mathematics of the Theory of Decision Functions. Even if this theory in its development so far had steered clear of confusion between the fields of tests of significance in the Natural Sciences, of policy or strategy in the Theory of Games, of rejection in Quality Control, and perhaps of other situations equally slightly related, it would still be true that the Natural Sciences can only be successfully conducted by responsible and independent thinkers applying their minds and their imaginations to the detailed interpretation of verifiable observations. The idea that this responsibility can be delegated to a giant computer programmed with Decision Functions belongs to the phantasy of circles rather remote from scientific research. The view has, however, really been advanced[19] (Neyman, 1938) that Inductive Reasoning does not exist, but only "Inductive Behaviour"!

A misconception having some troublesome consequences was introduced[18] by Neyman and Pearson in 1933, shortly after they had learnt of the possibility of deriving probability statements and therefore limits of significance by the fiducial argument, which had been published in the same journal, the *Proceedings of the Cambridge Philosophical Society*, in 1930.[5] Instead of perceiving that my method was appropriate to the absence of knowledge *a priori*, and, although I had not made this clear, would have been invalidated by the presence of such knowledge, Neyman and Pearson speak of my results as though they were a kind of "greatest common measure" of the inferences which could be made for all possible types of information *a priori*. In fact their paper opens as follows:

In a recent paper[17] we have discussed certain general principles underlying the determination of the most efficient tests of statistical hypothesis, but the method of approach did not involve any detailed consideration of the question of *a priori* probability. We propose now to consider more fully the bearing of the earlier results on this question and in particular to discuss what statements of value to the statistician in reaching his final judgment can be made from an analysis of observed data, which would not be modified by any change in the probabilities *a priori*. In dealing with the problem of statistical estimation,* R. A. Fisher has shown how, under certain conditions, what may be described as

---

* My paper had, however, been entitled "*Inverse Probability*."

rules of behaviour can be employed which will lead to results independent of these probabilities; in this connection he has discussed the important conception of what he terms fiducial limits.[5, 6] But the testing of statistical hypotheses cannot be treated as a problem in estimation, and it is necessary to discuss afresh in what sense tests can be employed which are independent of *a priori* probability laws.

This early misconception has led other writers to seek for inferences independent of *a priori* laws, whereas seeing that Bayes' theorem is based upon supposedly exact knowledge of probabilities *a priori*, and that these probabilities can be made to appear explicitly in the result, none but trivial conclusions can be common to all cases. It is perhaps some sort of recognition of this that makes these authors ascribe to me "rules of behaviour", which I had not mentioned at all, whereas I had written in quite conventional terms which refer to reasoning processes, such as "learning by experience" and the "probability of causes". The logical distinction must in any case be stressed between possessing no information of a certain kind, and possessing such information, although it may be provisionally expressed in a generalized notation. The confusion of these situations is a serious trap, especially for mathematicians without experience in the Sciences.

It is important that the scientific worker introduces no cost functions for faulty decisions, as it is reasonable and often necessary to do with an Acceptance Procedure. To do so would imply that the purposes to which new knowledge was to be put were known and capable of evaluation. If, however, scientific findings are communicated for the enlightenment of other free minds, they may be put sooner or later to the service of a number of purposes, of which we can know nothing. The contribution to the Improvement of Natural Knowledge, which research may accomplish, is disseminated in the hope and faith that, as more becomes known, or more surely known, a great variety of purposes by a great variety of men, and groups of men, will be facilitated. No one, happily, is in a position to censor these in advance. As workers in Science we aim, in fact, at methods of inference which shall be equally convincing to all freely reasoning minds, entirely independently of any intentions that might be furthered by utilizing the knowledge inferred.

REFERENCES

1. G. A. Barnard (1945). A new test for 2×2 tables. *Nature*, vol. 156, pp. 177.
2. G. A. Barnard (1949). Statistical inference. *J. Roy. Stat. Soc.*, B, vol. 11, pp. 115-139.
3. W.-U. Behrens (1929). Ein Beitrag zur Fehlen-Berechnung bei wenigen Beobachtungen. *Landw. Jb.*, vol. 68, pp. 807-837.
4. R. A. Fisher (1924). On a distribution yielding the error functions of several well-known statistics. *Proc. Inter. Math. Cong.*, Toronto, vol. 2, pp. 805-813.
5. R. A. Fisher (1930). Inverse probability. *Proc. Camb. Phil. Soc.*, vol. 26, pp. 528-535.
6. R. A. Fisher (1933). The concepts of inverse probability and fiducial probability referring to unknown parameters. *Proc. Roy. Soc.*, A, vol. 139, pp. 343-348.
7. R. A. Fisher (1925-1958). *Statistical Methods for Research Workers*. Oliver and Boyd, Edinburgh.
8. R. A. Fisher (1934). Randomization, and an old enigma of card play. *Math. Gazette*, vol. 18, pp. 294-297.

9. R. A. Fisher (1935).	The fiducial argument in statistical inference. *Ann. Eugen.*, vol. 6, pp. 391-398.
10. R. A. Fisher (1935-1966).	*The Design of Experiments.* Oliver and Boyd, Edinburgh. § 9 and 10.
11. R. A. Fisher (1937).	On a point raised by M. S. Bartlett on fiducial probability. *Ann. Eugen.*, vol. 7, pp. 370-375.
12. R. A. Fisher and F. Yates (1938-1963)	*Statistical Tables for Biological, Agricultural and Medical Research.* Oliver and Boyd, Edinburgh.
13. R. A. Fisher (1941).	The asymptotic approach to Behrens' integral, with further tables for the d test of significance. *Ann. Eugen.*, vol. 11, pp. 141-172.
14. R. A. Fisher (1945).	The logical inversion of the notion of the random variable. *Sankhya*, vol. 7, pp. 129-132.
15. R. A. Fisher (1956)	On a test of significance in Pearson's *Biometrika Tables* (No. 11). *J. R. Stat. Soc. B.*, vol. 18, pp. 36-40.
16. R. A. Fisher and M. J. R. Healy (1956).	New tables of Behrens' test of significance. *J. R. Stat. Soc. B.*, vol. 18, pp. 212-216.
17. F. R. Helmert (1875).	Ueber die Berechnung des wahrscheinlichen Fehlers aus einer endlichen Anzahl wahrer Beobachtungsfehler. *Z. Math. Phys.*, vol. 20, pp. 300-303.
18. J. Neyman and E. S. Pearson (1933).	The testing of statistical hypotheses in relation to probabilities *a priori*. *Proc. Camb. Phil. Soc.*, vol. 29, pp. 492-510.
19. J. Neyman (1938).	L'estimation statistique traité comme un problème classique de probabilité. *Actualités sci. industr.*, vol. 739, pp. 54-57.
20. E. S. Pearson (1947).	The choice of statistical tests illustrated on the interpretation of data classed in a 2×2 table. *Biometrika*, vol. 34, pp. 139-163.

21. E. S. Pearson and H. O. Hartley (1954). *Biometrika Tables for Statisticians.* Cambridge University Press.

22. K. Pearson (1900). On the criterion that a given system of deviations from the probable in the case of a correlated system of variables is such that it can be reasonably supposed to have arisen from random sampling.
Phil. Mag., Series V, vol. 1, pp. 157-175.

23. K. Pearson (1925). Further contributions to the theory of small samples.
Biometrika, vol. 17, pp. 176-199.

24. "Student" (1908). The probable error of a mean.
Biometrika, vol. 6, pp. 1-25.

25. P. V. Sukhatmé (1938). On Fisher and Behrens' test of significance for the difference in means of two normal samples.
Sankhya, vol. 4, pp. 39-48.

26. E. B. Wilson (1941). The controlled experiment and the fourfold table.
Science, vol. 93, pp. 557-560.

27. E. B. Wilson (1942). On contingency tables.
Proc. Nat. Acad. Sci., vol. 28, pp. 94-100.

28. E. B. Wilson and Jane Worcester (1942). Contingency tables.
Proc. Nat. Acad. Sci., vol. 28, pp. 378-384.

29. F. Yates (1934). Contingency tables involving small numbers and the χ^2 test.
Suppl. J. R. Stat. Soc. B., vol. 1, pp. 217-235.

CHAPTER V

SOME SIMPLE EXAMPLES OF INFERENCES INVOLVING PROBABILITY AND LIKELIHOOD

1. The logical consequences of uncertainty

The concepts sketched in Chapter III have arisen in the study of numerical observations in the Natural Sciences; they are intended for use in the inferences by which progress in the sciences is guided. Since the reasoning is quantitative it involves mathematical operations, which need not, however, be of a very complicated kind. Indeed, the examples may be confined to simple cases, though often at the expense of being scientifically trivial, for it is not the mathematics but the logical nature of these concepts that requires to be exemplified. Since the reasoning is inductive, the development for it of appropriate mathematical operations seems to run counter to the view that all mathematics can be reduced to a single and wholly deductive system. Admittedly deductive processes play a predominant part in mathematics, yet it is difficult to admit that mathematics is less than the whole art of exact quantitative reasoning, and as such must extend beyond the domain of deduction proper.

The theory that all mathematics could be reduced to a purely deductive system, which was popular about the beginning of this century, has, moreover, in the meantime suffered, with the development of axiomatic studies, some rather severe setbacks. It is common ground that the consistency of the axiomatic basis of a deductive system is essential for the reliability of its consequences. It has been formally demonstrated that a system admitting one contradiction must admit all, in the sense that any proposition whatever can be deduced from it, by formally rigorous processes. The non-existence of contradictory consequences is thus a burning question for the whole superstructure. Moreover, it has been proved that the non-existence of such contradictions can never be demonstrated on the basis of the axioms of the system themselves. It would be rather absurd, indeed, to imagine that any chain of theorems, derived from a given axiomatic basis, could disprove a possible property of that basis, when it is known that, if it had that property, these same theorems could certainly be deduced from it. For the possibility of proving such theorems does not depend upon the truth of what they assert. It would seem, therefore, that the validity of a purely deductive system has at best the same logical status as has a scientific theory, which has not yet been found in any case to be in conflict with the observations. As such it appears to be solidly based on a well-tested induction.

The axiomatic theory of mathematics has not been, and ought not to be; taken very seriously in those branches of the subject in which applications to real situations are in view. For, in applied mathematics, it is unavoidable that new concepts should from time to time be introduced as the cognate science develops, and any new definition having axiomatic implications is inevitably a threat to the internal consistency of the whole system of axioms into which it is to be incorporated. We have seen that the introduction of the concept of probability has caused just such an axiomatic disturbance which can only be remedied by a proper analysis of its meaning. In its applications, therefore, mathematics cannot easily be reduced to a closed and static system, but has to develop with the development of human thought, of which it is an important vehicle.

The purpose of deductive processes is to reveal, or uncover, the latent consequences of the axiomatic basis adopted. Nothing essentially new can be discovered, but the coherence of the whole structure can be usefully demonstrated, and its consistency to some extent tested. In the future these processes will perhaps be carried out better by machines than by men. The axiomatic basis, in any case, is tailored with a view to its deductive consequences, and it is this which gives it its real utility. Deductive arguments are, in fact, often only stages in an inductive process. For example, Bayes' theorem, on the data postulated, is strictly deductive, nevertheless we may include it among the processes of induction, on the ground that the probability statement *a priori* on which the argument is founded must in the real world have an inductive, and factual, rather than an axiomatic, basis.

On the contrary, the purpose of inductive reasoning, based on empirical observations, is to improve our understanding of the systems from which these observations are drawn. The appropriate mathematical forms for reasoning of this type have been becoming clear during the present century owing to the widespread application of statistical methods to scientific data, and of increasing understanding of the principles of the design of experiments. One of the obstacles which has had to be overcome is the tendency to impose on inductive thought the conventions and preconceptions appropriate only to deductive reasoning.

The governing characteristic of inductive reasoning is that it is always used to arrive at statements of uncertainty, and that logical situations are recognizable in which different types or degrees of uncertainty require to find rigorous expression. It has been thought that the Theory of Mathematical Probability, in spite of the fact that a probability statement is in reality a statement of a specific type of uncertainty, could be included among strictly deductive processes. This has seemed possible largely because many mathematical treatises have adopted a formal and abstract treatment in which the element of uncertainty is inoperative, just because applications to the real world are avoided.

The logical characteristic, which has been too much overlooked, of all inferences involving uncertainty is that the rigorous specification of the nature and extent of the uncertainty by which they are qualified must in general involve the whole of the data, quantitative and qualitative, on which they are based.

As soon as it is regarded realistically it is seen that the concept of Mathematical Probability shares this requirement. In a statement of probability the predicand, which may be conceived as an object, as an event, or as a proposition, is asserted to be one of a set of a number, however large, of like entities of which a known proportion, P, have some relevant characteristic, not possessed by the remainder. It is further asserted that no subset of the entire set, having a different proportion, can be recognized.

If, therefore, any portion of the data were to allow of the recognition of such a subset, to which the predicand belongs, a different probability would be asserted using the smallest such subset recognizable.

When no further subset is recognizable, which can be known only by an exhaustive scrutiny of the data, the predicand is spoken of as a random member of the ultimate set to which it belongs. An imagined process of sampling in which a succession of predicands are identified may be used to illustrate the relation between the proportion expected to be observed in the sample, and the primary proportion required to specify the set, now to be identified with the population sampled. Rather unsatisfactory attempts have been made to define the probability by reference to the supposed limit of such a random sampling process.

Difficulty has sometimes been expressed when the reference set, or the population sampled, is said to be infinite. The definition and consequent calculations can, however, be applied to any finite set however large, and the limit of these results, where the number in the set is increased indefinitely, is all that is meant by the results of sampling from an infinite population. The clarity of the subject has suffered from attempts to conceive of the "limit" of some physical process to be repeated indefinitely in time, instead of the ordinary mathematical limit of an expression of which some element is to be made increasingly great.

The following sections are intended to illustrate the kinds of reasoning, and concurrent mathematical operations, appropriate to various types of uncertainty.

2. Bayesian prediction

Expressed in terms of the hypothetical ratio $p : q$, Bayes' classical inference (from the uniform distribution *a priori* he used) is that the probability distribution of p is exactly

$$\frac{(a+b+1)!}{a!\,b!}\, p^a q^b\, dp , \qquad (61)$$

in the light of the empirical observation of a successes out of $(a+b)$ trials. Alternatively, the hypothetical parameter p (or q) can be eliminated, and the inference expressed wholly in terms of the probability of future observations.

For example, if $c+d$ further trials were to be made with the same causal system, the probability, for each possible value of p, of observing just c successes is

$$\frac{(c+d)!}{c!\,d!}\, p^c q^d . \qquad (62)$$

The average of this fraction, over all possible values of p, is then found by integration to be

$$\frac{(a+b+1)!}{a!\,b!} \cdot \frac{(a+c)!\,(b+d)!}{(a+b+c+d+1)!} \cdot \frac{(c+d)!}{c!\,d!} ; \qquad (63)$$

this represents the probability, in the light of the previous experience, of obtaining c successes in $(c+d)$ further trials, the hypothetical parameters p, q having been eliminated. It connects future rational anticipations directly with the experience on which they are based, without the mediation of hypothetical quantities. It is postulated only that the two samples are drawn from the same constant population of possibilities, and that Bayesian knowledge *a priori* is available.

It may be noticed that the last factor in the expression developed above,

$$\frac{(c+d)!}{c!\,d!} \tag{64}$$

stands only for the binomial coefficients forming the last line, or base, of Fermat's arithmetical triangle; but

$$\sum_{c=0}^{c+d} \frac{(c+d)!}{c!\,d!} p^c q^d \tag{65}$$

is not the only polynomial in p, q, the value of which is constantly equal to unity. If, in fact, the triangle is extended to any chosen boundary, as for example in the diagram (Fig. 3), the thirteen totals outside the boundary are the coefficients $\omega(c, d)$ of a polynomial

$$\Sigma\, \omega(c, d) p^c q^d = p^4 + 4p^6 q + 18 p^6 q^2 + \cdots \tag{66}$$

of which the value is unity for all values of p.

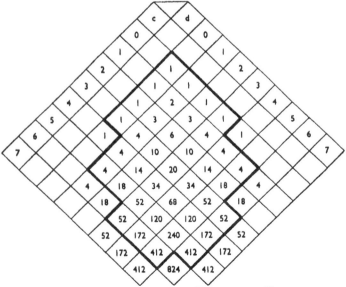

FIG. 3. THE ARITHMETICAL TRIANGLE EXTENDED

Then, based on previous experience of a successes out of $a+b$, we may infer the probability of reaching the terminal value (c, d) to be

$$\frac{(a+b+1)!}{a!\,b!} \cdot \frac{(a+c)!\,(b+d)!}{(a+b+c+d+1)!} \cdot \omega(c, d), \qquad (67)$$

if a subsequent trial were made with these end-points.

3. Fiducial prediction

When there are no data *a priori* and the fiducial argument is available, the parametric values may equally be eliminated, and the appropriate inferences expressed as probability statements about future observations. In the case of the radioactive source considered in Section 3.3, if the total length of N_1 measured time intervals were X_1, we previously drew the inference that

$$\chi^2_{2N_1} = 2\theta X_1, \qquad (68)$$

where θ is the true rate of emission, and χ^2 is distributed as is the sum of $2N_1$ independent squares of normal deviates each having unit variance.

If, now, a second series of N_2 time readings were to give a total time of X_2 it follows equally that

$$\chi^2_{2N_2} = 2\theta X_2. \qquad (69)$$

The ratio of X_2 to X_1 is distributed, therefore, in random samples, in a distribution independent of θ, and depending only on N_1 and N_2. This distribution is the basis of the analysis of variance. The distribution of X_2 given X_1 is, in fact,

$$\frac{(N_1+N_2-1)!}{(N_1-1)!\,(N_2-1)!} \cdot \frac{X_1^{N_1} X_2^{N_2-1}\,dX_2}{(X_1+X_2)^{N_1+N_2}}, \qquad (70)$$

for all values of X_2 from 0 to ∞. Without discussing the possible values of the parameter θ, therefore, the exact probability of the total time recorded in a second series of trials lying within any assigned limits is thus calculable on the basis of the total time observed in the first series.

The probability of the value X_2 to be observed, exceeding any chosen value x, is the integral of the frequency element for all values of X_2 exceeding x, and is expressible as the sum of the first N_2 terms of a negative binomial expansion, i.e.

$$P = \left(\frac{X_1}{X_1+x}\right)^{N_1}\left\{1 + \frac{N_1 x}{X_1+x} + \frac{N_1(N_1+1)}{2}\left(\frac{x}{X_1+x}\right)^2 + \cdots + \frac{(N_1+N_2-1)!}{(N_1-1)!(N_2-1)!}\left(\frac{x}{X_1+x}\right)^{N_2-1}\right\}. \quad (71)$$

It should be observed that such fiducial probability statements about future observations are verifiable by subsequent observations to any degree of precision required. This is not possible for probability statements inferred about parametric values, save on the supposition that they are capable in some other way of direct observation. Probability statements about the hypothetical parameters are, however, generally simpler in form, and, once their equivalence is understood to predictions in the form of probability statements about future observations, they are seen not to incur any logical vagueness by reason of the subjects of them being relatively unobservable.

In carrying out such a verification as that suggested above, it is to be supposed that the investigator is not deflected from his purpose by the fact that new data are becoming available from which predictions, better than the one he is testing, could at any time be made. For verification, the original prediction must be held firmly in view. This, of course, is a somewhat unnatural attitude for a worker whose main preoccupation is to improve his ideas. It is perhaps for this reason that some teachers assert that statements of fiducial probability cannot be tested by observations. It is also to be noted that future events or repetitions of the same event, which would be independent for a fixed value of a parameter, will not generally be independent when the parameter has a frequency distribution.

4. Predictions from a Normal Sample

A case of particular interest of the fiducial probability of future observations is offered by the process of sampling a normal distribution. From a sample of N observations the two statistics, the estimated mean

$$\bar{x} = \frac{1}{N} S(x), \qquad (72)$$

and the estimated variance of the mean,

$$s^2 = \frac{1}{N(N-1)} S(x-\bar{x})^2 \qquad (73)$$

subsume the whole of the information supplied by the sample about the population from which it was drawn. If μ is the true mean, the quantity,

$$\frac{\bar{x}-\mu}{s} = t, \qquad (74)$$

has a distribution independent of the unknown parameters, well known to be

$$\frac{(\frac{1}{2}(N-2))!}{(\frac{1}{2}(N-3))!\sqrt{\pi(N-1)}} \cdot \frac{dt}{\left(1 + \frac{t^2}{N-1}\right)^{\frac{1}{2}N}} \cdot \qquad (75)$$

for $N-1$ degrees of freedom.

This distribution has been very adequately tabulated so that the value of t is known for all levels of significance ordinarily required; the equation

$$\mu = \bar{x} - st \qquad (76)$$

expresses μ as a random variable having a distribution of "Student" 's type, with $N-1$ degrees of freedom, and a scale factor s, calculable from the sample observed by a rigorous fiducial argument, provided no information *a priori* is available.

Ignoring, however, the mean, μ, of the hypothetical population, it would equally have been possible by a direct fiducial argument to calculate the distribution, in the light of the N observations already made, of a further observation, x, so far unknown. For if it is to be drawn at random from the same population, x will be normally distributed with variance N times that of $\bar{x}$, about the same mean, and independently of it, so that

$$x - \bar{x}$$

has a normal distribution about zero with variance $(N+1)$ times as great as has $\bar{x}$, and consequently when the observed value s is used to eliminate σ has the distribution specified by

$$x - \bar{x} = st\sqrt{N+1} \qquad (77)$$

where, again, t has $N-1$ degrees of freedom. This prediction is evidently capable of verification to any degree of precision.

The same argument may be applied to predicting the values of a future sample of N', or, in particular of the statistics $\bar{x}'$ and s' derivable from it. In this case the value of the population variance could be estimated from the sum of the squares within both samples, that is as

$$\frac{1}{N+N'-2}\{S(x-\bar{x})^2+S'(x'-\bar{x}')^2\}, \quad (78)$$

and this quantity multiplied by

$$\frac{1}{N}+\frac{1}{N'} \quad (79)$$

would provide the estimated variance of the difference between the observed means

$$\bar{x}-\bar{x}'$$

based on $N+N'-2$ degrees of freedom. Consequently, for unselected samples

$$(\bar{x}-\bar{x}') \cdot \sqrt{\frac{NN'}{N+N'}} \cdot \sqrt{\frac{N+N'-2}{S+S'}} \quad (80)$$

will be distributed as is "Student" 's ratio t for $N+N'-2$ degrees of freedom. A second pivotal relationship is needed, since two values are to be predicted, and this is supplied by

$$2z = \log\frac{S}{N-1} - \log\frac{S'}{N'-1}, \quad (81)$$

the logarithm of the ratio of two independent estimates of the same variance, of which the distribution depends only on N and N', and is independent of the true mean.

To avoid misapplication of the method it should be noticed that for all pairs of primary observations $\bar{x}, S$, there is a one to one correspondence between pairs of "pivotal" values t, z, and corresponding pairs of predicted observations $\bar{x}'$, S'; equally for any pair of values $\bar{x}'$, S', a similar correspondence subsists between the pivotal values and the primary observations. Only on this condition the known frequency distribution of the pivotal values may be projected, or mapped, by direct substitution, to give the frequency distribution of the pair of unknowns to be predicted. The impossibility of statements of fiducial probability from discontinuous cases, such as the binomial distribution, is traceable to the fact that a single observational value corresponds, with any one parametric value, to a whole range in the values of the pivotal quantity, expressing the "probability integral" of the distribution.

In 1936,[4] addressing the Harvard tercentenary conference, I suggested that the condition for the further development of the use of fiducial inferences needed mathematical investigation, and would depend on the conditions of solubility of a type of problem, of which I gave, in general terms, an example, which has come to be known as the Problem of the Nile:

> The agricultural land of a pre-dynastic Egyptian village is of unequal fertility. Given the height to which the Nile will rise, the fertility of every portion of it is known with exactitude, but the height of the flood affects different parts of the territory unequally. It is required to divide the area, among the several households of the village, so that the yields of the lots assigned to each shall be in predetermined proportions, whatever may be the height to which the river rises.

The problem has not, I believe, in the meanwhile yielded up the conditions of its solubility, upon

PROBABILITY AND LIKELIHOOD 123

which, it would appear, the possibility of fiducial inference with two or more parameters, must depend.

It should, however, be noted, in the case of the two parameters of the normal distribution, that the sum of squares, the statistic S, not only yields a sufficient estimate for the true variance σ^2, but has a distribution independent of the true mean, μ, and so would supply also a solution of the type demanded in the Nile problem, if μ were the unknown variable.

An interesting application of the simultaneous distribution predicted for the two statistics of a future sample of N' values, is to allow N' to increase without limit, so that $\bar{x}'$ shall tend "in probability" to the population mean, μ, and the ratio

$$S'/(N'-1) \tag{82}$$

to the population variance, σ^2. We then have the simultaneous distribution, in the light of the first sample only, of the two parameters characterizing the population sampled. The frequency element of this simultaneous distribution is found to be the product

$$\sqrt{\frac{N}{2\pi\sigma^2}} \cdot e^{-\frac{N}{2\sigma^2}(\mu-\bar{x})^2} d\mu .$$

$$\frac{1}{\frac{N-3}{2}!} \cdot \left(\frac{S}{2\sigma^2}\right)^{\frac{1}{2}(N-1)} e^{-\frac{S}{2\sigma^2}} \cdot \frac{d\sigma^2}{\sigma^2}. \tag{83}$$

The rigorous step-by-step demonstration of the bivariate distribution by the fiducial argument would in fact consist first of the establishment of the second factor giving the distribution of σ given S, disregarding the other parameter, μ, and then of finding the first factor as the distribution of μ given $\bar{x}$ and σ. Several writers have adduced instances in which, when the formal requirements of the fiducial argument are ignored, the results of the projection of frequency elements using artificially constructed pivotal quantities may be inconsistent. When the fiducial argument itself is applicable, there can be no such inconsistency.

It will be noticed that in this simultaneous distribution (83) μ and σ^2 are not distributed independently. Integration with respect to either variable yields the unconditional distribution of the other, and these are naturally those obtainable by direct application of the fiducial argument, namely that

$$\frac{\mu-\bar{x}}{s} \tag{84}$$

is distributed as is t for $(N-1)$ degrees of freedom, while

$$\frac{S}{\sigma^2} \tag{85}$$

is distributed as is χ^2 for $(N-1)$ degrees of freedom. The distribution of any chosen function of μ and σ^2 can equally be obtained. The doubts expressed by Bartlett[1] on this point appear to be quite groundless.

It should, in general, be borne in mind that the population of parametric values, having the fiducial distribution inferred from any particular sample, does not, of course, concern any population of populations from which that sampled might have been in reality chosen at random, for the evidence available concerns one population only, and tells us nothing more of any parent population that might lie behind it. Being concerned with probability, not with history, the fiducial argument, when available, shows that the information provided by the sample about this one population is logically equivalent to the information, which we might alternatively have possessed, that it had been chosen at random from an aggregate specified by the fiducial probability distribution.

5. The fiducial distribution of functions of the parameters

A problem of this kind of some practical importance arises when it is desired to locate that value on a continuous scale which divides a hypothetical normal distribution in a given ratio. For example, to find the value which is only exceeded by one in forty of the population, of which we must judge by means of a randomly chosen sample of N measured individuals.

If μ and σ are the mean and standard deviation of the distribution, the point to be located may be represented by

$$\mu + a\sigma \tag{86}$$

where, for the chosen frequency of one in forty, the value of a must be about 1·96, and is always known in terms of the frequency specified.

If $\bar{x}$ and s stand for the mean and standard deviation as estimated from the sample of observations, using for this purpose

$$Ns^2 = S(x-\bar{x})^2,$$

we may put

$$\bar{x} + as = \mu + a\sigma, \tag{87}$$

and calculate the distribution of the quantity a in random samples, for, any sample must yield such a value. The distribution of a, which will depend on a, but not on the unknown parameters μ and σ, effectively exhibits in known terms the fiducial distribution of the particular linear function of the two parameters, chosen for examination. Equally, if some value a, were chosen with the intention of calculating from a succession of random samples,

126 STATISTICAL METHODS AND SCIENTIFIC INFERENCE

drawn perhaps from different populations, the value
$$\bar{x} + as, \tag{88}$$
then a knowledge of the sampling distribution of a for given α would display the distribution of the unknown deviate α, and therefore of the frequency ratio in which the true distribution had been partitioned.

The distribution of a for given α was first calculated for the Introduction to the *Mathematical Tables* (vol. 1)[2] of the British Association (1931) as an illustration of the appearance in statistical work of the function

$$I_n(x) = \frac{1}{\sqrt{2\pi}} \int_x^\infty \frac{(t-x)^n}{n!} e^{-\frac{1}{2}t^2} dt$$

$$= \frac{1}{\sqrt{2\pi}} \int_0^\infty \frac{t^n}{n!} e^{-\frac{1}{2}(t+x)^2} dt. \tag{89}$$

Using a sample of N observations as basis, it takes the form

$$\frac{(N-1)!}{2^{\frac{1}{2}(N-3)} \cdot \frac{N-3}{2}!} \cdot (1+a^2)^{-\frac{1}{2}N} \cdot e^{-\frac{N\alpha^2}{2(1+a^2)}} \cdot I_{N-1}\left(-\frac{a\alpha\sqrt{N}}{\sqrt{1+a^2}}\right) da,$$

$$\tag{90}$$

while the distribution of a for given α is easily expressed by putting

$$\alpha = \frac{u}{\sqrt{N}} + \frac{a\chi}{\sqrt{N}} \tag{91}$$

where u is normally distributed with unit variance, and χ is distributed *independently of u* in the familiar distribution for $N-1$ degrees of freedom.

$$\frac{1}{\frac{N-3}{2}!} (\tfrac{1}{2}\chi^2)^{\frac{1}{2}(N-3)} \cdot e^{-\frac{1}{2}\chi^2} \cdot \chi \, d\chi. \tag{92}$$

PROBABILITY AND LIKELIHOOD

The variable a has then the distribution of the sum of two variates distributed respectively in the normal distribution and in that of χ for $(N-1)$ degrees of freedom.

On a point of pure logic it may be noticed that the sampling distribution of a for given α, like that of α for given a, is entirely independent of the parameters of the distribution. In the fiducial distribution found for the parametric function

$$\mu + \alpha \sigma = \bar{x} + as$$

it is the introduction of the two sufficient statistics $\bar{x}$ and s, which brings in the requirement, that there must be no data *a priori* about the parameters, on which the fiducial argument relies. It requires that the observed statistics can be taken as random, and are unselected, and therefore representative values for the population from which they were obtained.

6. Observations of two kinds

It has been shown that observations of different kinds may justify conclusions involving uncertainty at different levels. It will be of some interest to consider the logical situation when observations of two such kinds are both available.

For example, let us suppose it to be possible to set a recorder to determine, for an exactly adjusted time interval, whether or no a charged particle has been received in that interval. It will not be supposed that the instrument will count them, if it receives more than one, but that it is capable of distinguishing the possible case of none, from the possible group of cases of "one or more."

Let p stand for the unknown probability of there being no particle, q the probability of there being one or more, then, if ξ is the time interval for which the instrument is set, and θ the unknown rate of delivery per unit of time, it appears that

$$p = e^{-\xi\theta}. \tag{93}$$

If then out of n trials it is observed that on a occasions no particle is recorded, while on b occasions there was one or more, we have a logical situation, without knowledge *a priori*, having a Likelihood Function,

$$e^L \propto p^a q^b, \tag{94}$$

in which expressions in θ can be substituted for p and q, to give the Mathematical Likelihood of any value of θ; but there is in the data no basis for making probability statements determining the probability that θ should lie between assigned limits.

Suppose, now, using the same supply of charged particles, it is possible to measure accurately a single randomly chosen time-interval between successive emissions. If this measured value is x_1, then for any given θ the distribution of x is

$$e^{-\theta x}\,\theta\,dx\,; \tag{95}$$

the probability of the random variable x exceeding the observed value, x_1, is

$$P = e^{-\theta x_1}, \tag{96}$$

and the fiducial distribution of θ is

$$dP = e^{-\theta x_1} x_1\, d\theta. \tag{97}$$

This fiducial distribution supplies information of exactly the same sort as that which Bayes showed how to obtain as a distribution of probability *a priori*. In fact

$$P = p^{x_1/\xi} = p^\lambda, \tag{98}$$

supplies the element of frequency
$$dP = \lambda p^{\lambda-1} dp, \qquad (99)$$
needed to complete Bayes' method. The simultaneous probability of p lying in this range, and of giving rise to the frequencies observed, is then
$$\frac{(a+b)!}{a!\,b!} \cdot \lambda p^{a-1+\lambda} q^b \, dp; \qquad (100)$$
the integral for all values of p over the range from 0 to 1 is
$$\frac{(a-1+\lambda)!\,b!}{(a+b+\lambda)!} \cdot \frac{(a+b)!}{a!\,b!} \lambda, * \qquad (101)$$
and the probability *a posteriori* is
$$\frac{(a+b+\lambda)!}{(a-1+\lambda)!\,b!} p^{a-1+\lambda} q^b \, dp. \qquad (102)$$
If the observation x_1 were made first and the recorder set so that
$$\xi = x_1, \qquad (103)$$
then λ would be unity, and we should have exactly Bayes' solution; but if it were thought better to set, for example,
$$\xi = 2x_1, \qquad (104)$$
then λ would be $\tfrac{1}{2}$, and the probability distribution *a posteriori* would be
$$\frac{(a+b+\tfrac{1}{2})!}{(a-\tfrac{1}{2})!\,b!} p^{a-\tfrac{1}{2}} q^b \, dp. \qquad (105)$$

* The factorial function, $x!$, has been generalized from positive integers only to all real numbers exceeding -1, by the Eulerian integral
$$x! = \int_0^\infty t^x e^{-t} \, dt.$$

130 STATISTICAL METHODS AND SCIENTIFIC INFERENCE

If the experimenter were bound to use a single setting, so that, after choosing ξ, the same time interval will be used for all tests with the same source, then the second choice, $\lambda=1/2$, will be on the average more informative (·4144 against ·4041) than the first proposal, $\lambda=1$. Between them there is an optimum with ξ about 51% above x; this makes the amount of information expected per trial about 2% higher still (·4232).

When the value of λ is open to choice, the question may arise as to what value it will be best to choose. There is here a problem in Experimental Design.

Supposing that the same value of ξ, and therefore of $\lambda(=x_1/\xi)$, is to be used in all subsequent trials, let

$$p = e^{-m},$$

so that $\qquad m = \xi\theta,$

or $\qquad \log m = \log \xi + \log \theta,$

and the amount of information supplied by each subsequent trial about $\log \theta$ will be equal to that supplied about $\log m$, since $\log \xi$ is a known constant. But the amount of information about $\log m$ given by each trial is known to be
(*Design of Experiments*, Section 68)

$$\frac{m^2}{e^m - 1} = i_{\log m}$$

The average value of this amount of information can be determined in terms of λ, and the value of λ which maximizes this average is the most profitable on the supposition adopted.

Since

$$P = p^\lambda = e^{-\lambda m}$$

the element of frequency is

$$\lambda e^{-\lambda m} dm,$$

and the function of λ to be maximized is

$$\lambda \int_0^\infty \frac{m^2 e^{-m\lambda}}{e^m - 1} dm$$

$$= \lambda \sum_{n=1}^\infty \int_0^\infty m^2 e^{-m(\lambda+n)} dm$$

$$= 2\lambda \sum_{n=1}^{\infty} \frac{1}{(n+\lambda)^3}$$

or
$$2\lambda\, s_3(\lambda),$$

where s_3 stands for the sum of the infinite series

$$\frac{1}{(1+\lambda)^3} + \frac{1}{(2+\lambda)^3} + \cdots$$

The equation for the value of λ which maximizes the amount of information expected is therefore

$$s_3 - 3\lambda s_4 = 0$$

where s_4 stands for the sum of the series

$$\frac{1}{(1+\lambda)^4} + \frac{1}{(2+\lambda)^4} + \cdots$$

I am indebted to Mr. R. H. Simpson of the Department of Statistics at Rothamsted for a rather exact solution of the equation, namely

$$\hat{\lambda} = \cdot 66142593,$$

or

$$\frac{1}{\lambda} = \frac{\xi}{x_1} = 1\cdot 5118851,$$

at which the amount of information about $\log \theta$ added by each further trial is nearly

$$\hat{i} = \cdot 41440,$$

which may be compared with the absolute maximum

$$\hat{i} = \cdot 64761,$$

obtainable with the most favourable value of p, namely

$$p = \cdot 203188,$$

as has been shown in Section 68 of *The Design of Experiments*.

Such a distribution *a posteriori*, whether expressed in terms of p or of θ, would be logically equivalent to a Bayesian probability *a posteriori*, or, equally, to one based exclusively on a fiducial argument, for the parameter is in each case a random variable of known distribution.

The percentile values of p in this distribution correspond each to each with the percentile values of the fiducial distribution of θ, through the relation

$$\theta = \frac{1}{\xi} \log 1/p \ .$$

The percentile values *a posteriori*, like those of the partial sum of the binomial expansion, may be found from the Table of z (Table V) of *Statistical Tables*, for the particular percentiles there tabulated, and their complements. The values of degrees of freedom (n_1, n_2) are equated to $2b+2$ and $2a+2\lambda$, so that if 2λ is not an integer, the latter would require interpolation; for this the Table of z is better suited than the associated Table of the Variance Ratio, e^{2z}.

In mathematical teaching the mistake is often made of overlooking the fact that Bayes obtained his probabilities *a priori* by an appropriate *experiment*, and that he specifically rejected the alternative of introducing them axiomatically on the ground that this "might not perhaps be looked at by all as reasonable"; moreover, he did not wish "to take into his mathematical reasoning anything that might admit dispute".

In many continental countries this distinction, which Bayes made perfectly clear, has been overlooked, and the axiomatic approach which he rejected has been actually taught as Bayes' method. The example of this Section exhibits Bayes' own method, replacing the billiard table by a radioactive source, as an apparatus more suitable for the 20th century.

7. Inferences from likelihoods

The mode of inference which takes the form of probability statements about parameters can lead to the alternative mode in the form of probability statements about future verifiable observations, by a general form of calculation; namely, if

$$F(\theta)\,d\theta$$

is the probability that the parameter lies in the range $d\theta$, and if for values within this range the probability of any future observable contingency, A, is

$$p_A(\theta)$$

then, eliminating θ, the probability of A is

$$P_A = \int p_A(\theta)\,F(\theta)\,d\theta\,, \tag{106}$$

taken over all possible values of the parameter.

The converse process of inferring the frequency distribution of θ from a knowledge of quantities such as P_A, for a sufficient variety of future possibilities A, is usually possible, and is extremely simple when A is taken to represent experience so ample as to confine the uncertainty of the parameter consistent with it, to an arbitrarily small range.

The cases in which the observations, together with other data, allow only of a statement of the Mathematical Likelihood, require separate consideration, to obtain a clear view of the rational prospect which future contingencies present, when knowledge of the Likelihood Function only is available.

In the previous Section it has been shown that, where data of two kinds are simultaneously available, one capable of supplying Likelihood statements only, while the other, however meagre and uninformative it may be quantitatively, is capable of leading to Probability statements, then the two kinds of data, available, as it were, in parallel, may, if exhaustive estimation is possible, supply inferences in terms of Mathematical Probability. In the case, also, in which the probability statement is based on an independent act of random sampling, this can be done, as Bayes had shown, by multiplying each element of the frequency distribution by a multiple of the corresponding Likelihood, this multiple being chosen to make the elements so formed add, or integrate, to unity.

Further, from the nature of the Likelihood Function it is evident that if data yielding Likelihood statements only are available from two independent sources, the aggregate of the two sources of data will supply simply a Likelihood Function found by multiplying together the two functions supplied by its parts.

In view of these various relationships it would be natural to expect that when the two types of nexus represented by Mathematical Likelihood and Mathematical Probability are connected, as it were, in series, so that we have the likelihood of an exhaustive set of possibilities, for each of which the probability of some event A, is known, the whole would yield statements, not of probability, but of likelihood only. This humbler status is not, however, incompatible with substantial utility.

Let us consider, from this standpoint, data of the Bayesian or Bernouillian type, without knowledge *a priori*, in which a successes have been observed out of $(a + b)$ trials, in relation to the rational bearing of such an observation on the prospect of observing c successes out of a subsequent $(c + d)$ trials. For example, if 3 successes have been observed out of 19 trials, what is the prospect of observing 14 successes in a subsequent set of 21 trials?

Considering the two-by-two table

3	16	19
14	7	21
17	23	40

we may recall that the relative likelihood of the first observation of three successes to sixteen failures is

$$p^3 q^{16} \cdot \frac{19^{19}}{3^3 \cdot 16^{16}}, \qquad (107)$$

which can be raised to unity by an appropriate choice of the ratio $p:q$. Similarly, with the second sample, if the probabilities of success and failure are p' and q', the relative likelihood is

$$p'^{14} q'^{7} \cdot \frac{21^{21}}{14^{14} \cdot 7^7}; \qquad (108)$$

now, if p and p' are to be the same, so that the two samples are drawn fairly from the same population, the most likely common value to give them is the total ratio of success, or $17/40$. Inserting this, and the complementary value, for p, q and for p', q', we find the likelihood of the whole table, namely the expression

$$\frac{17^{17} \cdot 23^{23}}{40^{40}} \cdot \frac{19^{19} \cdot 21^{21}}{3^3 \cdot 7^7 \cdot 14^{14} \cdot 16^{16}} \qquad (109)$$

in which the function

$$x^x \qquad (110)$$

has replaced the factorial function $(x!)$ used in the expression for the probability of the numerical table discussed, among others having the same margins. Taking logarithms to the base 10, the numerical values are

x	$\log x^x$	x	$\log x^x$
17	20·9176317	3	1·4313638
19	24·2963184	7	5·9156863
21	27·7666052	14	16·0457925
23	31·3197402	16	19·2659197
	104·3002955	40	64·0823997
	106·7411620		106·7411620
	997·5591335		

The relative likelihood of the two-by-two table under discussion is only about ·003623, and this is the likelihood assignable to the prospective contingency of fourteen successes out of twenty-one, in view of the data available. With the aid of a Table of x^x such as is given on page 143, it is easy to calculate such a series as the following:

Number of successes out of 21	Likelihood
14	·0036
13	·0093
12	·0216
11	·0460
10	·0904
9	·1642

Above 10 the likelihood is less than 1/15, and such future contingencies may be recognized in advance as definitely unlikely; above 8 the likelihood is still less than 1/5. The two values 9 and 10 thus lie in a zone in which the likelihood is still low. The test of significance discussed in Chapter IV, Section 4, could be used to confirm these judgements from another standpoint. The test of significance suffers, like the "confidence limits" calculated for a binomial distribution from some insensitivity due to the discontinuity of the distribution, and this may be thought to outweigh the rather formal advantage of asserting significance at a given level, when this yields no probability statement more definite than an inequality.

The likelihood assigned to a fourfold table is symmetrically related to the first (real) and second (conjectural) sample. The likelihood of observing 14 successes out of 21 as judged by data showing 3 successes out of 19, is exactly the same as the likelihood in prospect of observing 3 successes out of 19, judged on the basis of experience of 14 successes out of 21. Under another aspect, supposing both samples to have been observed, the same measure may be taken to be the likelihood of the hypothesis that they have been drawn from equivalent populations.

As in the cases in which probability rather than likelihood can be predicated, we may recover the likelihood statement appropriate to the parameters, by considering the likelihood in the limit of a very large sample showing in all pN successes out of N. The likelihood could then be written

$$\frac{(pN)^{pN}(qN)^{qN}}{N^N} \cdot \frac{19^{19}(N-19)^{N-19}}{3^3 16^{16}(pN-3)^{pN-3}(qN-16)^{qN-16}} \quad (111)$$

leading, when N is increased without limit to

$$\frac{19^{19}}{3^3 \cdot 16^{16}} p^3 q^{16} \quad (112)$$

the likelihood as inferred directly for the parameters.

It may be noted that the likelihood of a future trial yielding c successes and d failures does not involve the factor

$$\frac{(c+d)!}{c!\,d!}, \quad (113)$$

representing the number of ways in which such an outcome could occur. Nor, if a subsequent trial were to be made with chosen end-points, as in Fig. 3. would the likelihoods of these end-points involve the coefficients representing the number of paths in the extended triangle leading to each. Unlike a probability, the likelihood is independent of the number of ways in which the result could be brought about, for a statement of Likelihood does not involve a measureable reference set.

8. Variety of logical types

It is a noteworthy peculiarity of inductive inference that comparatively slight differences in the mathematical specification of a problem may have logically important effects on the inferences possible. In complicated cases, such effects may be very puzzling, as the conditions of solubility of problems of the Nile type have proved themselves to be. It may therefore be useful to consider some cases of extreme simplicity.

Let us suppose that x and y are two observed quantities, known each to be normally, and independently distributed with unit variance, x about an unknown value ξ, and y about η. It will be required to draw inferences about the pair of values (ξ, η) which may, of course, be represented as an unknown point, H, on a plane, on which (x, y) may be represented by an observed point, O.

If the data were as above, without further restriction, the probability distribution of the unknown point, formally demonstrable by the fiducial argument, is evidently a normal bivariate distribution, with unit variance in all directions, centred at the observed point (x, y). Additional data do, however, alter the character of the problem. It may be interesting to compare three cases:

(a) H is known to lie on a given straight line.
(b) H is known to lie on a circle.

(c) The given functional relationship between ξ and η does not confine this point either to a straight line, or to a circle, but to some other plane curve.

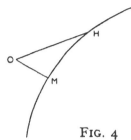

FIG. 4

In all cases the likelihood of any pair of parametric values is

$$e^{-\frac{1}{2}r^2},$$

where r is the distance from O to H. The nearest point on the curve to the observed point O then represents the solution of maximum likelihood: If M stands for this point, the relative likelihood of any other point is

$$exp\{-\tfrac{1}{2}(OH^2-OM^2)\}.$$

Setting this equal to any chosen series of conventional fractions we shall have defined zones around the point of maximal likelihood M, comprising all points satisfying the functional relationship, however these may be connected.

(a) In the particular case in which H is restricted to a given straight line, M satisfies the condition for a Sufficient estimate, for the relative likelihood of any point H on the line is simply

$$exp(-\tfrac{1}{2}HM^2) \tag{114}$$

and this is the same for all possible observation points on the line OM, produced if necessary. That is, for all observations leading to the same estimate.

The probability that HM should exceed any quantity u, positive or negative is therefore

$$\frac{1}{\sqrt{2\pi}}\int_u^\infty e^{-\frac{1}{2}t^2}dt\,; \tag{115}$$

otherwise stated, the fiducial distribution of H is Normal with unit variance, and centred at the estimate point M.

In view of the relation (115) it appears that if the data were modified so that instead of an unlimited straight line, the possible range for H had a terminus T, then necessarily the fiducial distribution extends only so far as T, and at T it has a probability condensation equal to

$$\frac{1}{\sqrt{2\pi}} \int_{TM}^{\infty} e^{-\frac{1}{2}t^2} dt, \qquad (116)$$

where TM is positive if M falls within the permitted range, but is taken to be negative if M lies outside it.

Knowing the frequency distribution of H, having coordinates (ξ, η), it is possible to calculate that of a second pair of random values (x', y'), the coordinates of a point O'. In the case of an unlimited straight line, it is easy to see that O' is distributed normally about M as centre, with variance unity in directions at right angle to the line, and twice as much in directions parallel with it. The contours of equal frequency density are ellipses with eccentricity $1/\sqrt{2}$, or ·707. The probability, on the data, of such a second trial lying within any defined area is then calculable.

(b)

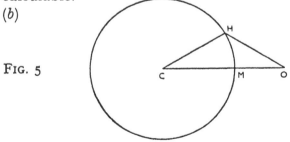

FIG. 5

The restriction which confines the unknown point H to the circumference of a given circle does not allow of a Sufficient estimate, for the relative likelihoods of the points on the circumference are not the same for different points O lying on the same radius CM, produced if necessary.

It is, however, a case in which an ancillary statistic can be recognized, namely the distance OC between the centre of the circle, and the point observed. For the sampling distribution of this distance must be the same for all points H on the circle. Consequently, we have only to determine the frequency distribution of the angle HCO, for given values of the distance OC, in order to obtain a fiducially determined distribution of the unknown point H.

If R stand for the radius of the circle, θ for the angle HCO, and a for the distance OC, then, since
$$HO^2 = R^2 + a^2 - 2aR\cos\theta, \qquad (117)$$
the factor of
$$e^{-\frac{1}{2}HO^2} \qquad (118)$$
which depends on θ is simply
$$e^{aR\cos\theta}. \qquad (119)$$

But
$$\int_0^{2\pi} e^{aR\cos\theta}\,d\theta$$
$$= 2\pi\left(1 + \frac{a^2R^2}{2^2} + \frac{a^4R^4}{2^2 \cdot 4^2} + \qquad \right) \qquad (120)$$
$$= 2\pi I_0(aR)$$
expressed as a Bessel function in Bassett's notation. Hence, the fiducial distribution of θ is
$$\frac{d\theta}{2\pi I_0(aR)}\, e^{aR\cos\theta}, \qquad (121)$$
the frequency density decreasing exponentially in the direction parallel to OC, at a rate, however, which depends on the ancillary statistic a. Here again we have a well-determined frequency distribution for the unknown point $H\,(\xi, \eta)$, from which fiducial limits

at all levels of probability can be calculated.

(c) In general, however, it is not to be expected either that a Sufficient statistic should exist, or that the most likely estimate could be made exhaustive by means of ancillary values. In such cases rational inference is effectively completed by the calculation of the Mathematical Likelihood for each plausible position of the unknown point. The practice of ignoring this quantity as a measure of rational belief appropriate to such cases would seem to leave the statistician who chooses this course without valid resource in respect of a great many subjects in which rational inference is possible. The fact that stronger inferences may be desired, and are certainly possible in other cases, seems to be no reason for not attuning our minds towards understanding the information actually available.

It is particularly to be noted in this example that the differences in the logical form of the available inferences flow from quite simple differences in the mathematical specification of the problem. Not to formal principles only, but there is also needed attention to particular analytic details.

REFERENCES

1. M. S. Bartlett (1937). Properties of sufficiency and statistical tests.
Proc. Roy. Soc., A, vol. 160, pp. 268-282.
2. R. A. Fisher (1931). The sampling error of estimated deviates.
Math. Tables, vol. 1, pp. xxvi-xxxv.
3. R. A. Fisher (1935). The fiducial argument in statistical inference.
Ann. Eugen., vol. 6, pp. 391-398.
4. R. A. Fisher (1936). Uncertain inference.
Proc. Amer. Acad. Arts and Sci., vol. 71, pp. 245-258.

PROBABILITY AND LIKELIHOOD

	$x \log_{10} x$	$x^x 10^{-x}$		$x \log_{10} x$	$x^x 10^{-x}$
1	0	1	41	66.124 1381	1.3309
2	0.602 0600	4	42	68.176 4702	1.5013
3	1.431 3638	2.7	43	70.239 1436	1.7344
4	2.408 2400	2.56	44	72.311 9178	2.0508
5	3.494 8500	3.125	45	74.394 5631	2.4806
6	4.668 9075	4.6656	46	76.486 8603	3.0680
7	5.915 6863	8.2354	47	78.588 5993	3.8779
8	7.224 7199	1.6777 ·	48	80.699 5794	5.0070
9	8.588 1826	3.8742	49	82.819 6079	6.6010
10	10.000 0000	1.0000	50	84.948 5002	8.8818
11	11.455 3195	2.8531	51	87.083 0190	1.2107
12	12.950 1750	8.9161	52	89.232 1739	1.7068
13	14.481 2636	3.0288	53	91.386 6211	2.4357
14	16.045 7925	1.1112	54	93.549 2630	3.5421
15	17.641 3689	4.3789	55	95.719 9479	5.2474
16	19.265 9197	1.8447	56	97.898 5295	7.9164
17	20.917 6317	8.2724	57	100.084 8668	1.2158
18	22.594 9051	3.9346	58	102.278 8236	1.9003
19	24.296 3184	1.9784	59	104.480 2687	3.0218
20	26.020 5999	1.0486	60	106.689 0750	4.8874
21	27.766 0652	5.8426	61	108.905 1199	8.0375
22	29.533 2990	3.4143	62	111.128 2847	1.3436
23	31.319 7402	2.0880	63	113.358 4546	2.2827
24	33.125 0698	1.3337	64	115.595 5183	3.9402
25	34.948 5002	8.8818	65	117.839 3682	6.9083
26	36.789 3070	6.1561	66	120.089 8997	1.2272
27	38.646 8216	4.4343	67	122.347 0118	2.2234
28	40.520 4249	3.3146	68	124.610 6061	4.0795
29	42.409 5419	2.5677	69	126.880 5873	7.5960
30	44.313 6376	2.0589	70	129.156 8628	1.4350
31	46.232 2125	1.7069	71	131.439 3428	2.7501
32	48.164 7993	1.4615	72	133.727 9397	5.3449
33	50.110 9600	1.2911	73	136.022 5688	1.0533
34	52.070 2832	1.1757	74	138.323 1473	2.1045
35	54.042 3816	1.1025	75	140.629 5948	4.2618
36	56.026 8900	1.0639	76	142.941 8330	8.7465
37	58.023 4638	1.0555	77	145.259 7858	1.8188
38	60.031 7767	1.0759	78	147.583 3790	3.8760
39	62.051 5197	1.1260	79	149.912 5402	8.1760
40	64.082 3997	1.2089	80	152.247 1990	1.7668

x	$x\log_{10}x$	$x^x 10^{-c}$	x	$x\log_{10}x$	$x^x 10^{-c}$
81	154·587 2865	3·8662	116	239·477 1267	3·0000
82	156·932 7359	8·5652	117	241·977 7458	9·5005
83	159·283 4817	1·9208	118	244·482 0769	3·0344
84	161·639 4600	4·3597	119	246·990 0884	9·7744
85	164·000 6087	1·0014	120	249·501 7496	3·1750
86	166·366 8668	2·3274	121	252·017 0298	1·0400
87	168·738 1750	5·4724	122	254·535 8993	3·4348
88	171·114 4752	1·3016	123	257·058 3287	1·1437
89	173·495 7106	3·1312	124	259·584 2890	3·8396
90	175·881 8258	7·6177	125	262·113 7516	1·2994
91	178·272 7667	1·8740	126	264·646 6887	4·4329
92	180·668 4801	4·1541	127	267·183 0726	1·5243
93	183·068 9142	1·1720	128	269·722 8761	5·2951
94	185·474 0182	2·9786	129	272·266 0726	1·8453
95	187·883 7425	7·6514	130	274·812 6358	6·4958
96	190·298 0384	1·9863	131	277·362 5397	2·3043
97	192·716 8582	5·2102	132	279·915 7589	8·2368
98	195·140 1554	1·3809	133	282·472 2683	2·9667
99	197·567 8843	3·6973	134	285·032 0430	1·0766
100	200·000 0000	1·0000	135	287·595 0587	3·9360
101	202·436 4588	2·7955	136	290·161 2915	1·4497
102	204·877 2175	7·5373	137	292·730 7177	5·3792
103	207·322 2341	2·1001	138	295·303 3139	2·0059
104	209·771 4673	5·9084	139	297·879 0572	7·5693
105	212·224 8764	1·6783	140	300·457 9250	2·8703
106	214·682 4217	4·8131	141	303·039 8949	1·0962
107	217·144 0642	1·3934	142	305·624 9449	4·2164
108	219·609 7656	4·0716	143	308·213 0534	1·6333
109	222·079 4883	1·2008	144	310·804 1989	6·3709
110	224·553 1954	3·5743	145	313·398 3603	2·5024
111	227·030 8506	1·0736	146	315·995 5169	9·8973
112	229·512 4185	3·2540	147	318·595 6482	3·9414
113	231·997 8641	9·9509	148	321·198 7339	1·5803
114	234·487 1530	3·0701	149	323·804 7540	6·3790
115	236·980 2516	9·5555	150	326·413 6889	2·5923

CHAPTER VI

THE PRINCIPLES OF ESTIMATION

1. Relations to other work

The logical principles of statistical reasoning, which it is my purpose in this book to set out for explicit consideration, have underlain and been implicitly required in the development of the two other main aspects of Statistical Science, namely (a) the mathematical methodology of the handling of bodies of observational data, so as to elicit what they have to tell us, and (b) the Design of the logical structure of an observational record, whether of an experiment, or of a survey, so as to ensure its completeness and cogency as a tool of research. In the two books that I have written with these ends in view it has not in either case seemed appropriate to enlarge upon purely logical considerations which had in fact found their fullest expression in earlier work on the Theory of Estimation. This theory is adverted to, therefore, in these books only with their particular ends in view. In *Statistical Methods* to exhibit the existence of competent and practical methods applicable to data of many types, to exemplify some of the kinds of complication which ordinarily arise, and to bring a wider class of cases into logical connection with the Analysis of Variance. In the *Design of Experiments*, I had chiefly in view, in this part of the book, the use of the concept of Amount of Information as a measurable characteristic by which the precision of an experiment could be anticipated, or confirmed, and compared with the expenditure of effort entailed.

In both books I hoped that the examples exhibited would not only indicate methods useful in themselves, but would also facilitate the development of principles of reasoning by which a body of data can be interpreted. And I believe they have indeed had this effect. Nevertheless, in both cases, it was my object to set out only what was immediately serviceable to a particular end, and to avoid the multiplicity of abstract concepts which an adequate discussion of the subject from a logical standpoint inevitably requires. I shall hope in this Chapter, on the contrary, to direct attention primarily to the logical aspects, and to develop these with a minimum of mathematical and technical complexity.

The Theory of Estimation discusses the principles upon which observational data may be used to estimate, or to throw light upon the values of theoretical quantities, not known numerically, which enter into our specification of the causal system operating. These principles have been more or less familiar for many years, but have been confused by a number of false starts due to insufficient appreciation of the nature of the problem.

A primary, and really very obvious, consideration is that if an unknown parameter θ is being estimated, any one-valued function of θ is necessarily being estimated by the same operation. The criteria used in the theory must, for this reason, be invariant for all such functional transformations of the parameters. This consideration would have eliminated such criteria as that the estimate should be "unbiased", meaning that the average value of the estimate should be equal to the true estimand; for if this were true of any parameter, it could not also be true of, for example, its square.

Although "unbiased" in this small-sample sense is thus a comparatively useless concept, a definition serviceable asymptotically in large-sample theory may be given in the form that

$$Pr\{N^\alpha |T - \theta| > \varepsilon\}$$

shall tend to zero as N is increased without limit, whenever $\alpha \leqslant \tfrac{1}{2}$, where ε is a positive quantity, however small, and T is the proposed estimate of the parameter θ. Another criterion in which the need for invariance in respect to functional transformations has been overlooked is that the Confidence Interval, or range within which the parameter is not rejected on some test of significance, shall be as short as possible. This is an inappropriate requirement since the relative lengths of any overlapping intervals may be adjusted arbitrarily by a functional transformation.

A distinction without a difference has been introduced by certain writers who distinguish "Point estimation", meaning some process of arriving at an estimate without regard to its precision, from "Interval estimation" in which the precision of the estimate is to some extent taken into account. "Point estimation" in this sense has never been practised either by myself, or by my predecessor Karl Pearson, who did consider the problem of estimation in some of its aspects, or by his predecessor Gauss of nearly one hundred years earlier, who laid the foundations of the subject. The distinction seems only to be made in order to support a claim, which is not indeed historical, to the effect that the authors have made in this matter an original contribution. It shows great confidence in the ignorance of students to put such a claim forward.

The following is not a complete exposition of the Theory of Estimation, but an outline emphasizing the origin and relevance of the logical concepts used elsewhere in this book.

2. Criteria of estimation

The fundamental criterion of estimation is known as the Criterion of Consistency, and is essentially a means of stipulating that the process of estimation is directed to the particular parameter under discussion, and not to some other function of the adjustable parameter or parameters. Of the attempts I have made to express this idea, one at least has been quite unsatisfactory, and all perhaps deserve restatement.

If a number, finite or infinite, of observable classes have probabilities of occurrence

$$p_j, \qquad S(p) = 1,$$

which are known functions of the parameters; and if out of N observations, the numbers observed to fall in these are

$$a_j, \qquad S(a) = N,$$

then the average value of a linear function of the observed frequencies

$$A = S(c_j a_j) \qquad (122)$$

when for each a is substituted its mean

$$\bar{a}_j = N p_j, \qquad (123)$$

is given by

$$\overline{A} = N S(c_j p_j), \qquad (124)$$

which is a known function of the probabilities, p, and therefore of the parameters.

The statistic A/N, calculable from the observations, will then be termed a Consistent estimate of the particular parametric function $S(c_j p_j)$. It will be noticed that it being linear in the frequencies is also

THE PRINCIPLES OF ESTIMATION 149

an unbiased estimate of this same function.

Now, if, for example, the value of
$$S(c_j p_i),\qquad(125)$$
summed over all possible observational classes, were the parametric function
$$\log \theta \qquad(126)$$
we might make the estimate
$$T = e^{A/N},\qquad(127)$$
and notice that when for all observed frequencies their expected values are substituted, the estimate T is such that it becomes identical with the estimand θ. This property is evidently invariant for transformations of the parameters, and does not imply the statement that T is an unbiased estimate of θ, for, in fact, it happens that $\log T$ is an unbiased estimate of $\log \theta$. It may be noted, for example, that the sufficient estimate T of Section 3, is not unbiased; and that this makes no difference to the fiducial distribution, of which the mean $\bar{\theta}$ is indeed equal to T.

In respect to bias it should be noted that no difficulty is usually experienced in adjusting an estimate so that the average of the adjusted value shall be equal to any particular parametric function; what has sometimes seemed to need emphasis is that the estimate is not necessarily improved by such an adjustment, which will introduce bias, not previously present, into the estimates of most functionally connected values. Before making any such adjustment, consideration should be given to its purpose.

The relations set out above are exact, and do not depend on the observed frequencies a being sufficiently large. It may well be that N of the frequencies are unity, and a very large, or infinite number, are zero; as, indeed will be the case when tolerably accurate measurements occur in the data, for these will be interpreted as single observations within very small ranges

$$x_1 \pm \tfrac{1}{2} \delta x_1 , \qquad (128)$$

and any function

$$S\{c(x_i)\} , \qquad (129)$$

where c is a continuous function of x, and the summation is taken over the observations and not over the classes, will be recognized as a linear function of the frequencies, and therefore as a suitable ingredient from which a consistent statistic can be built. The use of functions non-linear in the frequencies would in these cases introduce discontinuities whenever two measurements happen to coincide.

A Consistent Statistic may then be defined as:

> A function of the observed frequencies which takes the exact parametric value when for these frequencies their expectations are substituted.

This definition is applicable with exactitude to finite samples.

A much less satisfactory definition has often been used, namely that the probability that the error of estimation exceeds in absolute value any value ϵ, or, symbolically,

$$Pr\{|T-\theta|>\epsilon\} , \qquad (130)$$

shall tend to zero as the size of sample is increased, for all positive values ϵ, however small.

With respect to a function of the observations, T, defined for all possible sizes of sample this definition has a certain meaning. However, any particular method of treating a finite sample of N_1 observations may be represented as belonging to a great variety of such general functions. In particular, if T' stand for any function whatsoever of N_1 observations, and T_N for any function fulfilling the asymptotic condition of consistency, then

$$\frac{1}{N}\{N_1 T' + (N-N_1) T_{N-N_1}\} \tag{131}$$

is itself a statistic defined for all values of N, and tending asymptotically to the limit θ, yet it is recognizable when $N = N_1$ as the arbitrary function T' calculated from the finite sample.

In fact, the asymptotic definition is satisfied by any statistic whatsoever applied to a finite sample, and is useless for the development of a theory of small samples.

3. The concept of efficiency

An asymptotic or large-sample definition is, however, appropriate as a first step in defining the concept of efficiency. Consider the statistic

$$T = \frac{1}{N} S(a_i c_i), \tag{132}$$

in which the c_i are so far undetermined functions of θ, the parametric function of which T is to be an estimate. Then, T is consistent, when $\theta = \theta_0$, if

$$\theta_0 = S\{p_i c_i(\theta_0)\}, \tag{133}$$

and with the same coefficients c it will remain nearly consistent for small variations of θ if

$$1 = S\left\{\frac{\partial p_i}{\partial \theta_0} \cdot c_i(\theta_0)\right\}. \tag{134}$$

For large samples the variations to be expected may be made indefinitely small. Now the sampling variance of the linear function T is, exactly,

$$V(T) = \frac{1}{N}\{S(p_j c_j^2) - \theta_0^2\} \qquad (135)$$

and this variance may be minimized for variations of the coefficients c, subject to the limitations that the estimate shall be locally Consistent, by minimizing

$$S(p_j c_j^2) - \lambda S\left(c_j \frac{\partial p_j}{\partial \theta_0}\right) - \mu S(p_j c_j) \ . \qquad (136)$$

Varying any particular value c_i we find

$$2 p_j c_j - \lambda \frac{\partial p_j}{\partial \theta_0} - \mu p_j = 0 \ , \qquad (137)$$

from which μ may be determined by direct addition, giving

$$2\theta_0 = \mu \ , \qquad (138)$$

and λ by multiplying by c and adding for all classes, giving a second equation,

$$2 S(p_j c_j^2) = \lambda + \mu \theta_0 \ , \qquad (139)$$

so that substituting for λ and μ in (137) for each particular coefficient

$$p_j c_j = \frac{\partial p_j}{\partial \theta_0}\{S(p_j c_j^2) - \theta_0^2\} + p_j \theta_0 \ , \qquad (140)$$

or

$$p_j(c_j - \theta_0) = \frac{\partial p_j}{\partial \theta_0} S\{p_j(c_j - \theta_0)^2\} \ . \qquad (141)$$

Hence it easily follows that

$$V(T) = \frac{1}{N} S\{p_j(c_j - \theta_0)^2\} = \frac{1}{NS\left\{\frac{1}{p_j}\left(\frac{\partial p_j}{\partial \theta_0}\right)^2\right\}}, \qquad (142)$$

THE PRINCIPLES OF ESTIMATION 153

which we may write briefly as $1/I$. Moreover,

$$c_1 - \theta_0 = \frac{1}{p_s}\frac{\partial p_s}{\partial \theta_0} \div S\left\{\frac{1}{p_s}\left(\frac{\partial p_s}{\partial \theta_0}\right)^2\right\}. \quad (143)$$

For large samples, therefore, from populations having any particular parametric value θ, the linear function

$$T = \frac{1}{I} S\left(\frac{a_s}{p_s}\frac{\partial p_s}{\partial \theta_0}\right) + \theta_0, \quad (144)$$

is locally Consistent, and subject to this condition, has the least possible variance for samples of a given size. Since the p_s are known functions of θ, the general equation

$$S\left(\frac{a_s}{p_s}\frac{\partial p_s}{\partial \theta}\right) = 0 \quad (145)$$

or

$$S\left(\frac{a_s}{m_s}\frac{\partial m_s}{\partial \theta}\right) = 0, \quad (146)$$

if $m = pN$, being linear in the frequencies, though of any algebraic form in θ, will give, with large samples, such estimates of the highest precision for all values of θ.

In these limiting conditions the distribution of the estimate T tends to Normality, so that the specification of the variance as

$$\frac{1}{I} = \frac{1}{Ni}, \quad (147)$$

where, as above, i stands for

$$i = S\left\{\frac{1}{p_s}\left(\frac{\partial p_s}{\partial \theta}\right)^2\right\} = S\left\{p_s\left(\frac{\partial}{\partial \theta}\log p_s\right)^2\right\}, \quad (148)$$

is sufficient to specify fully the sampling distribution. The quantity I is the invariance of the estimate, and i

itself may be recognized as the amount of information to be expected for each observation made. The qualification "to be expected" is a reminder that the quantity i is itself a function of θ, of central importance also in the theory of small samples, and that with small samples the estimate of θ obtained from any sample will not be exactly correct.

The form of the Efficient equation of estimation,

$$S\left(\frac{a_j}{p_j}\frac{\partial p_j}{\partial \theta}\right) = 0, \qquad (149)$$

shows that it could be derived by maximizing

$$S(a_j \log p_j), \qquad (150)$$

or by maximizing that factor of the Mathematical Likelihood of θ which depends on θ, namely

$$\Pi(p_j^{a_j}). \qquad (151)$$

The solution of this equation is Consistent, and invariant for transformations of the parameter. Naturally, it is not generally unbiased in the small sample sense.

The equations of Maximum Likelihood are indeed the only equations of estimation, linear in the observed frequencies, which are efficient with large samples. The solutions of these equations are not generally linear functions of the frequencies.

4. Likelihood and Information

The connections between the Likelihood Function and the Amount of Information are worth noting. The Likelihood Function is determined by a particular sample, or *corpus* of observations, and shows for such observations the relative frequency with which different parametric values would yield such a sample. When the logarithm of the likelihood is used, different independent samples, of the same or of

THE PRINCIPLES OF ESTIMATION 155

different kinds, which throw light on the same parameter, may be combined merely by adding the log likelihoods for each value of the parameter.

For any particular value of the parameter, the probability of obtaining a given observational record may be represented by ϕ. Then using summation over all possible samples, it is clear that

$$S(\phi) = 1, \; S(\phi') = 0, \; S(\phi'') = 0, \qquad (152)$$

where differentiation with respect to a parameter is indicated by a prime.

Now

$$\frac{\partial^2}{\partial \theta^2}(\log \phi) = \frac{1}{\phi}\phi'' - \frac{\phi'^2}{\phi^2}, \qquad (153)$$

so that multiplying by ϕ and adding for all possible samples

$$E\left\{\frac{\partial^2}{\partial \theta^2}(\log \phi)\right\} = S(\phi'') - S\left(\frac{\phi'^2}{\phi}\right), \qquad (154)$$

or $\qquad 0 - I$,

since the last summation is merely the information expected for samples from populations having the particular value of the parameter chosen.

The average value of

$$-\frac{\partial^2}{\partial \theta^2}(\log \phi) = -\frac{\partial^2}{\partial \theta^2}L \qquad (155)$$

is thus equal to the amount of information expected. This relation shows very simply that the amount of information, like the likelihood function, is additive for independent bodies of data, even if of different sorts.

The value of the second differential coefficient of $(-L)$ with respect to θ is referred to as the amount of information realized at any value of θ. It is usually evaluated at that value for which L is maximized, since this has the highest likelihood of being the true value, but may be evaluated at any chosen value. At the maximum it measures the geometrical curvature. It is used in calculating the sampling variance of an estimate and is slightly more accurate, for this purpose, than the amount of information expected.

5. Grouping of samples

Ideally, we should like an estimate completely to replace the data from which it is drawn, so that no distinction need be made among different hypothetical lots of data which might yield the same estimate. This situation is sometimes, but not always, realizable.

If, for any given value of θ, the probabilities of observing the different types of samples which are to be grouped together are

$$\phi_1, \phi_2, \ldots \phi_s, \qquad (156)$$

such that

$$\sum_1^s (\phi) = \Phi \qquad (157)$$

Then

$$\sum_1^s \phi \left(\frac{\phi'}{\phi} - \frac{\Phi'}{\Phi} \right)^2 = \sum_1^s \left(\frac{\phi'^2}{\phi} \right) - \frac{\Phi'^2}{\Phi}, \qquad (158)$$

and since the expression on the left cannot ever be negative, it follows that the contribution to the amount of information is never increased by such grouping, and that the condition that it shall not be diminished is that, for all configurations yielding indistinguishable estimates,

$$\phi'/\phi \qquad (159)$$

shall be constant.

If this is so for all values of the parameter it follows that

$$\log \phi = L \tag{160}$$

shall be the same function of the parameter, apart from an additive constant. Samples to be grouped together must therefore have identical likelihood functions, if the grouping is not to be accompanied by loss of information. It is a characteristic of Sufficient statistics that the likelihood function of the statistic shall be the same as that of all samples from which such a statistic could have been calculated.

The actual loss incurred in other cases may be calculated by determining the frequency distribution of the statistic to be used in terms of its parameters, and calculating the amount of information supplied by a single observation from such a distribution. This supplies the criterion for efficiency in finite samples, and so completes the second stage of the theory. The information recovered will in general be less than the amount of information in the data from which the estimate was calculated, the differences having been lost through grouping the samples of different kinds which lead to the same estimate. With estimation by the Method of Maximum Likelihood, although the likelihood functions of samples leading to the same estimate may be different, yet the method of estimation has selected for grouping samples having likelihood functions so far alike as to have their maxima all at the same parametric value, and therefore with stationary ratios in the most important region. The method will necessarily lead to Sufficient Statistics when these exist.

In *The Theory of Estimation* (1925)[3] a good many examples have been given showing the loss of information for small samples of given size. When the likelihood function is everywhere differentiable these losses typically do not exceed the value of two or three observations, when the method of maximum likelihood is used in problems in which no sufficient estimate exists.

The concept of efficiency was introduced first in the foregoing discussion by a definition valid only in the limit for large samples. Such an approach is incomplete and must be taken only as a first step. In comparing estimates, all of which tend to be distributed Normally in the limit, a comparison of precision was immediately available; it is the evaluation of the maximum attainable that leads to the concept of the amount of information in an observation distributed in an error curve of any form, as is an estimate from a finite sample. It is the amount of information, and not the sampling variance, which completes the criterion of efficiency for finite samples, for when the distribution is not Normal the variance is an imperfect measure of the precision. The property, demonstrated above, that the amount of information may be diminished, or conserved, but cannot be increased by the processes of statistical reduction, guarantees the appropriateness of completing the definition by its means. Certain inequalities flow from the fact that a Normal distribution supplies the least possible information for an error curve of given variance, and has the least possible variance for a given amount of information.

6. Simultaneous estimation

In most practical situations there are a number of unknown parameters, not all of them necessarily of interest on their own account, but required in connection with the estimation of others. The likelihood is then a function of several variables, and

THE PRINCIPLES OF ESTIMATION 159

is maximized by the solution of simultaneous equations, all linear in the frequencies, obtained by differentiating the log likelihood with respect to a complete functionally independent set of parameters. The solution is an efficient set of simultaneous estimates.

In the practical procedure of solution, trial values of the parameters are chosen, principally with a view to their being near to the final solution, so far as this can be foreseen, but partly also for their simplicity as a basis for calculation, and the expressions for

$$S_1 = \partial L/\partial \theta_1,$$
$$S_2 = \partial L/\partial \theta_2,$$
(161)

etc., are evaluated by substituting this particular set of parametric values. These are known as the efficient scores, and it is by their means that the trial value may be adjusted.

For a single parameter the adjustment is, as has been seen, effected by dividing the score by the amount of information, that is, if $\delta\theta$ is the adjustment required by a trial value θ', then

$$I\delta\theta = S \qquad (162)$$

will supply an adjusted value $\theta' + \delta\theta$ which is an Efficient estimate on the data available; for this reason it is seldom necessary, though always possible as a check, to repeat the calculation with an improved trial value.

With many parameters further polishing of the solution is more often wanted, because in some types of data, the first appraisal is less likely to be correct. The amount of information, which, with one parameter is a simple scalar, becomes a symmetrical matrix, or square table of coefficients, in the general case, the coefficients being

$$\left\{\begin{array}{ll} S\left\{\dfrac{1}{p}\left(\dfrac{\partial p}{\partial \theta_1}\right)^2\right\} & S\left\{\dfrac{1}{p}\dfrac{\partial p}{\partial \theta_1}\dfrac{\partial p}{\partial \theta_2}\right\} \cdots \\ S\left\{\dfrac{1}{p}\dfrac{\partial p}{\partial \theta_1}\dfrac{\partial p}{\partial \theta_2}\right\} & S\left\{\dfrac{1}{p}\left(\dfrac{\partial p}{\partial \theta_2}\right)^2\right\} \cdots \\ \text{etc.} \end{array}\right\} \quad (163)$$

or, written more compactly,

$$i_{rs} = S\left\{\dfrac{1}{p}\dfrac{\partial p}{\partial \theta_r}\dfrac{\partial p}{\partial \theta_s}\right\}, \quad (164)$$

where r, s, are suffices specifying particular parameters θ_r, θ_s.

The series of adjustments $\delta\theta$ are calculated from the linear equations of which the coefficients are given by the information matrix, and the right hand sides, by the Efficient Scores. In matrix form

$$I\delta\theta = S, \quad (165)$$

in which now $\delta\theta$ and S stand for series of values corresponding with the series of parameters.

An outstanding advantage of this form of analysis is that the precision of the simultaneous estimate is, for large-sample theory, given by a covariance matrix which is the reciprocal of the information matrix. So that if the matrix product IV is equal to the identity, then V supplies the variances and covariances of the components of the efficient estimate. The matrix V can then be calculated simply by replacing the scores by the several series

$$\begin{array}{cccc} 1, & 0, & 0, & 0 \\ 0, & 1, & 0, & 0 \\ 0, & 0, & 1, & 0 \\ 0, & 0, & 0, & 1 \text{ etc.} \end{array} \quad (166)$$

THE PRINCIPLES OF ESTIMATION

and if this way is chosen the adjustments $\delta\theta$ are obtained by direct multiplication indicated by the matrix product

$$\delta\theta = VS.$$

It is to be noted that whereas the matrix I is exact for small samples, the identification of its reciprocal V with the covariance matrix of the simultaneous estimates is not exact; with small samples the sampling variation of the components will not generally be Normal, so that the covariance matrix cannot completely specify the distribution.

Properly speaking, therefore, the exact covariance matrix is not required; it could only be regarded as an approximation to V which does determine the simultaneous information matrix. To term V the covariance matrix is conventional and a useful reminder of an identity in large-sample theory.

With a single parameter, if θ were replaced by any known function $\phi(\theta)$, the amount of information, in each ingredient to be summed, is multiplied by

$$\left(\frac{\partial \theta}{\partial \phi}\right)^2$$

so that

$$I_\phi = \left(\frac{\partial \theta}{\partial \phi}\right)^2 I_\theta, \qquad (167)$$

where I_ϕ and I_θ stand for the amounts of information in any body of data in respect of ϕ and θ respectively. Similarly, with two or more parameters, when I has been replaced by a matrix which may still be written I_θ, if A is a matrix such that

$$a_{ij} = \frac{\partial \theta_i}{\partial \phi_j}, \qquad (168)$$

then I_ϕ is given by the matrix product[1]

$$I_\phi = A^* I_\theta A \qquad (169)$$

where A^* stands for the transpose of A.

This transformation system for the information matrices is exact with finite samples whereas the corresponding transformation of the covariance matrices could only be approximate.

Nevertheless, if

$$b_{ij} = \partial \phi_j / \partial \theta_i$$

the relation

$$v_\phi = B^* V_\theta B$$

is valid and exact, where V is the reciprocal of the information matrix.

7. Ancillary information

The study of the sampling errors, that is, of the precision, of statistical estimates, the core of the Theory of Errors as developed by Gauss, has led to the recognition among the multitude of Consistent estimates which can be invented, of a smaller class, such that, in the important class of cases in which the sampling distribution tends in large samples to the Normal form, then the limit of the product of the variance and the size of sample shall be as small as possible. The existence of such a limit has been demonstrated above, and its value has been expressed in terms of the relations between the unknown parameters and the frequencies of observations of all recognizable kinds. Such statistics of minimal limiting variances are termed Efficient, and thus

[†An inconsequential mistake involving matrices has been removed from this sentence.—J.H.B.]

THE PRINCIPLES OF ESTIMATION

satisfy a second rational criterion of what is required of a statistical estimate. It is easy to demonstrate[3] that any two estimates both efficient must have a correlation in random samples, and that their correlation coefficient must tend to a limit $+1$ as the size of the sample is increased. In fact, in the theory of "large samples" all efficient estimates are equivalent.

The theory of large samples can, however, never be more than a first step preliminary to the study of samples of finite size, although in fact a great many practical problems do not need for their effectual resolution, any further refinements. I do not think that this is a reason for not developing those concepts required for exact thought on small-sample problems. In such problems the different possible Efficient estimates must be distinguished. So far as the choice among them is concerned, a rational criterion is that we should prefer that estimate which conserves the most, or loses the least of the information supplied by the data. In practical terms, if from samples of 10 two or more different estimates can be calculated, we may compare their values by considering the precision of a large sample of such estimates each derived from a sample of only 10, and calculate for preference that estimate which would at this second stage give the highest precision. I do not know of a general proof, but no exception has been found to the rule that among Consistent Estimates, when properly defined, that which conserves the greatest amount of information is the estimate of Maximum Likelihood. The unique position of this method of estimation is also indicated by its being the only one in which the equations of estimation are linear in the frequencies. With small samples this obviates the irrationality of discontinuous changes in the estimates corresponding with minimal changes in the data.

A realistic consideration of the problem of estimation in small samples thus points unmistakably to the estimate of maximum likelihood as the uniquely appropriate single value for use in estimation, if any single value (i.e. with no ancillary values) is to be used. It indicates also that when using the maximum likelihood estimate, some loss of information may occur, and, although quantitatively this loss may be trifling, it cannot be unimportant logically, especially when an exhaustive treatment of the data is required in the calculation of probability statements.

The most important step which has been taken so far to complete the structure of the theory of estimation is the recognition of Ancillary statistics. The notion was first developed in a detailed study of the amount of information lost calculated exactly for a number of trivial but representative problems. It was shown not only that loss of information must vanish if all types of sample yielding the same estimate had identical Likelihood functions, and that for maximum likelihood estimates all must have functions stationary at the estimated value, but further that the loss of information suffered in the limit of large samples, was expressible directly in terms of the sampling variance of the second differential coefficient

$$\frac{\partial^2 L}{\partial \theta^2},$$

and indeed is equal in the limit to this variance multiplied by the variance of the estimate. When the likelihood function is repeatedly differentiable therefore, loss of information by simple estimation is due to the variance in the amount of information "realized". A simple remedy, appropriate for this asymptotic situation is merely to record, not only the estimate of maximum likelihood but the amount of information realized, or the "apparent precision", using this to supply a "weight" for the estimate more precise than that supplied by the size of the sample. Such a procedure, though merely asymptotic, and consisting of a refinement of the large-sample procedure, will guarantee that the loss of information in large samples shall tend to zero. The use of further differential coefficients of the Likelihood in the neighbourhood of the estimate can, by an extension of the same process, reduce the limit of N times the loss of information, or of N^r times its value to zero, as N tends to infinity. This is indeed an evasion of the problem of small samples properly speaking, but it does serve to show that it is the Likelihood function that must supply all the material for estimation, and that the ancillary statistics obtained by differentiating this function are inadequate only because they do not specify the function fully.

8. The location and scale of a frequency distribution of known form

If the probability of an observation falling into the range dx is given in the form

$$df = \exp\left\{\phi\left(\frac{x-a}{\beta}\right)\right\} \cdot \frac{dx}{\beta}, \qquad (170)$$

in which ϕ is a function of known form of the single real variable $(x-a)/\beta$, which may be taken to be differentiable almost everywhere, and a and β are two unknown parameters specifying the location and the scale of the distribution, then the logarithmic likelihood function is

$$L = -N \log \beta + S\left\{\phi\left(\frac{x-a}{\beta}\right)\right\}, \qquad (171)$$

where S stands for summation over the N values of a sample observed. It is not to be expected that the sum of the functions ϕ will be algebraically simple, and in consequence we shall not in general be able to express the likelihood in terms simply of the parameters and appropriate estimates of them. There will generally be no Sufficient pair of estimates.

The equations of maximum likelihood are easily seen to be

$$S\left\{\phi'\left(\frac{x-a}{\beta}\right)\right\} = 0$$

$$S\left\{\frac{x-a}{\beta}\phi'\left(\frac{x-a}{\beta}\right)\right\} + N = 0, \qquad (172)$$

in which ϕ' stands for the first differential coefficient of the known function ϕ.

Let us now suppose that A and B are values of a, β satisfying the equations of estimation, and that ϕ is such that these values are real and unique, then the sample observed will have supplied a set of N values of u such that

$$x_i = A + Bu_i, \qquad (173)$$

and the values of u satisfy the conditions

$$S\{\phi'(u)\} = 0 \qquad (174)$$

$$S\{u\phi'(u)\} + N = 0. \qquad (175)$$

The particular set of values u satisfying these equations, and derived from the sample observed,

THE PRINCIPLES OF ESTIMATION 167

may be said to specify the complexion of the sample. It is easy to see that if for the values x observed we had instead the values

$$X = \lambda + \mu x, \qquad (176)$$

then the complexion of the sample would be unchanged. To state the matter otherwise, the complexion depends only on the ratios of the $N-1$ successive differences among the N observations, when these are arranged in order of magnitude. It is evident, moreover, that the sampling variation of these ratios are severally and jointly independent of the parameters. The values u, specifying the complexion, thus define by their differences a set of $N-2$ functionally independent ancillary statistics, and the precision of the values A and B arrived at should be judged solely by reference to the variation of estimates among samples having the same complexion.

The precision of the estimates A and B may be specified by the measures of deviation

$$t_1 = \frac{A-a}{\beta}, \qquad t_2 = \frac{B}{\beta}, \qquad (177)$$

and since

$$\frac{x-a}{\beta} = t_1 + ut_2, \qquad (178)$$

the simultaneous frequency distribution of t_1 and t_2 is

$$df \propto e^{S\{\phi(t_1+ut_2)\}} t_2^{N-2} dt_1 dt_2. \qquad (179)$$

Moreover, the distribution of t_2 does not depend on a, being

$$df \propto \int_{-\infty}^{\infty} dt_1 \, e^{S\{\phi(t_1+ut_2)\}} t_2^{N-2} dt_2, \qquad (180)$$

so that the fiducial distribution of β may be found independently of α, and thence that of α for given β, as in the case of the Normal distribution; in fact the simultaneous distribution of α and β in the light of a given sample is

$$df \propto \exp\left[S\left\{\phi\left(\frac{A-\alpha}{\beta} + u\frac{B}{\beta}\right)\right\}\right]\frac{B^{N-1}}{\beta^{N+1}}\,d\alpha\,d\beta\ ,\quad (181)$$

for the actual set of u-values observed.

It can easily be verified that the distribution found in Chapter IV for the simultaneous frequency distribution of the mean (μ) and the standard deviation (σ) of a Normal distribution is a particular case of the solution given above, appropriate to the case

$$\phi(v) = c - \tfrac{1}{2}v^2\ ;\qquad (182)$$

the existence of Sufficient estimation, in that case, is replaced by the two estimates being in general rendered Exhaustive by taking account of $N-2$ independent Ancillary Statistics.

It is important to recognize the nature of the inversion that has been effected by the fiducial argument in this and in analogous cases. From the familiar form in which we have a frequency distribution of estimates such as A and B expressed in terms of the parameters, hypothetically supposed known, α and β, we have passed to a frequency distribution of α and β in a distribution specified by observable quantities including A and B. Since in reality A and B may be calculated from the observations, and are known in terms of them, while α and β are in reality unknown, the latter form of statement is the more realistic in representing the state of knowledge of a possible observer, while the former is a statement of what could be known in a hypothetical situation before any real observations had been made. By the fiducial argument we pass from the mathematical expression of a hypothetical to one of a realistic situation, in which the parameters are unknown, though exact probability statements can be made about them. Bartlett's criticism of the fiducial inference as "only to be regarded as a symbolic one" thus seems to be an example of mistaking the substance for the shadow. His statement that "there is no reason to suppose that from it we may infer the fiducial distribution of, say, $\mu + \sigma$" was presumably due to some analytic misapprehension. The problem involved no more than ordinary integration over the known bivariate frequency distribution, and its solution had been published before Bartlett wrote.

9. An example of the Nile problem

An example which illustrates well the connection between particular mathematical relationships on the one hand, and the existence of ancillary statistics, by means of which estimation can be made exhaustive, and exact probability statements inferred, as their consequences in mathematical logic, is as follows.

We suppose that pairs of observables (x, y) are distributed in a bivariate distribution

$$df = e^{-(\theta x + y/\theta)} dx\, dy , \qquad (183)$$

in which x and y take positive values only, then, if X, Y stand for the sums of the two coordinates over N pairs of observations, so that

$$S(x) = X , \qquad S(y) = Y , \qquad (184)$$

the Likelihood of any value θ in the light of the sample observed is

$$e^{-(\theta X + Y/\theta)} \tag{185}$$

so that its logarithm is determined by

$$L = -(\theta X + Y/\theta). \tag{186}$$

The equation of maximum likelihood is then

$$X = Y/\theta^2, \tag{187}$$

leading to the estimate

$$T = \sqrt{Y/X}. \tag{188}$$

The amount of information expected from each pair of observations will be the mean of the square of

$$-\frac{\partial}{\partial \theta}(\theta x + y/\theta) \tag{189}$$

$$= \frac{y}{\theta^2} - x,$$

but

$$E(x^2) = \frac{2}{\theta^2}, \quad E(xy) = 1, \quad E(y^2) = 2\theta^2, \tag{190}$$

hence

$$i_\theta = \frac{2}{\theta^2}, \tag{191}$$

or, considering the amount of information relative to $\log \theta$,

$$i_{\log \theta} = 2. \tag{192}$$

The amount of information supplied in a sample of N values is therefore

$$I_\theta = 2N/\theta^2$$
$$I_{\log \theta} = 2N. \tag{193}$$

THE PRINCIPLES OF ESTIMATION 171

Since the likelihood cannot be expressed in terms only of θ and T, there will be no Sufficient estimate, and some information will be lost if the sample is replaced by the estimate T only. This loss of information may now be calculated from the exact sampling distribution of T.

10. The sampling distribution of the estimate

Since $2\theta x$ and $2y/\theta$ are distributed independently in exponential distributions equivalent to χ^2 for 2 degrees of freedom, it follows that $2\theta X$ and $2Y/\theta$ are similarly distributed for $2N$ degrees of freedom, or, stated otherwise that their simultaneous distribution is

$$df = \frac{X^{N-1}}{(N-1)!} \cdot \frac{Y^{N-1}}{(N-1)!} e^{-(\theta X + Y/\theta)} dX\, dY\ . \quad (194)$$

Since we require the distribution of

$$T = \sqrt{Y/X} \quad (195)$$

we may substitute

$$X = U/T\ ,\quad Y = UT\ , \quad (196)$$

and obtain the simultaneous distribution of T and U, namely

$$df = 2e^{-U\left(\frac{T}{\theta} + \frac{\theta}{T}\right)} \cdot \frac{U^{2N-1} dU}{(N-1)!\,(N-1)!} \cdot \frac{dT}{T}\ . \quad (197)$$

The distribution of T alone is obtained by integration with respect to U from 0 to ∞, giving

$$df = 2\,\frac{(2N-1)!}{(N-1)!\,(N-1)!}\left(\frac{T}{\theta} + \frac{\theta}{T}\right)^{-2N}\frac{dT}{T}\ ; \quad (198)$$

the logarithm of the factor involving θ is

$$-2N \log\left(\frac{T}{\theta} + \frac{\theta}{T}\right), \quad (199)$$

its differential coefficient with respect to θ is

$$\frac{2N}{\theta}\left(\frac{T}{\theta}-\frac{\theta}{T}\right) \div \left(\frac{T}{\theta}+\frac{\theta}{T}\right) \qquad (200)$$

and the mean value of the square of this, giving the amount of information to be expected from a single observation, T, is to be evaluated.

Now from (198) it appears that

$$E\left(\frac{T}{\theta}+\frac{\theta}{T}\right)^{-2} = \frac{N^2}{2N(2N+1)}, \qquad (201)$$

but

$$\left(\frac{T}{\theta}-\frac{\theta}{T}\right)^2 = \left(\frac{T}{\theta}+\frac{\theta}{T}\right)^2 - 4, \qquad (202)$$

hence

$$\left(\frac{T}{\theta}-\frac{\theta}{T}\right)^2 \Big/ \left(\frac{T}{\theta}+\frac{\theta}{T}\right)^2 = 1 - 4\left(\frac{T}{\theta}+\frac{\theta}{T}\right)^{-2}. \qquad (203)$$

The amount of information is therefore

$$\frac{4N^2}{\theta^2}\left(1 - \frac{4N}{2(2N+1)}\right)$$
$$= \frac{2N}{\theta^2} \cdot \frac{2N}{2N+1}, \qquad (204)$$

being less than that supplied by the observations by one part in $(2N+1)$.

11. The use of an ancillary statistic to recover the information lost

The loss of information is less than half the value of a single pair of observations, and never exceeds one third of the total available. Nevertheless, its recovery does exemplify very well the mathematical processes required to complete the logical inference.

From the simultaneous distribution of U and T, we may find that of U only merely by integrating (197) with respect to T. The integral is in fact a standard form for the Bessel function K_0, and gives the distribution of U as

$$df = 4K_0(2U) \cdot \frac{U^{2N-1}\,dU}{(N-1)!\,(N-1)!}. \qquad (205)$$

As this distribution is independent of θ, U is available as an ancillary statistic. The sampling distribution of T, taking U into account, is found by dividing the bivariate element by the corresponding marginal frequency of U, and is evidently

$$\frac{1}{2K_0(2U)} e^{-U\left(\frac{T}{\theta} + \frac{\theta}{T}\right)} \frac{dT}{T}. \qquad (206)$$

From such an error distribution, having known U, the amount of information, calculated as usual, comes to

$$\frac{2U}{\theta^2} \cdot \frac{K_1(2U)}{K_0(2U)}, \qquad (207)$$

which, it will be observed, depends upon the value of U actually available, but has an average value, when variations of U are taken into account, of

$$2N/\theta^2 \qquad (208)$$

the total amount expected on the average from N observations; none is now lost.

The information is recovered and the inference completed by replacing the distribution of T for given size of sample N, by the distribution of T for given U, which indeed happens not to involve N at all. In fact, U has completely replaced N as a means of specifying the precision to be ascribed to the estimate. In both cases the estimate T is the same, the calculation of U enables us to see exactly how precise it is, not on the average, but for the particular value of U supplied by the sample.

In these circumstances it is possible to specify the precision by an exact statement of the probability of θ lying in any chosen range. Conveniently, if we write

$$\tau = \log T - \log \theta, \qquad (209)$$

then τ has the distribution

$$\frac{1}{2K_0(2U)} e^{-2U \cosh \tau} d\tau \qquad (210)$$

and the definite integral of this distribution between any chosen limits, τ_1 and τ_2 gives the probability that θ should lie between the corresponding limits

$$Te^{-\tau_1} \text{ and } Te^{-\tau_2}. \qquad (211)$$

It will be noticed that the distribution of τ is symmetrical. The success of the process by which the missing information was recovered, and the statement of probability *a posteriori* rendered exact, evidently depends on the distribution of

$$U = \sqrt{XY} \qquad (212)$$

being independent of θ. For a sample of one pair, it depends on the product xy having a distribution independent of θ, and therefore upon this being a solution of the Nile problem in the sense that the total frequency lying between any two rectangular hyperbolas

$$xy = c_1, \qquad xy = c_2, \qquad (213)$$

shall be independent of θ, and depend only on the chosen values c_1 and c_2. Such curves therefore divide the total frequency in fixed proportions independently of the value of the unknown parameter, representing in that case the unknown height to which the Nile will rise.

With a different mathematical specification of the problem, different logical consequences might ensue. If we take a more general distribution

$$df = \theta\phi e^{-\theta x - \phi y} dx\, dy , \qquad (214)$$

involving two parameters θ and ϕ, it may be seen that with any connection between θ and ϕ of the form

$$\phi = \theta' \qquad (215)$$

it will be possible to find an ancillary statistic and to derive probability statements about the parameter specifying θ and ϕ. However, the greater part of functional relationships which might subsist between two such positive quantities do not have this property, and apart from approximate statements appropriate to large samples, the totality of the information which the data supply is subsumed in the specification of the Likelihood function for all values of the unknown parameter.

12. Simultaneous distribution of the parameters of a bivariate Normal distribution

If from a Normal population with variances σ_1^2, σ_2^2 and correlation ρ, a sample yields the Sufficient estimates s_1, s_2 and r, then it was shown in 1915[2] that the sampling distribution of r was expressible in terms of ρ only, in the frequency element

$$\frac{1}{\pi(N-3)!}(1-\rho^2)^{\frac{1}{2}(N-1)} \cdot (1-r^2)^{\frac{1}{2}(N-4)} \cdot \frac{\partial^{N-2}}{\partial(\rho r)^{N-2}} \frac{\theta}{\sin \theta} \cdot dr \tag{216}$$

where $\cos \theta = -\rho r$, and $0 \leqslant \theta \leqslant \pi$.

Since the distribution of r does not depend on the parameters other than ρ, we have a known function of r and ρ,

$$P(r, \rho), \tag{217}$$

such that the distribution of r for given ρ is given by the frequency element

$$\frac{\partial P}{\partial r} dr, \tag{218}$$

and the frequency of ρ for given r is

$$-\frac{\partial}{\partial \rho}\{P(r, \rho)\} d\rho, \tag{219}$$

giving the marginal distribution of ρ in terms of r only. This was actually the first example of the derivation of a fiducial distribution (1930).[4]

For any given values of r, ρ the simultaneous distribution of s_1 and s_2 is

$$\frac{(N-1)^{N-1}}{(1-\rho^2)^{N-1}} \cdot \left(\frac{s_1 s_2}{\sigma_1 \sigma_2}\right)^{N-2}$$

$$\exp\left[-\frac{N-1}{2(1-\rho^2)}\left\{\frac{s_1^2}{\sigma_1^2} - 2r\rho \frac{s_1 s_2}{\sigma_1 \sigma_2} + \frac{s_2^2}{\sigma_2^2}\right\}\right]\frac{ds_1 ds_2}{\sigma_1 \sigma_2} \tag{220}$$

divided by the function of $r\rho$ only

$$\frac{\partial^{N-2}}{\partial(-\cos \theta)^{N-2}} \frac{\theta}{\sin \theta} \cdot \tag{221}$$

THE PRINCIPLES OF ESTIMATION 177

If we write

$$u = \sqrt{\frac{N-1}{1-\rho^2}} \cdot \frac{s_1}{\sigma_1}, \quad v = \sqrt{\frac{N-1}{1-\rho^2}} \cdot \frac{s_2}{\sigma_2}, \qquad (222)$$

the distribution for given values of r, ρ becomes

$$(uv)^{N-2} \exp\{-\tfrac{1}{2}(u^2 - 2r\rho uv + v^2)\} \, du \, dv$$
$$\div \frac{\partial^{N-2}}{(\sin\theta\partial\theta)^{N-2}} \frac{\theta}{\sin\theta} \cdot \qquad (223)$$

Now for any chosen values ξ, η expression (223) will supply a function $P(\xi, \eta)$ such that

$$Pr\{u > \xi, v > \eta\} = P(\xi, \eta), \qquad (224)$$

or, dividing each value into

$$s\sqrt{(N-1)/(1-\rho^2)}$$

$$Pr\left\{\sigma_1 < \sqrt{\frac{N-1}{1-\rho^2}} \cdot \frac{s_1}{\xi}, \quad \sigma_2 < \sqrt{\frac{N-1}{1-\rho^2}} \cdot \frac{s_2}{\eta}\right\} = P \qquad (225)$$

giving the simultaneous fiducial distribution of σ_1 and σ_2 with the frequency element

$$\frac{\partial^2}{\partial\sigma_1 \partial\sigma_2} P(\xi, \eta) \, d\sigma_1 \, d\sigma_2$$

$$= \frac{N-1}{1-\rho^2} \cdot \frac{s_1 s_2}{\sigma_1^2 \sigma_2^2} \cdot (\xi\eta)^{N-2} \exp\{-\tfrac{1}{2}(\xi^2 - 2r\rho\xi\eta + \eta^2)\} \, d\sigma_1 \, d\sigma_2$$

$$= (\xi\eta)^{N-1} \exp\{-\tfrac{1}{2}(\xi^2 - 2r\rho\xi\eta + \eta^2)\} \frac{d\sigma_1 d\sigma_2}{\sigma_1 \sigma_2}, \qquad (226)$$

divided by

$$\frac{\partial^{N-2}}{\partial(-\cos\theta)^{N-2}} \cdot \frac{\theta}{\sin\theta}, \qquad (227)$$

and multiplied by the marginal frequency

$$-\frac{\partial}{\partial\rho}\{P(r, \rho)\} \, d\rho, \qquad (228)$$

in which ξ and η are abbreviating symbols, standing for

$$\sqrt{\frac{N-1}{1-\rho^2}} \cdot \frac{s_1}{\sigma_1} \quad \text{and} \quad \sqrt{\frac{N-1}{1-\rho^2}} \cdot \frac{s_2}{\sigma_2}. \quad (229)$$

The simultaneous fiducial distribution of these three parameters with two more for the population means, may then be found from the consideration that the set of statistics, s_1, s_2, r is distributed independently of the means, so that we need only a further factor representing the fiducial distribution of the means for known values of σ_1, σ_2 and ρ. Namely,

$$\frac{N}{2\pi\sqrt{1-\rho^2}} e^{\frac{-N}{2(1-\rho^2)}\left\{\frac{(x_1-\mu_1)^2}{\sigma_1^2} - 2\rho \frac{(x_1-\mu_1)(x_2-\mu_2)}{\sigma_1\sigma_2} + \frac{(x_2-\mu_2)^2}{\sigma_2^2}\right\}} \frac{d\mu_1 d\mu_2}{\sigma_1 \sigma_2}.$$

$$(230)$$

It has been proposed that any set of functions having distributions independent of the parameters, such as

$$\left.\begin{array}{l} t_1 = \dfrac{s_1}{\sigma_1}\sqrt{N-1} \\[6pt] t_2 = \dfrac{s_2}{\sigma_2}\sqrt{\dfrac{(N-1)(1-r^2)}{1-\rho^2}} \\[6pt] t_3 = \sqrt{\dfrac{N-1}{1-\rho^2}}\left(r\,\dfrac{s_2}{\sigma_2} - \rho\,\dfrac{s_1}{\sigma_1}\right) \end{array}\right\} \quad (231)$$

can be used to transform the simultaneous frequency distribution of s_1, s_2, r in terms of σ_1, σ_2, ρ, into the simultaneous distribution of σ_1, σ_2, ρ in terms of s_1, s_2, r simply by multiplying by

$$\frac{\partial(s_1, s_2, r)}{\partial(t_1, t_2, t_3)} \times \frac{\partial(t_1, t_2, t_3)}{\partial(\sigma_1, \sigma_2, \rho)}. \quad (232)$$

The example has been chosen to illustrate the process of building up the simultaneous distribution of the parameters rigorously by a step by step process, since the short cut suggested by the use of Jacobians has no claim to validity unless it can be proved to be equivalent to a genuine fiducial argument. The expressions (231) cannot indeed be made to supply such a proof. Perhaps the change of sign of $\partial t_2/\partial \rho$ at $\rho = 0$ should be a sufficient warning.

The correct way of using the facts stated in (231) has been more recently demonstrated by D. A. S. Fraser and D. A. Sprott who eliminate s_1, s_2 and σ_1, σ_2, obtaining the equation in three random variables

$$w = \frac{r}{\sqrt{1-r^2}} \chi_{N-2} - \frac{\rho}{\sqrt{1-\rho^2}} \chi_{N-1} \qquad (233)$$

in which w is a normal variable with mean zero and unit variance, while the two others are χ-variables with $(N-2)$ and $(N-1)$ degrees of freedom respectively. Assigning any value to ρ the equation gives the distribution of r as it was given in my paper of 1915, while assigning any value to r, it gives the fiducial distribution of ρ as first given in 1930, and as used in this chapter.

Explicit forms for the distribution of ρ have been derived by C. R. Rao from Fraser's formula. For example the form corresponding with (216) is

$$\frac{1}{\pi(N-3)!} (1-\rho^2)^{\frac{1}{2}(N-3)} (1-r^2)^{\frac{1}{2}(N-2)}$$

$$\frac{\partial^{N-3}}{\partial(\rho r)^{N-3}} \left\{ \frac{\theta - \frac{1}{2}\sin 2\theta}{\sin^3 \theta} \right\} d\rho \qquad (234)$$

REFERENCES

1. M. S. Bartlett (1937). Properties of sufficiency and statistical tests.
 Proc. Roy. Soc., A, vol. 160, pp. 268-282.
2. R. A. Fisher (1915). Frequency distribution of the values of the correlation coefficient in samples from an indefinitely large population.
 Biometrika, vol. 10, pp. 507-521.
3. R. A. Fisher (1925). Theory of statistical estimation.
 Proc. Camb. Phil. Soc., vol. 22, pp. 700-725.
4. R. A. Fisher (1930). Inverse probability.
 Proc. Camb. Phil. Soc., vol. 26, pp. 528-535.
5. D. A. S. Fraser (1964). On the definition of fiducial probability.
 Bull. Int. Statist. Inst., vol. 40, pp. 842-856.

INDEX

Acceptance decisions, 79 seq.
Ancillary information, 112 seq.
Arithmetical triangle, 116
Atlas, 41
Axiomatic theory, 111

Barnard, G. A., 91, 92, 107
Bartlett, M. S., 100, 124, 169
Bassett, 141
Bateson, W., 3
Bayes, T., 8 seq., 38, 115, 132
Behrens, W.-U., 97, 99, 109
Bessel function, 141, 173
Biological assay, 27, 75
Bivariate normal distribution, 175
Boole, G., 17 seq., 38, 40
Boyle, Sir Robert, 9

Cards, 35, 93
Cauchy, 85
Chrystal, G., 29, 38
Composite hypothesis, 92 seq.
Confidence interval, 147
Confidence limits, 69, 74
Consistency, 148, 150
Cornish, E. A., and Fisher, R. A., 66, 78

Darwin, C., 4
Decision functions, 104
de Moivre, A., 9, 14, 39
De Morgan, A., 29, 30, 39
Dice, 34, 35, 46
Dictionary of National Biography, 8, 39
Discontinuous observations, 63 seq.
Disjunction, 42

Edgeworth, F. Y., 37, 38

Efficiency, 151 seq.
Estimation, 49, 145 seq.
Exhaustive estimation, 52
Experimental design, 6

Factorial function, 129
Fermat, 116
Fiducial prediction, 117
Fire, false alarm of, 25
Fisher, R. A., 7, 38, 78, 90, 92, 107, 108, 142, 177, 780
Fisher and Yates, 78, 108
Fraser, D. A. S., 179, 180

Galton, F., 1
Gauss, 82, 88
Goodness of Fit, 51
Gosset, W. S., 4, 82

Harvard, 122
Helmert, F. R., 83, 108

Inductive Behaviour, 105
Inductive Reasoning, 112
Information, 153

Jeffreys, H., 20, 59, 78

Keynes, J. M., 47, 78
Kolmogorov, A., 59, 78

Laplace, P. S., de, 14–16, 18, 20, 36, 38

Mahalanobis, P. C., 6, 7
Maia, 41
Mathematical Likelihood, 5, 71 seq., 110 seq., 132 seq., 154
Mice, 18
Michell, J., 21, 39, 40, 41
Montmort, 14, 39

Newton, Sir Isaac, 9
Neyman, J., 101, 103, 104
Neyman, J., and Pearson, E. S., 92, 95, 105, 108

Parapsychology, 43
Pearson, E. S., and Hartley, H. O., 3, 149
Pearson, K., 2, 37, 39, 50, 87, 109
Pleiades, 40
Precision, accurate statement of, 60
Price, R., 9
Probability, 14, 15, 33 *seq.*, 46, 47, 110 *seq.*, 113

Rain, 25
Randomization, 101 *seq.*, 102
Rao, C. R., 179
Roulettes, 35
Rule of Succession, 24 *seq.*

Simultaneous estimation, 158 *seq.*

Sprott, D. A., 179
Strychnine, 25
"Student", 4, 7, 82, 84, 109
Sufficient statistics, 139, 140, 157
Sukhatmé, P. V., 99, 109

Table of x^x, 143–144
Test of significance, 40 *seq.*, 79 *seq.*
The Fiducial Argument, 53, 54 *seq.*
Theory of Games, 102, 104
Theory of Testing Hypotheses, 4, 92
Todhunter, I., 31 *seq.*

Uncertainty, 110 *seq.*

Venn, J., 24 *seq.*, 39, 60

Wilson, E. B., and Worcester, J., 109
Wilson, E. B., 91, 92, 109

Yates, F., 6, 7, 90, 92, 109

Printed in the USA/Agawam, MA
May 1, 2018